AF411452

Polymer Biointerfaces

Polymer Biointerfaces

Special Issue Editors

Marián Lehocký
Petr Humpolíček

MDPI • Basel • Beijing • Wuhan • Barcelona • Belgrade • Manchester • Tokyo • Cluj • Tianjin

Special Issue Editors

Marián Lehocký
Tomas Bata University in Zlín
Czech Republic

Petr Humpolíček
Tomas Bata University in Zlín
Czech Republic

Editorial Office
MDPI
St. Alban-Anlage 66
4052 Basel, Switzerland

This is a reprint of articles from the Special Issue published online in the open access journal *Polymers* (ISSN 2073-4360) (available at: https://www.mdpi.com/journal/polymers/special_issues/polymer_biointerfaces).

For citation purposes, cite each article independently as indicated on the article page online and as indicated below:

LastName, A.A.; LastName, B.B.; LastName, C.C. Article Title. *Journal Name* **Year**, *Article Number*, Page Range.

ISBN 978-3-03928-977-6 (Hbk)
ISBN 978-3-03928-978-3 (PDF)

Contents

About the Special Issue Editors . . . . **vii**

Marián Lehocký and Petr Humpolíček
Polymer Biointerfaces
Reprinted from: *Polymers* **2020**, *12*, 793, doi:10.3390/polym12040793 . . . . **1**

Martina Klučáková
Agarose Hydrogels Enriched by Humic Acids as the Complexation Agent
Reprinted from: *Polymers* **2020**, *12*, 687, doi:10.3390/polym12030687 . . . . **7**

Andrea Catalina Villamil Ballesteros, Hugo Ramiro Segura Puello, Jorge Andres Lopez-Garcia, Andres Bernal-Ballen, Diana Lorena Nieto Mosquera, Diana Milena Muñoz Forero, Juan Sebastián Segura Charry and Yuli Alexandra Neira Bejarano
Bovine Decellularized Amniotic Membrane: Extracellular Matrix as Scaffold for Mammalian Skin
Reprinted from: *Polymers* **2020**, *12*, 590, doi:10.3390/polym12030590 . . . . **19**

Mehrnoush Narimisa, František Krčma, Yuliia Onyshchenko, Zdenka Kozáková, Rino Morent and Nathalie De Geyter
Atmospheric Pressure Microwave Plasma Jet for Organic Thin Film Deposition
Reprinted from: *Polymers* **2020**, *12*, 354, doi:10.3390/polym12020354 . . . . **37**

Alenka Vesel, Rok Zaplotnik, Gregor Primc and Miran Mozetič
Evolution of the Surface Wettability of PET Polymer upon Treatment with an Atmospheric-Pressure Plasma Jet
Reprinted from: *Polymers* **2020**, *12*, 87, doi:10.3390/polym120100875 . . . . **61**

Rushita Shah, Pavel Stodulka, Katerina Skopalova and Petr Saha
Dual Crosslinked Collagen/Chitosan Film for Potential Biomedical Applications
Reprinted from: *Polymers* **2019**, *11*, 2094, doi:10.3390/polym11122094 . . . . **87**

Pavel Stahel, Vera Mazankova, Klara Tomeckova, Petra Matouskova, Antonin Brablec, Lubomir Prokes, Jana Jurmanova, Vilma Bursikova, Roman Pribyl, Marian Lehocky, Petr Humpolicek, Kadir Ozaltin, and David Trunec
Atmospheric Pressure Plasma Polymerized Oxazoline-Based Thin Films—Antibacterial Propertiesand Cytocompatibility Performance
Reprinted from: *Polymers* **2019**, , 2069, doi:10.3390/polym11122069 . . . . **101**

Lucie Urbánková, Věra Kašpárková, Pavlína Egner, Ondřej Rudolf and Eva Korábková
Caseinate-Stabilized Emulsions of Black Cumin and Tamanu Oils: Preparation, Characterization and Antibacterial Activity
Reprinted from: *Polymers* **2019**, *11*, 1951, doi:10.3390/polym11121951 . . . . **115**

Kateřina Skopalová, Zdenka Capáková, Patrycja Bober, Jana Pelková, Jaroslav Stejskal, Věra Kašpárková, Marián Lehocký, Ita Junkar, Miran Mozetič and Petr Humpolíček
In-Vitro Hemocompatibility of Polyaniline Functionalized by Bioactive Molecules
Reprinted from: *Polymers* **2019**, *11*, 1861, doi:10.3390/polym11111861 . . . . **135**

Salma Habib, Marian Lehocky, Daniela Vesela, Petr Humpolíček, Igor Krupa and Anton Popelka
Preparation of Progressive Antibacterial LDPE Surface via Active Biomolecule Deposition Approach
Reprinted from: *Polymers* **2019**, *11*, 1704, doi:10.3390/polym11101704 . . . . **145**

Hana Dvořáková, JanČech, Monika Stupavská, Lubomír Prokeš, Jana Jurmanová, Vilma Buršíková, Jozef Ráheľ and Pavel Sťahel
Fast Surface Hydrophilization via Atmospheric Pressure Plasma Polymerization for Biological and Technical Applications
Reprinted from: *Polymers* **2019**, *11*, 1613, doi:10.3390/polym11101613 . . . . **161**

Lenka Vítková, Lenka Musilová, Eva Achbergerová, Antonín Minařík, Petr Smolka, Erik Wrzecionko and Aleš Mráček
Electrospinning of Hyaluronan Using Polymer Coelectrospinning and Intermediate Solvent
Reprinted from: *Polymers* **2019**, *11*, 1517, doi:10.3390/polym11091517 . . . . **177**

Andres Bernal-Ballen, Jorge-Andres Lopez-Garcia and Kadir Ozaltin
(PVA/Chitosan/Fucoidan)-Ampicillin: A Bioartificial Polymeric Material with Combined Properties in Cell Regeneration and Potential Antibacterial Features
Reprinted from: *Polymers* **2019**, *11*, 1325, doi:10.3390/polym11081325 . . . . **197**

Ji Yeon Lee, Ludwig Erik Aguilar, Chan Hee Park and Cheol Sang Kim
UV Light Assisted Coating Method of Polyphenol Caffeic Acid and Mediated Immobilization of Metallic Silver Particles for Antibacterial Implant Surface Modification
Reprinted from: *Polymers* **2019**, *11*, 1200, doi:10.3390/polym110712005 . . . . **211**

Ilkay Karakurt, Kadir Ozaltin, Daniela Vesela, Marian Lehocky, Petr Humpolíček and Miran Mozetič
Antibacterial Activity and Cytotoxicity of Immobilized Glucosamine/Chondroitin Sulfate on Polylactic Acid Films
Reprinted from: *Polymers* **2019**, *11*, 1186, doi:10.3390/polym11071186 . . . . **225**

Janina Belka, Joachim Nickel and Dirk G. Kurth
Growth on Metallo-Supramolecular Coordination Polyelectrolyte (MEPE) Stimulates Osteogenic Differentiation of Human Osteosarcoma Cells (MG63) and Human Bone Marrow Derived Mesenchymal Stem Cells
Reprinted from: *Polymers* **2019**, *11*, 1090, doi:10.3390/polym11071090 . . . . **237**

Pavel Mokrejš, Petr Mrázek, Robert Gál and Jana Pavlačková
Biotechnological Preparation of Gelatines from Chicken Feet
Reprinted from: *Polymers* **2019**, *11*, 1060, doi:10.3390/polym11061060 . . . . **253**

Jiří Smilek, Sabína Jarábková, Tomáš Velcer and Miloslav Pekař
Compositional and Temperature Effects on the Rheological Properties of Polyelectrolyte–Surfactant Hydrogels
Reprinted from: *Polymers* **2019**, *11*, 927, doi:10.3390/polym110509275 . . . . **267**

Kadir Ozaltin, Marian Lehocky, Petr Humpolicek, Jana Pelkova, Antonio Di Martino, Ilkay Karakurt and Petr Saha
Anticoagulant Polyethylene Terephthalate Surface by Plasma-Mediated Fucoidan Immobilization
Reprinted from: *Polymers* **2019**, *11*, 750, doi:10.3390/polym11050750 . . . . **287**

About the Special Issue Editors

Marián Lehocký is currently an Associate Professor in Physical Chemistry. After receiving his Masters degree in Physical and Applied Chemistry at the Faculty of Chemistry, Brno University of Technology, he moved on to a PhD in Chemistry and Technology of Macromolecular Substances at Tomas Bata University in Zlín. He spent time at the University of Aveiro, Portugal, as an Marie Curie Fellow, and he is currently a senior researcher at the Centre of Polymer Systems at Tomas Bata University in Zlin and a director of the Department of Lipids, Detergents and Cosmetics Technology at Faculty of Technology, Tomas Bata University in Zlín. His current scientific interests are polymer surface modification, emulsion and suspension phenomena, surface and interface science, colloid phenomena, antibacterial and atithrombotic surface modification and interaction of materials with cell systems and tissues, including STEM cells. He is an author or co-author of 80 original per-reviewed manuscripts in prestigious scientific journals (h-index 21) and inventor of 8 national and European patents. He is a member of Editorial board of Polymers, Materials and Design, and Marerials Science in Semiconductor Processing. In 2014, he won the "Molecules Best Paper Award" (2nd prize).

Petr Humpolíček is interested in biomaterials, genetics and interaction of cells and materials. He is actually a leader of the research group Bioactive polymer systems at research unit The Centre of Polymer Systems of Tomas Bata University in Zlín. His research interests centre around the evaluation of biocompatibility of materials, especially conducting polymers. Special attention pay to the preparation of biomimetic materials, structured surfaces and development of cultivation techniques mimicking the in vivo environment. Petr Humpolíček is author or co-author of 82 original peer-reviewed manuscripts in prestigious scientific journals (h-index 16) and the inventor of 6 national and European patents. He was or is the main investigator of four grants funded by the Czech Science Foundation.

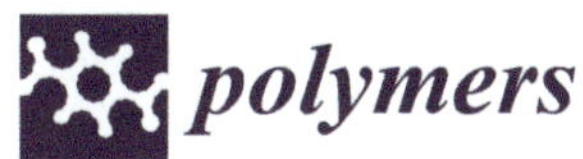 

Editorial

Polymer Biointerfaces

Marián Lehocký [1,2,*] **and Petr Humpolíček** [1,2]

1   Centre of Polymer Systems, University Institute, Tomas Bata University in Zlín, Nam. T.G.M. 5555, 76001 Zlín, Czech Republic; humpolicek@utb.cz
2   Faculty of Technology, Tomas Bata University in Zlín, Vavreckova 275, 76001 Zlín, Czech Republic
*   Correspondence: lehocky@post.cz

Received: 24 March 2020; Accepted: 30 March 2020; Published: 2 April 2020

Polymer biointerfaces are considered suitable materials for the improvement and development of numerous applications. The optimization of polymers' surface properties can control several biological processes, such as cell adhesion, proliferation, viability, and enhanced extracellular matrix secretion at biointerfaces [1–8].

Various routes of polymer biointerfaces preparation are targeted for numerous applications in the biomedical, biochemical, biophysical, biotechnological, food, pharmaceutical, and cosmetic fields [9–18].

This Special Issue, which consists of 18 articles written by research experts, reports on the most recent research on polymer biointerfaces. Several novel advanced methods related to polymer biointerfaces preparation, modification, analysis, and characterization are introduced.

Firstly, Ozaltin and co-workers treated a polyethylene terephthalate (PET) surface, for use in blood-contacting devices, by DC air plasma and immobilized fucoidan from *Fucus vesiculosus* (FU) on it, at different pH values in the range of 3–7. FU immobilization onto the PET surface after plasma treatment was found to be optimal at pH 5, as supported by FTIR, SEM, and XPS results, and provided the highest anticoagulant activity, more than 100 s, which indicates that the resulting FU-immobilized sample is an efficient anticoagulant, suitable for blood-contacting PET devices. Surface characterization was carried out by a wettability test, scanning electron microscopy, X-ray photoelectron spectroscopy, and Fourier-transform infrared spectroscopy. The anticoagulation activity of the samples was determined on the basis of prothrombin time, activated partial thromboplastin time, and thrombin time [19].

Smilek et al. prepared a hydrogel from oppositely charged biopolymer polyelectrolyte and surfactant in micellar form, at a surfactant/biopolymer charge ratio at least equal to one. Hyaluronan acted as a negatively charged biopolymer, whereas DEAED (amino-modified dextran) was used as a positively charged biopolymer. The former interacted with Septonex (carbethopendecinium bromide), whereas the latter interacted with sodium dodecylsulphate. The rheological properties of hyaluronan-based hydrogels were mainly dependent on the polymer molecular weight. Surfactant concentration (more precisely, the concentration of micelles at a surfactant/biopolymer charge ratio above 1) showed only a small effect. Surfactant concentration was found to have a much greater effect for DEAED-based hydrogels. The authors focused in more detail on the effects of various processing parameters on the properties of similar gels prepared using a slightly different cationic surfactant (approved for use in pharmaceutical formulations) and also investigated a reversely charged system—a positively charged polyelectrolyte (cationized dextran) and an anionic surfactant [20].

Mokrejš and co-workers introduced the preparation of gelatines from by-product collagen raw materials derived from the slaughter of chicken (chicken feet). Gelatine is a water-soluble protein that is obtained from collagenous raw materials by partial hydrolysis. The technological innovation consists in the biotechnological processing of (purified) feedstock by a commercial food endoprotease, which, in contrast to acidic (type A gelatines) or alkaline (type B gelatines) processing, has a variety of economic, technological, and environmental advantages [21].

Belka et al. contributed with current investigations employing MG63 cells grown on Fe-MEPE (metallo-supramolecular coordination polyelectrolyte) modified substrates, which suggest initiation of osteogenic differentiation by both high cell activity and altered morphology of the cells and/or cluster formation. The obtained results led to the conclusion that these surfaces individually support the specification of cell differentiation toward lineages that correspond to the natural commitment of the particular cell types. The authors, therefore, propose that Fe-MEPEs may be used as a scaffold for the treatment of defects in muscular or bone tissues [22].

Karakurt and collaborators investigated the antibacterial activity and cytotoxicity of immobilized glucosamine (GlcN)/chondroitin sulfate (ChS) on polylactic acid. The antibacterial surface modification of polylactic acid films was achieved through the immobilization of GlcN and ChS on film surfaces via plasma treatment, followed by acrylic acid grafting. It was found that the developed GlcN/ChS-coated polylactic acid films are excellent bactericide agents against representative Gram-positive and Gram-negative bacteria. Plasma-treated films immobilized with ChS and GlcN, separately and in combination, demonstrated bactericidal effect against strains of both bacterial types, and the results also revealed the absence of synergistic effect on antibacterial action for the combination. [23].

Lee and co-workers successfully coated Ti substrates with polycaffeic acid and metallic silver using a facile UV light-assisted method for implant applications. They confirmed that the coating process was successful by SEM and AFM analyses. At the same time, they verified the deposition of polycaffeic acid and metallic silver by confirming the elemental composition through XPS, EDS, and mapping methods, and the physical properties (hydrophilicity) of the samples were verified using water contact angle measurements. In vitro biocompatibility and antibacterial studies showed that polycaffeic acid with metallic silver can inhibit bacterial growth, while proliferation of MC-3T3 cells was observed. Therefore, the obtained results suggest that the introduced approach can be considered as a potential method for functional implant coating in the orthopedic field [24].

Bernal-Ballen et al. examined the development of a bioartificial polymeric material made of polyvinyl alcohol (PVA), chitosan (CHI), and fucoidan (FUC) and the incorporation of ampicillin as an antibacterial agent. The prepared films were tested, and it was elucidated that the bioartificial polymeric material has potential for inducing cell regeneration in vitro. The characterization techniques used in the manuscript indicated that PVA brings water resistibility to the system, whereas CHI and FUC are responsible for creating a porous microstructure, which allows the cells to adhere to and grow within the matrix. The obtained information indicated that PVA, CHI, and FUC are compatible, as evidenced by FTIR spectra and SEM images. The new material is an outstanding candidate for cell regeneration, as a result of the synergic effect that each component provides to the blend. [25].

Vitkova et al. obtained nanofibers containing hyaluronic acid (HA) by solution electrospinning. Two approaches were chosen: co-electrospinning of aqueous blend solutions of hyaluronic acid/polyvinyl alcohol and hyaluronic acid/polyethylene oxide and use of an intermediate solvent for electrospinning of pure hyaluronic acid solutions. The choice of materials was done with regard to their potential utilization for cell cultivation. The influences of polymer concentration, average molecular weight (M_w), viscosity, and solution surface tension were analyzed. HA and PVA were fluorescently labeled in order to examine the electrospun structures using fluorescence confocal microscopy [26].

Dvořáková and co-workers demonstrated a new hydrophilization technique based on plasma deposition of a thin film using mixtures of propane/butane with nitrogen at atmospheric pressure. Unlike a simple plasma treatment, the observed high-surface free energy values were due to the properties of the deposited plasma–polymer nanolayer. Therefore, the wettability improvement did not depend on the substrate material, and the aging of the surface modification was highly reduced. The thin layers of the prepared plasma–polymer exhibit highly stable wetting properties, are smooth, homogeneous, and flexible, and adhere well to the surface of polypropylene substrates. Moreover, they are constituted of essential elements only (C, H, N, O). This makes the presented modified plasma–polymer surfaces interesting for further studies in biological and/or technical applications [27].

Habib et al. grafted ascorbic acid onto a polyethylene surface via plasma treatment in order to improve its antimicrobial effects. Plasma treatment was effectively used as a radical initiator with subsequent incorporation of ascorbic acid, which served as an antimicrobial agent, on the polyethylene surface. This modification was confirmed by the enhanced wettability and adhesion properties. The results showed changes in the wettability, adhesion, and roughness of the polyethylene surface after plasma treatment as well as after ascorbic acid grafting. This is a positive indication of the possibility of grafting ascorbic acid onto polymeric materials using plasma pretreatment, enhancing its antibacterial activity [28].

Skopalová and co-workers studied polyaniline films modified by substances with anticipated anticoagulant activity, sodium dodecylbenzenesulfonate, 2-aminoethane-1-sulfonic acid, and *N*-(2-acetamido)-2-aminoethanesulfonic acid. The hemocompatibility tests conducted on these polyaniline films confirmed the absence of anticoagulation activity, though the functional groups typical of anticoagulation substances were present. Hemocompatibility is an essential prerequisite for the application of materials in the field of biomedicine and biosensing. In addition, mixed ionic and electronic conductivity of conducting polymers is an advantageous property for these applications. The results showed that the anticoagulation activity was highly affected by the presence of suitable functional groups originating from the used heparin-like substances and by the properties of the polyaniline polymer itself [29].

Urbánková et al. used sodium caseinate in order to stabilize emulsions containing bioactive tamanu and black cumin oils. The emulsions were prepared by ultrasound treatment or high-shear homogenization with Ultra-Turrax. The analysis of the oils' fatty acid composition revealed a higher degree of unsaturation for cumin oil, with higher content of linoleic acid C18:2, which corresponded to the higher iodine value determined for this oil. The antibacterial activities of both oils and of their emulsions were investigated with respect to the growth suppression of common spoilage bacteria, using the disk diffusion method. The oils and selected emulsions were proven to act against Gram-positive strains, mainly against *Staphylococcus aureus* and *Bacillus cereus*. Regrettably, Gram-negative species were fully resistant to their action [30].

Sťahel and co-workers deposited oxazoline-based thin films on glass substrates using atmospheric pressure dielectric barrier discharge. This study presents a new way to produce plasma-polymerized oxazoline polymers, which are a new promising class of polymers for biomedical applications, with antibiofouling properties and good biocompatibility. The authors describe the film preparation procedure. Nitrogen was used as the working gas for the discharge, 2-methyl-2-oxazoline vapors, used as the monomer, were admixed to the nitrogen flow. This gas composition made it possible to obtain a homogeneous discharge, which led to the deposition of homogeneous thin films. To improve the film properties, it was necessary to increase the substrate temperature during the deposition. All deposited films are cytocompatible and exhibited excellent antibacterial properties against *S. epidermidis*, *S. aureus*, and *Escherichia coli* [31].

Shah et al. developed a novel dual crosslinked film for promising future applications in ophthalmology, skin tissue engineering, and wound dressing. Firstly, a collagen/chitosan film was prepared by the solvent casting technique, utilizing two crosslinking agents together, i.e., tannic acid and genipin. The obtained final dual crosslinked film was translucent, thin, and greenish-blue in color. Enzymatic degradation of the films was performed with lysozyme and lipase. Cell adhesion and proliferation were tested using mouse embryonic cell lines, by culturing the cells on the dual crosslinked film. These dual crosslinked polymeric film find their application in ophthalmology, especially as implants for temporary injured cornea and skin tissue regeneration [32].

Vesel and co-workers presented results regarding systematic 2D mapping of surface wettability, which provide an insight in processes responsible for surface activation of the polymer polyethylene terephthalate using a simple atmospheric-pressure plasma jet. The discharge tube was only flushed with Ar gas to get rid of permanent gases, such as nitrogen, oxygen and CO_2, but the water remained on the surfaces, due to the humidity and laboratory atmosphere. In the case of a rapid activation, a very sharp interphase between the activated and the unaffected surface was observed and explained by the peculiarities of high-impedance discharges sustained in Ar with the presence of impurities from water vapor. The results obtained by X-ray photoelectron spectroscopy confirmed that the activation was a consequence of functionalization with oxygen functional groups [33].

Narimisa et al. applied a plasma source to deposit thin layers under atmospheric pressure. Using different techniques, the effect of process parameters such as applied power, carrier gas flow rate, distance from capillary to the substrate, and treatment time on the deposition efficiency was studied. A high level of monomer fragmentation observed with optical emission spectroscopy, together with the non-uniform distribution of the monomer observed by computational fluid dynamic simulations, was shown to be a reliable indicator of coating quality. By following this research strategy, crucial information regarding the effectiveness of atmospheric-pressure microwave plasma jet for thin film deposition can be revealed [34].

Villamil Ballesteros and co-workers decellularized bovine amniotic membranes using four different protocols, and the differences in terms of decellularization were considered as negligible. All membranes provided DNA concentrations <50 ng/mg, indicating that traces of the nucleic acid were present in the prepared material, although the obtained values were negligible, which implies that the decellularized membranes did not contain native cells from the bovine amniotic membranes. The in vitro biocompatibility of the studied samples was demonstrated. Hence, these matrices may be deemed as a potential scaffold for epithelial tissue regeneration [35].

Finally, Klučáková studied the influence of humic acids on the transport of metal ions and dyes in agarose hydrogels. It was confirmed that humic acids retarded the transport of diffusion probes. Humic acids' enrichment caused decreases in the values of effective diffusion coefficients, due to their complexation with the diffusion probes. The effect of complexation was selective for a particular diffusion probe. The aim of this study was to investigate the influence of the interactions between humic acids and probes on hydrogels' release ability, as hydrogels play an important role in the monitoring of the mobility of pollutants in nature as well as in their removal and in water treatment. They are usually based on materials able to absorb water and different pollutants in their structure [36].

Acknowledgments: First of all, we would like to express my deep gratitude to *Polymers*, especially to its Editorial office for continuous guide, care, and help during all steps of the preparation and production of this Special Issue titled "Polymer Biointerfaces". We would also like to extend our gratitude to all contributing authors for their valuable manuscripts as well as to all reviewers who have helped with valuable suggestions.

Conflicts of Interest: The authors declare no conflict of interest.

References

1. Lehocky, M.; Amaral, P.F.F.; Stahel, P.; Coelho, M.A.Z.; Barros-Timmons, A.M.; Coutinho, J.A.P. Preparation and characterization of organosilicon thin films for selective adhesion of Yarrowia lipolytica yeast cells. *J. Chem. Technol.* **2007**, *82*, 360–366.
2. Bilek, F.; Krizova, T.; Lehocky, M. Preparation of active antibacterial LDPE surface through multistep physicochemical approach: I. Allylamine grafting, attachment of antibacterial agent and antibacterial activity assessment. *Colloid Surf. B* **2011**, *88*, 440–447. [CrossRef]
3. Bilek, F.; Sulovska, K.; Lehocky, M.; Saha, P.; Humpolicek, P.; Mozetic, M.; Junkar, I. Preparation of active antibacterial LDPE surface through multistep physicochemical approach II: Graft type effect on antibacterial properties. *Colloid Surf. B* **2013**, *102*, 842–848. [CrossRef] [PubMed]
4. Asadinezhad, A.; Novak, I.; Lehocky, M.; Sedlarik, V.; Vesel, A.; Junkar, I.; Saha, P.; Chodak, I. A physicochemical approach to render antibacterial surfaces on medical-grade PVC. *Plasma Process. Polym.* **2010**, *7*, 504–514. [CrossRef]
5. Asadinezhad, A.; Novak, I.; Lehocky, M.; Sedlarik, V.; Vesel, A.; Junkar, I.; Saha, P.; Chodak, I. An in vitro bacterial adhesion assessment of surface modified medical-grade PVC. *Colloid Surf. B* **2010**, *77*, 246–256. [CrossRef] [PubMed]

6. Asadinezhad, A.; Novak, I.; Lehocky, M.; Bilek, F.; Vesel, A.; Junkar, I.; Saha, P.; Popelka, A. Polysaccharide coatings on medical-grade PVC: A probe into surface characteristics and bacterial adhesion extent. *Molecules* **2010**, *15*, 1007–1027. [CrossRef] [PubMed]

7. Mozetic, M.; Ostrikov, K.; Ruzic, D.N.; Curreli, D.; Cvelbar, U.; Vesel, A.; Primc, G.; Leisch, M.; Jousten, K.; Malyshev, O.B.; et al. Recent advances in vacuum sciences and applications. *J. Phys. D Appl. Phys.* **2014**, *47*, 153001:1–153001:23. [CrossRef]

8. Vanek, P.; Kolska, Z.; Luxbacher, T.; Garcia, J.A.L.; Lehocky, M.; Vandrovcova, M.; Bacakova, L.; Petzelt, J. Electrical activity of ferroelectric biomaterials and its effects on the adhesion, growth and enzymatic activity of human osteoblast-like cells. *J. Phys. D Appl. Phys.* **2016**, *49*, 175403. [CrossRef]

9. Lehocky, M.; Amaral, P.F.F.; Stahel, P.; Coelho, M.A.Z.; Barros-Timmons, A.M.; Coutinho, J.A.P. Deposition of Yarrowia lipolytica on Plasma Prepared Teflonlike Thin Films. *Surf. Eng.* **2008**, *24*, 23–27. [CrossRef]

10. Karbassi, E.; Asadinezhad, A.; Lehocky, M.; Humpolicek, P.; Vesel, A.; Novak, I.; Saha, P. Antibacterial performance of alginic acid coating on polyethylene film. *Int. J. Mol. Sci.* **2014**, *15*, 14684–14696. [CrossRef]

11. Kováčová, M.; Markovic, Z.M.; Humpolíček, P.; Mičušík, M.; Švajdlenková, H.; Kleinová, A.; Danko, M.; Kubát, P.; Vajďák, J.; Capáková, Z.; et al. Carbon quantum dots modified polyurethane nano-composite as effective photocatalytic and antibacterial agents. *ACS Biomater. Sci. Eng.* **2018**, *4*, 3983–3993. [CrossRef]

12. Adamczyk, Z.; Szyk-Warszynska, L.; Zembala, M.; Lehocky, M. In situ studies of particle deposition on non-transparent substrates. *Colloids Surf. A* **2004**, *235*, 65–72. [CrossRef]

13. Mozetic, M.; Primc, G.; Vesel, A.; Zaplotnik, R.; Modic, M.; Junkar, I.; Recek, N.; Klanjsek-Gunde, M.; Guhy, L.; Sunkara, M.K.; et al. Application of extremely non-equilibrium plasmas in the processing of nano and biomedical materials. *Plasma Sources Sci. Technol.* **2015**, *24*, 015026:1–015026:12. [CrossRef]

14. Mozetic, M.; Vesel, A.; Primc, G.; Eisenmenger-Sittner, C.; Bauer, J.; Eder, A.; Schmid, G.H.S.; Ruzic, D.N.; Ahmed, Z.; Barker, D.; et al. Recent developments in surface science and engineering, thin films, nanoscience, biomaterials, plasma science, and vacuum technology. *Thin Solid Films* **2018**, *660*, 120–160. [CrossRef]

15. Lehocky, M.; Stahel, P.; Koutny, M.; Cech, J.; Institoris, J.; Mracek, A. Adhesion of *Rhodococcus* sp. S3E2 and *Rhodococcus* sp. S3E3 to plasma prepared Teflon-like and organosilicon surfaces. *J. Mater. Process. Technol.* **2009**, *209*, 2871–2875. [CrossRef]

16. Junkar, I.; Cvelbar, U.; Lehocky, M. Plasma treatment of biomedical materials. *Mater. Tehnol.* **2011**, *45*, 221–226.

17. Kalachyova, Y.; Guselnikova, O.; Hnatowicz, V.; Postnikov, P.; Švorčík, V.; Lyutakov, O. Flexible Conductive Polymer Film Grafted with Azo-Moieties and Patterned by Light Illumination with Anisotropic Conductivity. *Polymers* **2019**, *11*, 1856. [CrossRef]

18. Fajstavr, D.; Neznalová, K.; Švorčík, V.; Slepička, P. LIPSS Structures Induced on Graphene-Polystyrene Composite. *Materials* **2019**, *12*, 3460. [CrossRef]

19. Ozaltin, K.; Lehocky, M.; Humpolicek, P.; Pelkova, J.; Di Martino, A.; Karakurt, I.; Saha, P. Anticoagulant Polyethylene Terephthalate Surface by Plasma-Mediated Fucoidan Immobilization. *Polymers* **2019**, *11*, 750. [CrossRef]

20. Smilek, J.; Jarábková, S.; Velcer, T.; Pekař, M. Compositional and Temperature Effects on the Rheological Properties of Polyelectrolyte-Surfactant Hydrogels. *Polymers* **2019**, *11*, 927. [CrossRef]

21. Mokrejš, P.; Mrázek, P.; Gál, R.; Pavlačková, J. Biotechnological preparation of gelatines from chicken feet. *Polymers* **2019**, *11*, 1060. [CrossRef] [PubMed]

22. Belka, J.; Nickel, J.; Kurth, D.G. Growth on metallo-supramolecular coordination polyelectrolyte (MEPE) stimulates osteogenic differentiation of human osteosarcoma cells (MG63) and human bone marrow derived mesenchymal stem cells. *Polymers* **2019**, *11*, 1090. [CrossRef] [PubMed]

23. Karakurt, I.; Ozaltin, K.; Vesela, D.; Lehocky, M.; Humpolíček, P.; Mozetič, M. Antibacterial activity and cytotoxicity of immobilized glucosamine/chondroitin sulfate on polylactic acid films. *Polymers* **2019**, *11*, 1186. [CrossRef]

24. Lee, J.Y.; Aguilar, L.E.; Park, C.H.; Kim, C.S. UV Light assisted coating method of polyphenol caffeic acid and mediated immobilization of metallic silver particles for antibacterial implant surface modification. *Polymers* **2019**, *11*, 1200. [CrossRef] [PubMed]

25. Bernal-Ballen, A.; Lopez-Garcia, J.-A.; Ozaltin, K. (PVA/Chitosan/Fucoidan)-ampicillin: A bioartificial polymeric material with combined properties in cell regeneration and potential antibacterial features. *Polymers* **2019**, *11*, 1325. [CrossRef]

26. Vítková, L.; Musilová, L.; Achbergerová, E.; Minařík, A.; Smolka, P.; Wrzecionko, E.; Mráček, A. Electrospinning of hyaluronan using polymer coelectrospinning and intermediate solvent. *Polymers* **2019**, *11*, 1517. [CrossRef]

27. Dvořáková, H.; Čech, J.; Stupavská, M.; Prokeš, L.; Jurmanová, J.; Buršíková, V.; Ráheľ, J.; Sťahel, P. Fast Surface hydrophilization via atmospheric pressure plasma polymerization for biological and technical applications. *Polymers* **2019**, *11*, 1613. [CrossRef]

28. Habib, S.; Lehocky, M.; Vesela, D.; Humpolíček, P.; Krupa, I.; Popelka, A. Preparation of progressive antibacterial LDPE surface via active biomolecule deposition approach. *Polymers* **2019**, *11*, 1704. [CrossRef]

29. Skopalová, K.; Capáková, Z.; Bober, P.; Pelková, J.; Stejskal, J.; Kašpárková, V.; Lehocký, M.; Junkar, I.; Mozetič, M.; Humpolíček, P. In-vitro hemocompatibility of polyaniline functionalized by bioactive molecules. *Polymers* **2019**, *11*, 1861. [CrossRef]

30. Urbánková, L.; Kašpárková, V.; Egner, P.; Rudolf, O.; Korábková, E. Caseinate-stabilized emulsions of black cumin and Tamanu oils: Preparation, characterization and antibacterial activity. *Polymers* **2019**, *11*, 1951. [CrossRef]

31. Sťahel, P.; Mazánková, V.; Tomečková, K.; Matoušková, P.; Brablec, A.; Prokeš, L.; Jurmanová, J.; Buršíková, V.; Přibyl, R.; Lehocký, M.; et al. Atmospheric pressure plasma polymerized oxazoline-based thin films—antibacterial properties and cytocompatibility performance. *Polymers* **2019**, *11*, 2069. [CrossRef] [PubMed]

32. Shah, R.; Stodulka, P.; Skopalova, K.; Saha, P. Dual crosslinked collagen/chitosan film for potential biomedical applications. *Polymers* **2019**, *11*, 2094. [CrossRef] [PubMed]

33. Vesel, A.; Zaplotnik, R.; Primc, G.; Mozetič, M. Evolution of the surface wettability of PET polymer upon treatment with an atmospheric-pressure plasma jet. *Polymers* **2020**, *12*, 87. [CrossRef] [PubMed]

34. Narimisa, M.; Krčma, F.; Onyshchenko, Y.; Kozáková, Z.; Morent, R.; De Geyter, N. Atmospheric pressure microwave plasma jet for organic thin film deposition. *Polymers* **2020**, *12*, 354. [CrossRef]

35. Villamil Ballesteros, A.C.; Segura Puello, H.R.; Lopez-Garcia, J.A.; Bernal-Ballen, A.; Nieto Mosquera, D.L.; Muñoz Forero, D.M.; Segura Charry, J.S.; Neira Bejarano, Y.A. Bovine decellularized amniotic membrane: Extracellular matrix as scaffold for mammalian skin. *Polymers* **2020**, *12*, 590. [CrossRef]

36. Klučáková, M. Agarose hydrogels enriched by humic acids as the complexation agent. *Polymers* **2020**, *12*, 687. [CrossRef]

Article

Agarose Hydrogels Enriched by Humic Acids as the Complexation Agent

Martina Klučáková

Faculty of Chemistry, Brno University of Technology, Purkyňova 118/464, 612 00 Brno, Czech Republic; klucakova@fch.vutbr.cz; Tel.: +42-054-114-9410

Received: 26 February 2020; Accepted: 17 March 2020; Published: 19 March 2020

Abstract: The transport properties of agarose hydrogels enriched by humic acids were studied. Methylene blue, rhodamine 6G and Cu(II) ions were incorporated into hydrogel as diffusion probes, and then their release into water was monitored. Cu(II) ions as well as both the dyes studied in this work have high affinity to humic substances and their interactions strongly affected their diffusion in hydrogels. It was confirmed that humic acids retarded the transport of diffusion probes. Humic acids' enrichment caused the decrease in the values of effective diffusion coefficients due to their complexation with diffusion probes. In general, the diffusion of dyes was more affected by the complexation with humic acids in comparison with Cu(II) ions. The effect of complexation was selective for the particular diffusion probe. The strongest effect was obtained for the diffusion of methylene blue. It was assumed that metal ions interacted preferentially with acidic functional groups. In contrast to Cu(II) ions, dyes can interact with acidic functional groups, and the condensed cyclic structures of the dye probes supported their interactions with the hydrophobic domains of humic substances.

Keywords: hydrogel; agarose; humic acid; reactivity; diffusion

---

1. Introduction

Hydrogels play an important role in the monitoring of the mobility of pollutants in nature as well as in their removal and water treatment. They are usually based on the materials able to absorb water and different pollutants in their structure. There are many studies and reviews dealing with bio-polymeric and polymeric hydrogels for different environmental applications. Agarose, pectin, alginate or lignin may be mentioned as examples of materials eligible for these purposes [1–9]. This work is focused on the agarose and its hydrogels as a medium for the investigation of their transport properties and effect of humic substances on the release of metal ions and dyes from them. Agarose hydrogels have a number of practical utilizations. They can be, e.g., used as separation media in column chromatography and act as bacterial culture support [1,2]. Their gelling [3], rheological and thermal properties [7–9], and internal structure [10–12] have been widely studied, but the effect of micro-scale structural factors of porous media on effective mass diffusion is not well understood [11], and the molecular structure of agarose is still a matter of debate [3,7]. Some studies stated that the hydrogels consist of thick bundles of agarose chains and large pores of water [10,13], which constitute the main paths for diffusing particles. The hydrogel is widely used as a transport medium for the determination of diffusion characteristics of different molecules and ions [2,13–16]. Pluen et al. [14] studied the diffusion of several different macromolecules through 2% agarose hydrogel. Data were analysed by means of Yimm-Rouse model [17] and the reptation model [13,18]. Gong et al. [13] investigated the effect of aspect ratio of protein on its diffusion in hydrogel and the effect of electrostatic interaction betarween protein and hydrogel on its transport through hydrogel. The influence of of electrostatic and specific interactions on the diffusion and partitioning of various solutes in agarose hydrogels was studied by Fation-Rouge et al. [2]. Gutenwik et al. [15] determined diffusion coefficients of proteins at different pH values and ionic strengths by means of diffusion cells. Golmohamadi et al. [16] measured the self and mutual diffusion of different cations and correlated them with Donnan potentials of hydrogels. Wang et al. [19] characterized the diffusion of cations and anions in thin films of agarose hydrogel. Agarose hydrogel is often used as transport medium passive samplers based on diffusion gradients in thin hydrogel films [19,20].

As can be seen, agarose hydrogels are widely studied and utilized as transport media for different diffusing particles. In our previous studies [21–24], agarose hydrogels were enriched by humic acids as an active component in order to support the complexation of diffusing particles in hydrogels. Humic-agarose hydrogels were characterized by means of their viscoelastic properties [21,24], and the transport of different ions through the hydrogels was studied by means of diffusion cells [21,23,24] and non-stationary transient diffusion [22,23]. The effect of acidic functional groups of humic acids on the complexation and transport of metal and dye ions was investigated by means of the selective blocking of carboxylic groups by methylation [23,24]. Our experiments showed that the addition of humic acids in agarose hydrogel can strongly influence the complexation and diffusion of metal ions and dyes, which resulted in changes in effective diffusion coefficients.

In this study, a different type of experiment was performed. The study is focused on transport properties of agarose hydrogels enriched with humic acids, especially the release of different diffusion probes from the hydrogels. Agarose hydrogel was enriched by humic acids and also by metal or dye ions. The transport out of hydrogel was monitored and the diffusion characteristics of the transport were determined. Simultaneously, the degree of immobilization of ions was calculated, and the ratio between mobile and complexed ions was calculated. The aim was to investigate the influence of interactions of humic acids with probes on their release ability.

2. Experimental

2.1. Materials

Agarose (AG; routine use class), $CuCl_2 \cdot 2H_2O$ (p.a.), methylene blue hydrate (MB; CI basic blue 9) and rhodamine 6G (RH; CI basic red 1) were purchased from Sigma-Aldrich (St. Luis, MO, USA).

Samples of humic acids were purchased from the International Humic Substances Society (IHSS, St. Paul, MN, USA). Elliot soil humic acids (ESHA), Pahokee peat humic acids (PPHA), Suwannee river humic acids (SRHA) and Leonardite humic acids (LEHA) were used in this study. The main characteristics such as elemental composition and the contents and properties of acidic functional groups can be found on the website of the International Humic Substances Society (IHSS).

2.2. Preparation of Hydrogels

The preparation of hydrogels was based on the thermo-reversible gelation of AG aqueous solution. An accurately weighed amount of AG was dissolved in deionized water or in an aqueous solution of humic acids. The mixture was slowly heated with continuous stirring up to 80 °C and stirred at this temperature in order to obtain a transparent solution, and finally sonicated (1 min) to remove gasses. AG hydrogels were prepared using 1 wt % AG solution [21–24]. Afterwards, the AG solution was slowly poured into the polymethylmethacrylate (PMMA) spectrophotometric cuvette (inner dimensions: 10 mm × 10 mm × 45 mm). The cuvette orifice was immediately covered with pre-heated plate of glass to prevent drying and shrinking of gel. Flat surface of the boundary of resulting hydrogels was provided by wiping an excess solution away. Gentle cooling of cuvettes at the laboratory temperature led to the gradual gelation of the mixture [22,23]. AG–HA hydrogels were prepared from 1 wt % AG solution containing 0.01 wt % of HA. The AG and humic contents in final hydrogels were chosen on the basis of our previous results and experimental experiences [21–24]. Images of pure agarose hydrogel and hydrogel enriched by humic acids in cuvettes are shown in Figure S1 (from the Supplementary Materials).

Aqueous solutions of CuCl$_2$, MB, and RH were used as the donor solutions for the incorporation of the diffusion probes into hydrogels. Their initial concentrations were equal to 0.1 mol.dm^{-3} for Cu(II) salt and 1 mg.dm^{-3} for dyes. The incorporation of the probes into hydrogels was based on diffusing of Cu(II) ions and dyes into the hydrogels. The cuvettes filled with hydrogels were placed into stirred donor solutions (4 cuvettes in 200 cm^3). The diffusion probes have been diffusing into the hydrogel until a constant concentration throughout the whole hydrogel was achieved [25,26].

2.3. Diffusion-Release Experiments

The cuvettes with AG and AG–HA hydrogels enriched by diffusion probes were placed in stirred distilled water (4 cuvettes in 200 cm^3). The release of diffusion probes into water was monitored over time. The concentrations of probes in leachates were measured by means of UV-VIS spectrometer Hitachi U3900H (Hitachi, Tokyo, Japan). The data were used for the calculation of diffusion fluxes from the hydrogels into water through the square orifices of the cuvettes.

Simultaneously, the distributions of diffusion probes in hydrogels were determined in selected time intervals. The cuvettes were taken out of the leachates and the UV-VIS spectra were measured at various distances from the orifice by means of Varian Cary 50 UV-VIS spectrophotometer equipped with the special accessory providing controlled fine vertical movement of the cuvette in the spectrophotometer. Using the collected UV-VIS spectra, the concentrations of the probes were determined at different positions in gels [22]. The obtained data were used to compute the concentration profiles of probes in the cuvettes. The diffusion fluxes determined as the differences between the total contents of probes in hydrogels before diffusion experiments and the contents in hydrogels at given times should be the same as the values calculated on the basis of the concentrations measured in leachates; therefore their values were determined by two different measurements and averaged. All experiments were performed at laboratory temperature (25 ± 1 °C). Data are presented as average values with standard deviation bars. Schematic illustration of release experiment is shown in Figure S2 (from the Supplementary Materials).

3. Results and Discussion

In this work, the effect of standard humic acids as the complexation agents added in agarose hydrogels was studied by means of so-called diffusion-release experiments. Humic acids are known as substances which complex effectively with metal ions [25–33] and dyes [21–24]. Carboxylic functional groups, as well as aromatic structures and π–π interactions, are important in their reactivity. The amounts of diffusion probes in hydrogels differed slightly according to type of added humic acids (Table 1).

Table 1. Concentrations of diffusion probes in the AG and AG–HA hydrogels before diffusion-release experiments.

Hydrogel	Cu: $c_{0,h}$ (mmol.dm^{-3})	MB: $c_{0,h}$ (mmol.dm^{-3})	RH: $c_{0,h}$ (mmol.dm^{-3})
AG	85.3 ± 7.1	7.6 ± 0.4	6.4 ± 0.1
AG–ESHA	91.2 ± 7.6	6.9 ± 0.6	6.0 ± 0.2
AG–PPHA	90.8 ± 5.9	5.7 ± 0.3	4.6 ± 0.1
AG–SRHA	88.6 ± 6.6	5.2 ± 0.2	4.4 ± 0.2
AG–LEHA	96.2 ± 8.2	7.3 ± 0.2	5.7 ± 0.1

The contents differed more in the case of organic dyes. Their amounts in hydrogels without humic acids seem to be higher in comparison with enriched hydrogels. In contrast, the content of copper is slightly higher. It is well known that humic acids have very high affinity to Cu(II) ions [24–28,30–33]. Therefore, copper is a traditional model metal used to study humic reactivity. The increase in the content of Cu(II) ions in agarose hydrogels enriched by humic acids can be considered as the result of this humic affinity observed also in our previous studies [24–28]. If we compare the contents of Cu(II) ions in hydrogels containing different humic acids with the contents of their acidic functional groups declared by IHSS [34,35], we can find that the content of Cu(II) ions in hydrogels increases with the increasing total acidity of studied humic acids. This confirmed that the acidic functional groups play the most important role in the interactions of humic acids with metal ions. Nevertheless, we must take account of the strengths (dissociation abilities) of functional groups and the fact that metal ions can be bound by other active centres, as studied in detail in [27]. It should be noted that IHSS published the content of acidic functional groups related to the content of carbon in humic acids and it is necessary to re-calculate the data on the whole samples of humic acids. On the other hand, this increase is not high, which means that only a smaller portion of metal present in hydrogel can be bonded by humic acids which corresponds with the low content of humic acids in hydrogel. No relationship exists between content of dyes in hydrogels and amounts of functional groups. There are more possibilities for the binding of dyes by humic acids. Apart from dissociable functional groups, the unsaturated and aromatic structures are more asserted due to the aromatic structure of studied dyes.

The knowledge of contents of diffusion probes was necessary for the mathematical description of their release from hydrogels. The effective diffusion coefficients $D_{\text{ef,h}}$ of Cu(II) ions and dyes in hydrogels were calculated on the basis of the following equation [26,36,37]:

$$m_{\text{h}\rightarrow\text{s}} = 2\frac{\varepsilon c_{0,\text{h}} - c_{0,\text{s}}}{1 + \varepsilon\sqrt{D_\text{s}/D_{\text{ef,h}}}}\sqrt{\frac{D_\text{s}t}{\pi}} \tag{1}$$

where $m_{\text{h}\rightarrow\text{s}}$ is the total diffusion flux at time t; $c_{0,\text{h}}$ and $c_{0,\text{s}}$ are the initial concentrations of the probe in the hydrogel and aqueous solution (equal to zero in this case); $D_{\text{ef,h}}$ and D_s are the effective diffusion coefficient of the probe in the hydrogel and the diffusion coefficient of the probe in the supernatant; ε is the ratio between concentrations of the probe in the supernatant (c_s) and hydrogel (c_h) in given time, i.e., $\varepsilon = c_\text{s}/c_\text{h}$.

The values of D_s for Cu(II) ions are tabulated [38]: 1.43×10^{-9} m$^2\cdot$s^{-1}. The values of D_s for dyes were determined in our previous study [22]. They were extrapolated for 25 °C and used in this work as: 8.42×10^{-10} m$^2\cdot$s^{-1}for MB and 8.93×10^{-10} m$^2\cdot$s^{-1}for RH [24]. These results are in agreement with values determined using other methods [16,19,39–41].

Experimental data fitted by Equation (1) are shown in Figure 1. We can see that they are in good agreement with the mathematical model. The slopes of the lines were used for the calculation of effective diffusion coefficients $D_{\text{ef,h}}$. Their values are listed in Table 2.

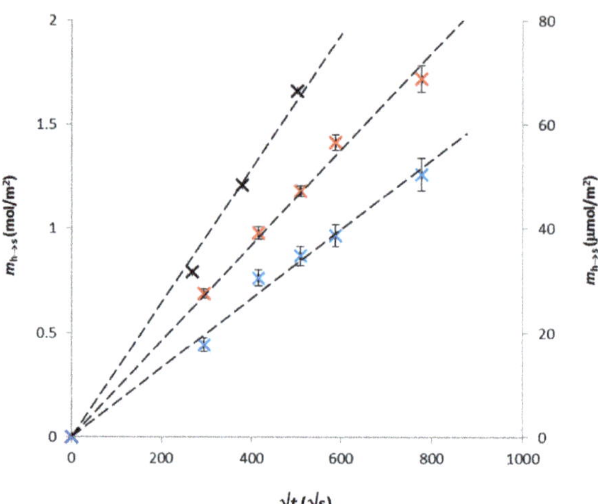

Figure 1. Experimental data obtained for Cu (**black**), MB (**blue**) and RH (**red**) fitted by Equation (1).

Table 2. The values of effective diffusion coefficients of diffusion probes in the AG and AG–HA hydrogels.

Hydrogel	Cu: $D_{ef,h}$ (10^{-10} m^2·s^{-1})	MB: $D_{ef,h}$ (10^{-11} m^2·s^{-1})	RH: $D_{ef,h}$ (10^{-11} m^2·s^{-1})
AG	12.21 ± 0.72	4.16 ± 0.13	5.21 ± 0.16
AG–ESHA	6.40 ± 0.30	1.25 ± 0.10	1.33 ± 0.06
AG–PPHA	7.34 ± 0.36	1.86 ± 0.21	3.30 ± 0.25
AG–SRHA	8.39 ± 0.30	0.93 ± 0.04	3.76 ± 0.25
AG–LEHA	8.76 ± 0.38	1.26 ± 0.06	1.49 ± 0.07

The highest values of effective diffusion coefficients were determined for pure AG hydrogel. The obtained values can be compared with the results published in other studies [16,19,40,42–44]. Wang et al. [19] characterized the agarose hydrogel used in so-called DGT technique (diffusive gradients in thin films) for monitoring of different substances in natural environments (waters, soils, sediments). They investigated diffusivities of several ions including Cu(II) in 1.5 wt % agarose hydrogel and determined the value diffusion coefficient equal to 6.59×10^{-10} m^2·s^{-1}. The value obtain in other study [43] for 2% agarose hydrogel was slightly lower (6.59×10^{-10} m^2·s^{-1}). In this study, practically double value of $D_{ef,h} = 1.22 \times 10^{-9}$ m^2·s^{-1}. This difference is partially caused by lower content agarose in our hydrogel and partially by different methods used for the determination of diffusivity, which is principal for the resulting value of the diffusivity [44]. The published values of diffusion coefficients of MB in 1.5 wt % agarose hydrogel (enriched by 3 wt % CaCl$_2$) were between 2.9 and 3.9×10^{-10} m^2·s^{-1} depending on the MB concentration and pH [40]. Similarly, the diffusion coefficients of RH in in 1.5 wt % agarose hydrogel were between 2 and 3.5×10^{-10} m^2·s^{-1} depending on the concentration and pH [16]. Variations in obtained values of diffusion coefficients of RH were observed also for its diffusion in water (2.8–4.3×10^{-10} m^2·s^{-1}) [44].

In the case of release of Cu(II) ions, the highest value of D_{ef} was determined for the AG–LEHA hydrogel. The same hydrogel achieved the highest initial content of Cu(II) ions. The LEHA sample can be characterized by the highest total acidity, and C/H and C/N ratios; aromaticity; and the lowest O/C ratio [34]. The lowest value of D_{ef} was determined for the AG–ESHA hydrogel which can be characterized by the lowest C/N ratio, but C/H and O/C are comparable with LEHA. Simultaneously, ESHA has relatively high total acidity and aromaticity. The mobility of dyes in AG–ESHA and AG–LEHA hydrogels were similar. Both humic acids have high C/H and low O/C ratios. They are also more aromatic in comparison with other two samples. In contrast, the highest diffusivity of MB in the AG-PPHA hydrogel is probably caused by common impact of low content of acidic functional groups, C/H and C/N ratios and low aromaticity.

Chakraborty et al. [42] combined DGT with the CLE (competing ligand exchange) technique in order to investigate diffusion of metal ions in the presence of humic substances. They observed the increase in the diffusion coefficients from 6.06×10^{-10} m^2·s^{-1} (obtained for Cu(II) ion) to 6.2×10^{-11} m^2·s^{-1} (obtained for Cu–NLHA complex), 8.0×10^{-11} m^2·s^{-1} (obtained for Cu–NLFA complex), and 8.5×10^{-11} m^2·s^{-1} (obtained for complex of Cu with Suwannee River natural organic matter). Similarly, the decrease in diffusion coefficient of Cu–HA in comparison with free Cu(II) ions in the hydrogel based on polyacrylamide cross-linked with an agarose derivative was from 5.48×10^{-10} to 5.70×10^{-11} m^2·s^{-1} [43]. In contrast, the effect of humic acids on the diffusion of RH in water was much weaker: from 2.88×10^{-10} to 2.22×10^{-10} m^2·s^{-1} for RH-PPHA complex and 2.15×10^{-10} m^2·s^{-1} for RH-SRHA complex. Their values of diffusion coefficient in 1.35 wt % AG hydrogels achieved 91%, 87% and 88% of diffusion coefficients in water, respectively [5]. It means that the effect of humic substances on the diffusion in agarose hydrogels observed by different authors differed. This finding showed that we must be very careful in the comparison of diffusion characteristics obtained by different authors and different methods [44].

The decrease of effective diffusion coefficients obtained for the hydrogels enriched by humic acids can be the result of two effects. The first is a possible change in hydrogel structure (see SEM of lyophilized hydrogels in Figure S3, from the Supplementary Materials). In spite of the fact that the content of humic acids in hydrogel is relatively low, their incorporation into AG hydrogel can influence its inner structure, including the distribution, size and shape of hydrogel pores [24]. The structure of humic acids is very dynamic and sensitive to circumstances such as concentration, pH and ionic strength [24,45,46]. They can be characterized by a supramolecular arrangement of relatively small particles in co-existence with bigger macromolecules [24,45–51], which makes it possible to respond to changes in their surroundings.

The second effect is a possible interaction between the diffusion probes and HA and humic acids during the transport of the probes through the hydrogel. In comparison with our previous works [21–28], the diffusion probes are in equilibrium with humic acids at the beginning of the release experiments. It means that the immobilization of the probe has the same rate as its liberation from binding sites. It is known that diffusion probes occurring in the hydrogels can be divided into three fractions: free mobile particles without chemical binding to humic acids, ion exchangeable ions bound by electrostatic forces, and strongly (covalently) bound particles in humic complexes [25,28,52]. These fractions are in a dynamic equilibrium and can convert to other forms as a result of changes in circumstances.

In the case of release experiments, the mobile fraction can easily diffuse out of the hydrogel. It results in the displacement from equilibrium, and particles of probe can be liberated from the exchangeable and strongly bound fractions. These processes can strongly affect the release of probe from AG–HA hydrogels and are dependent on the character of humic acids. In Figure 2, the ratios between effective diffusion coefficients $D_{ef,h}$ and the diffusion coefficients of probes in aqueous solutions D_s and the ratios between effective diffusion coefficients $D_{ef,h}$ for AG–HA hydrogels and the values obtained for pure AG hydrogel are shown. The differences between dyes and Cu(II) ions were observed. While Cu(II) ions amount to 40%–80% of their diffusion coefficients in solution, that proportion is only 1%–6% in the case of dyes, mainly because of their sizes. The liberation of dyes from their ion-exchangeable and strongly bound fractions has an influence comparable with Cu(II) ions (AG–PPHA and AG–SRHA) or lower (AG–ESHA and AG–LEHA). The values of $D_{ef,h}$ obtained for hydrogels enriched by humic acids amount to 20%–70% of the values obtained for pure AG hydrogel.

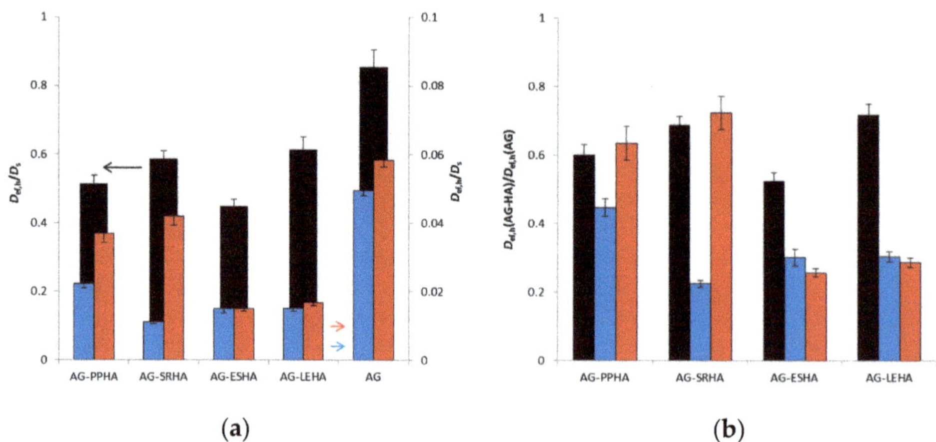

(a) (b)

Figure 2. The ratios between effective diffusion coefficients $D_{ef,h}$ and the diffusion coefficients of probes in aqueous solutions D_s (**a**); the ratios between effective diffusion coefficients $D_{ef,h}$ for AG-HA hydrogels and the values obtained for pure AG hydrogel (**b**): Cu (black); MB (blue); and RH (red).

The influence of inner structure of hydrogel on the diffusion can be characterized as so-called structure fraction μ, which is the ratio between the porosity ϕ and tortuosity τ:

$$\mu = \phi/\tau \tag{2}$$

The value of μ can be determined as the ratio between the diffusion coefficient of the probe in AG hydrogel and the diffusion coefficient of the probe in water (D_s). If we focus on the ratios obtained for pure AG hydrogel, we can state that their values are lower for the diffusion of dyes (5%–6%) in comparison with Cu(II) ions (> 80%). In the case of AG–HA hydrogels, the situation is more complex. The release of diffusion probes from hydrogels can be described as the non-stationary diffusion based on Fick's equation [36,37]:

$$\frac{\partial c}{\partial t} = D_{\text{ef,h}} \frac{\partial^2 c}{\partial x^2} \tag{3}$$

where c represents the concentration of the diffusing compound at time t and position x (the coordinate parallel to the direction of the diffusion movement). The diffusion coefficient $D_{\text{ef,h}}$ is the main parameter characterizing the rate of the transport. The diffusion coefficient is an "effective" characteristic which reflects the influence of chemical interactions of diffusion probe with humic acids in their transport through the hydrogel and the influence of inner structure of hydrogel. Mathematically, the effects of the chemical reaction can be described by the following equation based on the conservation of mass:

$$\frac{\partial c}{\partial t} = D^* \frac{\partial^2 c}{\partial x^2} - \dot{r} \tag{4}$$

where D^* is the diffusion coefficient affected only by the porous structure of the hydrogel and $\dot{r}$ is the rate of chemical reaction. In this case, the value of D^* is equal to $D_{\text{ef,h}}$ for pure AG hydrogel. If a fast chemical reaction in the presence of local equilibrium between free mobile probes (c) and immobilized ones (c_{im}) is presumed (K is the equilibrium constant), then

$$c_{\text{im}} = Kc \tag{5}$$

and Equation (4) can be written as

$$\frac{\partial c}{\partial t} = D^* \frac{\partial^2 c}{\partial x^2} - K \frac{\partial c}{\partial t}, \tag{6}$$

and consequently,

$$\frac{\partial c}{\partial t} = \frac{D^*}{1 + K} \frac{\partial^2 c}{\partial x^2} = D_{\text{ef,h}} \frac{\partial^2 c}{\partial x^2} \tag{7}$$

Since the diffusion coefficient in the hydrogel D^* is dependent on its porosity and tortuosity expressed by the structural factor μ according to the Equation (2), the following relation can be written:

$$D_{\text{ef,h}} = \frac{D^*}{1 + K} = \frac{\mu D_s}{1 + K} \tag{8}$$

in which the effects of the tortuous movement of the diffusing matter in the hydrogel and the chemical reaction between diffusion probe and humic acids are involved [25–28,36,37].

The values of K can be calculated only assuming that the inner structure of hydrogel was not changed by the addition of humic acids and they should be proportional to the ratios between effective diffusion coefficients $D_{\text{ef,h}}$ for AG–HA hydrogels and the values obtained for pure AG hydrogel shown in Figure 2b. As it was described in [24], rheological measurements showed that the AG hydrogel is more resistant to applied stress than hydrogels enriched with humic substances and the networks of the AG–HA hydrogels can easily collapse. The behaviour of hydrogels enriched with humic substances shifted towards that of viscoelastic liquids. This means that hydrogels containing humic substances had a lower ability to resist mechanical stresses, which can be connected with their higher permeability.

Therefore the effect of interactions between diffusion probes and humic acids is probably higher than the values of K listed in Table 3.

Table 3. Values of the apparent equilibrium constants K determined on the basis of Equation (8).

Hydrogel	Cu: K (-)	MB: K (-)	RH: K (-)
AG–ESHA	0.64 ± 0.03	2.32 ± 0.19	2.90 ± 0.13
AG–PPHA	0.22 ± 0.01	1.23 ± 0.07	0.58 ± 0.04
AG–SRHA	0.37 ± 0.01	3.45 ± 0.15	0.38 ± 0.02
AG–LEHA	0.89 ± 0.01	2.29 ± 0.11	2.49 ± 0.12

The concentration profile in hydrogel during the release of Cu(II) ions and dyes can be described as [25,36,37]:

$$c = \frac{1}{2}c_{0,h}\left[erf\frac{l-x}{2\sqrt{D_{ef,h}t}} + erf\frac{l+x}{2\sqrt{D_{ef,h}t}}\right] \qquad (9)$$

where l is the length of hydrogel and x is the distance from the interface between hydrogel and solution. This model is in a good agreement with data obtained for Cu(II) ions (see Figure 3). The small differences were observed close to the interface between hydrogels and solutions. The agreement between mathematical model and experimental data showed on the fact that the release of Cu(II) ions corresponded with our presumptions and they were accumulated on the interface. Similar results were obtained for all studied AG–HA hydrogels and Cu(II) ions. In contrast, the agreement of the Equation (8) with experimental data obtained for dyes was worse. It seems that dyes are accumulated in a certain distance from the interface. This was observed mainly in the case of RH (see Figure 3b). It is not easy to explain it. It is known that MB and RH can form bigger aggregates [53–57]. This formation together with general bigger size of dye probes (in comparison with metal ions) can support the observed accumulation.

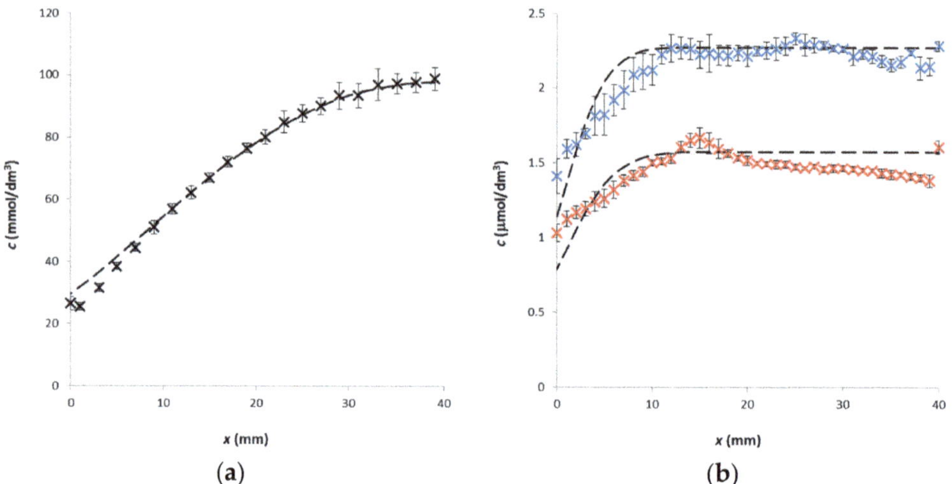

Figure 3. The concentration profile of Cu(II) ions in AG–LEHA hydrogel after 40 h from the start of release (**a**), and the concentration profiles of MB (blue) and RH (red) in AG–LEHA hydrogel after 168 h from the start of release (**b**) fitted by Equation (8) – dashed curves.

As mentioned above, the transport through hydrogel can be affected by two factors: the tortuous movement of the diffusing particles in the porous structure of hydrogel and the interactions of diffusing particles with hydrogel. On the condition that pure agarose hydrogel cannot interact with a diffusion probe, we can determine the influence of the porous structure on the diffusion. The decrease in the diffusivity of probes in pure agarose hydrogels (in comparison with the diffusivity in water) can be attributed fully to the tortuosity effect. In general the movement of diffusing particles can be suppressed by their sizes. Particles of dyes are generally bigger than metal ions; therefore their Brownian motion is less intensive. On the other hand, pore size of agarose hydrogel exceed significantly the Stokes hydrodynamic radius of dyes [21,58]. The decrease is much stronger for dyes which can be connected with their sizes. The accumulation of dyes in a certain distance from the interface is more likely connected with the humic acids contained in enriched hydrogel. No accumulation was observed in the case of diffusion in pure agarose hydrogel. The most intensive accumulation was observed for the hydrogel enriched by LEHA, the weakest for ESHA. It is not easy to explain this finding. The phenomenon was observed only in release experiments. It means that dye was homogeneously distributed in hydrogel and (partially) complexed with humic acids and the equilibrium between humic acids, dye and formed complexes in the beginning is assumed. When the release of dye from hydrogel started, the equilibrium was distorted and some dye can be liberated from humic complexes. We assumed that the observed accumulation can be connected with the disruption of equilibrium and an effort of the system to attain new equilibrium. It seems that humic-dye complexes are (partially) able to diffuse towards the interface between hydrogel and water, but their movement is much slower in comparison with free dye particles. It resulted in the situation wherein an excess of free dye particles arose in the hydrogel closer to interface and their depletion in the hydrogel far from the interface; therefore, the equilibrium must be attained again and again as the release proceeds. This means that a part of free movable dyes can be complexed in the hydrogel closer to interface and a part of complexes can be disintegrated in the hydrogel far from the interface. Other effect is that the pores in hydrogel are filled by solution containing both free dyes and probably also by their complexes with humic acids which can obstruct the movement of smaller free dye particles. Both these effects probably resulted in the described state and the maximum observed on the concentration profile of dye in hydrogel. Different types of humic acids (extracted from different matrices) were used in order to compare their abilities to interact with diffusion probes and influence their release out of hydrogel. It is necessary to realize further experiments in order to investigate our findings in detail.

4. Conclusions

The influence of humic acids on the transport of metal ions and dyes in agarose hydrogel was studied. It was confirmed that humic acids retarded the transport of diffusion probes. Humic acids' enrichment caused decreases in the values of effective diffusion coefficients due to their complexation with diffusion probes. The effect of complexation was selective for the particular diffusion probe. The strongest effect was obtained for the diffusion of MB in the AG–SRHA hydrogel, the lowest one for the diffusion of Cu(II) ions in the AG–PPHA hydrogel. In general, the diffusion of dyes was more affected by the complexation with humic acids in comparison with metal ions. We assume that metal ions interacted preferentially with acidic functional groups. In contrast, dye can interact with acidic functional groups and the condensed cyclic structure of the dye probes supported their interactions with the hydrophobic domains of humic substances.

The results can be used in the investigation of the functioning of natural organic matter in the transport of pollutants in natural systems. Humic acids, as important constituents of soil organic matter, are able to affect, significantly, the migration and bioavailability of some pollutants in nature. In this study, agarose hydrogel was used as a model of a system with a homogeneous distribution of humic acids contaminated by metal ions and dyes. This model hydrogel was very wet in order to study the release of pollutants out of hydrogel. The purpose was to assess the effect of humic acids (as the constituent of soil organic matter) on the mobility of pollutants in wet soil. It means that pollutants present in soil can be partially complexed by humic substances and the movements of complexed and free pollutants are generally restricted if the soil is dry. In contrast, pollutants can diffuse relatively fast in wet soils, and their movement can be supported also by a convection in soggy soils. This study was focused on the diffusion of pollutants in a model of wet soil (e.g., after rain). Results described in this study showed that some pollutants complexed by humic acids are (partially) able to diffuse through pore structure, but their movement is slower and can cause an accumulation of dyes in a certain position (distance from interface). This accumulation can influence the ensuing release from the pore structure into water (e.g., it can reduce the effective size of pores). Therefore, the results obtained in this study can help in the investigation of the functioning of natural organic matter in the transport of pollutants in natural systems. The effective diffusion coefficient determined on the basis of this study included the influence of pore structure and the interactions between humic substances and pollutants. Both pores and affinity of organic matter to pollutants differ with the type and quality of soil. Therefore, the methods and technics presented in this study can be used for predictions of the mobility and bioavailability of pollutants. The mathematical models for different diffusion processes are "universal" and can be used for different hydrogel materials (inert and reactive) and different diffusion probes. One of them could be the use of humic hydrogels as the material having controlled release of nutrients in agriculture.

Supplementary Materials: The following are available online at http://www.mdpi.com/2073-4360/12/3/687/s1, Figure S1: Pure agarose hydrogel (left) and agarose hydrogel enriched by humic acids (right) in cuvettes, Figure S2: Schematic illustration of release experiment, Figure S3. SEM of lyophilized hydrogels (ZEISS EVO LS 10): pure agarose hydrogel (left) and agarose hydrogel enriched by humic acids (right).

Author Contributions: M.K. performed the experiments, processed and analysed the experimental data and wrote the paper. All authors have read and agreed to the published version of the manuscript.

Funding: This research was funded by the National Programme for Sustainability I (Ministry of Education, Youth and Sports), grant number REG LO1211, Materials Research Centre at FCH BUT-Sustainability and Development.

Conflicts of Interest: The authors declare no conflict of interest.

References

1. Djabourov, M.; Clark, A.H.; Rowlands, D.W.; Rossmurphy, S.B. Small-angle X-ray-scattering characterization of agarose sols and gels. *Macromolecules* **1989**, *22*, 180–188. [CrossRef]
2. Fatin-Rouge, N.; Milon, A.; Buffle, J.; Goulet, R.R.; Tessier, A. Diffusion and partitioning of solutes in agarose hydrogels: The relative influence of electrostatic and specific interactions. *J. Phys. Chem. B* **2003**, *107*, 12126–12137. [CrossRef]
3. Singh, T.; Meena, R.; Kumar, A. Effect of sodium sulfate on the gelling behavior of agarose and water structure inside the gel networks. *J. Phys. Chem. B* **2009**, *113*, 2519–2525. [CrossRef]
4. Sourbh, T.; Jyoti, C.; Vinod, K.; Vijay, K.T. Progress in pectin based hydrogels for water purification: Trends and challenges. *J. Environ. Manag.* **2019**, *238*, 210–223.
5. Sourbh, T.; Bhawna, S.; Anki, V.; Jyoti, C.; Sigitas, T.; Vijay, K.T. Recent progress in sodium alginate based sustainable hydrogels for environmental applications. *J. Clean. Prod.* **2018**, *198*, 143–159.
6. Sourbh, T.; Penny, P.G.; Messai, A.M.; Sigitas, T.; Yogendra, K.M.; Vijay, K.T. Progress in lignin hydrogels and nanocomposites for water purification: Future perspectives. *Vacuum* **2017**, *146*, 342–355.
7. Fernandez, E.; Lopez, D.; Mijangos, C.; Duskova-Smrckova, M.; Ilavsky, M.; Dusek, K. Rheological and thermal properties of agarose aqueous solutions and hydrogels. *J. Polym. Sci. B* **2008**, *46*, 322–328. [CrossRef]
8. Barrangou, L.M.; Daubert, C.R.; Foegeding, E.A. Textural properties of agarose gels. I. Rheological and fracture properties. *Food Hydrocolloid.* **2006**, *20*, 184–195. [CrossRef]
9. Barrangou, L.M.; Drake, M.; Daubert, C.R.; Foegeding, E.A. Textural properties of agarose gels. II. Relationships between rheological properties and sensory texture. *Food Hydrocolloid.* **2006**, *20*, 196–203. [CrossRef]
10. Aymard, P.; Martin, D.R.; Plucknett, K.; Foster, T.J.; Clark, A.H.; Norton, I.T. Influence of thermal history on the structural and mechanical properties of agarose gels. *Biopolymers* **2001**, *59*, 131–144. [CrossRef]
11. Kim, H.; Kim, H.J.; Huh, H.K.; Hwang, H.J.; Lee, S.J. Structural design of a double-layered porous hydrogel for effective mass transport. *Biomicrofluidisc* **2015**, *9*, 024104. [CrossRef] [PubMed]
12. Narayanan, J.; Xiong, J.Y.; Liu, X.Y. Determination of agarose gel pore size: Absorbance measurements vis a vis other techniques. *J. Phys. Conf. Ser.* **2006**, *28*, 83–86. [CrossRef]

13. Gong, J.P.; Hirota, N.; Kakugo, A.; Narita, T.; Osada, Y. Effect of aspect ratio on protein diffusion in hydrogels. *J. Phys. Chem. B* **2000**, *104*, 9904–9908. [CrossRef]

14. Pluen, A.; Netti, P.A.; Jain, R.K.; Berk, D.A. Diffusion of macromolecules in agarose gels: Comparison of linear and globular configurations. *Biophys. J.* **1999**, *77*, 542–552. [CrossRef]

15. Gutenwik, J.; Nilsson, B.; Axelsson, A. Determination of protein diffusion coefficients in agarose gel with a diffusion cell. *Biochem. Eng. J.* **2004**, *19*, 1–7. [CrossRef]

16. Golmohamadi, M.; Davis, T.A.; Wilkinson, K.J. Diffusion and partitioning of cations in an agarose hydrogel. *J. Phys. Chem. A* **2012**, *116*, 6505–6510. [CrossRef]

17. Doi, M.; Edwards, S.F. *The Theory of Polymer Dynamics*; Oxford University Press: Oxford, UK, 1986.

18. De Gennes, P.G. *Sealing Concepts in Polymer Physics*; Cornell University Press: Ithac, NY, USA, 1979.

19. Wang, Y.; Ding, S.; Gong, M.; Xu, S.; Xu, W.; Zhang, C. Diffusion characteristics of agarose hydrogel used in diffusive gradients in thin films for measurements of cations and anions. *Anal. Chim. Acta* **2016**, *945*, 47–56. [CrossRef]

20. Urík, J.; Vrána, B. An improved design of a passive sampler for polar organic compounds based on diffusion in agarose hydrogel. *Environ. Sci. Pollut. Res.* **2019**, *26*, 15273–15284. [CrossRef]

21. Sedláček, P.; Smilek, J.; Klučáková, M. How interactions with polyelectrolytes affect mobility of low molecular ions - Results from diffusion cells. *React. Funct. Polym.* **2013**, *73*, 1500–1509. [CrossRef]

22. Sedláček, P.; Smilek, J.; Klučáková, M. How interactions with polyelectrolytes affect mobility of low molecular ions – 2. Non-stationary diffusion experiments. *React. Funct. Polym.* **2014**, *75*, 41–50. [CrossRef]

23. Smilek, J.; Sedláček, P.; Kalina, M.; Klučáková, M. On the role of humic acids' carboxyl groups in the binding of charged organic compounds. *Chemosphere* **2015**, *138*, 503–510. [CrossRef] [PubMed]

24. Klučáková, M.; Smilek, J.; Sedláček, P. How humic acids affect the rheological and transport properties of hydrogels. *Molecules* **2019**, *24*, 1545. [CrossRef] [PubMed]

25. Klučáková, M.; Pekař, M. Transport of copper (II) ions in humic gel—New results from diffusion couple. *Colloid. Surf. A* **2009**, *349*, 96–101. [CrossRef]

26. Klučáková, M.; Jarábková, S.; Velcer, T.; Kalina, M.; Pekař, M. Transport of a model diffusion probe in polyelectrolyte-surfactant hydrogels. *Colloid. Surf. A* **2019**, *573*, 73–79. [CrossRef]

27. Klučáková, M.; Pekař, M. Study of Structure and properties of humic and fulvic acids. III. Study of complexation of Cu^{2+} ions with humic acid in sols. *J. Polym. Mater.* **2003**, *20*, 145–154.

28. Klučáková, M.; Pekař, M. Study of structure and properties of humic and fulvic acids. IV. Study of interactions of Cu^{2+} ions with humic gels and final comparison. *J. Polym. Mater.* **2003**, *20*, 155–162.

29. Klučáková, M.; Kalina, M.; Smilek, J.; Laštůvková, M. The transport of metal ions in hydrogels containing humic acids as active complexation agent. *Colloid. Surf. A* **2018**, *557*, 116–122. [CrossRef]

30. Manceau, A.; Matynia, A. The nature of Cu bonding to natural organic matter. *Geochim. Cosmochim. Acta* **2010**, *74*, 2556–2580. [CrossRef]

31. Xu, J.; Tan, W.; Xiong, J.; Wang, M.; Fang, L.; Koopal, L.K. Copper binding to soil fulvic and humic acids: NICA-Donnan modeling and conditional affinity spectra. *J. Colloid Interface Sci.* **2016**, *473*, 141–151. [CrossRef]

32. Sierra, J.; Roig, N.; Gimenez Papiol, G.; Perez-Gallego, E.; Schuhmacher, M. Prediction of the bioavailability of potentially toxic elements in freshwaters. Comparison between speciation models and passive samplers. *Sci. Total Environ.* **2017**, *605–606*, 211–218. [CrossRef]

33. Baek, K.; Yang, J.-W. Humic-substance-enhanced ultrafiltration for removal of heavy metals. *Sep. Sci. Technol.* **2005**, *40*, 699–708. [CrossRef]

34. Humic-Substances.Org. Available online: https://www.humic-substances.org (accessed on 24 February 2020).

35. Ritchie, J.D.; Perdue, E.M. Proton-binding study of standard and reference fulvic acids, humic acids, and natural organic matter. *Geochim. Cosmochim. Acta* **2003**, *67*, 85–96. [CrossRef]

36. Crank, J. *The Mathematics of Diffusion*, 2nd ed.; Clarendon Press: Oxford, UK, 1975.

37. Cussler, E.L. *Diffusion: Mass Transfer in Fluid Systems*, 2nd ed.; Cambridge University Press: Cambridge, UK, 1984.

38. Haynes, W.M. *Handbook of Chemistry and Physics*, 93rd ed.; CRC Press: Boca Raton, FL, USA, 2012.

39. Leaist, D.G. The effect of aggregation, counterion binding, and added NaCl on diffusion of aqueous methylene blue. *Can. J. Chem.* **1988**, *66*, 2452–2456. [CrossRef]

40. Samprovalaki, K.; Robbins, P.T.; Fryer, P.J. Investigation of the diffusion of dyes in agar gels. *J. Food Eng.* **2012**, *111*, 537–545. [CrossRef]

41. Gendron, P.O.; Avaltroni, F.; Wilkinson, K.J. Diffusion coefficients of several rhodamine derivatives as determined by pulsed field gradient-nuclear magnetic resonance and fluorescence correlation spectroscopy. *J. Fluoresc.* **2008**, *18*, 1093–1101. [CrossRef] [PubMed]

42. Chakraborty, P.; Manek, A.; Nizogi, S.; Hudson, J. Determination of dynamic metal complexes and their diffusion coefficients in the presence of different humic substances by combining two analytical techniques. *Anal. Lett.* **2014**, *47*, 1224–1241. [CrossRef]

43. Zhang, H.; Davison, W. Diffusional characteristics of hydrogels used in DGT and DET techniques. *Anal. Chim. Acta* **1999**, *398*, 329–340. [CrossRef]

44. Majer, G.; Melchior, J.P. Characterization of the fluorescence correlation spectroscopy (FCS) standard Rhodamine 6G and calibration of its diffusion coefficient in aqueous solutions. *J. Chem. Phys.* **2014**, *140*, 094201. [CrossRef]

45. Klucakova, M.; Veznikova, K. The role of concentration and solvent character in the molecular organization of humic acids. *Molecules* **2016**, *21*, 1410. [CrossRef]

46. Klucakova, M.; Veznikova, K. Micro-organization of humic acids in aqueous solutions. *J. Mol. Struct.* **2017**, *1144*, 33–40. [CrossRef]

47. Piccolo, A. The supramolecular structure of humic substances. *Soil Sci.* **2001**, *166*, 810–832. [CrossRef]

48. Ramirez Coutino, V.A.; Torres Bustillos, L.G.; Godinez Mora Tovar, L.A.; Guerra Sanchez, R.J.; Rodriguez Valadez, F.J. pH effect on surfactant properties and supramolecular structure of humic substances obtained from sewage sludge composting. *Rev. Int. Contam. Ambie.* **2013**, *29*, 191–199.

49. Fischer, T. Humic supramolecular structures have polar surfaces and unpolar cores in native soil. *Chemosphere* **2017**, *183*, 437–443. [CrossRef] [PubMed]

50. Tarasevich, Y.I.; Dolenko, S.A.; Trifonova, M.Y.; Alekseenko, E.Y. Association and colloid-chemical properties of humic acids in aqueous solutions. *Colloid J.* **2013**, *75*, 207–213. [CrossRef]

51. Baalousha, M.; Motelica-Heino, M.; Galaup, S.; Le Coustumer, P. Supramolecular structure of humic acids by TEM with improved sample preparation and staining. *Microsc. Res. Technol.* **2005**, *66*, 299–306. [CrossRef]

52. Klucakova, M.; Kalina, M. Diffusivity of Cu (II) ions in humic gels—influence of reactive functional groups of humic acids. *Colloid. Surface. A* **2015**, *483*, 162–170. [CrossRef]

53. Terdale, S.; Tantray, A. Spectroscopic study of the dimerization of rhodamine 6G in water and different organic solvents. *J. Mol. Liq.* **2017**, *225*, 662–671. [CrossRef]

54. Talap, P.D. Self-aggregation of Rhodamine—6G in aqueous medium and aqueous solution of Bu4NBr. *Arch. Appl. Sci. Res.* **2014**, *6*, 183–187.

55. Florence, N.; Naorem, H. Study on the effect of an electrolyte on the self-aggregation and the geometry of the dye aggregates of methylene blue in aqueous media. *J. Surf. Sci. Technol.* **2016**, *32*, 28–34. [CrossRef]

56. Moreno-Vasilda, I.; Torres-Gallegos, C.; Araya-Hermosilla, R.; Nishide, H. Influence of the linear aromatic density on methylene blue aggregation around polyanions containing sulfonate groups. *J. Phys. Chem. B* **2010**, *114*, 4151–4158. [CrossRef]

57. Marras-Marquez, T.; Pena, J.; Veiga-Ochoa, M.D. Agarose drug delivery systems upgraded by surfactants inclusion: Critical role of the pore architecture. *Carbohydr. Polym.* **2014**, *103*, 359–368. [CrossRef] [PubMed]

58. Li, X.; Li, Y.; Chen, C.; Zhao, D.; Wang, X.; Zhao, L.; Shi, H.; Ma, G.; Su, Z. Pore size analysis from low field NMR spin–spin relaxation measurements of porous microspheres. *J. Porous Mater.* **2015**, *22*, 11–20. [CrossRef]

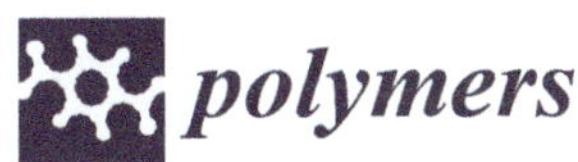 

Article

Bovine Decellularized Amniotic Membrane: Extracellular Matrix as Scaffold for Mammalian Skin

Andrea Catalina Villamil Ballesteros [1,*], Hugo Ramiro Segura Puello [1],
Jorge Andres Lopez-Garcia [2], Andres Bernal-Ballen [3], Diana Lorena Nieto Mosquera [1],
Diana Milena Muñoz Forero [1], Juan Sebastián Segura Charry [1] and Yuli Alexandra Neira Bejarano [1]

[1] Laboratorio de Investigaciones en Salud, Universidad Manuela Beltrán, Avenida Circunvalar No. 60-00, Bogotá 110231, Colombia; hugo.segura@umb.edu.co (H.R.S.P.); lorena.nieto@umb.edu.co (D.L.N.M.); diana.munoz@umb.edu.co (D.M.M.F.); reumatologiasegura@hotmail.com (J.S.S.C.); yuli.neira@umb.edu.co (Y.A.N.B.)

[2] Centre of Polymer Systems, University Institute, Tomas Bata University in Zlín, Trida Tomase Bati 5678, 76001 Zlín, Czech Republic; vextropk@gmail.com

[3] Grupo de Investigación en Ingeniería Biomédica, Vicerrectoría de Investigaciones, Universidad Manuela Beltrán, Avenida Circunvalar No. 60-00, Bogotá 110231, Colombia; andres.bernal@docentes.umb.edu.co

* Correspondence: gemecata15@hotmail.com

Received: 31 August 2019; Accepted: 23 November 2019; Published: 5 March 2020

Abstract: Decellularized membranes (DM) were obtained from bovine amniotic membranes (BAM) using four different decellularization protocols, based on physical, chemical, and mechanical treatment. The new material was used as a biological scaffold for in vitro skin cell culture. The DM were characterized using hematoxylin-eosin assay, scanning electron microscopy (SEM), Fourier transform infrared spectroscopy (FTIR-ATR), and differential scanning calorimetry (DSC). The in vitro cytotoxicity of DM was evaluated using MTT. The efficacy of decellularization process was assessed through DNA quantification and electrophoresis. All the used protocols showed a high effectiveness in terms of elimination of native cells, confirmed by DNA extraction and quantification, electrophoresis, and SEM, although protocol IV removes the cellular contents and preserve the native extracellular matrix (ECM) architecture which it can be considered as the most effective in terms of decellularization. FTIR-ATR and DSC on the other hand, revealed the effects of decellularization on the biochemical composition of the matrices. There was no cytotoxicity and the biological matrices obtained were a source of collagen for recellularization. The matrices of protocols I, II, and III were degraded at day 21 of cell culture, forming a gel. The biocompatibility in vitro was demonstrated; hence these matrices may be deemed as potential scaffold for epithelial tissue regeneration.

Keywords: decellularization; biological scaffolding; bovine amniotic membrane; extracellular matrix; tissue regeneration

---

1. Introduction

Tissue engineering aims to regenerate damaged tissues, developing biological substitutes which along with a thriving cell growth, may restore, maintain, or improve a functional tissue [1–3]. This field has undergone rapid development in the last quarter of the twentieth century, although this science is devoted to skin regeneration it is still a major scientific and clinical challenge [4,5], and the healing response to chronic wounds is poorly understood and a matter of debate [6]. The skin can be considered as the largest organ, which covers the entire surface of the body and its main function is to serve as protective barrier against chemical, mechanical, and infectious damage. Nonetheless, injuries from trauma or skin-burns result in large-scale tissue loss, therefore, autografts, allografts, and xenografts are traditionally used. However, these kinds of treatments have limitations, such as immune rejection, and primary contraction [7].

Ideal skin substitutes should mimic the natural functions of the skin and the structural properties of the extracellular matrix; moreover, it has to protect the organism from protein loss, and it should improve the aesthetic appearance of the wound as well as inhibit the growth of exogenous microorganisms [7,8].

As an innovative treatment for skin injuries, biological substitutes have appeared, and they have the function of supporting growth, differentiation, and cell migration, and may come from different substrates either natural or synthetic, such as, collagen, gelatin, hyaluronic acid, fibronectin/fibrin, chitosan, alginate, polyglycolic acid, polylactic acid, and polycaprolactone; likewise, biological scaffolds composed of extracellular matrix (ECM) of decellularized tissues may be also used as biological substitutes [9–13]. Within this context, decellularized tissues and organs have successfully been used in a variety of tissue engineering/regenerative medicine applications, and the used decellularization methods vary as widely as the tissues and organs of interest [14]. The importance of ECM stems from its three-dimensional ultrastructure and its composition provides a microenvironment that guides the organization, growth, and differentiation of skin cells [15–21]. From all the mentioned substrates, collagen as a part of the ECM has been extensively employed as a biomaterial in cellular therapies and tissue engineering [14,21–24], and its relevance as a candidate for tissue engineering has been described in great extent [25,26].

Despite of the recent breakthroughs in terms of tissue regeneration, wound healing is a complex process that involves activation and synchronization of intracellular, intercellular, and extracellular mechanisms, including coagulation and inflammatory events, fibrous tissue accumulation, collagen deposition, epithelialization, contraction of the wound, tissue granulation, and remodeling [14,15,27,28]. Therefore, grafting materials must exhibit biodegradable, biocompatible, and adequate mechanical properties as well as support normal tissue regeneration [3,29]. In that regard, skin substitutes for wound healing from biological materials based on animal ECM have been developed and the decellularization process has reached an important level of success [14,15,30–33]. Over the years, xenografts have been obtained from various animal species, including birds, rodents, felines, canines, bovines, and swine [34,35]. To this day, the available acellular ECM scaffolds include swine and bovine equine substrates as well as human amniotic membranes [14,15,30,35–40].

Unfortunately, despite the numerous investigations in this area, clinical wound treatment remains unsatisfactory in many cases [15]; chronic wounds require long term and intensive care, and the associated cost are high [31]. Specialists agree that there is still no ideal skin substitute available [31,41,42] and the high costs and time required for the preparation of biological substitutes are crucial factors for developing new materials [35]. Moreover, the risk of zoonotic infections that might be transferred from the graft to the patient is latent, and allergic reactions is the main contraindication for using these kinds of materials [3,34,43]. On the other hand, the human amniotic membrane presents relevant disadvantages, such as that it is scarce due to its high cost, it has poor mechanical properties, and it may be a source of infectious diseases spreading [44,45].

The need for skin substitutes is of paramount importance specifically for large defects of burns, congenital diseases, traumas, and infections [46]. In this frame, acellular amniotic membrane might have potential as a matrix for tissue regeneration or as a substrate to facilitate autologous/allogeneic cell transfer [47]. Human decellularized amniotic membrane has been widely shown as a biodegradable and bioactive matrix for regenerative tissue repair [48]. Thus, this study proposes a candidate that brings the inherent attributes of bovine amniotic membranes (BAM), which can be useful for being used as a matrix for skin regeneration. For this purpose, four distinct protocols were proved (details are shown in methodology section). The obtained decellularized membranes (DM) were used as an alternative scaffold for skin regeneration, and as a sources of collagen IV and VII, elastin, laminin one and five, fibronectin, and entactin [46,49,50]. This material has similar properties to other matrixes obtained from skin [51,52]. Therefore, DM were obtained from four different methods and the efficiency of these methods were evaluated. The cell cultures were carried out on samples and there is evidence of the material does not behave cytotoxically. The obtained DM demonstrated potential in the use of skin regeneration, which is a valuable alternative for tissue engineering and the prospectives of its applications are a new challenge in the field of biomaterial science.

2. Materials and Methods

2.1. Decellularization of the BAM

The bovine amniotic membrane was obtained from a vaginal birth of a bovine female with no infectious-contagious diseases within aseptic conditions. Samples were collected and transported at 4 °C in centrifuge tubes containing a solution of phosphate buffered saline (PBS) that included antibiotics (penicillin, streptomycin, and amphotericin B). Thereupon, the samples were washed with cold PBS and dissected in sections of 16 cm^2 sections in the biological class II biosafety cabinet (ESCO, IDN).

Four protocols were used to decellularize the BAM (Table 1). A control sample was kept without frozen treatment at −20 °C. For protocols I, II, and III, the separation of two layers, fetal and maternal, was performed as it is reported in most of the investigations [52–55]. On the other hand, this procedure was not performed in protocol IV, in order to observe variations in the properties of the membrane regarding exposure with the chemical solutions used and in cell culture.

Table 1. Decellularization protocols for the bovine amniotic membranes (BAM).

No.	Protocols
I	SDS 0.1% for 4 h NaOH 0.1 M for 1 h PAA + ascorbic acid 0.1 for 12 h Ethanol 70% for 1 h PBS for 2 h
II	SDS 0.1% for 4 h NaOH 0.1 M for 1 h PAA 0.15% + EtOH for 12 h NaOH 0.1 M for 1 h PAA for 1 h Ethanol 70% for 1 h PBS for 2 h
III	Tween 80 for 4 h NaOH 0.1 M for 1 h, PAA + ascorbic acid 0.1 for 12 h Ethanol 70% for 1 h PBS for 2 h
IV	Tween 80 for 4 h NaOH 0.1 M for 1 h PAA 0.15% + EtOH for 12 h NaOH 0.1 M for 1 h PAA for 1 h Ethanol 70% for 1 h PBS for 2 h

All BAM were subjected to a freezing cycle in liquid nitrogen (−196 °C) for 22 h and unfreezing in a serological bath (Polyscience, Niles, Illinois, USA) at 37 °C for two hours. Then, BAM were treated with strong and weak detergents (sodium dodecyl sulphate (SDS) 0.1% or Tween 80) for 4 h followed by being soaked in a base solution (NaOH 0.1 M) for 1 h and acid solution (peracetic acid (PAA) and ascorbic acid or ethanol). After, as a final wash, ethanol at 70% was applied for 1 h to remove residual nucleic acids and phospholipids from the tissue and finally, PBS as a buffer solution was pertained for 2 h. The membranes were mechanically stirred throughout the process using an orbital shaker (Camlab, Cambridge, UK) to ensure a homogeneous wash and a minimal damage to the tissue ultrastructure [5–7,30].

Once the abovementioned process was finished, protocols II and IV needed a new acid/basic treatment in order to wipe out the remains of color in the membranes; therefore, it was necessary to immerse them again in NaOH for one hour and PAA for another hour. For BAM treated in protocols I and III, it was not necessary to carry out more washings, and samples were stirred in ethanol at 70%.

After each decellularization step, BAM were washed with deionized water for 30 min in a shaker to eliminate tissue remnants and the used substances. Finally, all the membranes were washed four times with PBS for 30 min and stored at −22 °C.

2.2. Determination of DNA Content

2.2.1. Extraction of DNA

To ensure the removal of all cellular and nuclear material in the decellularized BAM, the DNA extraction process was carried out using a PureLink® kit (Invitrogen). <25 mg of BAM and DM were placed into a micro centrifuge tube. It was added to 180 μL of genomic digestion buffer and 20 μL of proteinase K to remove lipids and digest proteins. Then, the treated samples were incubated at 55 °C in a serological bath with vortex every 10 min for one hour and centrifuged at 13,000 rpm for 3 min at room temperature. Each supernatant was transferred to a new sterile micro centrifuge tube and 20 μL of RNase A was added, mixed, and incubated for two minutes. Subsequently, 200 μL of lysis buffer and 200 μL of 99.9% ethanol were added to each lysate to precipitate the DNA by vortexing for five seconds.

Once the DNA was extracted, it was purified by a series of washes, placing the previous preparations in collector tubes with a column and centrifuging at 12,000 rpm for one minute. Washes were carried out with 500 μL of wash buffer, 1500 μL of wash buffer two, and 50 μL of buffer elution, centrifuging after each addition for one, two, and three minutes, respectively. Finally, the microcentrifuge tubes containing DNA were stored at 4 °C.

In the wells of an agarose gel, samples of the extracted DNA were placed in the horizontal electrophoresis chamber (Thermo EC, Holbrook, NY, USA). Afterwards, the movement of the bands was observed in the transilluminator (Fisher Biotech, Pittsburgh, PA, USA).

2.2.2. DNA Quantification

DNA concentrations were obtained using a QuantiFluor® dsDNA System Kit (Promega, Madison, Wisconsin, USA). After the DNA extraction, the DNA samples were prepared by the addition of 1–20 μL to 200 μL of working solution in 0.5 mL PCR tubes and vortexing, and incubated at room temperature for five minutes, in a dark condition. Finally, fluorescence was measured in the calibrated Quantus™ fluorometer (Promega, Madison, Wisconsin, USA). The effectiveness of each decellularization protocol was evaluated by triplicated.

2.3. Cell Culture

Cells were obtained from a full thickness ovine skin biopsy and cultured with RPMI-1640 culture medium, supplemented with 5% fetal bovine serum (SFB) and 1% antibiotic penicillin, streptomycin, and amphotericin B (Sigma-Aldrich, Bogota, Colombia) in an incubator at 37 °C in a humidified atmosphere of 5% CO_2. Monitoring culture was carried out every three days with an inverted microscope (Olympus, New York, NY, USA). After cell confluence, cells were sub-cultured through trypsinization and used in the second pass.

Cell Seeding on DM and 3-(4,5-d imethylthiazol-2-yl)-2,5-diphenyltetrazoliumbromide (MTT) Assay

The in vitro cytotoxicity of DM was evaluated using 3-(4,5-dimethylthiazol-2-yl)-2,5-diphenyltetrazoliumbromide (MTT) assay (Cell Biolabs Inc., Bogota, Colombia). Specimens of 10 mm in diameter were cut and placed at the bottom of a 24-well-plate (Corning, New York, NY, USA). Before the culture, samples underwent a sterilization process with ultraviolet light for 15 min each side. Then, cells were seeded on DM at a density 5×10^4 cells/well in the previous mentioned incubation conditions.

After culturing for 24, 48, and 72 h, 50 µL of the CytoSelect™ MTT Cell Proliferation Assay Reagent was added to each well and incubated for 4 h, until purple precipitate was visible. Then, 500 µL detergent solution was added and incubated at room temperature for two hours. A specific culture media (RPMI) was also considered as control. The absorbance of solution was measured using a microplate reader 800 ^{TS} (Biotek, Winousky, Vermont, USA) at 490 nm. Cell viability was determined using Equation (1). For histological analysis with hematoxylin-eosin, a DM sample was cultured until day 21.

$$\text{Cell viability (\%)} = \text{Absorbance sample/Absorbance control (untreated)} \times 100 \tag{1}$$

2.4. Histological Analysis with Hematoxylin-Eosin

Control sample BAM, DM, and recellularized DM (at 21 days culture) tissues were fixed in 10% formaldehyde, embedded in paraffin, cut into sections of 5 µm, stained with hematoxylin-eosin, and observed under the optical microscope (Olympus, Tokyo, Japan) to evaluate the presence of nuclear material.

2.5. Scanning Electron Microscopy (SEM)

Micrographs of the prepared samples were taken by the scanning electron microscope Nova NanoSEM 450 (FEI, Brno, Czech Republic) with a Schottky field emission electron source operated at an acceleration voltage ranging from 200 V to 30 kV and a low-vacuum SED (LVD) detector. A coating with a thin layer of gold was performed by a sputter coater SC 7640 (Quorum Technologies, Newhaven, East Sussex, UK).

2.6. FTIR-ATR Spectroscopy

FITR spectroscopy analysis was carried out on NICOLET 6700 FTIR spectrometer device (Thermo Scientific, Waltham, MA, USA) equipped with attenuated total reflectance (ATR) accessory utilizing the Zn–Se crystal and software package OMNIC over the range of wavelengths from 4000 to 600 cm^{-1} at room temperature under a resolution of 4 cm^{-1}. Each spectrum represents 64 co-added scans referenced against an empty ATR cell spectrum.

2.7. Differential Scanning Calorimetry (DSC)

Calorimetric measurements were carried out in a differential scanning calorimetry (DSC) 1 calorimeter, Mettler Toledo (Greifensee, Zurich, Switzerland), under nitrogen flowing at a rate of 30 mL min^{-1}. The specimens were pressed in sealed aluminum pans. A heating cycle was performed in order to acquire the glass transition temperature (T_g) and melting temperature (T_m). The samples were cooled down by nitrogen at an exponentially decreasing rate. The heating of the cycle was performed from 25 to 240 °C at a rate of 20 °C/min. The T_g was determined as the midpoint temperature by standard extrapolation of the linear part of DSC curves using Mettler-Toledo Stare software and the T_m as the maximum value of the melting peak.

2.8. Statistic Analysis

MTT measurements were performed in triplicate. All experimental values were expressed in form of average ± standard deviation. Results were statistically compared using one-way analysis of variance (ANOVA) with $p < 0.05$.

3. Results and Discussion

3.1. Decellularization of BAM and DNA Content

The main purpose of decellularization of xenogeneic matrices is to effectively eliminate cells and nucleic acid residues, as well as preserve the composition of the ECM [11]. In this frame, the DNA content analysis in DM indicated total cell absence whereas in BAM is easily observable (Figure 1). Moreover, the four protocols achieve cell removal until the detection limit of the test.

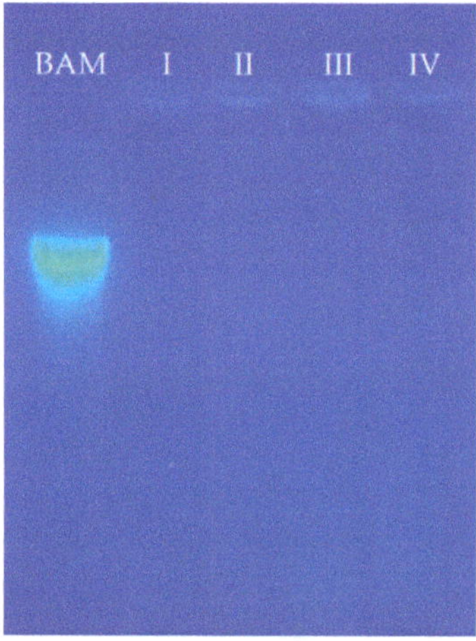

Figure 1. Electrophoresis pattern obtained in agarose gel for the used protocols (I, II, III, and IV), and for the control sample (BAM).

In the electrophoresis technique, the DNA moieties are so small that they cannot be observed while they migrate through the gel, as it is possible in the control membrane (BAM). This technique allowed the separation, identification and isolation of DNA fragments, which cannot be separated by other methods. However, the quantification of DNA allows for the detection of small amounts of the nucleic acid, that is, the actual value of respective DNA for each protocol. From smallest to largest value, so it is highly sensitive. DNA content analysis of DM was conducted to compare the efficiency of the previously developed decellularization protocols. The obtained results showed that the DNA levels decreased with each protocol in comparison to BAM (Figure 2). ANOVA test showed that no significant differences were evidenced in the tested protocols. It has been reported that a lower concentration of <50 ng/mg in a membrane implies that the matrix can be considered as decellularized [32,56]. Therefore, the chemical and mechanical methods used were effective at eliminating the DNA content from the DM. Other studies have shown that higher degrees of decellularization measured by DNA content are associated to a better tissue remodeling in vivo and macroscopic response in the host [57,58]; therefore, protocol II would be considered as the most suitable for decellularization process.

The decellularization protocols consisted of the application of physical freeze-unfreeze method to lyse cells through the formation of microcrystals. This technique requires smaller amounts of chemical agents, which do not significantly alter the ECM properties [10,59,60]. With liquid nitrogen, a lower number of cycles and shorter time were required compared to freezing-unfreeze protocol at −20 or −80 °C [60].

Sodium dodecyl sulphate used as an ionic detergent has the ability to efficiently remove cells and genetic material [58,61] as it was observed in the electrophoresis and confirmed by the DNA content of protocols I and II. Likewise, SDS contributes to the inhibition of collagen calcification processes [62]. However, it can alter the ultrastructure and the elimination of growth factors [58,59,61]. Tween-80 is a non-ionic detergent, considered mild, that has the property of solubilizing proteins while maintaining the structure of the native protein [10]. In the electrophoresis of samples prepared using protocols III and IV, a slight sweep was observed due to protein residues most likely associated with the use of this detergent and DNA content analysis corroborated the presence of DNA in low concentration after decellularization. DM from protocol IV contained double layer (amnion and chorion), therefore, the surface area of exposure to chemical agents was smaller and consisted of even more DNA residues. For this reason, it contains more DNA; however, the obtained value for this protocol is lower in comparison to the reported value of <50 ng/mg for a membrane which is considered decellularized.

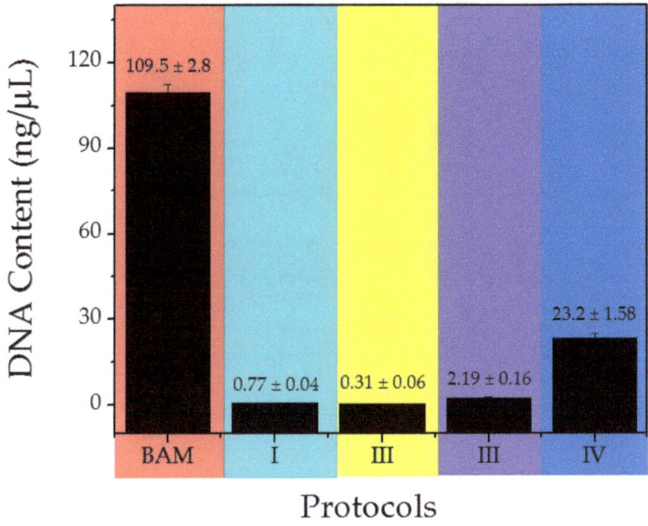

Figure 2. DNA content obtained for the control sample and decellularized membranes (DM).

It is imperative to emphasize that the use of alkaline or acidic solutions in excess may cause serious alteration on the ECM [10]. Alkaline solutions denature chromosomal DNA and plasmid; however, they degrade collagen to a certain extent and eliminate growth factors from the resulting DM and reducing its mechanical properties. The exposure of the matrices to NaOH was performed for one hour, since a prolonged exposure may disintegrate the tissues and interrupt the formation of collagen crosslinks [3]. On the other hand, acids dissociate the DNA of the ECMs via solubilization of cytoplasmic components and the disruption of nucleic acids. PAA with hydrogen peroxide or ethanol was effective for the disinfection and removal of cellular debris from the BAM [3]; ethanol was used for the final wash to eliminate the residual nucleic acids, and delipidize the tissues in addition to its microbicide action, necessary for the manipulation to which the membranes were exposed [10]. Washes with PBS were indeed effective to remove chemical traces and to neutralize the pH of the samples for cell culture. Cells need strict culture conditions to survive, and variation in those conditions can trigger apoptosis [63].

In comparison to other human amniotic membrane decellularization studies, no antibiotics or enzymes were used, which are usually associated with bacterial resistance and irreparable damage to the matrices [53]. Finally, the low obtained standard deviation in this process is an indicative of the reproducibility of the decellularization protocols.

3.2. MTT Assay

The viability of skin cells seeded on DM of different protocols was measured in terms of cellular mitochondrial dehydrogenase activity using MTT assay. The viability of seeded cells for 24, 48, and 72 h are depicted in Figure 3. It was observed that the cells sustained their metabolic activity in culture on the DM, and that activity was increased during the time, showing a considerable biocompatibility of DM. At 72 h, the activity was not observable as a consequence of the detergent did not longer dissolve MTT.

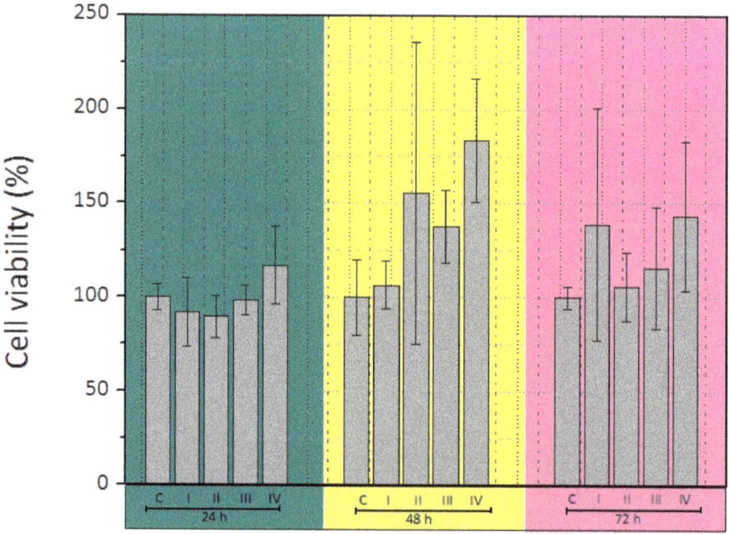

Figure 3. 3-(4,5-dimethylthiazol-2-yl)-2,5-diphenyltetrazoliumbromide (MTT) assay of skin cells growing on different DM for 24, 48, and 72 h. * $p < 0.05$.

The obtained data underwent ANOVA testing and the results indicated that no significant differences were evidenced in all the tested protocols. This was likely due to the nature of DM and and its composition (mainly collagen). Moreover, it was revealed that detergent, acids, and bases removal are critical for generating optimal acellular scaffolds with potencial clínical uses. In this way, any of the tested protocols show cytotoxic effects on the seeded cells. The cell number augmented with increasing the incubation time, which is an indicator for improving the effect of DM on the metabolic activity of the cells compared to a control culture. DM of protocol IV exhibited a visible increase in metabolic activity, associated to the fact that it retained its biochemical properties to a greater extent. These results are in a good agreement with other reports which indicate that scaffolds made of decellularized amniotic membrane, did not exhibit cytotoxicity [64].

The results of the previous studies suggest that the vast majority of current decellularization protocols are detergent-based and incompletely removed residual detergents may have a deleterious impact on subsequent scaffold recellularization [10,29,58,65]. Residual SDS within biomaterials has severe cellular toxicity and may be responsible for the decrease in cell growth [10,29]. Therefore, the success of subsequent recellularization is based on the removal of the lysed cellular material and cytotoxic detergent after the decellularization process [65].

The progress of cell cultures is shown in Figure 4. On day 21 in protocols I, II, and III, degradation was observed and the membrane of protocol IV remained intact. Furthermore, protocols I, II, and III were degraded and it was not possible to carry out histological analysis. The histological findings corroborated the cell growth on the DM.

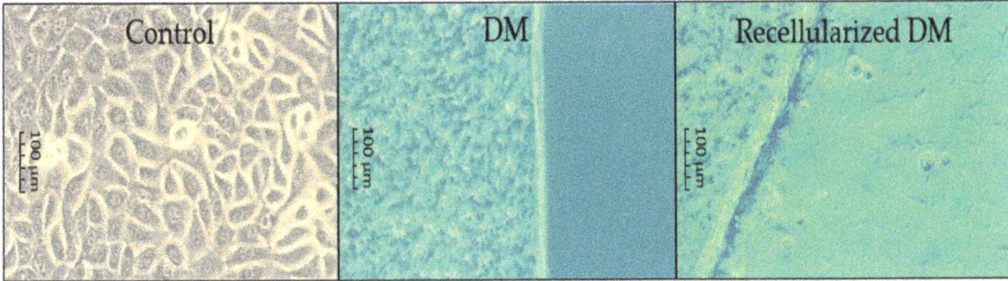

Figure 4. Images of cell culture on DM for mammalian skin cells obtained using an inverted microscope (40×).

3.3. Histological Analysis

In the histological study of BAM, a simple cubic epithelium was observed (Figure 5) that included large binucleated (basophilic) cells and native collagenous (eosinophilic) fibers [49,54] of normal bovine tissue. In this technique, the efficiency of the decellularization protocols was substantiated by cellular absence. Hematoxylin-eosin assay (H&E) disclosed abundant mammalian skin cells adhered to the recellularized BAM as it was observed during cell culture monitoring until day 21. The microphotographs are shown in Figure 5. No cells were observed in the DM as a consequence of the acidophilic matrix. Moreover, cells were present after 21 days of culture in the recellularized DM.

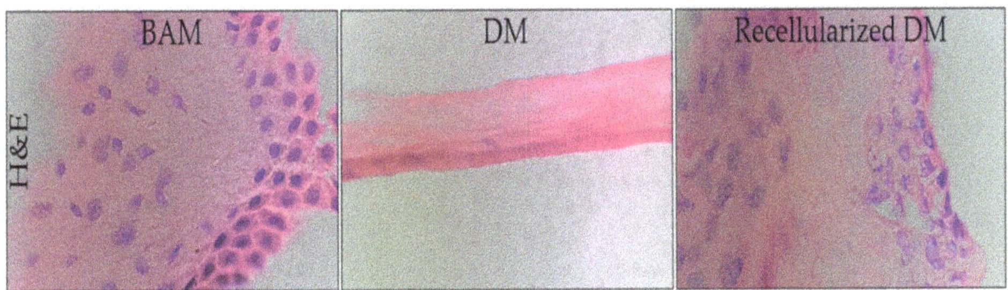

Figure 5. Microphotographs for BAM (**left**), DM (**middle**), and recellularized DM (**right**) obtained from the hematoxylin-eosin (H&E) assay (100×).

Although several studies recommend the use of cell lines in this kind of experiments [66–71], it is of paramount importance to indicate that in vitro studies have evidenced that in a standard cell culture, fibroblast positively influence keratinocyte growth, most likely due to the fact that these cells secrete soluble growth factors. In natural skin, the interaction is relevant as well. Without fibroblasts, the keratinocyte differentiation is severely affected. Moreover, keratinocytes have also a positive effect on the proliferation of fibroblasts. Based on these findings, it is possible to affirm that in order to gain meaningful data from toxicological in vitro studies, the isolated focus on a keratinocyte-containing epidermal layer alone is not sufficient, making the use of a full-thickness skin model essential [14,72–74].

The amniotic membrane has structures, which are histologically similar to the skin, i.e., composed of a multilayer epithelium and the basic membrane, and the structure might be considered as a good support for wound healing, reepithelialization and inhibition of scar formation and bacterial growth [50,52,75,76].

3.4. Scanning Electron Microscopy

Topographical analysis shows that the native BAM contained collagen fibers with tissue cells on an irregular surface (Figure 6). This result is in a good agreement with the obtained by electrophoresis and histology studies, where the control membranes presented the DNA band.

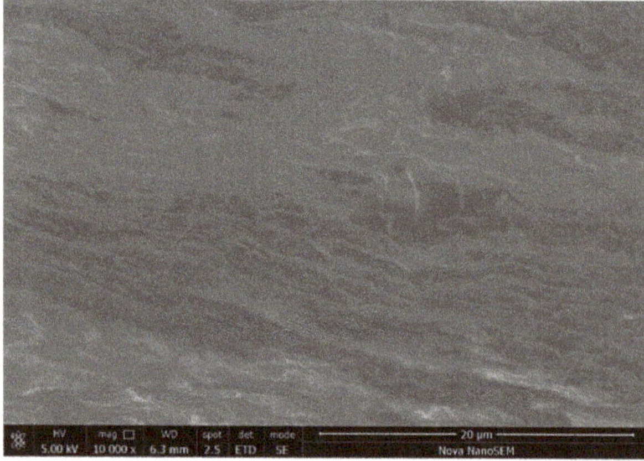

Figure 6. SEM image for BAM.

The micrographs of the studied membranes also confirm that the processes were efficacious for the elimination of the cells in all the tested protocols. There are differences in the surface of each membrane; for instance, image from protocol I depicts a surface where the collagen fibers are very similar to the native ones, whereas DM for protocol II is a smoother surface. DM obtained using protocol III showed tissue wear along with some crystalline residues and the sample of protocol IV is the most homogeneous of the appraised surfaces (Figure 7).

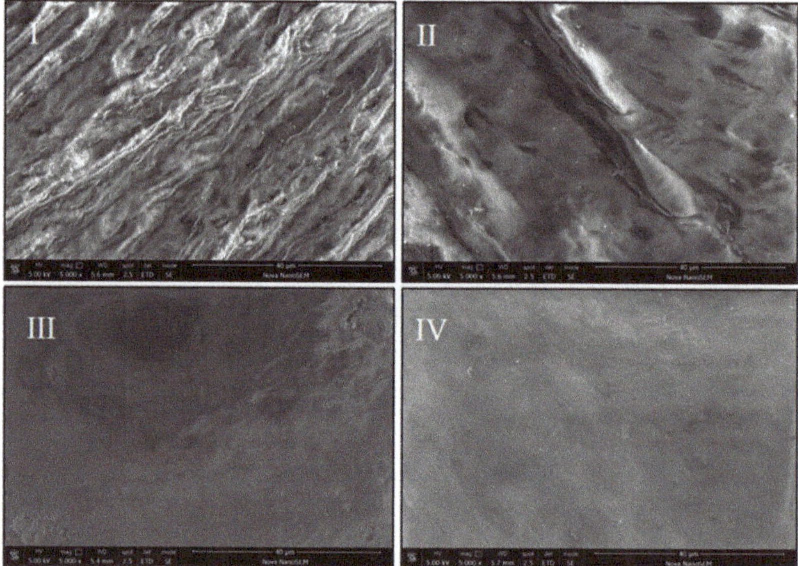

Figure 7. SEM images for DM obtained from the tested protocols.

3.5. FTIR-ATR Spectroscopy

Attenuated total reflectance Fourier transform infrared (ATR-FTIR) spectra of the assessed samples are shown in Figure 8. The peptide characteristic bands at approximately 3300, 3000, 1630, 1545, 1240, and 690 cm^{-1} are identifiable. For example, the amide I is a broad band around 1640–1630 cm^{-1} originated from C=O stretching vibrations coupled to N–H bending vibration. The amide II band, which is located at around 1550 cm^{-1} arises from N–H bending vibrations coupled to C–N stretching vibrations. Finally, the amide III characteristics bands, that usually appear within the range of 1300–1200 cm^{-1} result from the interaction between N–H bending and C–N stretching. The band locate at 690 cm^{-1} is an usual amide vibration which emerges from out of plane N–H wagging [77–80].

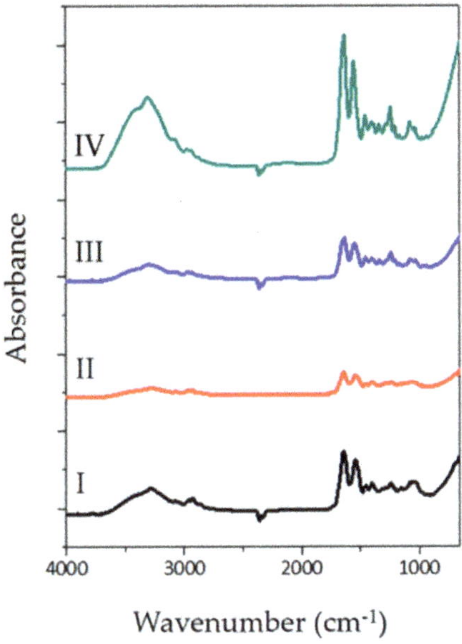

Figure 8. FTIR spectra for DM obtained from the four protocols.

The absorption peaks within the 3000–2800 cm^{-1} spectral range are attributed to aliphatic C–H stretching; on the other hand, the bands around 1500 cm^{-1} are associated with C–H bending. The studied spectra possess the typical features of collagen-like proteins, which have been extensively studied in previous scientific works [77,78,81]. Nevertheless, the characteristic collagen bands are visible, which implies that collagen is retained upon each decellularization process, there are visible differences in the intensity of spectral bands, which may be ascribed to the interaction of the membranes with the solutions, the duration, and harshness of each decellularization protocol. In fact, there is a triple helix denaturation, which may be evinced on the intensity bands.

3.6. Differential Scanning Calorimetry (DSC)

The thermograms of the decellularized membranes of protocols II, III, and IV are shown in Figure 9, and they correspond to the first heating scan. It was not possible to obtain the thermogram from the DM of protocol I because of the sample decomposition. Different endothermic denaturation peaks may be seen viz.: II, 185 °C; III, 200 °C; and IV, 152 °C. Collagen materials exposed to high temperatures endure irreversible denaturation process [81–86]. Previous thermo-analytical studies of denaturation of collagen report that the denaturation temperature for bovine skin is 55 °C [87], bovine intramuscular connective tissue 90 °C [88], rat tail collagen 65 °C [89], bovine skin 50–55 °C [84], and type I collagen from bovine skin soluble in acid 117 °C [90]. As the water content is higher, collagen denaturation temperature gets higher, and this phenomenon may be observed in this study. It should be noted that collagen denatures; therefore, there was no second heating scan; furthermore the treated samples also showed another endothermic peak (105–115 °C) which is most likely related to gelatinous structures by the denaturation that were obtained at day 21 of culture for protocols II and III [84].

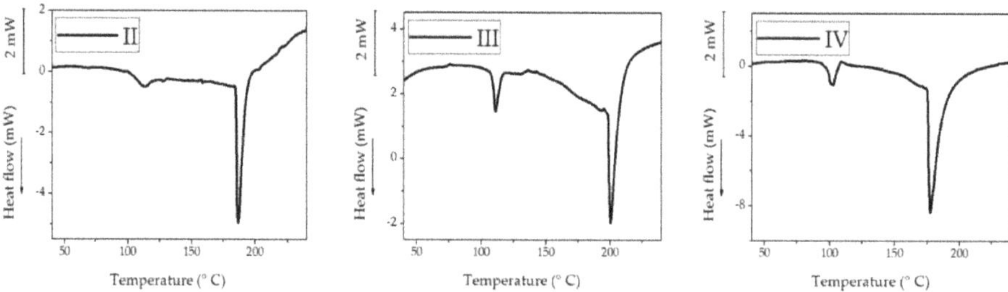

Figure 9. Thermograms for DM obtained from four different protocols.

4. Conclusions

Decellularized membranes have attracted the attention of the scientific community since through tissue engineering it is possible to develop biological scaffolds that aim to deliver cells and proteins to damaged tissue, and at the same time, gradually degrade to make room for regenerated tissue. Within this frame, BAM were decellularized using four different protocols and the differences in terms of decellularization can be considered as negligible. All membranes obtained DNA concentrations <50 ng/mg, indicating that traces of the nucleic acid were present in the prepared material, although the obtained values are negligible which implies that DM do not have presence of native cells from the BAM. Nonetheless, protocol II proved to be the best method in terms of eliminating DNA content.

In the biological test, the obtained matrices from BAM were not cytotoxic for the cells (confirmed by MTT) and consisted of a source of collagen for recellularization. The mammalian skin cells adhered and conducted a remodeling effect on the BAM.

Each protocol may damage the ultrastructure of the tissue in different grade, mainly related to the chemical substances that were used. The extent of denaturation depends upon the interaction of the chemical substances with the molecules present in the tissue, and this analysis was supported by spectroscopic, thermal and topographical techniques.

Results showed that protocol IV (SDS 0.1%, NaOH 0.1 M, PAA + ascorbic acid 0.1, ethanol 70%, and PBS) could efficiently remove the cellular contents and preserve the native ECM architecture (confirmed for FTIR-ATR spectroscopy). Therefore, double layer bovine amniotic membranes (fetal and maternal) retained its biochemical properties after decellularization in comparison with the other membranes. Moreover, the mentioned double layer membrane exhibits a very low DNA concentration which is below to 50 ng/mg; for this reason, DM of protocol IV might be used as a possible biological substitute for skin.

The membranes of protocols I, II, and III, being single layer (stromal), had a greater surface area of exposure to the chemical agents used and, therefore, degraded further in terms of their composition. However, degradation was observed in culture, a semi-transparent gel was formed that may have potential biomedical applications, which may be part of later studies, underlining the potential applications of this matrices for tissue engineering.

The in vitro biocompatibility was demonstrated in this study, and it is indeed of pivotal importance, since this matrix may be considered as a potential source for the regeneration of epithelial tissue.

Author Contributions: A.C.V.B. and H.R.S.P. designed the study and carried out the biological experiments. J.A.L.-G. and A.B.-B. were responsible for the characterization techniques. D.L.N.M., D.M.M.F., and Y.A.N.B. collaborated in the biological test. A.C.V.B., J.A.L.-G., and A.B.-B. wrote and edited the manuscript. J.S.S.C. collaborated by reviewing the manuscript. All authors have read and agreed to the published version of the manuscript.

Funding: This research received no external funding.

Acknowledgments: We would like to thank to Juan Villamil Fraile for donating the BAM; to Andrés Gutierrez Beltrán, Eng. Sergio Andrés Pineda, and Clínica Veterinaria Potencia Animal for their support in data analysis. Special thanks go to the program *Reconocimiento Docente* for facilitating a posdoctoral visit to Tomas Bata University in Zlin and to Universidad Distrital Francisco José de Caldas, where part of the characterization techniques was carried out.

Conflicts of Interest: The authors declare no conflict of interest.

References

1. Taylor, D.A.; Sampaio, L.C.; Ferdous, Z.; Gobin, A.S.; Taite, L.J. Decellularized matrices in regenerative medicine. *Acta Biomater.* **2018**, *74*, 74–89. [CrossRef] [PubMed]
2. O'Brien, F.J. Biomaterials & scaffolds for tissue engineering. *Mater. Today* **2011**, *14*, 88–95.
3. Parmaksiz, M.; Elcin, A.E.; Elcin, Y.M. Decellularization of bovine small intestinal submucosa and its use for the healing of a critical-sized full-thickness skin defect, alone and in combination with stem cells, in a small rodent model. *J. Tissue Eng. Regen. Med.* **2017**, *11*, 1754–1765. [CrossRef] [PubMed]
4. Greenwood, J.E. Hybrid Biomaterials for Skin Tissue Engineering. In *Skin Tissue Engineering and Regenerative Medicine*; Elsevier: Winston-Salem, NC, USA, 2016; pp. 185–210.
5. Downes, S.; Mishra, A.A. Tissue-biomaterial interactions. In *Advanced Wound Repair Therapies*; Elsevier Inc.: Cornwall, UK, 2011; pp. 174–185. ISBN 9781845697006.
6. Naasani, L.S.; Damo Souza, A.F.; Rodrigues, C.; Vedovatto, S.; Azevedo, J.G.; Santin Bertoni, A.P.; Da Cruz Fernandes, M.; Buchner, S.; Wink, M.R. Decellularized human amniotic membrane associated with adipose derived mesenchymal stromal cells as a bioscaffold: Physical, histological and molecular analysis. *Biochem. Eng. J.* **2019**, 107366. [CrossRef]
7. Ding, J.; Zhang, J.; Li, J.; Li, D.; Xiao, C.; Xiao, H.; Yang, H.; Zhuang, X.; Chen, X. Electrospun polymer biomaterials. *Prog. Polym. Sci.* **2019**, *90*, 1–34. [CrossRef]
8. Subramanian, A.; Krishnan, U.M.; Sethuraman, S. Skin tissue regeneration. In *Electrospinning for Tissue Regeneration*; Elsevier: Cornwall, UK, 2011; pp. 298–316.
9. Somuncu, Ö.S.; ßak Ballica, B.; Furkan Temiz, A.; Somuncu, D. Experimental study In vitro artificial skin engineering by decellularized placental scaffold for secondary skin problems of meningomyelocele. *J. Clin. Neurosci.* **2019**, *59*, 291–297. [CrossRef]
10. Keane, T.J.; Swinehart, I.T.; Badylak, S.F. Methods of tissue decellularization used for preparation of biologic scaffolds and in vivo relevance. *Methods* **2015**, *84*, 25–34. [CrossRef]
11. Park, S.M.; Yang, S.; Rye, S.-M.; Choi, S.W. Effect of pulsatile flow perfusion on decellularization. *Biomed. Eng. Online* **2018**, *17*, 15. [CrossRef]
12. Singh, D.; Singh, D.; Han, S. 3D Printing of Scaffold for Cells Delivery: Advances in Skin Tissue Engineering. *Polymers* **2016**, *8*, 19. [CrossRef]
13. Feng, X.; Li, J.; Zhang, X.; Liu, T.; Ding, J.; Chen, X. Electrospun polymer micro/nanofibers as pharmaceutical repositories for healthcare. *J. Control. Release* **2019**, *302*, 19–41. [CrossRef]
14. Catalano, E.; Cochis, A.; Varoni, E.; Rimondini, L.; Azzimonti, B. Tissue-engineered skin substitutes: An overview. *J. Artif. Organs* **2013**, *16*, 397–403. [CrossRef] [PubMed]
15. Kim, H.S.; Sun, X.; Lee, J.-H.; Kim, H.-W.; Fu, X.; Leong, K.W. Advanced drug delivery systems and artificial skin grafts for skin wound healing. *Adv. Drug Deliv. Rev.* **2018**, *146*, 209–239. [CrossRef] [PubMed]
16. Simões, D.; Miguel, S.P.; Ribeiro, M.P.; Coutinho, P.; Mendonça, A.G.; Correia, I.J. Recent advances on antimicrobial wound dressing: A review. *Eur. J. Pharm. Biopharm.* **2018**, *127*, 130–141. [CrossRef]
17. Theocharis, A.D.; Skandalis, S.S.; Gialeli, C.; Karamanos, N.K. Extracellular matrix structure. *Adv. Drug Deliv. Rev.* **2016**, *97*, 4–27. [CrossRef]
18. Nyström, A.; Bernasconi, R.; Bornert, O. Therapies for genetic extracellular matrix diseases of the skin. *Matrix Biol.* **2018**, *71–72*, 330–347. [CrossRef] [PubMed]
19. Iozzo, R.V.; Gubbiotti, M.A. Extracellular matrix: The driving force of mammalian diseases. 2017, 71, 1–9. *Matrix Biol.* **2017**, *71*, 1–9.
20. Waldeck, H.M.; Guerra, A.D.; Kao, W.J. Extracellular Matrix: Inspired Biomaterials. *Compr. Biomater. II* **2017**, *2*, 132–146. [CrossRef]
21. Chalikias, G.K.; Tziakas, D.N. Biomarkers of the extracellular matrix and of collagen fragments. *Clin. Chim. Acta* **2015**, *443*, 39–47. [CrossRef]
22. Badylak, S.F.; Freytes, D.O.; Gilbert, T.W. Reprint of: Extracellular matrix as a biological scaffold material: Structure and function. *Acta Biomater.* **2015**, *23*, S17–S26. [CrossRef]

23. De Castro Brás, L.E.; Ramirez, T.A.; Deleon-Pennell, K.Y.; Chiao, Y.A.; Ma, Y.; Dai, Q.; Halade, G.V.; Hakala, K.; Weintraub, S.T.; Lindsey, M.L. Texas 3-Step decellularization protocol: Looking at the cardiac extracellular matrix. *J. Proteomics* **2013**, *86*, 43–52. [CrossRef]

24. Vyas, K.; Vasconez, H.; Vyas, K.S.; Vasconez, H.C. Wound Healing: Biologics, Skin Substitutes, Biomembranes and Scaffolds. *Healthcare* **2014**, *2*, 356–400. [CrossRef] [PubMed]

25. Li, A.; Wei, Y.; Hung, C.; Vunjak-Novakovic, G. Chondrogenic properties of collagen type XI, a component of cartilage extracellular matrix. *Biomaterials* **2018**, *173*, 47–57. [CrossRef] [PubMed]

26. Kawecki, M.; Łabuś, W.; Klama-Baryla, A.; Kitala, D.; Kraut, M.; Glik, J.; Misiuga, M.; Nowak, M.; Bielecki, T.; Kasperczyk, A. A review of decellurization methods caused by an urgent need for quality control of cell-free extracellular matrix' scaffolds and their role in regenerative medicine. *J. Biomed. Mater. Res. Part B Appl. Biomater.* **2018**, *106*, 909–923. [CrossRef] [PubMed]

27. Han, S.-K. *Innovations and Advances in Wound Healing*; Springer: Berlin/Heidelberg, Germany, 2016; ISBN 978-3-662-46586-8.

28. Milan, P.B.; Kargozar, S.; Joghataie, M.T.; Samadikuchaksaraei, A. Nanoengineered biomaterials for skin regeneration. In *Nanoengineered Biomaterials for Regenerative Medicine*; Elsevier: Amsterdam, Netherlands, 2019; pp. 265–283.

29. Hany Hussein, K.; Park, K.-M.; Kang, K.-S.; Woo, H.-M. Biocompatibility evaluation of tissue-engineered decellularized scaffolds for biomedical application. *Mater. Sci. Eng.: C* **2016**, *67*, 766–788. [CrossRef] [PubMed]

30. Dunckel, A.P. Acellular bovine-derived matrix used on a traumatic crush injury of the hand: A case study. *Ostomy Wound Manag.* **2009**, *55*, 44–49.

31. Wurzer, P.; Keil, H.; Branski, L.K.; Parvizi, D.; Clayton, R.P.; Finnerty, C.C.; Herndon, D.N.; Kamolz, L.P. The use of skin substitutes and burn careda survey. *J. Surg. Res.* **2016**, *201*, 293–298. [CrossRef]

32. Scarritt, M.; Murdock, M.; Badylak, S.F. Biologic Scaffolds Composed of Extracellular Matrix for Regenerative Medicine. *Princ. Regen. Med.* **2019**, 613–626. [CrossRef]

33. Barreto, R.d.S.N.; Romagnolli, P.; Mess, A.M.; Miglino, M.A. Decellularized bovine cotyledons may serve as biological scaffolds with preserved vascular arrangement. *J. Tissue Eng. Regen. Med.* **2018**, *12*, e1880–e1888. [CrossRef]

34. Nyame, T.T.; Chiang, H.A.; Orgill, D.P. Clinical Applications of Skin Substitutes. *Surg. Clin. NA* **2014**, *94*, 839–850. [CrossRef]

35. Yamamoto, T.; Iwase, H.; King, T.W.; Hara, H.; Cooper, D.K.C. Skin xenotransplantation: Historical review and clinical potential. *Burns* **2018**, *44*, 1738–1749. [CrossRef]

36. Macleod, T.M.; Sarathchandra, P.; Williams, G.; Sanders, R.; Green, C.J. Evaluation of a porcine origin acellular dermal matrix and small intestinal submucosa as dermal replacements in preventing secondary skin graft contraction. *Burns* **2004**, *30*, 431–437. [CrossRef] [PubMed]

37. Schallberger, S.P.; Stanley, B.J.; Hauptman, J.G.; Steficek, B.A. Effect of Porcine Small Intestinal Submucosa on Acute Full-Thickness Wounds in Dogs. *Vet. Surg.* **2008**, *37*, 515–524. [CrossRef]

38. Brown-Etris, M.; Milne, C.T.; Hodde, J.P. An extracellular matrix graft (Oasis ®wound matrix) for treating full-thickness pressure ulcers: A randomized clinical trial. *J. Tissue Viability* **2018**, *28*, 21–26. [CrossRef] [PubMed]

39. Middelkoop, E.; Sheridan, R.L. Skin Substitutes and 'the next level'. *Total Burn Care* **2018**, 167–173. [CrossRef]

40. Chawla, R.; Seifalian, A.; Moiemen, N.S.; Butler, P.E.; Seifalian, A.M. The Use of Skin Substitutes in the Treatment of Burns. *Regen. Med. Appl. Organ Transplant.* **2014**, 771–782. [CrossRef]

41. Haddad, A.G.; Giatsidis, G.; Orgill, D.P.; Halvorson, E.G. Skin Substitutes and Bioscaffolds Temporary and Permanent Coverage. *Clin. Plast. Surg.* **2017**, *44*, 627–634. [CrossRef]

42. Halim, A.; Khoo, T.; Shah, J.Y. Biologic and synthetic skin substitutes: An overview. *Indian J. Plast. Surg.* **2010**, *43*, 23. [CrossRef]

43. Aamodt, J.M.; Grainger, D.W. Extracellular matrix-based biomaterial scaffolds and the host response. *Biomaterials* **2016**, *86*, 68–82. [CrossRef]

44. Saffle, J.R. Closure of the Excised Burn Wound: Temporary Skin Substitutes. *Clin. Plast. Surg.* **2009**, *36*, 627–641. [CrossRef]

45. Park, M.; Kim, S.; Kim, I.S.; Son, D. Healing of a porcine burn wound dressed with human and bovine amniotic membranes. *Wound Repair Regen.* **2008**, *16*, 520–528. [CrossRef]

46. Valladares, M. *Estudio de Tres Diferentes Métodos de Preparación y Conservación de la Membrana Amniótica Para Usos Oftalmológicos*; Universidade de Santiago de Compostela: Sandriago de Compostela, Spain, 2008.

47. Francisco, J.C.; Correa Cunha, R.; Cardoso, M.A.; Baggio Simeoni, R.; Mogharbel, B.F.; Picharski, G.L.; Silva Moreira Dziedzic, D.; Guarita-Souza, L.C.; Carvalho, K.A.T. Decellularized Amniotic Membrane Scaffold as a Pericardial Substitute: An In Vivo Study. *Transplant. Proc.* **2016**, *48*, 2845–2849. [CrossRef] [PubMed]

48. Zhou, Z.; Long, D.; Hsu, C.-C.; Liu, H.; Chen, L.; Slavin, B.; Lin, H.; Li, X.; Tang, J.; Yiu, S.; et al. Nanofiber-reinforced decellularized amniotic membrane improves limbal stem cell transplantation in a rabbit model of corneal epithelial defect. *Acta Biomater.* **2019**, *97*, 310–320. [CrossRef] [PubMed]

49. da Anunciação, A.; Mess, A.; Orechio, D.; Aguiar, B.; Favaron, P.; Miglino, M. Extracellular matrix in epitheliochorial, endotheliochorial and haemochorial placentation and its potential application for regenerative medicine. *Reprod. Domest. Anim.* **2017**, *52*, 3–15. [CrossRef] [PubMed]

50. Min, S.; Yoon, J.Y.; Park, S.Y.; Kwon, H.H.; Suh, D.H. Clinical effect of bovine amniotic membrane and hydrocolloid on wound by laser treatment: Prospective comparative randomized clinical trial. *Wound Repair Regen.* **2014**, *22*, 212–219. [CrossRef]

51. Leonel, L.; Miranda, C.; Coelho, T.; Ferreira, G.; Cañada, R.; Miglino, M.; Lobo, S. Decellularization of placentas: Establishing a protocol. *Braz. J. Med. Biol. Res.* **2018**, *51*. [CrossRef]

52. Kang, M.; Choi, S.; Cho Lee, A.-R. Effect of freeze dried bovine amniotic membrane extract on full thickness wound healing. *Arch. Pharm. Res.* **2013**, *36*, 472–478. [CrossRef]

53. Sanluis-Verdes, A.; Yebra-Pimentel Vilar, M.T.; García-Barreiro, J.J.; García-Camba, M.; Ibáñez, J.S.; Doménech, N.; Rendal-Vázquez, M.E. Production of an acellular matrix from amniotic membrane for the synthesis of a human skin equivalent. *Cell Tissue Bank.* **2015**, *16*, 411–423. [CrossRef]

54. Favaron, P.; Carvalho, R.; Borghesi, J.; Anunciação, A.; Miglino, M. The Amniotic Membrane: Development and Potential Applications - A Review. *Reprod. Domest. Anim.* **2015**, *50*, 881–892. [CrossRef]

55. Milan, P.B.; Amini, N.; Joghataei, M.T.; Ebrahimi, L.; Amoupour, M.; Sarveazad, A.; Kargozar, S.; Mozafari, M. Decellularized human amniotic membrane: From animal models to clinical trials. *Methods* **2019**. [CrossRef]

56. Bruyneel, A.A.N.; Carr, C.A. Ambiguity in the Presentation of Decellularized Tissue Composition: The Need for Standardized Approaches. *Artif. Organs* **2017**, *41*, 778–784. [CrossRef]

57. Gilbert, T.W.; Freund, J.; Badylak, S.F. Quantification of DNA in Biologic Scaffold Materials. *J. Surg. Res.* **2009**, *152*, 135–139. [CrossRef]

58. Crapo, P.M.; Gilbert, T.W.; Badylak, S.F. An overview of tissue and whole organ decellularization processes. *Biomaterials* **2011**, *32*, 3233–3243. [CrossRef] [PubMed]

59. Morales-Valencia, M.; Patiño-Vargas, M.; Correa-Londoño, L.; Restrepo-Múnera, L. Evaluación del método químico-enzimático de descelularización para la obtención de matrices extracelulares de tráquea en el modelo porcino. *Iatreia* **2016**, *29*, 144–156. [CrossRef]

60. Wu, L.-C.; Kuo, Y.-J.; Sun, F.-W.; Chen, C.-H.; Chiang, C.-J.; Weng, P.-W.; Tsuang, Y.-H.; Huang, Y.-Y. Optimized decellularization protocol including α-Gal epitope reduction for fabrication of an acellular porcine annulus fibrosus scaffold. *Cell Tissue Bank.* **2017**, *18*, 383–396. [CrossRef]

61. Gilpin, A.; Yang, Y. Decellularization Strategies for Regenerative Medicine: From Processing Techniques to Applications. *Biomed. Res. Int.* **2017**, *2017*, 1–13. [CrossRef]

62. Olgierd, B.; Sklarek, A.; Siwek, P.; Waluga, E. Methods of Biomaterial-Aided Cell or Drug Delivery: Extracellular Matrix Proteins as Biomaterials. *Stem Cells Biomater. Regen. Med.* **2019**, 163–189. [CrossRef]

63. Vargas, N.; González, C. *Técnicas de Cultivos Celulares e Ingeniería de Tejidos*; México, 2016; ISBN 978-607-28-0688-7.

64. Ganjibakhsh, M.; Mehraein, F.; Koruji, M.; Aflatoonian, R.; Farzaneh, P. Three-dimensional decellularized amnion membrane scaffold as a novel tool for cancer research; cell behavior, drug resistance and cancer stem cell content. *Mater. Sci. Eng. C* **2019**, *100*, 330–340. [CrossRef] [PubMed]

65. Zvarova, B.; Uhl, F.E.; Uriarte, J.J.; Borg, Z.D.; Coffey, A.L.; Bonenfant, N.R.; Weiss, D.J.; Wagner, D.E. Residual Detergent Detection Method for Nondestructive Cytocompatibility Evaluation of Decellularized Whole Lung Scaffolds. *Tissue Eng. Part C Methods* **2016**, *22*, 418–428. [CrossRef] [PubMed]

66. Fernando, I.P.S.; Jayawardena, T.U.; Kim, H.-S.; Vaas, A.P.J.P.; De Silva, H.I.C.; Nanayakkara, C.M.; Abeytunga, D.T.U.; Lee, W.; Ahn, G.; Lee, D.-S.; et al. A keratinocyte and integrated fibroblast culture model for studying particulate matter-induced skin lesions and therapeutic intervention of fucosterol. *Life Sci.* **2019**, *233*, 116714. [CrossRef]

67. Celik, S.B.; Dominici, S.R.; Filby, B.W.; Das, A.A.; Madden, L.A.; Paunov, V.N. Fabrication of Human Keratinocyte Cell Clusters for Skin Graft Applications by Templating Water-in-Water Pickering Emulsions. *Biomimetics* **2019**, *4*, 50. [CrossRef]

68. Sapru, S.; Das, S.; Mandal, M.; Ghosh, A.K.; Kundu, S.C. Nonmulberry silk protein sericin blend hydrogels for skin tissue regeneration - in vitro and in vivo. *Int. J. Biol. Macromol.* **2019**, *137*, 545–553. [CrossRef]

69. Izadyari Aghmiuni, A.; Heidari Keshel, S.; Sefat, F.; Akbarzadeh Khiyavi, A. Quince seed mucilage-based scaffold as a smart biological substrate to mimic mechanobiological behavior of skin and promote fibroblasts proliferation and h-ASCs differentiation into keratinocytes. *Int. J. Biol. Macromol.* **2019**. [CrossRef] [PubMed]

70. Duan, H.; Feng, B.; Guo, X.; Wang, J.; Zhao, L.; Zhou, G.; Liu, W.; Cao, Y.; Zhang, W.J. Engineering of epidermis skin grafts using electrospun nanofibrous gelatin/polycaprolactone membranes. *Int. J. Nanomed.* **2013**, *8*, 2077–2084.

71. Wolf, M.T.; Daly, K.A.; Brennan-Pierce, E.P.; Johnson, S.A.; Carruthers, C.A.; D'Amore, A.; Nagarkar, S.P.; Velankar, S.S.; Badylak, S.F. A hydrogel derived from decellularized dermal extracellular matrix. *Biomaterials* **2012**, *33*, 7028–7038. [CrossRef] [PubMed]

72. Groeber, F.; Holeiter, M.; Hampel, M.; Hinderer, S.; Schenke-Layland, K. Skin tissue engineering—In vivo and in vitro applications. *Adv. Drug Deliv. Rev.* **2011**, *63*, 352–366. [CrossRef]

73. Wood, F.M. Therapeutic Applications: Tissue Engineering of Skin. *Princ. Regen. Med.* **2019**, 1281–1295. [CrossRef]

74. Petreaca, M.; Martins-Green, M. Cell–Extracellular Matrix Interactions in Repair and Regeneration. In *Principles of Regenerative Medicine*; Elsevier: London, UK, 2019; pp. 15–35.

75. Leonel, L.C.P.C.; Miranda, C.M.F.C.; Coelho, T.M.; Ferreira, G.A.S.; Caãada, R.R.; Miglino, M.A.; Lobo, S.E. Decellularization of placentas: Establishing a protocol. *Braz. J. Med Biol. Res.* **2017**, *51*, e6382. [CrossRef] [PubMed]

76. Kim, H.; Son, D.; Choi, T.H.; Jung, S.; Kwon, S.; Kim, J.; Han, K. Evaluation of an Amniotic Membrane-Collagen Dermal Substitute in the Management of Full-Thickness Skin Defects in a Pig. *Arch. Plast. Surg.* **2013**, *40*, 11–18. [CrossRef] [PubMed]

77. Barraza-Garza, G.; de la Rosa, L.A.; Martínez-Martínez, A.; Castillo-Michel, H.; Cotte, M.; Alvarez-Parrilla, E. La microespectroscopia de infrarrojo con t ransformada de Fourier (FTIRM) en el estudio de sistemas biológicos. *Revista Latinoamericana de Química RLQ.* **2013**, *41*, 125–148.

78. Bernal, A.; Balkova, R.; Kuritka, I.; Saha, P. Preparation and characterisation of a new double-sided bio-artificial material prepared by casting of poly(vinyl alcohol) on collagen. *Polym. Bull.* **2013**, *70*, 431–453. [CrossRef]

79. Zhou, Y.; Chen, C.; Guo, Z.; Xie, S.; Hu, J.; Lu, H. SR-FTIR as a tool for quantitative mapping of the content and distribution of extracellular matrix in decellularized book-shape bioscaffolds. *BMC Musculoskelet. Disord.* **2018**, *19*, 220. [CrossRef] [PubMed]

80. Chi Ting Au-Yeung, G.; Sarig, U.; Sarig, H.; Bogireddi, H.; Bronshtein, T.; Baruch, L.; Spizzichino, A.; Bortman, J.; Freddy, B.Y.C.; Machluf, M.; et al. Restoring the biophysical properties of decellularized patches through recellularization. *Biomater. Sci.* **2017**, *5*, 1183–1194. [CrossRef] [PubMed]

81. Shi, D.; Liu, F.; Yu, Z.; Chang, B.; Douglas Goff, H.; Zhong, F. Effect of aging treatment on the physicochemical properties of collagen films. *Food Hydrocoll.* **2018**, *87*, 436–447. [CrossRef]

82. Li, M.; Han, M.; Sun, Y.; Hua, Y.; Chen, G.; Zhang, L. Oligoarginine mediated collagen/chitosan gel composite for cutaneous wound healing. *Int. J. Biol. Macromol.* **2018**, *122*, 1120–1127. [CrossRef]

83. Zouhair, S.; Aguiari, P.; Iop, L.; Vásquez-Rivera, A.; Filippi, A.; Romanato, F.; Korossis, S.; Wolkers, W.F.; Gerosa, G. Preservation strategies for decellularized pericardial scaffolds for off-the-shelf availability. *Acta Biomater.* **2019**, *84*, 208–221. [CrossRef]

84. Badea, E.; Della Gatta, G.; Usacheva, T. Effects of temperature and relative humidity on fibrillar collagen in parchment: A micro differential scanning calorimetry (micro DSC) study. *Polym. Degrad. Stab.* **2012**, *97*, 346–353. [CrossRef]

85. Latorre, M.E.; Velázquez, D.E.; Purslow, P.P. Differences in the energetics of collagen denaturation in connective tissue from two muscles. *Int. J. Biol. Macromol.* **2018**, *113*, 1294–1301. [CrossRef]

86. Schroepfer, M.; Meyer, M. DSC investigation of bovine hide collagen at varying degrees of crosslinking and humidities. *Int. J. Biol. Macromol.* **2017**, *103*, 120–128. [CrossRef]

87. Zhang, Y.; Snow, T.; Smith, A.J.; Holmes, G.; Prabakar, S. A guide to high-efficiency chromium (III)-collagen cross-linking: Synchrotron SAXS and DSC study. *Int. J. Biol. Macromol.* **2019**, *126*, 123–129. [CrossRef]
88. Rochdi, A.; Foucat, L.; Renou, J.-P. NMR and DSC studies during thermal denaturation of collagen. *Food Chem.* **2000**, *69*, 295–299. [CrossRef]
89. Bozec, L.; Odlyha, M. Thermal denaturation studies of collagen by microthermal analysis and atomic force microscopy. *Biophys. J.* **2011**, *101*, 228–236. [CrossRef] [PubMed]
90. Sarti, B.; Scandola, M. Viscoelastic and thermal properties of collagen/poly(vinyl alcohol) blends. *Biomaterials* **1995**, *16*, 785–792. [CrossRef]

Article

Atmospheric Pressure Microwave Plasma Jet for Organic Thin Film Deposition

Mehrnoush Narimisa [1,*], František Krčma [2], Yuliia Onyshchenko [1], Zdenka Kozáková [2], Rino Morent [1] and Nathalie De Geyter [1]

[1] Research Unit Plasma Technology, Department of Applied Physics, Faculty of Engineering and Architecture, Ghent University, Sint-Pietersnieuwstraat 41, B4, 9000 Ghent, Belgium; Yuliia.Onyshchenko@UGent.be (Y.O.); Rino.Morent@UGent.be (R.M.); Nathalie.DeGeyter@UGent.be (N.D.G.)

[2] Faculty of Chemistry, Brno University of Technology, Purkyňova 118, 612 00 Brno, Czech Republic; krcma@fch.vut.cz (F.K.); kozakova@fch.vut.cz (Z.K.)

* Correspondence: Mehrnoush.Narimisa@UGent.be; Tel.: +32-9-264-38-25

Received: 29 November 2019; Accepted: 28 January 2020; Published: 6 February 2020

Abstract: In this work, the potential of a microwave (MW)-induced atmospheric pressure plasma jet (APPJ) in film deposition of styrene and methyl methacrylate (MMA) precursors is investigated. Plasma properties during the deposition and resultant coating characteristics are studied. Optical emission spectroscopy (OES) results indicate a higher degree of monomer dissociation in the APPJ with increasing power and a carrier gas flow rate of up to 250 standard cubic centimeters per minute (sccm). Computational fluid dynamic (CFD) simulations demonstrate non-uniform monomer distribution near the substrate and the dependency of the deposition area on the monomer-containing gas flow rate. A non-homogeneous surface morphology and topography of the deposited coatings is also observed using atomic force microscopy (AFM) and SEM. Coating chemical analysis and wettability are studied by XPS and water contact angle (WCA), respectively. A lower monomer flow rate was found to result in a higher C–O/C–C ratio and a higher wettability of the deposited coatings.

Keywords: atmospheric pressure plasma jet (APPJ); microwave (MW) discharge; thin film deposition; optical emission spectroscopy (OES); Comsol MultiPhysics; methyl methacrylate (MMA); styrene

---

1. Introduction

Atmospheric pressure non-thermal plasma deposition as an advantageous coating technique has been a continually growing research field for over a few decades [1–3]. This method has already been frequently used to deposit a wide range of convenient thin films that were previously fabricated by various other methods such as chemical synthesis, electrochemical polymerization, and low pressure plasma-enhanced chemical vapor deposition (PECVD) [4–6]. When using the latter methods, controlling the deposition rate, optimizing the structure and adhesion of the produced polymers, reducing the energy consumption, and the enormous cost of vacuum techniques were remaining concerns. Consequently, the field of thin film deposition remains open to alternative methods that can overcome these issues such as atmospheric pressure plasma deposition. Among atmospheric pressure plasma sources suitable for coating deposition, atmospheric pressure plasma jets (APPJs) have gained recognition due to their straightforward design and operation together with their ability to effectively coat 3D complex substrates and their capacity to produce a localized deposition [7–11]. Furthermore, due to its construction, a plasma jet offers the possibility of transporting large amounts of charged and active species while the sample is in indirect exposure to the active discharge, which in turn is helpful for biomedical purposes as well as for the treatment of other sensitive materials [12–15]. The concentration of reactive species in the plasma directly correlates with the discharge excitation method. The majority of studied atmospheric pressure plasma jets are driven by audio- (typically around 10 kHz) or radiofrequency (usually 13.54 or 27.2 MHz) power supplies and contain a large amount of reactive species [16–19]. Discharges, generated with even higher excitation frequency (e.g., microwave (MW)), contain an even higher concentration of these reactive species [20,21]. In the case of an MW discharge, the plasma can be (as in the present case) created using the surface wave that propagates along the plasma–capillary or plasma–open-air boundary, and, thus, the active discharge propagates out of the capillary and can also interact directly with the surface. Due to this fact, in an MW-induced discharge, the radicals are present at a high density, thereby contributing to a raised level of fragmentation of the introduced monomer while maintaining room temperature. The plasma torch length can be varied by gas flow and supplied energy [22] and the monomer can be mixed with plasma at different positions with respect to the end of an active discharge. The high radical density in an MW-induced discharge is very advantageous as a high rate of chemical dissociation accompanied by subsequent recombination processes is a crucial reaction for thin film deposition. Furthermore, such a relatively high frequency can also minimize the plasma instability and its transition to arc mode during film deposition, which can bypass the degradation of molecules and chemical bonds during the deposition stage. Nevertheless, despite all the above-listed advantages of MW discharge sources, only a few works have been conducted which have focused on the performance of MW plasma jets in thin film deposition under different operational conditions [8,23,24]. Additional advantages of MW plasma application for thin film deposition are its simplicity and its generally cheap scalability up to a couple of meters. Thus, laboratory achievements could be generally simply extended to industrial scale practice.

A thorough investigation of a complete experiment is required to give a full insight into the processes involved in plasma jet polymerization and properties of the deposited films. Plasma diagnostics in this work are used as a powerful tool to establish the connection between monomer fragmentation/distribution and coating properties, as it is crucial to demonstrate correlations between the plasma discharge and the resultant coatings. This information is profoundly beneficial for future plasma source development and for tuning the experimental conditions in order to obtain plasma-polymerized coatings with specific requirements suitable for a particular application.

Many different monomers can be used for organic thin film deposition by an MW plasma jet. However, each particular monomer has a specific chemical structure, and, thus, requires a particular energy of the plasma reactive species to be successfully deposited as a coating on a substrate. Among the available monomers, methyl methacrylate (MMA) and styrene are very interesting from a technological point of view. A wide range of applications such as humidity sensors, optical biosensors, protective coatings, biomedical applications, and anti-reflection optical coatings require MMA and/or styrene-based coatings because of their ideal plasticity, transparency, electric insulation, and cushioning properties [25–28]. Taking this into account, this work focuses on the optimization of styrene and MMA coating deposition by an APPJ operating at an MW frequency of 2.45 GHz. This plasma source has been recently developed and has never been applied for thin film deposition before [20,29]. Consequently, this work is a unique example of using an MW source to polymerize styrene and MMA. Initially, the reactive species in the plasma source are characterized using optical emission spectroscopy (OES). In the next step, gas flow dynamic simulations with Comsol MultiPhysics 5.3 are performed to demonstrate the distribution of the monomer during the deposition process [30–32]. In doing so, the gas mixing and propagation in the working volume can be prognosticated, which can in turn help to understand the processes that are taking place during plasma deposition. Moreover, the results of the gas dynamic simulations extensively help to interpret other experimentally obtained data such as temperature and incorporation of other species, which can significantly influence the final coating properties [33,34]. In the final part of the study, the physical and chemical properties of the deposited films are also characterized by water contact angle (WCA) analysis and XPS, while the surface morphology and topography are analyzed using SEM and atomic force microscopy (AFM), respectively. By following this research strategy, crucial information regarding atmospheric pressure MW plasma jet effectiveness for thin film deposition will be revealed.

2. Materials and Methods

2.1. Materials

Plasma deposition experiments were carried out using styrene (99.5%, Brenntag CZ) and MMA (99.9%, Pliska, Brno, Czech Republic) as precursors. Cover slip glasses 25 mm in diameter were selected as substrates. Furthermore, to have a better visualization of the coating lateral dimensions, silicon wafers (Siegert Wafers, Aachen, Germany) were also used as substrates. Argon of 99.996% purity purchased from Linde was used as the main and carrier gas in the discharge.

2.2. Experimental Plasma Set-Up

A schematic representation of the experimental plasma set-up is depicted in Figure 1. Similarly to a previous study [29], a high voltage power source (Sairem, GMS 200 W) operating with low power at a 2.45 GHz frequency was used for plasma generation. This power supply was connected to a surfatron surface wave (SW) launcher (Sairem, SURFATRON 80, Décines-Charpieu, France) using a coaxial cable. A continuous flow of water was utilized to cool down the surfatron resonator to avoid the influence of elevated temperature. Plasma was generated in a flow of 1 standard liter per minute (slm) argon gas using an Omega FMA-A2408 mass flow controller (Omega, Norwalk, CT, USA) inside a quartz capillary with a 3.06 and 8.0 mm inner and outer diameter, respectively. To introduce the monomer into the plasma, a Macor disc (60 mm diameter and 6 mm thickness) with a cylindrical opening 2 mm in diameter at one side for the monomer carrier gas inlet was placed at the end of the capillary edge. In the middle of this disc there was also a circular opening 4 mm in diameter which ensured that the plasma effluent could freely pass through this opening. Due to the geometry of the MW plasma jet set-up, the precursor was inserted in the plasma afterglow region perpendicular to the main gas stream and towards its core. The Macor disc was also able to improve the efficiency of deposition due to the fact that it alters the gas dynamics in the volume close to the substrate, as previously reported by Onyshchenko et al. [35]. To deliver different amounts of evaporated monomer molecules into the plasma plume, a Bronkhorst F201 mass flow controller was used to regulate an additional argon gas flow, as a monomer carrier gas, with different flow rates (125, 250, 500, and 750 standard cubic centimeters per minute (sccm)) through a glass bubbler filled with MMA or styrene. The bubbler was kept in a water–ice mixture to keep the temperature of the liquid monomers constant. During the plasma deposition experiments, the distance between the bottom edge of the disc and the sample was varied between 3, 5, and 7 mm. For these three distances, the active discharge part of the plasma effluent did not touch the substrate. The other plasma operational parameters that were varied in this work were applied power and deposition time; these were 7, 10, and 13 W and 1, 3, and 5 min, respectively. This work aimed to demonstrate the influence of each of these working parameters on plasma jet performance in terms of thin film deposition. For sake of clarity, one set of experimental parameters ("Main") was selected as a reference point for all other parameters that were varied. The 10 sets of plasma operational parameters used in this study are summarized in Table 1 and their corresponding labels are also indicated. All deposition experiments were performed in stationary mode, producing only a single plasma treated spot on the substrate to properly verify the deposited area as well as its homogeneity.

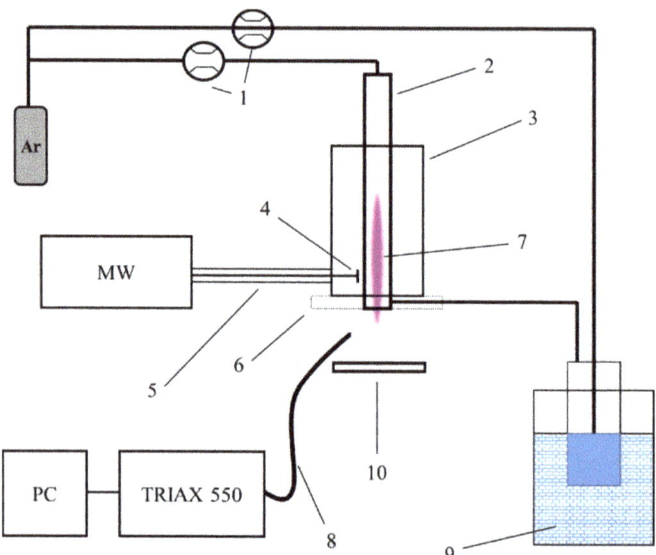

Figure 1. Scheme of the experimental plasma set-up: 1—mass flow controllers; 2—quartz capillary; 3—surfatron resonator; 4—microwave (MW) antenna; 5—MW coaxial cable; 6—Macor plate with precursor injection; 7—plasma torch; 8—optical fiber; 9—bubbler system in water–ice bath; 10—substrate.

Table 1. Summary of the selected experimental conditions. Legend: sccm, standard cubic centimeters per minute.

Sample	Power (W)	Time (s)	Distance to Substrate (mm)	Carrier Gas Flow Rate (sccm)
Main	10	3	5	250
Main + t1	10	1	5	250
Main + t2	10	5	5	250
Main + p1	7	3	5	250
Main + p2	13	3	5	250
Main + f1	10	3	5	125
Main + f2	10	3	5	500
Main + f3	10	3	5	750
Main + d1	10	3	3	250
Main + d2	10	3	7	250

2.3. OES

OES was used for plasma diagnostics in an effort to reveal more information on the radiative plasma species present in the MW plasma jet. The light emitted by the discharge was integrally collected by a quartz optical cable mounted 5 cm away from the plasma output under an angle of 30° with respect to the plasma jet axis as shown in Figure 1. No additional optical components were used in this case, so no space-resolved OES spectra were collected. For the OES spectra acquisition, a Jobin Yvon TRIAX 550 spectrometer (Longjumeau, France) with an LN2 cooled back-illuminated CCD (1025 × 256, pixel size 26 × 26 μm^2) was used. Using the standard Ocean Optics DT-MINI-2-GS source (Largo, FL, USA), the system was calibrated with respect to its spectral response. These calibration results were subsequently used for measured spectra correction. The OES spectra were recorded in the range 300–850 nm using a 1200 gr/mm grating at a fixed spectrometer entrance slit of 30 μm. Four integration times (0.002, 0.01, 0.1, and 1 s) were applied for an appropriate signal/noise ratio for each studied line and band.

2.4. Computational Fluid Dynamic Simulations

To obtain information on the impact of the MW plasma jet configuration on the gas dynamics in the plasma region, computational fluid dynamic (CFD) modeling using Comsol MultiPhysics 5.3 was carried out [36]. Considering the design of the experimental set-up and the interaction of the gases with the substrate, simulations were performed under the assumption of turbulent flow [35]. Transport of concentrated species was also applied to obtain information on the distribution of the mass fraction of argon and the precursor. Capillary edge–substrate distances and carrier gas flow rates were selected as variables for computation of the mole fraction and the velocity field in the observation region. Three-dimensional geometry was used for these simulations since there was no axial symmetry in the experimental plasma set-up. Figure 2 represents the constructed geometry of the plasma set-up for the XZ- and XY-plane views. The simulations were carried out in a cylindrical surrounding region with a 150 mm radius and 130 mm height, which was assumed to be filled with ambient air, and the dimensions of all the plasma set-up parts were in accordance with the experimental set-up description of Section 2.2. Argon and a monomer-containing Ar carrier gas were injected from the main gas and side gas inlets, respectively, as shown in Figure 2b, and incompressible flows were considered. For the main gas inlet, a flow rate corresponding to 1 slm was set. For the sideway inlet, the gas flow rate was varied between 125, 250, 500, and 750 sccm. The Navier–Stokes equation and the continuity equation were solved with a fine physics controlled mesh in the volume of interest, i.e.,

$$\rho(\boldsymbol{u}.\nabla)\boldsymbol{u} = \nabla.\left[-p\boldsymbol{I} + (\mu + \mu_T(\nabla\boldsymbol{u} + (\nabla\boldsymbol{u})^T)\right] + \boldsymbol{F} \tag{1}$$

$$\rho\nabla.(\boldsymbol{u}) = 0 \tag{2}$$

where ρ is the density (kg/m^3), $\boldsymbol{u}$ the gas velocity (m/s), $\boldsymbol{I}$ the identity matrix, p the gas pressure, μ the dynamic viscosity (Pa·s), μ_T the turbulent dynamic viscosity (Pa·s), $\boldsymbol{F}$ the volume force (N/m^3), and ∇ the differential operator.

As previously mentioned, transport of the concentrated species was also introduced, making use of the following equations, i.e.,

$$\nabla.\mathrm{j}_i + \rho(\boldsymbol{u}.\nabla)\omega_i = R_i \tag{3}$$

$$N_i = \mathrm{j}_i + \rho\boldsymbol{u}\omega_i \tag{4}$$

where j_i kg(m^2·s) is the relative mass flux component and ω_i a mass fraction.

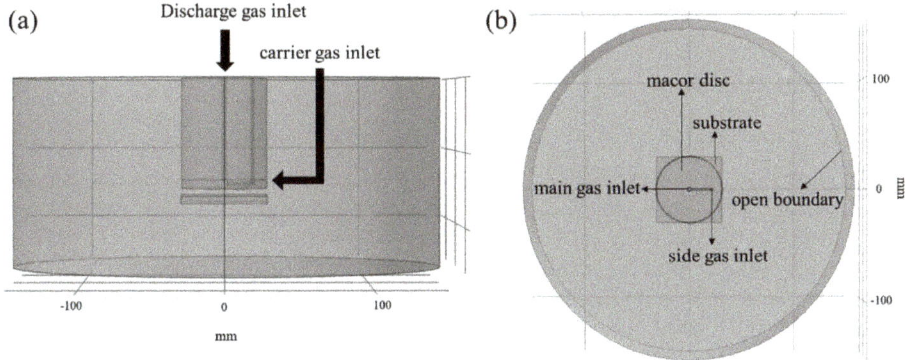

Figure 2. (**a**) XZ-plane and (**b**) XY-plane views of the MW plasma jet geometry.

2.5. AFM

The surface topography and the thickness of the deposited coatings were obtained by means of an XE-70 AFM apparatus (Park Scientific Instruments, Suwon, South Korea). In this work, the non-contact tapping mode using a highly-doped single crystal silicon cantilever with a spring constant of approximately 40 N/m was used to collect AFM images. Subsequently, using XEP software, the roughness of the surfaces was calculated. Additionally, to measure the thickness of the deposited films, a scratch was first made with a sharp blade on the surface of the cover slip glass covered by the coating. Subsequently, the scratch region was visualized using AFM and the difference in height formed on the substrate by the scratch was determined using XEP software. Each reported result is an average value calculated from three different measurements and presented with standard deviation.

2.6. SEM

After gold sputtering of the coated substrates with a JEOL JFC-1300 Auto Fine Coater (Peabody, MA, USA) for 20 s, the surface morphology of the coatings was visualized using a JEOL JSM-6010 PLUS/LV SEM device (Peabody, MA, USA) which applied an accelerating voltage of 7 kV. The magnifications used to acquire the SEM images in this work were 1000× and 30,000×.

2.7. Static WCA Analysis

The wettability of the deposited layers was measured using a Data Physics OCA 10 WCA instrument (Filderstadt, Germany). Droplets of distilled water 1 µL in volume were first placed on top of the deposited coating, after which an image of the water droplet resting on the surface was captured by a CCD video camera (Filderstadt, Germany). Afterwards, by fitting the profile of the water drop using Laplace–Young curve fitting, the WCA value was obtained. The experiments were repeated three times and the measurements were also reported as an average value with an error of less than $3.0°$.

2.8. XPS

Chemical analysis of the surface of the deposited films was assessed by performing XPS experiments. The measurements were conducted using a PHI Versaprobe II spectrometer (Kanagawa, Japan) with a monochromatic Al K_α X-ray source ($h\nu = 1486.6$ eV) operating at 25 W, resulting in a beam size of 100 µm. Survey scans and high-resolution C1s spectra were acquired with 187.85 and 23.5 eV pass energy, respectively. All XPS measurements were conducted in a vacuum of at least 10^{-6} Pa with an angle of 45° between the sample surface and the analyzer. By processing the survey scans using Multipak software (v 9.6) and applying a Shirley background, the elemental composition of the samples under investigation was determined. In addition, the high-resolution C1s peaks were curve-fitted by utilizing Gaussian–Lorentzian peak shapes with a 1.4 eV limitation of the full width at half maximum (FWHM) after a calibration of the energy scale was performed with respect to the hydrocarbon component of the C1s spectrum (285.0 eV).

3. Results and Discussion

3.1. Discharge Characterization

In the first stage of this study, the composition of the plasma during the coating deposition process was investigated. In an effort to obtain better insight into the quantity and quality of excited plasma species and their influence on the properties of the deposited coatings, OES was performed. The following bands and lines were observed in the obtained OES spectra: bands of OH (A→X); N_2 first negative and second positive systems; and argon, carbon, hydrogen, and oxygen atomic lines. Additionally, a weak radiation of atomic hydrogen was also noticed in the OES spectra. As mentioned above, all spectra were collected using four integration times (1, 0.1, 0.01, and 0.002 s). Nevertheless, in this work, the intensity of each selected component has been presented using an integration time which was sufficiently long to provide a good intensity of signal but which avoided signal saturation at the same time. The evolutions of OH, N_2 second positive system, N, Ar, Ar$^+$, C, CH, and O as some representative lines were selected to assess information on the discharge excited species in correlation with the experimental conditions. Table 2 provides an overview of these selected components with the wavelengths, corresponding transitions, and integration times used for their acquisition. The influence of power and carrier gas flow rate on the intensity of these lines was investigated, since these two parameters had the most significant impact on the concentration of excited species.

Table 2. Emission bands/lines and their corresponding transitions, wavelengths, and selected integration time.

Band/Line	Transition	Wavelength (nm)	Integration Time (s)
OH	A-X, band 0-0	305.0–310.0	0.1
N_2 2nd positive	C-B, bands 0-0, 2-4, 1-3, 0-2	337.1, 370.9, 375.4, 380.4	0.1
N	$^4P_{3/2}$-$^4P_{5/2}$	824.2	1
Ar^+	$^2D_{3/2}$-$^2D_{3/2}$	404.3	1
Ar	$^2[^3/_2]$-$^2[^3/_2]°$	714.7	0.1
C	$^1P°$-^1S	247.9	0.1
CH	$^2\Delta$-$^2\Pi$, band 0-0	431.3	1
O	^5P-$^5S°$	777.0	0.01

The selected spectral lines can be divided into two different groups: one group (OH, N, N_2 second positive, Ar, Ar^+, and O) features the discharge itself, and those in the other (C and CH) are representatives of the injected monomers MMA and styrene. The first evident trend observed in all emission spectra was the increase in intensity with increasing MW power, which was due to the extra dissociation and fragmentation of the monomer molecules that subsequently occurred with increasing input power. In fact, higher energy delivered to the plasma resulted in an increased amount of excited species generation. The dependence of the carbon line emission intensity on the carrier gas flow rate at different input powers is presented in Figure 3 for the two monomers under study (styrene and MMA). Based on these graphs, it can be concluded that precursor fragmentation was almost twice as significant when using MMA as a monomer instead of styrene. This difference can be explained by the different levels of dissociation energy required for breaking the bonds in each monomer molecule. Immediately after introducing the monomer to the system, the intensity of the carbon line also increased for both monomers under study. The intensity line sharply increased with increasing carrier gas flow rate until it reached its maximum value at a carrier gas flow rate of 250 sccm. Further increasing the monomer input, however, resulted in a decrease in the intensity of the atomic carbon line, demonstrating that a higher monomer concentration present at the same power level results in a decrease in available fragmentation energy per monomer molecule. Moreover, for a higher carrier gas flow rate, the interaction time between monomer molecules and plasma reactive species was shorter, which in turn decreased the concentration of the carbon species. The same trend with maximum intensity at a 250 sccm carrier gas flow rate was also observed for the CH compound, as can be seen in Supplementary Materials Figure S1.

The optical emission of the Ar and Ar^+ lines is proportional to the concentration of reactive species in the argon discharge. Thus, the dependency of these line intensities on the carrier gas flow rate and applied power can provide insight into the processes occurring during thin film deposition. The evolution of the Ar and Ar^+ line emission intensity as a function of carrier gas flow rate at different discharges powers is presented in Figure 4a,b for styrene and MMA, respectively. A comparison between Figure 4a,b suggests that for a higher applied power, the intensity of Ar and Ar^+ were almost independent of the type of monomer which was present in the discharge, while for lower power, an admixture of MMA resulted in an enhanced intensity of Ar and Ar^+ lines in comparison with styrene. In addition, the signal from these two argon lines increased with increasing carrier gas flow rate and again reached its maximal level at a carrier gas flow rate of 250 sccm for both monomers under study. The growing trend can be explained by the higher amount of overall argon gas in the system which flowed with a higher speed. At a carrier gas flow rate of above 250 sccm, a steady state of the intensity of the Ar and Ar^+ lines appeared, which can be seen to be related to the air admixing and the presence of a higher concentration of monomer molecules in the discharge. The same trend of intensity profiles was also observed for the N_2 second positive, N, OH, and O bands (see Figures S2–S5, Supplementary Materials). The similarity in the optical emission intensity between all these mentioned bands and the argon lines thus indicates that the processes standing behind the excitation were strongly interconnected.

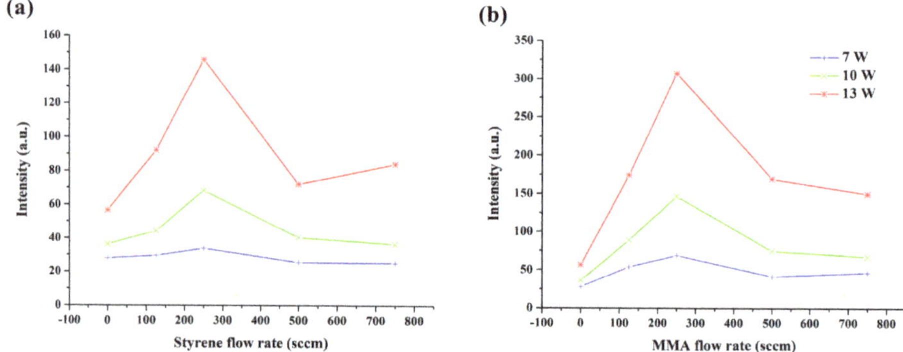

Figure 3. Dependence of the C line emission intensity on the carrier gas flow rate at different powers for the (**a**) styrene and (**b**) methyl methacrylate (MMA) monomers.

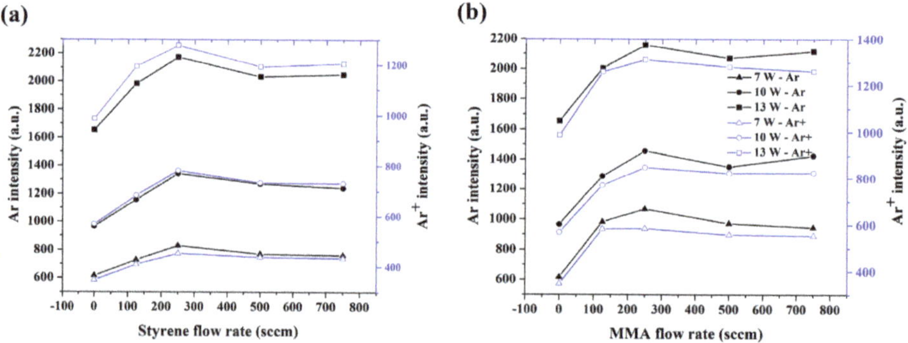

Figure 4. Dependence of the Ar and Ar^+ line emission intensities on carrier gas flow rate at different applied powers for the (**a**) styrene and (**b**) MMA monomers.

Making use of the integral intensity of a few selected argon lines, namely the lines at 603.2, 675.3, 687.1, and 714.7 nm, the excitation temperature of argon was also calculated using corresponding constants from the NIST (National Institute of Standards and Technology) database [37]. The calculations resulted in excitation temperatures in the range 4300–5000 K under all experimental conditions and for both monomers. In general, no dependence on the carrier gas flow rate was observed for the excitation temperature. However, the excitation temperature was slightly higher with increasing power. In the next step, a Boltzmann plot from the N_2 second positive 0-2, 1-3, and 2-4 band heads was used to determine the vibrational temperature. This temperature was found to be dependent on discharge power and monomer type and was rather stable when changing the Ar carrier gas flow rate. For the MMA precursor, the vibrational temperature varied between 2100 and 2900 K depending on the applied power while this temperature range was slightly lower for styrene, being 2000–2800 K. Furthermore, according to the Boltzmann distribution, using the N_2 second positive 0-2 band with J = 27–32, the rotational temperatures were also calculated [38]. For this calculated rotational temperature, no significant impact of power, monomer type, and carrier gas flow rate variation on the obtained value was observed. The range of rotational temperature variation for both monomers was found to be 1250–2500 K, with a mean uncertainty of 250 K. Additionally, the rotational temperature calculated from the OH radical lowest lines (details in [20]) was 550–700 K, with a mean uncertainty of 70 K.

3.2. CFD Simulations

Additional to the plasma characterization, 3D numerical simulations of fluid dynamics were also carried out in this study to obtain a basic perception of how the mixing of the main Ar and the Ar carrier gas stream occurred. Moreover, information on the distribution of the monomer in the volume close to the substrate was also able assist in evaluating the thin film growth during the deposition process. For example, it has already been shown that the gas velocity and mass fraction can be correlated to the deposition pattern on a substrate [39] and the plasma behavior [35]. In the latter study it was also revealed that placing an additional plate at the edge of the plasma jet capillary intensified the gas purity near the surface when working with a small gap distance to the substrate [35]. In this study, the experimental set-up was quite similar to that used in [35], as the additional disc for the purpose of monomer introduction also played the role of an additional plate and thus constrained the gas in a small volume, which in turn influenced the gas diffusion and propagation near the substrate. All computations in this work were only performed for the styrene monomer admixture, since its deposition pattern on the substrate differed from a simple circular shape. This was not the case when using MMA as monomer, as will be shown in the following sections.

The first stage of the simulations was focused on defining the velocity distribution in the MW plasma jet system. In this step, the main gas flow rate was fixed at 1 slm and the distance between the capillary and the substrate was set to 5 mm. Computations were performed for different monomer-containing gas flow rates (125, 250, 500, and 750 sccm), and Figure 5 demonstrates the gas velocity for these different gas flow rates in the XZ-plane. Figure 5a shows that for the smallest value of the carrier gas flow, there was no disturbance to the main gas stream, as a higher velocity was present in the larger channel compared to the sideway gas inlet. Figure 5b also shows that when the additional gas reached a flow rate of 250 sccm, the velocity of the two gas streams became comparable at their meeting point. Moreover, when using even higher carrier gas flow rates (Figure 5c,d), the impact of the sideway gas inlet became more pronounced. Under these experimental conditions, the side gas flow could not easily deviate from its direction and could not mix well with the main gas at their meeting point. The higher velocity of the sideway gas pushed the monomer-containing gas and the main gas flow on the side of the capillary opposite to the sideway gas inlet, as shown in Figure 5c,d. Changes in the velocity of the gas near the substrate can be seen as well in Figure 5. The homogeneous velocity distribution of the gas in the case of small side gas flow rates changed to an uneven distribution when injecting sideway gas with higher flow rates. It is also worth mentioning that the diameters of the two gas inlets were different (3.1 mm for the main gas stream and 2 mm for the sideway gas inlet), which influenced the velocity magnitude. The high velocity of the main argon gas stream in the center could also prove that there should be a negligible amount of deposition in the center of the substrate, especially when using low side gas flow rates.

In a next step, the mass fraction of the monomer precursor and the way the monomer propagated on the surface were simulated. The results are shown in Figure 6 for the XY-plane (top view) and in Figure 7 for the XZ-plane of the used 3D model. The XY-plane crossed the volume at the position where the substrate surface was located during the experiments. The shown series of graphs indicates that for the smallest carrier gas flow rate (125 sccm), only a small amount of monomer was present on the substrate surface and that the monomer completely diverged to the right-hand side of the plasma set-up. An increase in the monomer-containing gas flow rate to 250 sccm intensified the presence of the monomer at the right side of the set-up with only the presence of a negligible amount of monomer on the other side of the substrate. On the other hand, when the flow rate of the side gas was further increased to 500 sccm, a small propagation of the monomer to the left side of the substrate was observed. Moreover, a 750 sccm side gas flow rate even led to the monomer spreading over almost the complete area of the substrate. In the XZ-plane of the 3D model, depicted in Figure 7, the majority of the monomer also deviated to the right-hand side below the opening of the Macor disc on the substrate surface when the carrier gas flow rate was small (Figure 7a,b). For these studied conditions, the velocity of the main gas stream was much higher than the velocity of the gas stream coming from the smaller side channel. Consequently, the monomer only slightly mixed with the main stream of the system and remained on the same side when the gas outflowed the capillary. On the other hand, in the cases where carrier gas flow rates of 500 and 750 sccm were used (Figure 7c,d), the mixing process of the two streams was significantly improved and the monomer was more directed to the center, since it was entering the channel with a higher velocity. Moreover, at the highest flow of the side gas, the volume between the Macor disc and the substrate was almost entirely filled with monomer, which was not the case for the other analyzed carrier gas flow rates. It can therefore be concluded that increasing the carrier gas flow rate resulted in a more uniform monomer dispersion on the substrate surface based on the mass fraction of the precursor in the volume between the plasma jet and the substrate. These results of the monomer propagation region shape can thus provide a preliminary prediction of the footprint of the deposited layer on the substrate.

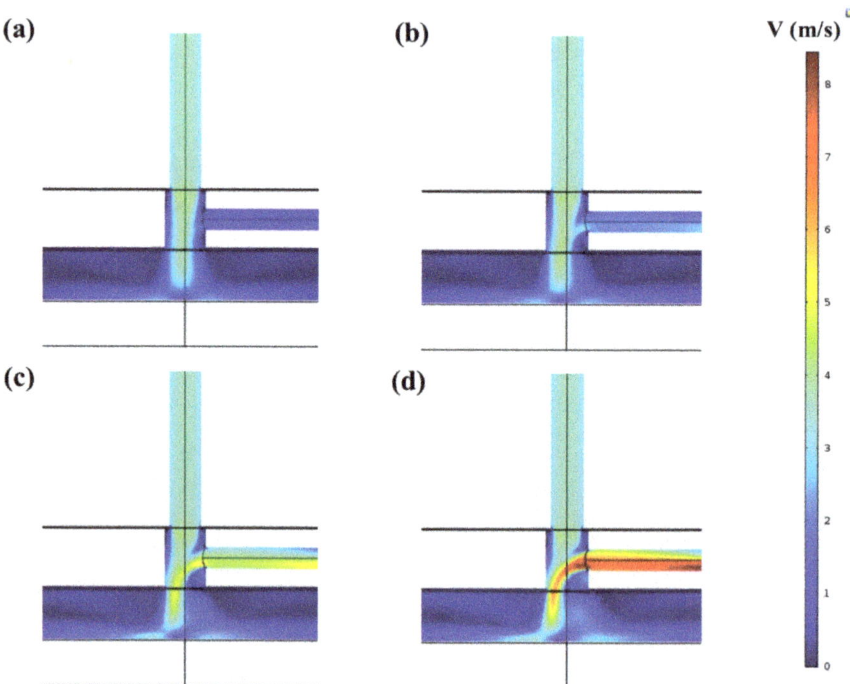

Figure 5. Velocity distribution (m/s) in the XZ-plane at a 5 mm gap size between the capillary and the substrate for a sideway gas flow rate of (**a**) 125 sccm, (**b**) 250 sccm, (**c**) 500 sccm, and (**d**) 750 sccm.

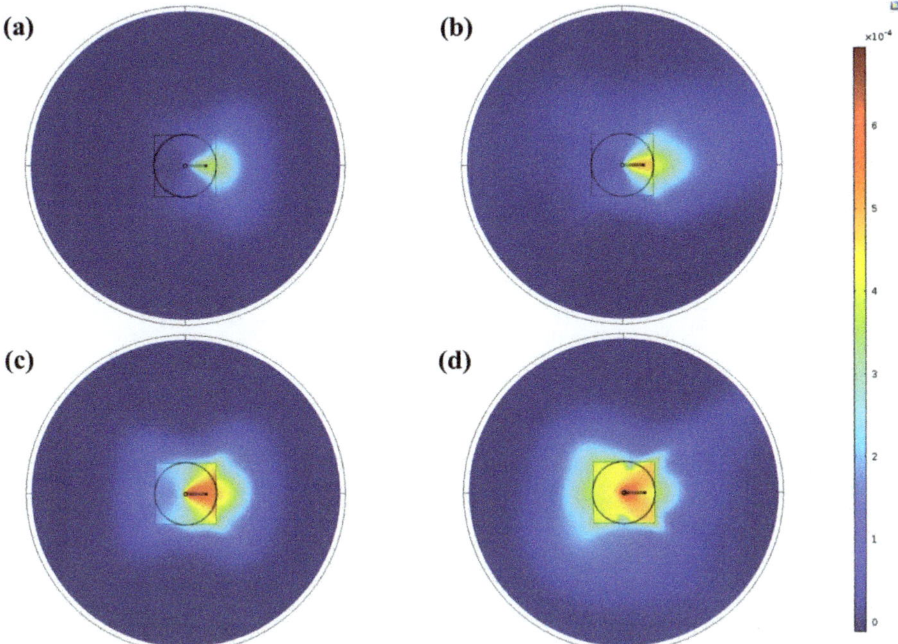

Figure 6. Top view (XY-plane) of the monomer mass fraction distribution at 5 mm gap size for (**a**) 125 sccm, (**b**) 250 sccm, (**c**) 500 sccm, and (**d**) 750 sccm side gas flow rates.

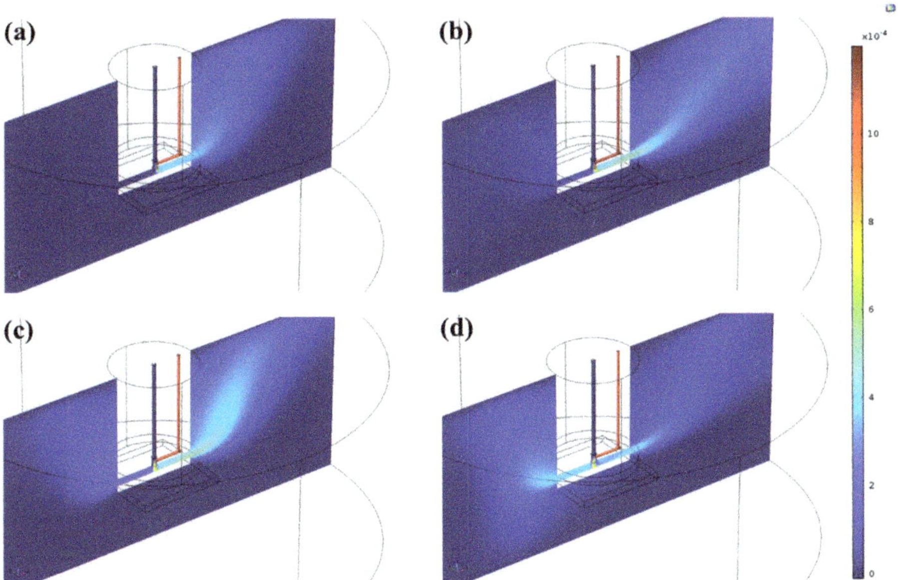

Figure 7. XZ-plane cross-sections of the monomer mass fraction distribution at 5 mm gap size for (**a**) 125 sccm, (**b**) 250 sccm, (**c**) 500 sccm, and (**d**) 750 sccm side gas flow rates.

The size of the volume between the Macor disc and the sample was regulated by the gap size between these two planes. Consequently, the working distance also had a significant influence on gas propagation at the capillary outlet. Figure 8 shows the influence of the gap size on the monomer distribution for distances of 3, 5, and 7 mm between the bottom of the Macor disc and the substrate plane. In these simulations, the sideway gas flow rate was fixed at 250 sccm. For a smaller distance (Figure 8a), the volume of propagation was too small to provide a sufficient distance for efficient mixing of the two gas streams. As a result, the pattern of the monomer fraction was in this case mainly controlled by the main gas flow. A similar behavior was also observed when using a 5 mm distance, yet a more uniform gas propagation on the right side of the sample was present due to a longer available distance for mixing of the two gas streams. Finally, for a 7 mm distance between the disc and the substrate (Figure 8c), the mass fraction of the monomer was more evenly distributed on the surface of the substrate, which might in turn have resulted in a better homogeneity of the deposited coating.

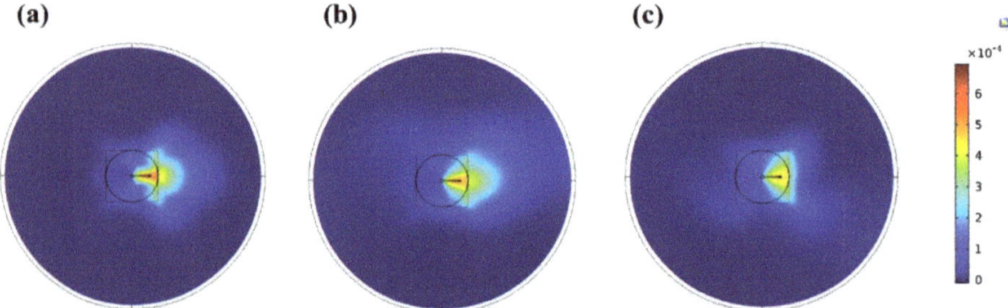

Figure 8. Top view (XY-plane) of the monomer mass fraction distribution for (**a**) 3 mm, (**b**) 5 mm, and (**c**) 7 mm gap sizes. The monomer-containing gas flow rate was fixed at 250 sccm.

3.3. Physical and Chemical Surface Characterization

3.3.1. Coating Appearance and Homogeneity

In the next step of this study the physical and chemical properties of the deposited coatings were investigated. For this purpose, the MW plasma jet was used for thin film deposition of MMA and styrene coatings. Initially, both monomers were deposited on a silicon wafer, making use of the main conditions (power: 10 W, monomer-containing gas flow rate: 250 sccm, deposition time: 3 min, and gap size: 5 mm (see Table 1)) to get an idea of the deposited thin film pattern on a flat substrate. The obtained plasma-deposited thin films were studied in terms of surface morphology and topography by utilizing SEM and AFM, respectively, and the obtained results have been depicted in Figure 9. As can be seen in this figure, each coated silicon wafer had a unique appearance depending on the used monomer and the way it was mixed into the MW plasma jet effluent. The diameter of the coated area for the plasma-deposited styrene was approximately 20 ± 5 mm, while for MMA, the diameter was reduced to 3.0 ± 0.5 mm. However, in the case of styrene, the deposition pattern on the substrate was not entirely symmetrical. This result can be explained by the non-homogeneous mixing of the monomer into the main stream of the plasma, as was shown in the previous section. On the other hand, when MMA was added to the plasma, a coating located near the center of the substrate was deposited, as demonstrated in Figure 9b. Figure 9 thus clearly shows that the styrene coating had larger dimensions compared to the MMA-based film deposited under the same experimental conditions. A possible explanation for this dissimilarity may be the difference in deposition rate of the two monomers under study. For instance, organic compounds containing double bonds and aromatic rings are known to have a higher deposition rate in comparison to branched-chain or non-aromatic structures [40]. Hence, styrene is expected to show a higher deposition rate than MMA, which can in turn result in a larger deposition area.

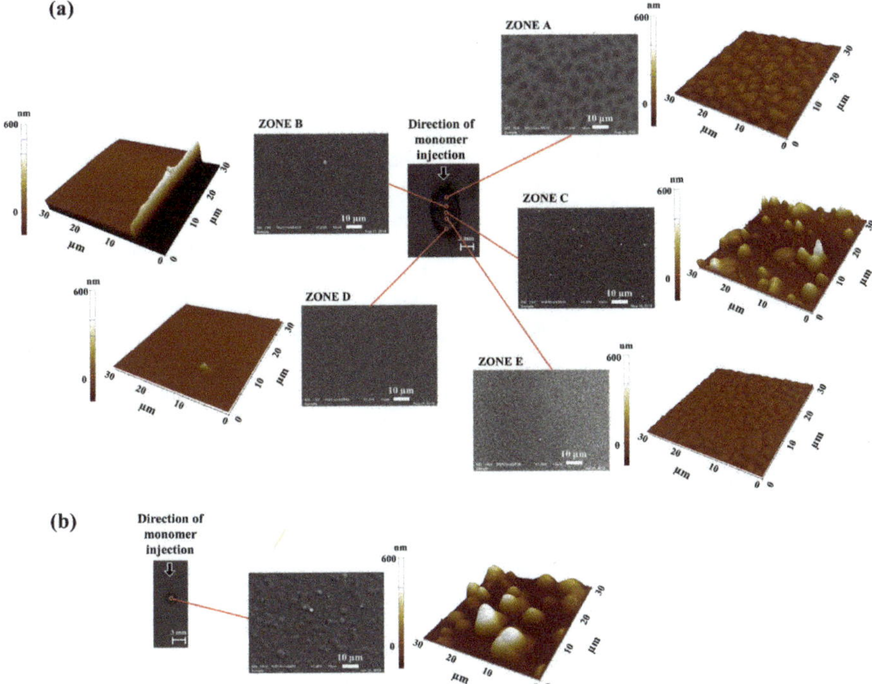

Figure 9. Overview of the appearance, surface morphology, and surface topography of thin films deposited with the MW plasma jet by injection of the (**a**) styrene and (**b**) MMA monomers. The observed area of the 3D atomic force microscopy (AFM) image is $30 \times 30 \ \mu m^2$.

On the styrene-based coating area, five different zones have been determined which all showed a different surface topography and morphology. Based on Figure 9a, the following zones were selected: zone C—directly under the MW plasma jet capillary where the coating appeared in the form of dispersed droplets with round shapes and sizes reaching up to 1 μm; zone B—located above zone C, consisting of a very smooth and thick coating; zone D—located below zone C where no deposition at all could be observed; and zone A and zone E—located at the edges of the coated area. In these last two regions, the deposits were characterized by a very thin non-uniform coating that contained plenty of vacant spots. The location of the MMA coating, as shown in Figure 9b, matched the position of zone B of the styrene-based sample, suggesting similarity in gas dynamics for both monomers as the same plasma jet device was used. Indeed, a higher amount of deposited precursor was situated above the center of the substrate as the monomer gas injection point was located in this zone. This not-centered position of the deposited coating can be explained by the suppressive behavior of the main gas stream towards the injected monomer. Since the monomer is injected in the region where the plasma jet is fully developed, well-structured, and close to the outlet, it does not have enough residence time as it propagates towards the core of the plume. Thus, the majority of the monomer is not deposited in the central area but in the zone shifted towards the monomer injection side. Figure 9b also shows that the coating morphology was different when MMA instead of styrene was used as a monomer, which could be attributed to the lighter weight of MMA molecules compared to styrene. The OES results suggest that the plasma jet with injected MMA contained a higher amount of excited species which could have interacted with the monomer molecules. This could in turn have led to nucleation in the gas phase, which is followed by condensation and finally deposition in the form of droplets on the surface. This effect was also valid for styrene but was less pronounced, as shown in Figure 9a.

3.3.2. Coating Morphology

In this part of the study, the morphology of the coatings was examined using SEM imaging. As mentioned above, the majority of the coated sample surfaces contained droplets distributed all across of the deposit. Figure S8 (Supplementary Materials) depicts a cross-sectional view of the coatings on the glass substrate under the main conditions. The formation of droplets and clusters on the sample surface is a sign that monomer plasma polymerization actually occurs during the coating deposition [25,41,42]. Based on the used experimental conditions, the size and distribution of these deposited droplets were found to fluctuate. Figure 10 shows the SEM images of both styrene (zone C) and MMA samples prepared with different carrier gas flow rates. Generally, deposited styrene in the center contains droplets with a smaller size compared to the MMA monomer for all experimental conditions. A common trend can also be noticed for the SEM images presented in Figure 10: increasing the monomer-containing gas flow rate caused an increase in the number of droplets formed on the surface of the coatings. Additionally, by adding monomer to the system the size and shape of the droplets also varied. For example, when using 750 sccm MMA, the large amount of injected monomer resulted in a surface coverage by a high amount of droplets which appeared as a smooth coating. The influence of other working parameters (time, power, and distance to substrate) on the surface morphology can be observed in Supplementary Materials Figures S6 and S7 for MMA and styrene, respectively. A short deposition time (1 min) and lower carrier gas flow rate resulted in coatings possessing a smaller amount of droplets, while higher values of these parameters produced deposits with a higher amount of droplets across the investigated area for both monomers. In addition, the distance between the MW plasma jet disc and the substrate influenced the size of the droplets: increasing this distance to 7 mm made the size of the deposited droplets larger. This can be explained by the gas dynamics in the volume between the Macor disc and the substrate. A smaller distance can provoke vortex creation at the center [14], which can affect the location of droplet creation on the substrate surface, while for a larger distance this effect is less prevalent, resulting in deposits with larger-sized droplets. To prove that a plasma-polymerized cross-linked coating was deposited on the substrates in this work, the coatings were submerged in water for a short duration. This was tested for the main condition deposits and Figure S9 (Supplementary Materials) confirms that the coating was still present on the substrate after immersion in water for 24 h. This result clearly evidences that a plasma-polymerized cross-linked coating was deposited in this work [43,44].

3.3.3. Coating Thickness

As previously mentioned, to measure the thickness of the deposited film, a scratch was first made with a sharp blade along the zones of the samples that were defined on the surface of the coating deposited on the cover slip glass. Subsequently, the scratch region was visualized using AFM and the difference in height formed on the substrate by the scratch was determined. Within this context, it is important to mention that during the making of the scratches, the cover slip glasses remained untouched, making the thickness measurement more reliable. As previously shown in Figure 9, it had already been observed that a thick homogeneous coating was obtained in zone B of the plasma-deposited styrene sample. As a result, in the case of styrene, the thickness of the films was measured in zone B. Coating thickness measurements were, however, not performed for samples prepared using MMA as a monomer, as in this case the coating was only spread across the substrate in the form of droplets. Consequently, Table 3 only contains the coating thickness results for styrene samples prepared under different experimental conditions.

Table 3. Thickness of the styrene-based coating under different experimental conditions.

Condition	Main	Main + t1	Main + t2	Main + p1	Main + p2	Main + f1	Main + f2	Main + f3	Main + d1	Main + d2
Thickness (nm)	229 ± 8	58 ± 3	486 ± 12	210 ± 5	217 ± 6	110 ± 5	220 ± 4	190 ± 10	140 ± 6	248 ± 8

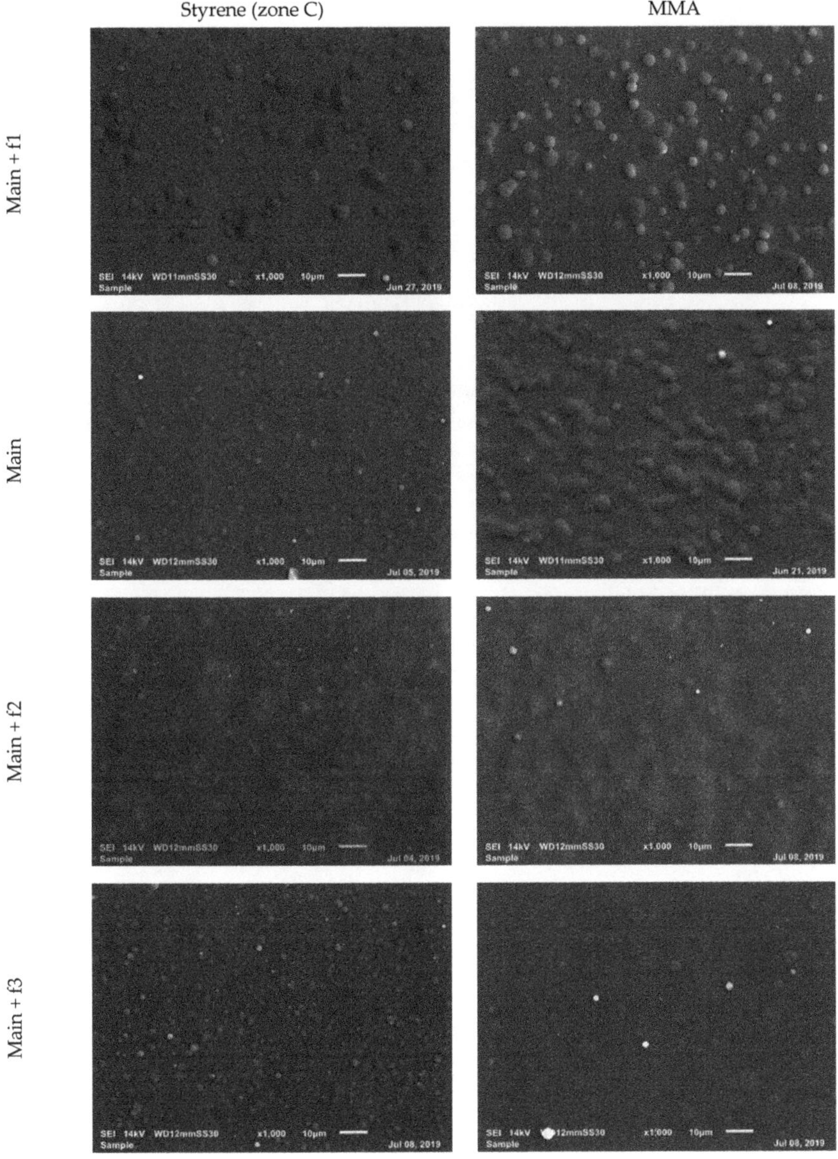

Figure 10. SEM images of the central zone of styrene (zone C) and MMA coatings deposited with different carrier gas flow rates.

These results demonstrate that the lowest measured thickness corresponded to the lowest deposition time (1 min), and, conversely, that the highest deposition time resulted in the thickest coating. Another parameter that had a direct influence on the coating thickness was the carrier gas flow rate. As is reported in Table 3, the smallest Ar flow-carrying styrene (125 sccm) produced a coating with a 110 nm thickness after a deposition time of 3 min. When a higher Ar carrier gas flow rate of 250 sccm was used, the coating thickness reached its maximum level of 229 nm after the same deposition time. Interestingly, any further increase of the carrier gas flow rate did not result in a thicker coating but resulted in a more homogenous film deposition over the substrate area as other measurements along the scratch in this zone showed only a small deviation from the reported coating thickness. Based on the simulation results reported in Section 3.2, it was concluded that at the highest carrier gas flow rate (750 sccm), the carrier gas was injected into the system with a considerably high velocity, which could in turn spread the monomer all over the substrate surface due to efficient mixing with the main plasma jet gas stream. Besides the gas propagation behavior, which only predicted a better uniformity of the coating at the highest carrier gas flow rate, the available energy for monomer fragmentation should also be taken into account to explain why the coating thickness did not increase when increasing the Ar carrier gas flow rate to 500 and 750 sccm. Since the applied power remained the same while the carrier gas flow rate was varied, the average energy per monomer molecule reduced with increasing monomer injection flow. As a result, the efficiency of precursor fragmentation was also less pronounced at higher sideway gas flow rates, which is in agreement with the OES results shown in this work. Although one would expect a thicker coating with increasing carrier gas flow rate as a higher amount of monomer was injected in the system, this increase in monomer amount was in this study counteracted by the lower amount of monomer fragmentation, ultimately leading to the deposition of a thinner coating.

Table 3 also reveals that the distance between the disc and the substrate also influenced the thickness of the coating: a higher distance to the substrate yielded a thicker and more uniform coating, while a smaller gap size provided a wider coated area with a lower coating thickness and less uniformity. The latter conclusion concerning coated area and coating uniformity was drawn based on thickness measurements which were performed along the scratch. These results can again be well correlated with the simulations performed in this work, showing the effect of distance on the monomer distribution. Moreover, at a lower distance, a developed turbulent flow of main and carrier gas can occur, which can in turn influence the size and homogeneity of the coated area. Finally, Table 3 also shows that the coating thickness was also affected by the applied power. An increase in power to 10 W resulted in a higher coating thickness, resulting from more monomer fragmentation at elevated power. However, at a power of 13 W, the coating thickness was not increased further, which could be explained by the fact that in this case monomer fragmentation was too pronounced to efficiently deposit a coating.

3.3.4. Coating Roughness

Another critical coating parameter that can be measured by AFM is surface roughness, and the obtained roughness results are presented in Table 4. Generally, for styrene samples, the highest surface roughness was observed in the center of the deposit (zone C), as in this area the monomer was mostly deposited in the form of dispersed droplets with varying sizes. This roughness, however, decreased in zone B in comparison to central zone C, as indicated by the results shown in Table 4. Noting that the roughness for the pristine glass is 5 nm, zone B did contain a coating, and this coating was quite smooth, as the roughness varied between 14 ± 2 and 32 ± 6 nm, depending on the experimental conditions. These results are thus consistent with the AFM image of zone B, shown in Figure 9a. After this smooth coating region, an area covered by a non-homogeneous coating with different vacant sites was observed to follow (zone A). In this region, the roughness was found to be rather similar to or slightly higher than the roughness values obtained in zone B, depending on the applied experimental conditions. Table 4 also reveals the influence of deposition time on the surface roughness: for both monomers, an increase in deposition time resulted in a higher surface roughness. This finding can be attributed to a higher number of accumulations of the monomer on the surface at random locations when using longer treatment duration. Consequently, this monomer distribution in the form of droplets resulted in an increase in coating surface roughness. Another experimental parameter that was found to influence the surface roughness was the applied power: for both monomers, an increase in power resulted in a rougher deposit. This can be explained as follows: at a higher power, a higher level of monomer fragment reactions occurred in the gas phase instead of on the gas–substrate interface, thereby leading to a rougher coating surface. In contrast to deposition time and power, the influence of carrier gas flow rate on the surface roughness was dependent on the type of injected precursor (MMA or styrene).

For MMA, the highest surface roughness was obtained when using the lowest carrier gas flow rate, and this roughness decreased with increasing flow of monomer into the system. As stated above, with an increase in the amount of injected monomer to the system the droplets on the substrate recombined with each other and decreased the surface roughness and finally completely covered the surface at the higher flow rate, producing a smooth coating. In the case of styrene, the smallest carrier gas flow rate of 125 sccm produced films with a comparatively small roughness in all zones. These roughness values in the center of the sample (zone C) increased when a 250 sccm monomer-containing gas flow rate was used, and then again decreased when this gas flow rate was further increased to 500 and 750 sccm. This behavior can be explained by a combination of monomer distribution and gas velocity during the deposition process. Zone C is just below the plasma jet capillary, and, thus, with a 125 sccm carrier gas flow rate, the monomer was not well mixed with the main stream, leaving this area almost without a coating (just a few deposited droplets). A higher amount of droplets was able to be observed for the next value of Ar carrier gas, since in this case more monomer fragments were delivered to the surface by the faster gas stream. Lastly, a smooth layer in zone C was observed when deposition was performed at higher flow rates (500 and 750 sccm) due to good monomer mixing with the main gas stream and its uniform distribution under the plasma jet capillary, as was shown in Section 3.2. In zones A and B, which were located at the side where the monomer was injected, the values of surface roughness increased with increasing monomer-containing gas flow rate. These findings can be explained by the simulation results obtained earlier. It was shown that at a lower flow rate the monomer was not well mixed with the main stream and mostly remained on one side of the plasma jet, while higher values of carrier gas flow rate caused direct injection of the monomer into the core of the MW jet. Thus, the nucleation of styrene near zones A and B proceeded at different radial positions of the plasma jet: near its edge or in the core for smaller and higher side gas flow rates, respectively. This different behavior resulted in coatings with different roughness values at these positions on the substrate, depending on the amount of injected monomer. Finally, the distance between the MW plasma jet and the sample was found to have an insignificant influence on surface roughness.

Table 4. Surface roughness of styrene and MMA deposited thin films under different experimental conditions.

Condition/Roughness (nm)	Main	Main + t1	Main + t2	Main + p1	Main + p2	Main + f1	Main + f2	Main + f3	Main + d1	Main + d2
Styrene (zone A)	22 ± 2	23 ± 3	38 ± 5	17 ± 2	29 ± 4	22 ± 3	41 ± 2	45 ± 4	19 ± 2	20 ± 3
Styrene (zone B)	20 ± 2	14 ± 2	30 ± 4	18 ± 3	25 ± 5	15 ± 3	29 ± 6	32 ± 6	19 ± 3	20 ± 3
Styrene (zone C)	128 ± 6	63 ± 7	173 ±5	75 ± 8	149 ± 10	64 ± 4	79 ± 5	102 ± 8	125 ± 5	136 ± 8
MMA	137 ± 7	110 ± 10	156 ± 11	122 ± 10	185 ± 9	236 ± 12	110 ± 9	54 ± 5	122 ± 6	136 ± 8

3.3.5. Coating Wettability

In this part of the study, the wettability of the deposits prepared in this work was determined utilizing WCA analysis. The measurements were again performed on the styrene and MMA samples in the zones that were introduced earlier in this work. The wettability of an uncoated cover slip glass was measured to be 79° ± 2°. After plasma deposition, the WCA values of the coated samples significantly decreased, suggesting that the coated layers were hydrophilic. Summarized results of WCA measurements for both monomers and under different experimental conditions are presented in Table 5. These results indicate that shorter deposition times and smaller carrier gas flow rates resulted in coatings with higher hydrophilicity for both precursors. Consequently, a higher flow rate of the monomer-containing gas produced deposits with higher WCA values. This latter behavior can be explained by a reduction in monomer molecule fragmentation, and, therefore, an incorporation of more non-oxidized long-chain monomers on the surface with increasing carrier gas flow rate. Furthermore, Table 5 also shows that an increase in applied power resulted in more hydrophobic coatings. A possible explanation for this can be that there was lower fragmentation with lower power. In the case of styrene, the highest wettability was for all experimental conditions observed in zone C, the location precisely under the plasma exposure point, which might be due to the fact that this point had the best interactions with the plasma jet during the deposition process. However, while radially proceeding to the edge of the sample, in the direction of zone A, a reduction in coating wettability was observed for all experimental conditions under which styrene coatings were deposited. Only a small difference in WCA was observed for different distances between the MW plasma jet and the substrate, which can be explained by variations in the gas flow dynamics in the volume above the sample.

Table 5. Water contact angle (WCA) angle values for styrene and MMA coatings prepared under different experimental conditions.

Condition	Zone A	Zone B	Zone C	MMA
Main	45.0 ± 2.0	34.2 ± 1.8	28.2 ± 3.1	26.5 ± 3.1
Main + t1	41.3 ± 3.5	31.0 ± 1.5	22.3 ± 2.3	23.7 ± 3.5
Main + t2	45.7 ± 2.1	36.5 ± 2.6	31.1 ± 2.7	30.8 ± 2.8
Main + p1	35.3 ± 3.2	20.5 ± 2.4	18.0 ± 3.3	24.4 ± 3.1
Main + p2	50.5 ± 3.5	36.3 ± 2.4	32.4 ± 3.6	35.4 ± 2.9
Main + f1	18.0 ± 1.9	12.1 ± 1.8	8.1 ± 2.1	10.8 ± 1.9
Main + f2	55.4 ± 2.5	45.8 ± 1.2	30.0 ± 2.2	37.6 ± 2.0
Main + f3	61.6 ± 2.4	52.9 ± 1.0	35.2 ± 1.7	62.0 ± 1.7
Main + d1	45.0 ± 3.6	36.1 ± 2.0	23.0 ± 3.0	28.5 ± 3.2
Main + d2	42.5 ± 2.5	33.6 ± 1.5	27.3 ± 2.3	38.0 ± 2.3

3.3.6. Coating Chemical Composition

Chemical surface analysis is one of the most valuable characterization tools for deposited coatings, as it provides information on thin film surface composition. The elemental compositions of the coatings prepared under various experimental conditions were determined by XPS in addition to the concentration of the carbon-containing chemical bonds. Similarly to WCA, the coatings were again analyzed in the same zones as where all other measurements were performed. From XPS survey spectra it was found that the coatings produced from styrene only consisted of carbon and oxygen, except for the central point (zone C), at which a small amount of nitrogen was also detected. It is well known that APPJ treatment can induce the incorporation of oxygen by post-treatment exposure of the sample to the atmosphere or by reactions with oxygen which are present in the discharge due to gas impurities and ambient air admixture [45,46]. OES measurements demonstrated that in this work a considerable amount of oxygen and nitrogen species were present in the plasma jet during the thin film deposition when using styrene as a monomer. This can explain why in case of styrene, which contains no oxygen in its chemical structure, it was possible to detect oxygen and nitrogen in the deposited coatings. In the case of MMA, the surface of the thin films only consisted of carbon and oxygen, regardless of the location or the applied experimental conditions, which is in agreement with the chemical structure of the original monomer. Based on the elemental composition results, O/C ratios were calculated for MMA and all regions of the styrene samples, and the obtained values have been presented in Supplementary Materials Table S1. These results were found to be in good agreement with the earlier obtained WCA results, since a higher oxygen incorporation (higher O/C ratio) corresponded to a more hydrophilic region. Moreover, for the styrene samples, the O/C ratio followed the same trend across the surface as was noticed before for the coating wettability properties. A higher amount of oxygen was incorporated in the central area (zone C), and each subsequent area towards the edge of the coating contained less oxygen.

Deconvolution of high-resolution C1s spectra was also performed to obtain more detailed information on the chemical structure of the deposited thin films. The C1s high-resolution spectra for deposited styrene coatings were fitted with five curves, namely, (1) aromatic C–C/C–H bonds at 284.8 eV; (2) aliphatic C–C/C–H bonds at 285 eV; (3) the oxygen-containing group C–O at a binding energy of 286.6 eV; (4) O–C–O/C=O groups at a binding energy of 288.1 eV; and (5) a shake-up peak at 291.5. For MMA samples, curve-fitting of C–C/C–H, C–COO, C–O, O–C–O/C=O, and O–C=O were used at binding energies of 284.8, 285.4, 286.6, 287.9, and 289.1 eV, respectively. Figure S10

(Supplementary Materials) shows a typical deconvolution of the detailed C1s spectra for styrene and MMA deposits with the indication of the reference peaks ascribed to conventional PS (polystyrene) and PMMA (poly(methyl methacrylate)) as well as new peaks induced during plasma polymerization [47]. The discrimination of the standard peaks attributed to the mentioned polymers is evidence that the plasma polymerization process was happening during the coating deposition with MW APPJ for both monomers. For the MMA coatings, the appearance of the peak at 287.9 eV could have been due to the scission of C–OCH$_3$ bonds during plasma exposure, as was reported in the literature [48]. In the case of styrene, the presence of two extra peaks attributed to oxygen-containing groups suggests the breakage of C–C/C–H bonds, followed by recombination with oxygen from the surrounding environment [49]. The presence of the minor shake-up bond revealed that by using the MW plasma jet for thin film deposition, the benzene ring of the styrene monomer was at least partly preserved in the deposits. Figure 11a,b show the C–O/C–C ratio as a function of carrier gas flow rate for styrene and MMA, respectively. From these figures, it is clear that for both monomers the amount of C-O functional groups was highly dependent on the rate of monomer injection into the system. In all regions of the styrene coating and in case of the MMA coating, the smallest flow rate of the carrier gas (125 sccm) resulted in the highest values of C–O/C–C ratio. Increasing the carrier gas flow rate up to 500 sccm led to the incorporation of a higher amount of carbon in the form of C–C/C–H bonds and, consequently, the C–O/C–C ratio decreased. Further increasing the carrier gas flow rate to 750 sccm, on the other hand, did not significantly affect the C–O/C–C ratio anymore for both monomers under study. This trend can be explained by the concentration of the monomer in the discharge during its fragmentation: at a lower carrier gas flow rate, fragmented monomer molecules mostly interacted with oxygen molecules present in the plasma rather than with other monomer fragments which occurred at higher rates of monomer injection. Hence, the C–O/C–C ratio was lower in the case of higher carrier gas flow rate due to more pronounced recombination processes between the monomer fragments. The influence of the applied power, the distance between the plasma jet disc and substrate, and the deposition time on the coating chemical composition was also investigated and the results have been reported in Figure S11 (Supplementary Materials) together with the carrier gas flow rate results for both styrene and MMA. This figure reveals that the deposition time and the jet-substrate distance had a mostly insignificant impact on the C–O/C–C ratio: generally speaking, all these samples only varied in a small range of C–O/C–C ratio with varying time and distance. By contrast, the discharge power was found to have an impact on the C–O/C–C ratio: in the case of both monomers, a higher C–O/C–C ratio was obtained for a lower applied power. This result could be explained by the presence of other oxygen-containing functional groups such as C=O and O-C=O at higher applied powers. Finally, Figure 11 and Figure S11(Supplementary Materials) also reveal that for all applied experimental conditions, zones B and C had the highest C–O/C–C ratios, suggesting that the C–O functionalities were mainly present in the region close to the MW plasma jet.

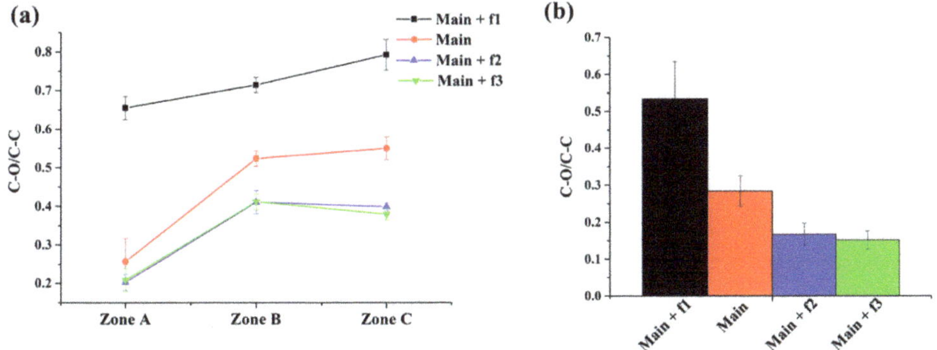

Figure 11. Influence of carrier gas flow rate on C-O/C-C ratio for (**a**) styrene and (**b**) MMA samples.

4. Conclusions

In this study, the feasibility of MMA and styrene deposition by means of a MW APPJ has been investigated. For the first time, this type of plasma source has been applied to deposit thin layers under atmospheric pressure. Using different techniques, the effect of process parameters such as applied power, carrier gas flow rate, distance from capillary to the substrate, and treatment time on the deposition efficiency has been studied. Firstly, using OES, a MW plasma jet was characterized in terms of excited species. The results demonstrated that a higher amount of reactive species was present in the plasma containing MMA as the monomer compared to the plasma jet containing styrene. It was also observed that the Ar and C bands had a well-defined maximum of OE spectra intensity which was reached when adding 250 sccm of the monomer-containing gas into the system. It can therefore be concluded that the optimal fragmentation of the introduced monomers and argon atom excitation occurred under mid-experimental conditions. CFD simulations were also employed to determine the pattern of precursor propagation in the volume close to the substrate surface. The obtained results of the CFD simulations displayed a remarkable resemblance between the pattern obtained in the experimental stage and the CFD simulations. It was shown that the injection of the monomer from one side caused non-uniformity of the coating due to the non-homogeneous precursor distribution in the examined system. Additionally, an increase in carrier gas flow rate also influenced the pattern of the deposited layer on the substrate and enforced the monomer to propagate further into the main gas stream, thereby providing better mixing. As a conclusion, CFD simulations thus predicted the distribution of the coating on the substrate during the experiments. Coating thickness, roughness, and topography were evaluated using AFM measurements and SEM imaging. The results demonstrated the appearance of droplets in the central area of the deposited styrene coatings while the areas further away from the center showed almost empty regions in one direction and a comparatively thick coating layer in the other direction due to monomer distribution in the volume close to the substrate, as was demonstrated by Comsol simulations. Finally, the edges of the coated area contained a thin coating with vacant spots across their surface. In the case of MMA monomer deposition, the formation of droplets was only observed in the center within a small deposition area. Finally, WCA and XPS measurements provided information about the surface chemical properties of the deposits. Good agreement between surface wettability and oxygen incorporation was demonstrated. Higher hydrophilicity and higher O/C ratios were observed in zones closer to the center where better contact with the MW plasma jet was established. Based on C–O/C–C ratios calculated for all experimental conditions for both monomers it was concluded that the highest ratios were obtained at low carrier gas flow rates due to the effective fragmentation and oxidation. The obtained results with support from the literature allow for us to conclude that the styrene and MMA monomers were plasma-polymerized by the MW APPJ used. In this work, a well-established correlation between the plasma diagnostic results and coating surface properties was achieved. A high level of monomer fragmentation observed with OES, together with the non-uniform distribution of the monomer presented by CFD simulations, was shown to be a reliable indicator of coating quality. In conclusion, it can be asserted that this study gives useful insights into the properties of deposited films using an atmospheric pressure MW plasma jet with different working parameters. The carrier gas flow rate was found to have the most significant impact on the properties of the thin deposited films due to its effect on the gas flow dynamics in the system.

Supplementary Materials: The following are available online at http://www.mdpi.com/2073-4360/12/2/354/s1.

Author Contributions: M.N. wrote the manuscript, conducted most of the experimental work, collected and interpreted the data, and performed the simulations; F.K. also carried out deposition experiments, prepared samples, and performed OES measurements and analysis; Y.O. supervised the work, performed data interpretation, and conducted XPS measurements; Z.K. helped conduct experimental work and additional analysis; and R.M. and N.D.G. supervised and financed the work, aided in the interpretation of the results, and worked on the manuscript. All authors have read and agreed to the published version of the manuscript.

Funding: This work was funded by a Starting Grant of the Special Research Fund of Ghent University (project number 01N00516).

Acknowledgments: The authors acknowledge the support from the Special Research Fund of Ghent University.

Conflicts of Interest: The authors declare no conflict of interest.

References

1. Desmet, T.; Morent, R.; De Geyter, N.; Leys, C.; Schacht, E.; Dubruel, P. Nonthermal plasma technology as a versatile strategy for polymeric biomaterials surface modification: A review. *Biomacromolecules* **2009**, *10*, 2351–2378. [CrossRef] [PubMed]

2. Liston, E.; Martinu, L.; Wertheimer, M. Plasma surface modification of polymers for improved adhesion: A critical review. *J. Adhes. Sci. Technol.* **1993**, *7*, 1091–1127. [CrossRef]

3. Fanelli, F.; Fracassi, F. Atmospheric pressure non-equilibrium plasma jet technology: General features, specificities and applications in surface processing of materials. *Surf. Coat. Technol.* **2017**, *322*, 174–201. [CrossRef]

4. Paunovic, M.; Schlesinger, M. *Fundamentals of Electrochemical Eeposition*; John Wiley & Sons: New York, NY, USA, 1998.

5. Sherman, A. *Chemical Vapor Deposition for Microelectronics: Principles, Technology, and Applications*; Elsevier Science: Park Ridge, NJ, USA, 1987.

6. Smith, D.L. *Thin-Film Deposition: Principles and Practice*; McGraw Hill Professional: New York, NY, USA, 1995.

7. Kumar, V.; Kim, J.H.; Pendyala, C.; Chernomordik, B.; Sunkara, M.K. Gas-phase, bulk production of metal oxide nanowires and nanoparticles using a microwave plasma jet reactor. *J. Phys. Chem. C* **2008**, *112*, 17750–17754. [CrossRef]

8. Li, S.-Z.; Hong, Y.C.; Uhm, H.S.; Li, Z.-K. Synthesis of nanocrystalline iron oxide particles by microwave plasma jet at atmospheric pressure. *Jpn. J. Appl. Phys.* **2004**, *43*, 7714. [CrossRef]

9. Van Vrekhem, S.; Morent, R.; De Geyter, N. Deposition of a PMMA coating with an atmospheric pressure plasma jet. *J. Coat. Technol. Res.* **2018**, *15*, 679–690. [CrossRef]

10. Sankaran, R.M.; Giapis, K.P. Hollow cathode sustained plasma microjets: Characterization and application to diamond deposition. *J. Appl. Phys.* **2002**, *92*, 2406–2411. [CrossRef]

11. Schäfer, J.; Foest, R.; Quade, A.; Ohl, A.; Weltmann, K. Local deposition of SiOx plasma polymer films by a miniaturized atmospheric pressure plasma jet (APPJ). *J. Phys. D Appl. Phys.* **2008**, *41*, 194010. [CrossRef]

12. Carton, O.; Ben Salem, D.; Bhatt, S.; Pulpytel, J.; Arefi-Khonsari, F. Plasma polymerization of acrylic acid by atmospheric pressure nitrogen plasma jet for biomedical applications. *Plasma Process. Polym.* **2012**, *9*, 984–993. [CrossRef]

13. Donegan, M.; Dowling, D.P. Protein adhesion on water stable atmospheric plasma deposited acrylic acid coatings. *Surf. Coat. Technol.* **2013**, *234*, 53–59. [CrossRef]

14. Ben Salem, D.; Carton, O.; Fakhouri, H.; Pulpytel, J.; Arefi-Khonsari, F. Deposition of water stable plasma polymerized acrylic acid/MBA organic coatings by atmospheric pressure air plasma jet. *Plasma Process. Polym.* **2014**, *11*, 269–278. [CrossRef]

15. Farhat, S.; Gilliam, M.; Rabago-Smith, M.; Baran, C.; Walter, N.; Zand, A. Polymer coatings for biomedical applications using atmospheric pressure plasma. *Surf. Coat. Technol.* **2014**, *241*, 123–129. [CrossRef]

16. Horvath, G.; Moravsky, L.; Krcma, F.; Matejcik, Š. Characterization of a low-cost kilohertz-driven plasma pen operated in Ar gas. *IEEE IEEE Trans. Plasma Sci.* **2013**, *41*, 613–619. [CrossRef]

17. Guaitella, O.; Sobota, A. The impingement of a kHz helium atmospheric pressure plasma jet on a dielectric surface. *J. Phys. D Appl. Phys.* **2015**, *48*, 255202. [CrossRef]

18. Chen, X. Modelling of a radio-frequency plasma torch including a self-consistent electromagnetic field formulation. *J. Phys. D Appl. Phys.* **1989**, *22*, 361. [CrossRef]

19. van der Mullen, J.; Jonkers, J. Fundamental comparison between non-equilibrium aspects of ICP and MIP discharges. *Spectrochim. Acta Part B* **1999**, *54*, 1017–1044. [CrossRef]

20. Krčma, F.; Tsonev, I.; Smejkalová, K.; Truchlá, D.; Kozáková, Z.; Zhekova, M.; Marinova, P.; Bogdanov, T.; Benova, E. Microwave micro torch generated in argon based mixtures for biomedical applications. *J. Phys. D Appl. Phys.* **2018**, *51*, 414001. [CrossRef]

21. Wang, C.; Srivastava, N. OH number densities and plasma jet behavior in atmospheric microwave plasma jets operating with different plasma gases (Ar, Ar/N$_2$, and Ar/O$_2$). *Eur. Phys. J. D* **2010**, *60*, 465–477. [CrossRef]

22. Tsonev, I.; Atanasov, N.; Atanasova, G.; Krčma, F.; Bogdanov, T. Atmospheric Pressure Microwave Plasma Torch for Biomedical Applications. *Plasma Med.* **2018**, *8*. [CrossRef]

23. Schäfer, J.; Hnilica, J.; Šperka, J.; Quade, A.; Kudrle, V.; Foest, R.; Vodák, J.; Zajíčková, L. Tetrakis (trimethylsilyloxy) silane for nanostructured SiO_2-like films deposited by PECVD at atmospheric pressure. *Surf. Coat. Technol.* **2016**, *295*, 112–118. [CrossRef]

24. Ting, J.A.S.; Rosario, L.M.D.; Lee, H.V., Jr.; Ramos, H.J.; Tumlos, R.B. Hydrophobic coating on glass surfaces via application of silicone oil and activated using a microwave atmospheric plasma jet. *Surf. Coat. Technol.* **2014**, *259*, 7–11. [CrossRef]

25. Choudhury, A.; Kakati, H.; Pal, A.; Patil, D.; Chutia, J. Synthesis and characterization of plasma polymerized styrene films by rf discharge. *Proc. J. Phys. Conf. Ser.* **2010**, *208*, 012104. [CrossRef]

26. Kinzig, B.; Smardzewski, R. Plasma-polymerized thin coatings from methyl-methacrylate, styrene and tetrafluoroethylene. *Surf. Technol.* **1981**, *14*, 3–16. [CrossRef]

27. Geckeler, K.; Gebhardt, R.; Grünwald, H. Surface modification of polyethylene by plasma grafting with styrene for enhanced biocompatibility. *Naturwissenschaften* **1997**, *84*, 150–151. [CrossRef]

28. Morita, S.; Tamano, J.; Hattori, S.; Ieda, M. Plasma polymerized methyl-methacrylate as an electron-beam resist. *J. Appl. Phys.* **1980**, *51*, 3938–3941. [CrossRef]

29. Bogdanov, T.; Tsonev, I.; Marinova, P.; Benova, E.; Rusanov, K.; Rusanova, M.; Atanassov, I.; Kozáková, Z.; Krčma, F. Microwave plasma torch generated in argon for small berries surface treatment. *Appl. Sci.* **2018**, *8*, 1870. [CrossRef]

30. Chen, W.-K.; Huang, J.-C.; Chen, Y.-C.; Lee, M.-T.; Juang, J.-Y. Deposition of highly transparent and conductive Ga-doped zinc oxide films on tilted substrates by atmospheric pressure plasma jet. *J. Alloys Compd.* **2019**, *802*, 458–466. [CrossRef]

31. Belmonte, T.; Henrion, G.; Gries, T. Nonequilibrium atmospheric plasma deposition. *J. Therm. Spray Technol.* **2011**, *20*, 744. [CrossRef]

32. Obrusník, A.; Jelínek, P.; Zajíčková, L. Modelling of the gas flow and plasma co-polymerization of two monomers in an atmospheric-pressure dielectric barrier discharge. *Surf. Coat. Technol.* **2017**, *314*, 139–147. [CrossRef]

33. Yusoff, H.M.; Abrahamson, J.; Shastry, R. Influence of anode surface temperature in a continuously-fed arc discharge depositing carbon nanotubes. In Proceedings of the 2006 International Conference on Nanoscience and Nanotechnology, Brisbane, Australia, 3–7 July 2006.

34. Juang, J.-Y.; Lin, H.-T.; Liang, C.-T.; Li, P.-R.; Chen, W.-K.; Chen, Y.-Y.; Pan, K.-L. Effect of ambient air flow on resistivity uniformity of transparent Ga-doped ZnO film deposited by atmospheric pressure plasma jet. *J. Alloys Compd.* **2018**, *766*, 868–875. [CrossRef]

35. Onyshchenko, I.; De Geyter, N.; Morent, R. Improvement of the plasma treatment effect on PET with a newly designed atmospheric pressure plasma jet. *Plasma Process. Polym.* **2017**, *14*, 1600200. [CrossRef]

36. *COMSOL Multiphysics Reference Guide*; COMSOL AB: Stockholm, Sweden, 2011.

37. Kramida, A.; Ralchenko, Y.; Reader, J. *Team 2018 NIST Atomic Spectra Database (ver. 5.5. 3)*; National Institute of Standards and Technology: Gaithersburg, MD, USA, 2018.

38. Gilmore, F.R.; Laher, R.R.; Espy, P.J. Franck–Condon factors, r-centroids, electronic transition moments, and Einstein coefficients for many nitrogen and oxygen band systems. *J. Phys. Chem. Ref. Data* **1992**, *21*, 1005–1107. [CrossRef]

39. Cools, P.; Sainz-García, E.; Geyter, N.D.; Nikiforov, A.; Blajan, M.; Shimizu, K.; Alba-Elías, F.; Leys, C.; Morent, R. Influence of DBD Inlet Geometry on the Homogeneity of Plasma-Polymerized Acrylic Acid Films: The Use of a Microplasma–Electrode Inlet Configuration. *Plasma Process. Polym.* **2015**, *12*, 1153–1163. [CrossRef]

40. Yasuda, H. Glow discharge polymerization. *J. Polym. Sci. Macromol. Rev.* **1981**, *16*, 199–293. [CrossRef]

41. Thompson, L.; Mayhan, K. The plasma polymerization of vinyl monomers. II. A detailed study of the plasma polymerization of styrene. *J. Appl. Polym. Sci.* **1972**, *16*, 2317–2341. [CrossRef]

42. Mackie, N.M.; Castner, D.G.; Fisher, E.R. Characterization of pulsed-plasma-polymerized aromatic films. *Langmuir* **1998**, *14*, 1227–1235. [CrossRef]

43. Liguori, A.; Pollicino, A.; Stancampiano, A.; Tarterini, F.; Focarete, M.L.; Colombo, V.; Gherardi, M. Deposition of Plasma-Polymerized Polyacrylic Acid Coatings by a Non-Equilibrium Atmospheric Pressure Nanopulsed Plasma Jet. *Plasma Process. Polym.* **2016**, *13*, 375–386. [CrossRef]

44. Dong, B.; Manolache, S.; Somers, E.B.; Lee Wong, A.C.; Denes, F.S. Generation of antifouling layers on stainless steel surfaces by plasma-enhanced crosslinking of polyethylene glycol. *J. Appl. Polym. Sci.* **2005**, *97*, 485–497. [CrossRef]
45. France, R.M.; Short, R.D. Plasma treatment of polymers: The effects of energy transfer from an argon plasma on the surface chemistry of polystyrene, and polypropylene. A high-energy resolution X-ray photoelectron spectroscopy study. *Langmuir* **1998**, *14*, 4827–4835. [CrossRef]
46. Noeske, M.; Degenhardt, J.; Strudthoff, S.; Lommatzsch, U. Plasma jet treatment of five polymers at atmospheric pressure: Surface modifications and the relevance for adhesion. *Int. J. Adhes. Adhes.* **2004**, *24*, 171–177. [CrossRef]
47. Beamson, G.; Briggs, D. *High Resolution XPS of Organic Polymers: The Scienta ESCA300 Database*; Wiley: Hoboken, NJ, USA, 1992.
48. Casserly, T.B.; Gleason, K.K. Effect of substrate temperature on the plasma polymerization of poly (methyl methacrylate). *Chem. Vap. Depos.* **2006**, *12*, 59–66. [CrossRef]
49. Merche, D.; Poleunis, C.; Bertrand, P.; Sferrazza, M.; Reniers, F. Synthesis of polystyrene thin films by means of an atmospheric-pressure plasma torch and a dielectric barrier discharge. *IEEE Trans. Plasma Sci.* **2009**, *37*, 951–960. [CrossRef]

Article

Evolution of the Surface Wettability of PET Polymer upon Treatment with an Atmospheric-Pressure Plasma Jet

Alenka Vesel *, Rok Zaplotnik, Gregor Primc and Miran Mozetič

Department of Surface Engineering, Jozef Stefan Institute, Jamova cesta 39, 1000 Ljubljana, Slovenia; rok.zaplotnik@ijs.si (R.Z.); gregor.primc@ijs.si (G.P.); miran.mozetic@ijs.si (M.M.)
* Correspondence: alenka.vesel@guest.arnes.si

Received: 20 November 2019; Accepted: 1 January 2020; Published: 3 January 2020

Abstract: A useful technique for pre-treatment of polymers for improved biocompatibility is surface activation. A method for achieving optimal wettability at a minimal thermal load and unwanted modifications of the polymer properties is elaborated in this paper. Samples of polyethylene terephthalate polymer were exposed to an atmospheric-pressure plasma jet created by a high-impedance low-frequency discharge in wet argon. Different treatment times and distances from the end of the glowing discharge enabled detailed investigation of the evolution of surface activation. A rather fast saturation of the surface wettability over the area of the order of cm^2 was observed upon direct treatment with the glowing discharge. At a distance of few mm from the glowing discharge, the activation was already two orders of magnitude lower. Further increase of the distance resulted in negligible surface effects. In the cases of a rapid activation, very sharp interphase between the activated and unaffected surface was observed and explained by peculiarities of high-impedance discharges sustained in argon with the presence of impurities of water vapor. Results obtained by X-ray photoelectron spectroscopy confirmed that the activation was a consequence of functionalization with oxygen functional groups.

Keywords: atmospheric-pressure plasma jet; polymer; surface patterns; wettability mapping; XPS mapping

---

1. Introduction

Polymers are nowadays widely used in industry, as well as in medicine because of their excellent chemical inertness and mechanical properties. The biocompatibility of most polymers with appropriate mechanical properties is, however, below requirements; therefore, they should be modified prior to incubation with a biological matter. The biocompatibility of materials depends predominantly on surface properties. The best surface properties are obtained by grafting biocompatible molecules, preferably by forming covalent bonds between a monolayer of grafted molecules and the substrate [1–6]. Usually, such bonding cannot occur because of the inertness of most polymers of appropriate chemical and mechanical properties. The surface properties of polymers should be, therefore, altered to enable at least reasonable if not excellent biocompatibility. A common method for tailoring surface properties of polymers is a brief exposure to non-equilibrium gaseous plasma. A review of plasma techniques for achieving desired properties has been published recently [7].

Any attempt of grafting biocompatible coatings on polymer substrates require alternation of the surface wettability. The surface composition needs to be modified to achieve desired hydrophilicity, and thus, adhesion properties. Many different plasma configurations can be used to achieve better wettability ranging from low-pressure to atmospheric-pressure plasmas. Nowadays, atmospheric-pressure plasma jets (APPJ) are particularly popular, and their application in polymer surface treatment is still increasing [8–11]. Such plasmas enable localized treatment of what is beneficial in various biomedical applications [12–15]. Details about the surface finish, in particular, on gradients of surface properties, however, are rarely reported in the scientific literature. Such details are of crucial importance to understand the surface chemistry and prepare the surface conditions which are regarded as optimal for a particular application. Below are selected papers that have addressed the surface finish, in particular, two-dimensional mapping and/or determination of gradients likely to occur on polymer surfaces upon treatment with atmospheric plasma jets.

Birer [16] performed a very interesting investigation regarding the reactivity zones formed on polyethylene (PE) surface after treating with an APPJ using helium and argon gas or its mixture with oxygen or nitrogen. The sample was positioned 1 cm away from the nozzle of the discharge tube. Reactivity zones were investigated by two-dimensional X-ray photoelectron spectroscopy (XPS) mapping. He found that oxidation started at the center hit by the plasma jet and then expanded outwards forming a ring-shape pattern. Formation of ring patterns of –NO, –COO, –CO and –NO$_3$ groups with diameters increasing with treatment time was detected. Surface modified by APPJ treatment expanded several millimeters from the center. Thus, a diameter of the zone modified by APPJ was found to be up to about 2 cm from the plasma jet axis. Kostov et al. [17] also investigated the wettability as well as surface morphology by atomic force microscopy (AFM) of polymers polyethylene (PE), polypropylene (PP) and polyethylene terephthalate (PET) treated by APPJ operating in argon. Samples were placed at various distances between 2 and 3.5 cm from the nozzle of the APPJ. He provided radial water contact angle profiles and found the similar size of the modified area as Bierer [1] found by XPS mapping-approximately 2 cm of diameter.

Jofre-Reche et al. [18] investigated the effect of the nozzle distance to the polydimethylsiloxane (PDMS) sample surface (in the range up to 10 mm) and found an optimum surface wettability at the distance of 6.6 mm. This correlated well with the lower gas temperature at this distance and higher optical emission intensity (OES) of O (777 nm) line in the plasma jet. Moreover, Deynse et al. [19] investigated the effect of the nozzle distance on the surface wettability of PE treated with Ar plasma jet. For distances up to 15 mm, 70% of increase of surface wettability was found, whereas, at longer distances a sharp knee appeared on the curve of the surface wettability versus the distance. Such variation of the surface wettability corresponded well with the changes in the chemical composition regarding XPS O/C ratio and concentration of various oxygen functional groups. Wagenaars et al. [20] investigated the wettability of PP treated with APPJ in helium with various oxygen admixtures. They found a minimum in the water contact angle (WCA) when 0.5% of oxygen was added to helium. The WCA also depended on treatment time and a distance between the APPJ nozzle and PP sample. A significant decrease of WCA was observed in the first 10 s of treatment for distances up to 20 mm, whereas, later WCA slowly stabilized. For very short distances 3 and 5 mm the WCA stabilized in 40 s. The minimum achievable WCA decreased with increasing distance. Because the O-atom density is lower at longer distances, applying longer treatment time would give the same total flux as in the case of a low distance (high O-atom density) and low treatment time. Therefore, the same final effect should be expected. However, this was not observed; therefore, the authors concluded that O-atom density is important and not the total flux.

Dowling et al. [21] investigated the influence of DC pulse plasma cycle time (PCT) on the activation of polypropylene (PP), polystyrene (PS) and polycarbonate (PC) polymers. They found that optimized PCT was specific for a given polymer and related to the polymer thermal properties. Lommatzsch [22] was investigating differences in the surface reactions when treating polymer PE with APPJ created in air or nitrogen.

Some authors have also investigated the effect of water vapor on surface modification of polymers [23–25]. Oehrlein et al. [24] have investigated the effect of water vapor in Ar/H$_2$O plasma on the etching rates of polymers. He found that OH radicals formed in plasma play a dominant role in the etching process. The etching rates dropped exponentially when a distance between polymer and APPJ was increased what was consistent with a density of OH radicals that also exponentially decreased with a distance. The exponential decay constant for Ar/H_2O plasma was 3.40 and 6.07 mm for the case of air or nitrogen environment, respectively. Contrary, regarding an oxygen content on the etched surface, a maximum was observed at intermedium distance of a polymer from the APPJ. Moreover, Sarani et al. [25] investigated the effect of water vapor in Ar plasma. They found a higher oxygen content on a polymer surface as measured by XPS for the case when the polymer was treated with Ar/H_2O plasma jet, what was explained with a higher density of radicals in the jet afterglow. Nevertheless, a difference in XPS oxygen concentration for the case of Ar/H_2O plasma in comparison to pure Ar plasma is small.

Foest et al. [26] investigated vacuum ultraviolet (VUV) emission (115–135 nm) from Ar plasma jet and the effect of nitrogen addition. The intensity of VUV radiation over the radius of the plasma jet was measured. They found that the addition of 5% of N_2 reduces the integral of VUV emission to approximately 10% of the original value. The radial dependence of VUV emission showed a ring shape with about 1 mm diameter and a minimum in the center. Moreover, Oehrlein et al. [27] investigated the effect of VUV-induced surface modification using an optical window made from pure MgF_2 that transmits VUV down to the suitable wavelength. They investigated modification of polymethyl methacrylate (PMMA) based 193 nm photoresist (PR193) with 300 nm film thickness and polystyrene (PS) based 248 nm photoresist (PR248) with 400 nm film thickness. Surface modification was investigated by attenuated total reflection Fourier transform infrared spectroscopy (ATR–FTIR) whereas the thickness loss rate measured by ellipsometry. For this investigation different kHz- or MHz-driven argon APPJ plasma sources were used: (1) kHz driven ring-APPJ source, (2) kHz driven pin-APPJ source, (3) MHz driven pin-APPJ source, and (4) kHz driven surface microdischarge source. It was found that the type of the APPJ source is the crucial factor regarding the effect of VUV photons relative to other reactive plasma species to surface modification. Ar fed kHz-driven ring-APPJ source caused the largest VUV surface modification—the highest thickness loss rate when using MgF_2 filter. If no filter was used, a MHz driven pin-APPJ source caused the highest polymer thickness loss rate. If oxygen was added to Ar feed gas, a reduced VUV effect was observed and explained by the absorption of VUV photons by oxygen molecules. The importance of VUV has also been stressed by Schneider et al. [28]. He has exposed a model a-C: H films to an affluent of helium APPJ with a small admixture of oxygen and found that VUV/UV photons caused hardening of a model a-C: H films on the area in line-of-sight to the jet nozzle, which resulted in slower etching rates of the area directly under the nozzle.

Onyshchenko et al. [29] performed a two-dimensional mapping of both water contact angle and oxygen concentration on the surface of PET polymer after treatment with the atmospheric pressure plasma jet sustained in high-purity argon. Numerous parameters were varied to estimate the influence of the plasma treatment on the surface composition and wettability: The distance between the dielectric tube and the sample, the exposure time, the discharge power and the gas flow rate. Because of four variable parameters, a limited number of experiments was feasible: Three distances, three exposure times, two powers and two gas flows. Little differences in the surface finish were observed between experiments at different gas flows and discharge powers, whereas, the distance and the treatment time exhibited more pronounced variations of the surface wettability and functionalization. The width of the water contact angle footprints saturated at a rather short treatment time of 20 s at a fixed distance, power and gas flow. The minimal water contact angle of approximately 20° was observed at the shortest distance and the result correlated well with the oxygen concentration as determined by XPS. Detailed investigation on the surface phenomena was not feasible in the work of Onyshchenko et al. because of the large number of independent parameters. In another work, Onyshchenko et al. [30] used a modified design of APPJ with an additional plate at the end of the quartz capillary discharge tube. They found significant improvement in wettability (i.e., much wider hydrophilic footprint area) at short distances between the sample and the nozzle of the capillary discharge tube. At large distances between the sample and the nozzle of the capillary discharge tube, there was only a minor effect on the hydrophilic region created on the PET surface.

The brief survey of the most relevant literature, as shown in Table 1 indicates a variety of techniques, as well as reported results on plasma activation of polymers at ambient pressure. Because of the wide spread of the techniques, it is difficult to draw any correlations. In general, all authors agree that the treatment of polymers with atmospheric pressure plasmas causes significant surface modification, but little explanation for the observed effects is provided. Understanding the reaction mechanisms requires basic knowledge on reactive gaseous species and detailed two-dimensional mapping of the surface wettability at various treatment conditions. A brief study on the influence of the distance between the dielectric tube and the sample, the exposure time, the discharge power and the gas flow rate was already disclosed by Onyshchenko et al. [29,30]. Because there are numerous variable parameters, it is not feasible to vary all of them like the type of gas, the concentration of any additional gases, including impurities, the discharge power, intensity of electrical field, the gas flow and velocity gradients, the distance between a sample and the jet tip, influence of any conductivity of the substrate etc. In this paper, we addressed the wettability at various treatment times and distances between the plasma jet and the samples at practically steady other parameters. The broad ranges of both treatment times and distances enabled a detailed investigation of the evolution of the surface finish. We used a common argon APPJ which contains water vapor as an impurity gas.

2. Materials and Methods

2.1. Plasma Treatment

Biaxially oriented PET foil (125 µm in thickness) from Goodfellow was cut to squares with a size of 5×5 cm^2 and placed to a wooden substrate holder. Samples were treated in the center of the square with an atmospheric-pressure plasma jet operated in Ar gas at the flow of 1 slm. Schematic illustration of the experimental setup is shown in Figure 1. Plasma was generated along a Pyrex tube with an outer diameter of 4 mm, but the most luminous discharge was observed at the nozzle. A high-voltage electrode (a copper wire) was placed inside the Pyrex tube, as shown in Figure 1. A diameter of the electrode was 0.3 mm, and its length was the same as the length of the Pyrex tube, i.e., 15 cm. A peak-to-peak voltage of 7 kV was applied to the electrode. Plasma was generated with an excitation frequency of 25 kHz using an almost sinusoidal power supply. A visible part of plasma jet extended less than 3 cm from the nozzle of the Pyrex tube (Figure 2).

Table 1. Literature overview. APPJ, atmospheric-pressure plasma jets

Reference	APPJ Parameters	Gas	Sample Distance	Variables	Methods for Plasma Characterisation	Methods for Sample Characterisation	Polymer	Wettability before and after Treatment
[16]	5 kH, 20 W, 12 kV DC, treatment time: 1, 5, 10 min	He, Ar or mixture with 1% O_2 or N_2	10 mm	Gas type	/	XPS mapping SEM	PE	
[17]	AC 37 kHz, 4.4 W, gas flow 1.3 L/min, treatment time: Up to 90	Ar	2–3.5 cm	Treatment time, distance	Current-voltage waveforms	WCA line profile, XPS AFM	PET PP PE	$78° \rightarrow 25°$ $102° \rightarrow 52°$ $94° \rightarrow 36°$
[20]	13.56 MHZ, gas flow 1 slm, treatment time up to 30	He with up to 1% of O_2 admixture	3–50 mm	Treatment time, distance, oxygen admixture	/	WCA	PP	$95° \rightarrow 50°$
[18]	Rotating plasma jet (Openair system, Plasmatreat GmbH), torch speed: 10–30 m/min, DC arc plasma, 270–300 V, 19–22 kHz, pulsed mode, stage moving speed 5–50 m/min	Compre-ssed air	5–10 mm, optimum at 6.6 mm	Distance, torch speed	OES, gas temperature	WCA, XPS	PDMS	$109° \rightarrow {<}10°$
[21]	Openair system (Plasmatreat GmbH), 25 kHz, flow rate 76,6 L/min, pulse peak height 2–5 kV, stage moving speed 30 m/min, treatment time ~5 ms	Compre-ssed air	16 mm	Pulse plasma cycle time	OES	WCA, aging, AFM, XPS, polymer temperature	PP PS PC	$90° \rightarrow 70°$ $93° \rightarrow 20°$ $83° \rightarrow 30°$
[22]	Openair system (Plasmatreat GmbH), 17–22 kHz, pulse peak height 5 kV, flow rate 17 L/min, stage moving speed 100 m/min, treatment time < 0.1 s	Air or nitrogen	3 mm	Gas type	OES	WCA, XPS	PE	$93° \rightarrow 22°$ (for air) $93° \rightarrow 41°$ (for N_2)
[19]	Flow 1.25 slm, scanning velocity 0.08-4.5 m/min, treatment time 78 ms	Ar	5–45 mm	Distance	Current-voltage waveforms	WCA, XPS, aging	PE	$104° \rightarrow 28°$
[27]	Different APPJ sources: (1) kHz driven ring-APPJ source, (2) kHz driven pin-APPJ source, (3) MHz driven pin-APPJ source, (4) kHz driven surface microdischarge source	Ar with or without O_2 addition	Up to 3 cm	Different APPJ sources	optical window for VUV, COMSOL modelling	ATR-FTIR, ellipsometry	PMMA based 193 nm photore-sist (PR193) and PS based 248 photore-sist (PR248)	
[29]	AC 60 kHz, peak-to-peak voltage 7 or 10 kV, discharge power 3.1–5.7 W, gas flow: 1 and 3 slm, treatment time 5 to 20 s	Ar	2, 10 and 15 mm	Treatment time, distance, discharge power, gas flow	OES	WCA, XPS	PET	$87° \rightarrow 22°$

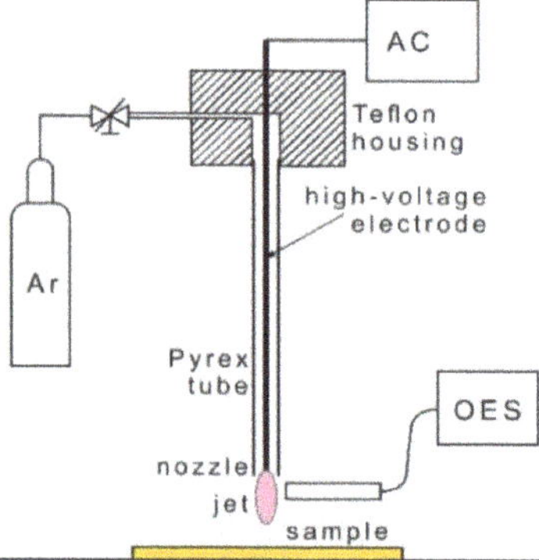

Figure 1. Schematic drawing of the APPJ system used for polymer surface modification.

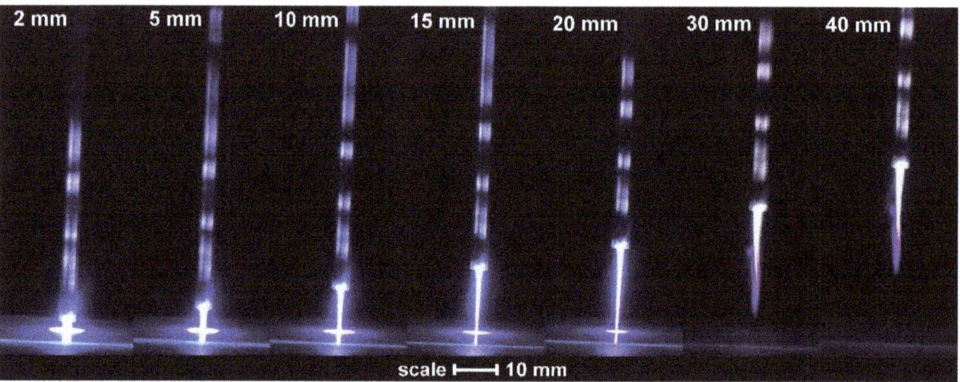

Figure 2. Photos of the plasma jet at various distances of the nozzle from the sample.

As-received polymer samples were placed at different distances to the plasma jet. A distance between the nozzle and the polymer sample was varied between 2 and 40 mm. Individual samples were treated for various treatment times between 0.5 s and 10 min.

Plasma is a source of various reactive species. To see the role of UV/VUV radiation, an additional experiment was performed where one of the samples was covered with MgF_2 optical window to transmit only UV/VUV radiation and eliminate other reactive plasma species.

2.2. Optical Emission Spectroscopy

AvaSpec-3648 Fiber Optic Spectrometer (Avantes, Apeldoorn, The Netherlands) was used for characterisation of our APPJ. The spectrometer resolution was 0.5 nm in the range of wavelengths between 200 to 1100 nm. The spectra acquisition time was set to 100 ms. Optical spectra were measured during the treatment of PET foil at various distances of the sample from the nozzle. A collimating lens of the OES spectrometer was placed 2 mm below the nozzle.

2.3. Polymer Temperature Measurements

The average temperature of the polymer surface during plasma treatment was measured with the infra-red IR camera Optris PI 160 (Optris GmbH, Berlin, Germany) working at a wavelength range from 7.5 to 13 μm. The temperature was averaged on the area approximately 7×7 mm^2 around the center of the impact point of the plasma jet with the polymer surface. The surface temperature was measured for two distances of the sample from the discharge nozzle, i.e., 5 and 30 mm. The albedo was fixed at 0.95.

2.4. Wettability Measurements

Mapping of the surface wettability on polymer samples was performed with Drop Shape Analyser DSA-100 (Krüss GmbH, Hannover, Germany). A static contact angle was measured using a sessile drop method. An array of distilled water drops with a volume of 1 µL was applied to the surface with a distance of 5 mm between individual drops. The whole polymer surface was mapped what enabled 2D images of the surface wettability. Each sample of a size of 5×5 cm^2 was, therefore, probed with 73 droplets placed at different spots. A photo of such deposited drops on the sample is shown in Figure 3. The device for measuring the wettability had a stage with fully automatically controlled movement in X and Y direction. By setting the X, Y grid coordinates of the spots on the surface that needed to be measured, it was possible to simultaneously deposit a drop, record an image of this drop and immediately determine the contact angle. Later, all images were manually checked for correct contact angle determination. A time elapsed between the first and the last measured drop was less than 13 min. An ellipse-tangent fitting method was used to obtain contact angles from the shape of the drop.

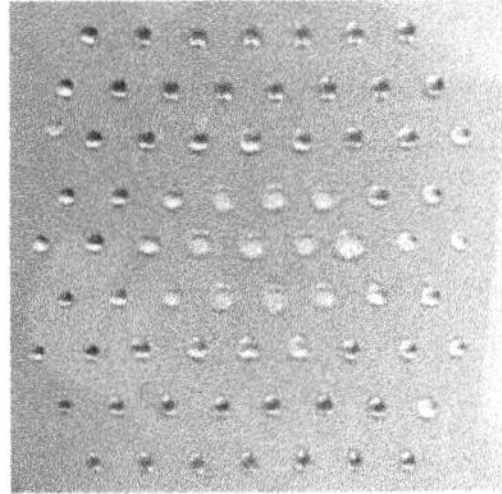

Figure 3. A photo of a grid of water droplets deposited on the polymer surface. In the center, where surface wettability was higher, we can observe wider droplets. The size of the image is 5×5 cm^2.

2.5. X-ray Photoelectron Spectroscopy

XPS characterization of the polymer surface was performed to determine changes in the chemical composition after APPJ treatment using an XPS (TFA XPS Physical Electronics, München, Germany). The samples were excited with monochromatic Al K$\alpha_{1,2}$ radiation at 1486.6 eV over an area with a diameter of 400 µm. Photoelectrons were detected with a hemispherical analyser positioned at an angle of 45° with respect to the normal of the sample surface. To determine the variation of the oxygen concentration over the sample surface, carbon C1s and oxygen O1s spectra were measured in the middle of the treated sample, as well as at various positions over the sample surface (an array of measured points with a distance of 5 mm). In such a way a similar 2D mapping was performed as for wettability measurements. The spectra were measured at a pass-energy of 23.5 eV using an energy step of 0.1 eV. An additional electron gun was used for surface neutralization during the XPS measurements. The measured spectra were analyzed using MultiPak v8.1c software (Ulvac-Phi Inc., Kanagawa, Japan, 2006) from Physical Electronics, which was supplied with the spectrometer. Because of time-consuming experiments, the XPS spectra were acquired only on the limited number of the samples.

2.6. Atomic Force Microscopy

The surface morphology was examined by Atomic Force Microscopy (AFM). An AFM (Solver PRO, NT-MDT, Moscow, Russia) was used to determine variations of the surface morphology and roughness around the center of the impact point of the plasma jet with the polymer surface. The measurements were performed in the semi-contact mode. Images with a size of 2×2 µm^2 were recorded.

3. Results and Discussion

Individual samples were mounted, as shown in Figure 1, and treated by APPJ. Before performing detailed 2D mapping of the surface wettability, the optical spectra of gaseous plasma 2 mm from the nozzle were acquired. A typical OES spectrum is shown in Figure 4. As expected, the OES spectrum is dominated by Ar lines which correspond to transitions of Ar atoms from highly excited states to metastable states. Apart from Ar ions one can observe a small oxygen line at 777 nm and nitrogen molecular band in the near UV range of spectra. The appearance of these spectral features is attributed to mixing of surrounding air with highly excited species of the plasma jet, particularly Ar metastables. A very intensive band is also observed at the bandhead of 309 nm. This band corresponds to the transition of OH radicals from excited to the ground state. The origin is obviously water vapor, which is presented in the ambient air, as well as inside the discharge tube. The intensity of radiation at the same spot, i.e., 2 mm from the nozzle, depends on the distance between the nozzle and the polymer substrate. Figure 5 represents behavior of spectral features at various distances of the sample from the nozzle. The integral intensity is the largest at the smallest distance which is explained by the fact that the light reflected from the sample and the plasma spread across the surface is also captured by the acceptance angle of the collimating lens. This can be seen in Figure 2.

More interesting is the behavior of the relative intensities of particular spectral features. Figure 6 represents the intensity of selected spectral features versus the distance between the nozzle and the sample, normalized to the main Ar line at 763 nm. One can observe large differences between nitrogen and OH lines. While nitrogen lines practically vanish over the moderate distance, the OH line still persists even at the largest distance. In fact, the relative intensity of the OH band remains comparable for all distances. The differences in the behavior of the nitrogen and OH lines can be explained by the fact that water retains on any walls of the discharge system throughout the measurement; therefore, the origin of the OH radicals is in the source gas, and not only in the effusion area. Opposite to OH, nitrogen is quickly removed from the discharge tube because it does not condense on the surfaces.

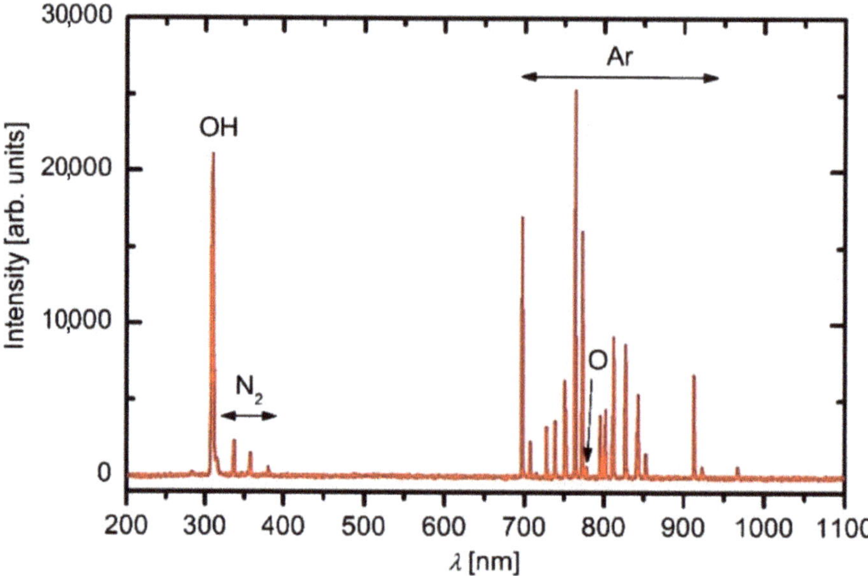

Figure 4. An optical spectrum acquired at a distance between the nozzle and the sample of 2 mm.

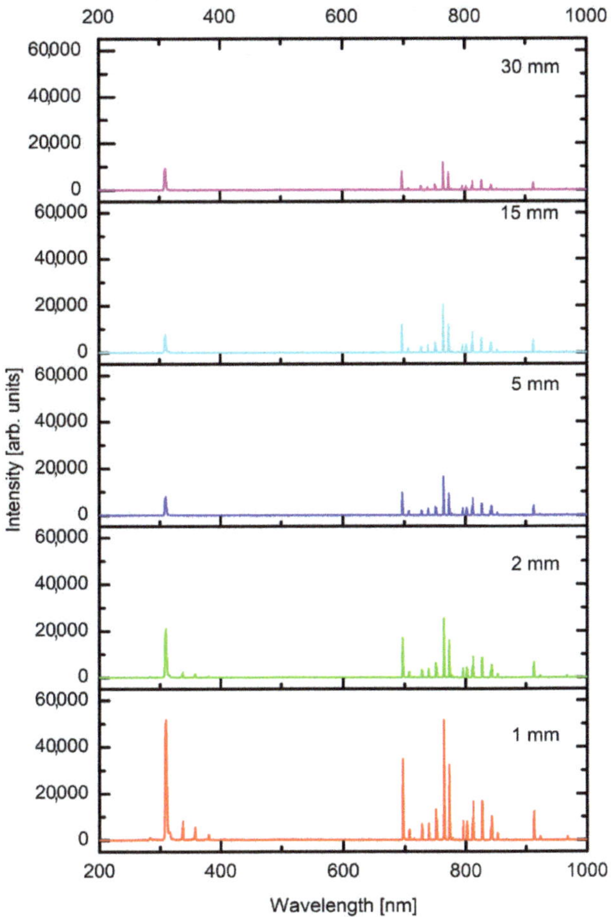

Figure 5. Optical spectra acquired at various distances between the nozzle and the sample. All spectra were acquired with the optical fibre tip placed 2 mm below the nozzle and lenses of an acceptance angle of approximately 3°.

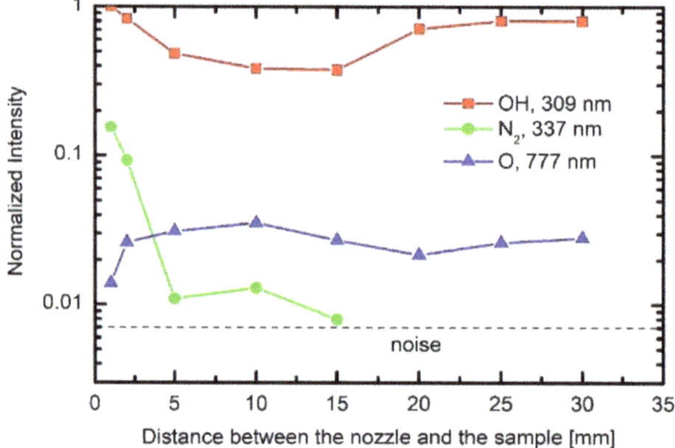

Figure 6. The behavior of normalized spectral features versus the distance between the nozzle and the sample.

Figure 7 represents results of systematic measurements of the water contact angles on the surface of samples mounted 5 mm below the nozzle. As explained above, the luminous plasma jet was in physical contact with the polymer surface in this case (Figure 2). Plasma jet was focused on the center of the samples. The samples were treated for various periods, and then the surface wettability was measured within less than 15 min after the treatment.

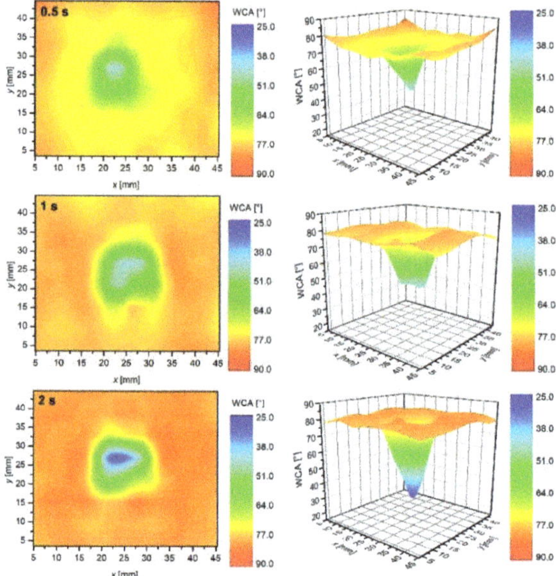

Figure 7. *Cont.*

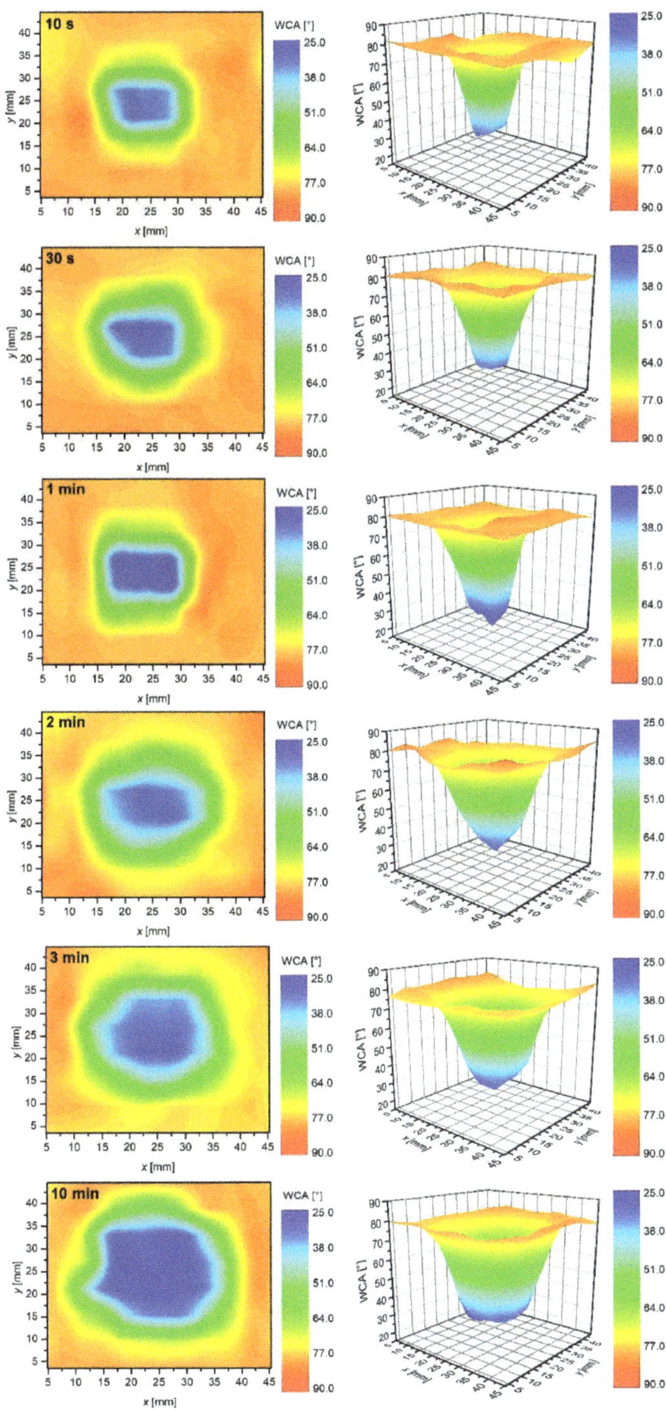

Figure 7. The time evolution of water contact angles on samples placed 5 mm below the APPJ nozzle.

The uppermost image of Figure 7 corresponds to the plasma treatment time for 0.5 s. We can observe the highest wettability at the axis of the plasma jet. The water contact angle for the sample treated for 0.5 s reaches the minimum of approximately 47° in a small spot in the center of the sample. Away from the spot, the water contact angle increases rather monotonically. The distribution of the water contact angles is best viewed in 3D graphs which are presented in Figure 7. The extreme radial distribution of the surface wettability indicates that the most extensive reactions leading to the improved wettability occur in the center of the plasma jet. The center of a plasma jet in our case is free from nitrogen or oxygen that could diffuse from surrounding atmosphere into the gas jet. The rapid activation of the sample at the center, therefore, cannot be explained by nitrogen or oxygen radicals that appear in the effusion zone of the atmospheric pressure plasma jet. The activation of the sample after 0.5 s treatment should, therefore, be explained by other factors. Among them, and consistent with the literature survey presented in Introduction to this paper, there were energetic Ar particles, OH radicals and UV/VUV radiation. Atmospheric pressure plasma jets sustained at rather low frequencies, and high impedances do not form continuous plasma, but rather appear as numerous streamers. The streamers propagate from the powered electrode in the form of electron avalanches. The driving force of such avalanches is the ionization wavefront. The wavefront propagates in the direction of the highest electrical field, i.e., along the axis of the gas jet. As a result, there is a strong radial gradient of both free electrons and Ar metastables. The flux of Ar ions and metastables on the polymer surface in the case when the polymer is in direct contact with the plasma jet is, therefore, the largest at the axis and decreases rapidly with increased radial distance from the jet axis. The shape of the 3D plot after treatment for 0.5 s (the uppermost image in Figure 7) is in agreement with the radial distribution of the fluxes. From this point of view, we can attribute the initial stage of polymer activation to the effects of energetic Ar particles, in particular, metastables and ions. Both particles transfer their potential energies to the solid material upon impinging.

While the upper explanation is highly visible from Figure 7, one should also consider the results presented in Figures 4–6, in particular, the behavior of the normalized OH "line" in Figure 6. As explained above, such a behavior (almost independent of the distance between the nozzle and the sample) indicates that the water vapor is already presented in the source gas. The origin is desorption of water condensed on any surfaces inside the discharge tube. The molecules dissociate upon interaction with energetic Ar metastables. The OH radicals are clearly visible in the spectra (Figures 4–6), whereas, the H atoms are not observed. One explanation for this observation is a rather low electron temperature, which does not allow for significant excitation of hydrogen atoms to radiative states. Another one is the association of H atoms to H_2 molecules, which are rather poor emitters in the visible range, therefore, they cannot be distinguished from the background of the measured spectra either. The OH radicals are renowned for their oxidation potential. The shape of the 3D graph of the surface wettability after plasma treatment for 0.5 s could, therefore, be explained by chemical activation and formation of oxygen functional groups, due to the existence of OH radicals in the plasma jet.

The third mechanism of surface activation is a bond scission upon the interaction between UV and/or VUV radiation and the surface polymer film. According to Oehrlein et al. [27] and Schneider et al. [28], especially VUV radiation is particularly relevant. VUV radiation in the Ar plasma jet is assigned to Ar_2^* excimer continuum [31,32]. As reported by Oehrlein et al. [27] and explained in Section 1, the type of the APPJ source is a significant factor in the importance of VUV radiation relative to other plasma species in its contribution to surface modification. They found that the relative importance of VUV effects for a kHz ring-APPJ source was over 20%, whereas, for a kHz pin-APPJ and a MHz pin-APPJ it was only about 6% and 1%, respectively. The distribution of any radiation arising from a source is spatially uniform: The photons are emitted in all directions uniformly. However, as mentioned by various authors, VUV effects are reduced in the presence of molecular oxygen [27,28,33], as well as nitrogen [26]. As reported by Schneider et al. [28], the flux of VUV is the most intense just underneath the jet and is absorbed by air at any radial distance from the jet larger than few millimeters. This can also be observed in Figure 8a where the O/C ratio measured by XPS at various radial distances from the axis of the plasma jet is shown. In this case, we have covered one sample with MgF_2 optical window that transmits only VUV radiation [34], and exposed it to APPJ, at a distance of 5 mm. For comparison, also the XPS composition of the uncovered sample is shown. As observed in Figure 8a, the sample covered with MgF_2 window was modified only in the center (at the axis of the plasma jet). Because the MgF_2 window was in contact with the sample surface, the increased oxygen concentration can be explained either by reaction of oxygen adsorbed on the surface with the polymer or by post-treatment reactions of dangling bonds with the ambient atmosphere. With increasing radial distance from the center, a sharp drop in the O/C concentration appeared. This is not observed for the uncovered sample, where a diameter of the modified surface is much larger and comparable with that shown in Figure 8a. In Figure 8b is also shown mapping of the surface wettability of the sample covered with the MgF_2 optical window (the wettability of the corresponding uncovered sample is shown in Figure 7 (1 min)). The results of surface wettability are in agreement with XPS results–only the small spot on the axis was activated. The minimum contact angle is higher compared to the uncovered sample. This is expected—because the wettability is also influenced by the surface roughness. The uncovered sample was, of corse, exposed to etching effects that were absent in the case of the sample covered with MgF_2 optical window.

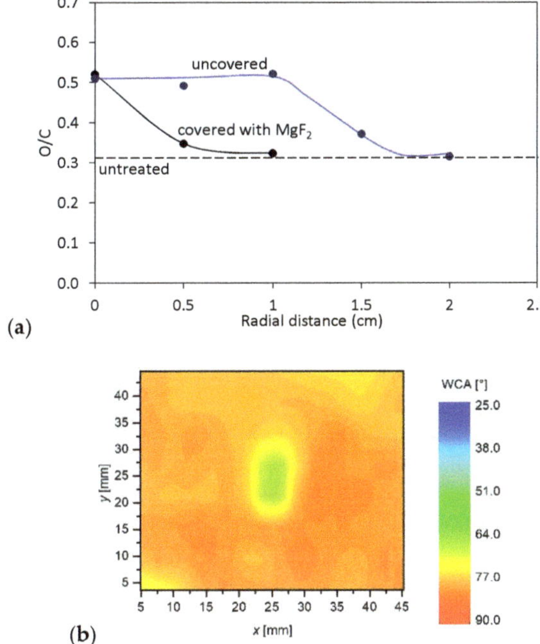

Figure 8. (**a**) Comparison of XPS ratio O/C for the sample covered with MgF_2 optical window, and thus, exposed to radiation only with the uncovered sample exposed to all reactive species. The sample was placed 5 mm from the nozzle. Treatment time was 1 min. (**b**) Mapping of the surface wettability of the polymer surface covered with a MgF_2 window. The surface wettability of the corresponding uncovered sample is shown in Figure 7 (1 min).

Therefore, at experimental conditions in this work, the VUV radiation is regarded to not play a dominant role in the activation of the polymers, because it cannot explain the formation of such large spots as observed in Figure 7. Therefore, we can conclude that, in our case, the activation by functionalization with OH radicals and/or interaction with long-living energetic Ar species prevails the activation induced by VUV.

The polymer samples were treated at a distance of 5 mm also for longer times. Figure 7 reveals the temporal evolution of the surface wettability. As expected, the area of a low water contact angle expands with increasing treatment time. Less expected, however, is the fact that the surface activation is limited to a rather small spot even for prolonged treatment times. For example, the sample treated for 10 min at the distance of 5 mm exhibits a surprisingly sharp distribution of the surface wettability. For this sample, the central area of a diameter of about 2 cm is saturated, whereas, the edges of the sample are not affected by plasma treatment at all. In between, there is a rather steep change of the surface wettability, where the water contact angle increases from approximately 25° to over 70° at a radial distance of approximately 0.5 cm. Such a sharp transition is a consequence of several effects, which will be briefly presented below.

As already discussed, the reactive species presented at the axis of the plasma jet cause rapid activation of the polymer sample. The concentration of free electrons, as well as other species formed in the streamers (in particular, within the ionization wavefront) away from the axis, decreases rapidly because of the gas phase collisions. The highest concentration of reactive species is, therefore, concentrated at the axis and decreases with increasing radial distance from the axis. There should be huge radial gradients, so the air molecules effusing the Ar jet, but not reaching the axis cannot be excited to any state that causes polymer activation. If the concentration of reactive species, such as O, NO and N, let alone molecular metastables several cm from the axis were detectable, the polymer sample exposed to plasma for a long time would have been activated entirely. The graphs presented in Figure 7, however, show that the plasma activation is radially limited even for treatment times as long as 10 min.

Another important doubtless conclusion that can be drawn from results presented in Figure 7 is that the contact angle saturates in the center of the spot even at the treatment time as low as a few seconds. The results show that a rather large spot of a low contact angle appears in the center of the sample already after treating the polymer for 10 s. After saturation is reached, some expansion of the spot with a minimal contact angle is observed with increasing treatment time, but the contact angle does not become lower than about 25° even after minutes of plasma treatment. Such a saturation is typical for many polymers and has been reported by numerous authors using both high and low-pressure discharges, as well as plasma afterglows [35,36]. The saturation is usually explained by a balance between functionalization, etching, as well as possible thermal degradation. Dowling et al. reported that some thermal decomposition might occur at polymers having low glass transition temperature [21].

Treatment of this polymer by the plasma jet, therefore, allows for a rapid increase of the surface wettability which is limited to a spot of typical diameter of few cm. Such a surface finish is achievable providing the plasma jet is in direct contact with the polymer sample. In many cases, however, it is not feasible to provide direct contact, for example, in the case of a polymer product of complex geometry. In such cases, the surface wettability may or may not be altered, depending on particularities of the experimental conditions. If a polymer surface is far away from the end of the plasma jet, the concentration of any reactive species is too low to allow for any effect. Particularly interesting are surface alternations in the case where the polymer sample is placed close to the visible end of the glowing plasma jet. Such experiments were also performed, and the results are shown in Figure 9.

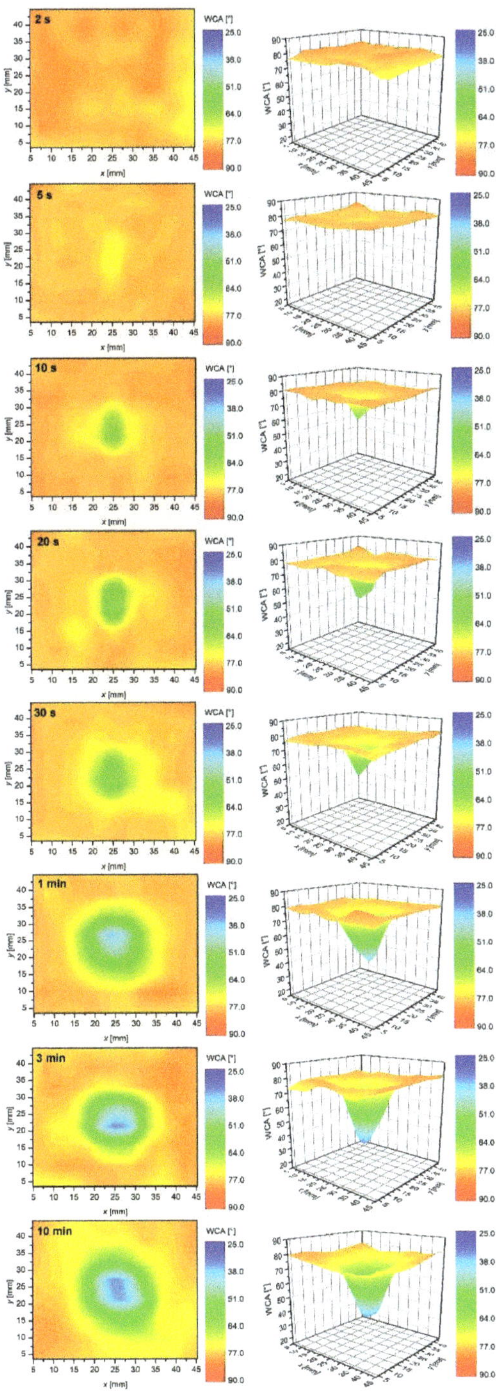

Figure 9. The time evolution of water contact angles on samples placed 30 mm below the APPJ nozzle.

The graphs presented in Figure 9, where acquired exactly in the same manner as those in Figure 7 except the distance between the sample and the nozzle was much larger (30 mm); therefore, the visible part of the plasma jet did not touch the samples (Figure 2). The shortest treatment time, shown in Figure 9, is 2 s. No statistically significant modification in the surface wettability is observed for this sample, meaning that a dose of reactive species was too low. For the sample treated for 5 s, however, we can already observe a statistically significant pattern. Right in the center of the sample, there is a spot of a diameter of ~1 mm, where the water contact angle is ~68°. This spot is surrounded by a larger area of a contact angle of ~78°. The observed feature indicates the initial stage of the sample activation. A double time (10 s) already reveals a rather large spot of the activated material of the contact angle ~52°. The spot diameter increases with increasing treatment time, but the contact angle in the center of the spot does not drop below 39° even after 1 min of plasma treatment. Contact angles lower than 39° are detectable only after prolonged treatment. The results presented in Figure 9, therefore, reveal that the activation not only takes longer time than in the case of direct exposure to the glowing plasma, but the minimal contact angle of 25° appears only after prolonged treatment time (10 min). Because the visible plasma jet expands only up to approximately 25 mm from the nozzle, i.e., 5 mm from the sample surface (Figure 2), the concentration of reactive species that owe their existence to electron impact events is negligible as compared to the glowing jet. The reactive species found in the early afterglow are definitely Ar metastables, and there could also be some long living chemically reactive species, such as nitric oxides and ozone [10,37]. There is a strong gradient in the concentration of OH radicals at the edge of the plasma jet [38].

The sample treated at a distance of 30 mm from the nozzle for 10 min exhibits a similar wettability as a sample treated directly with the plasma jet for 10 s. The ability for activation of the PET surface just outside the plasma jet is, therefore, about two orders of magnitude smaller than in the glowing plasma. The practical consequence of the observed effects is that any treatment of products of complex geometry using such a simple device as our APPJ is impractical because the required treatment time to saturate the surface wettability is too long even for a short distance between the end of the glowing plasma jet and the substrate.

The upper discussion is proved by additional results presented in Figure 10. In that set of measurements, the plasma-on-time was fixed to 30 s, but the distance between the nozzle and the sample was varied between 2 and 40 mm. Figure 10 reveals little difference between the samples treated at various distances up to 20 mm, i.e., when the plasma jet was in direct contact with a sample surface. For the case of 30 mm, there was a reasonable activation, but at 40 mm we observe no statistically significant modifications of the polymer surface. The huge difference between the last two images, shown in Figure 10, indicates that the concentration of reactive species capable of activation of the polymer surface about 1 cm away from the glowing plasma jet is negligible.

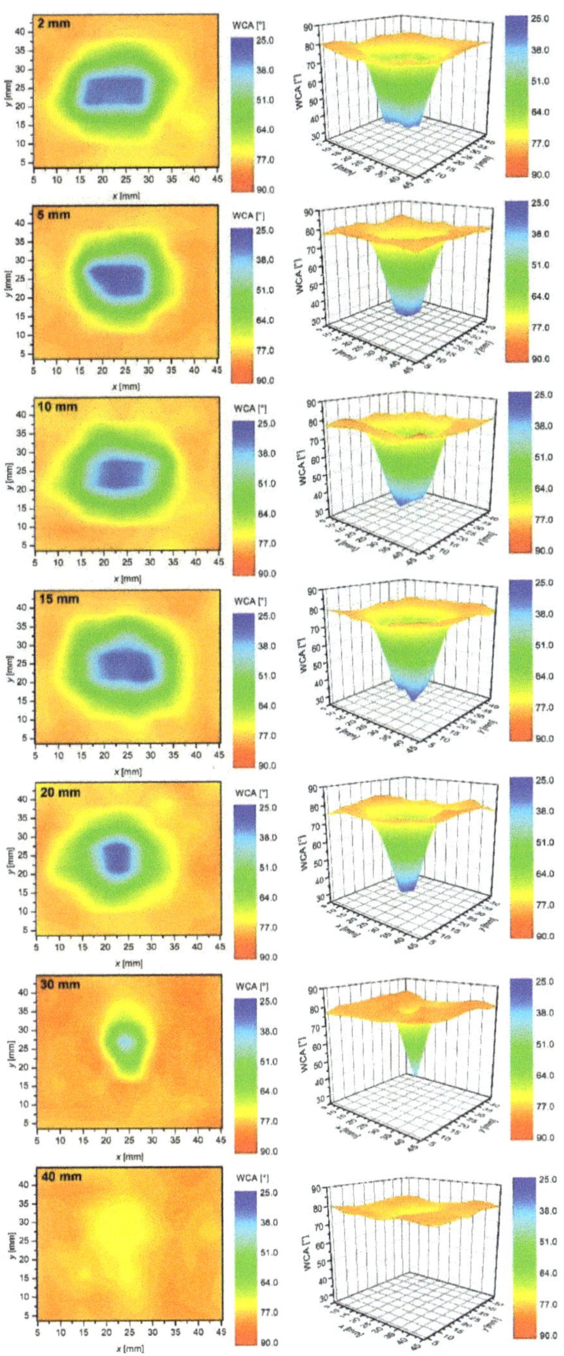

Figure 10. The water contact angles on samples treated for 30 s at different distances between the APPJ nozzle and samples. The visual length of plasma jet is below 30 mm from the nozzle.

Polymers like PET are moderately hydrophobic; therefore, any hydrophilicity is thermodynamically unstable. The hydrophilicity decreases spontaneously upon aging. The effect is known as "hydrophobic recovery". The aging of a sample treated at a distance of 5 mm, and for 3 min was, thus, measured and is shown in Figures 11 and 12. Figures 11 and 12 represents the evolution of water contact angles upon aging at room temperature for different periods. The area of a low contact angle in the middle of the modified zone ages preferentially. It is clearly visible that the zone of originally low contact angles ages rather quickly, whereas, the edges remain almost perfectly intact. After several days, the water contact angle assumes the same value over the entire area of the modified zone. The area of a higher wettability, therefore, ages faster than the area of a moderate wettability. This observation is in agreement with several reports on the hydrophobic recovery of various polymers: The aging is faster for highly activated polymers. The aging roughly follows the logarithmic dependence as revealed from Figure 13, which shows the water contact angle in the center of the modified zone versus the aging time.

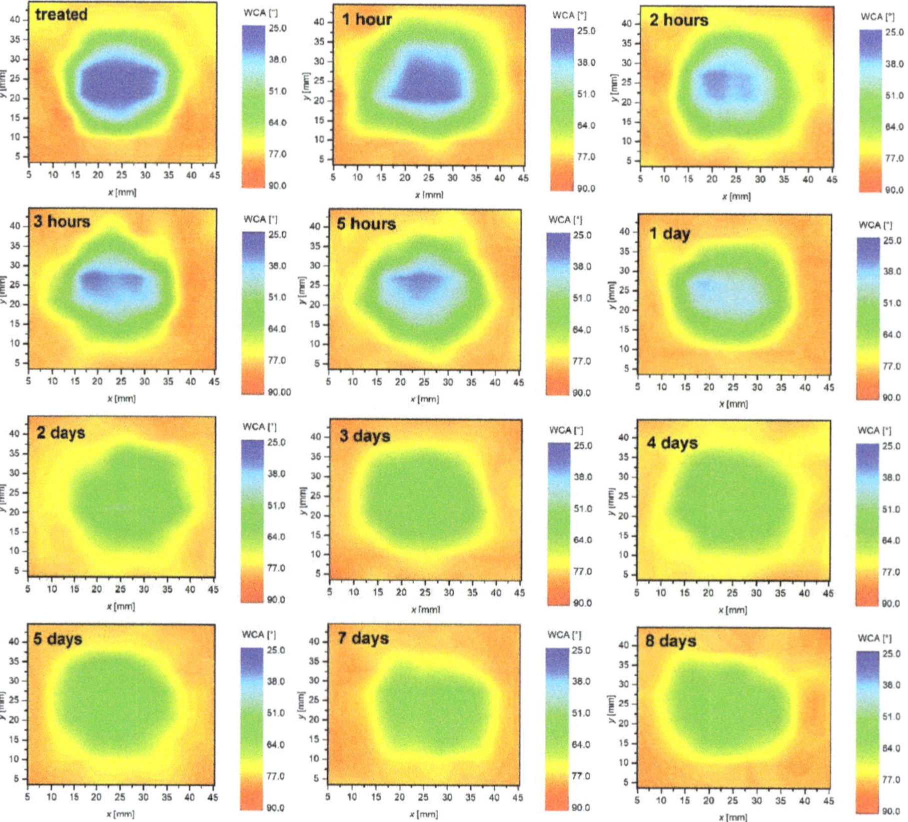

Figure 11. 2D evolution of water contact angles versus aging time.

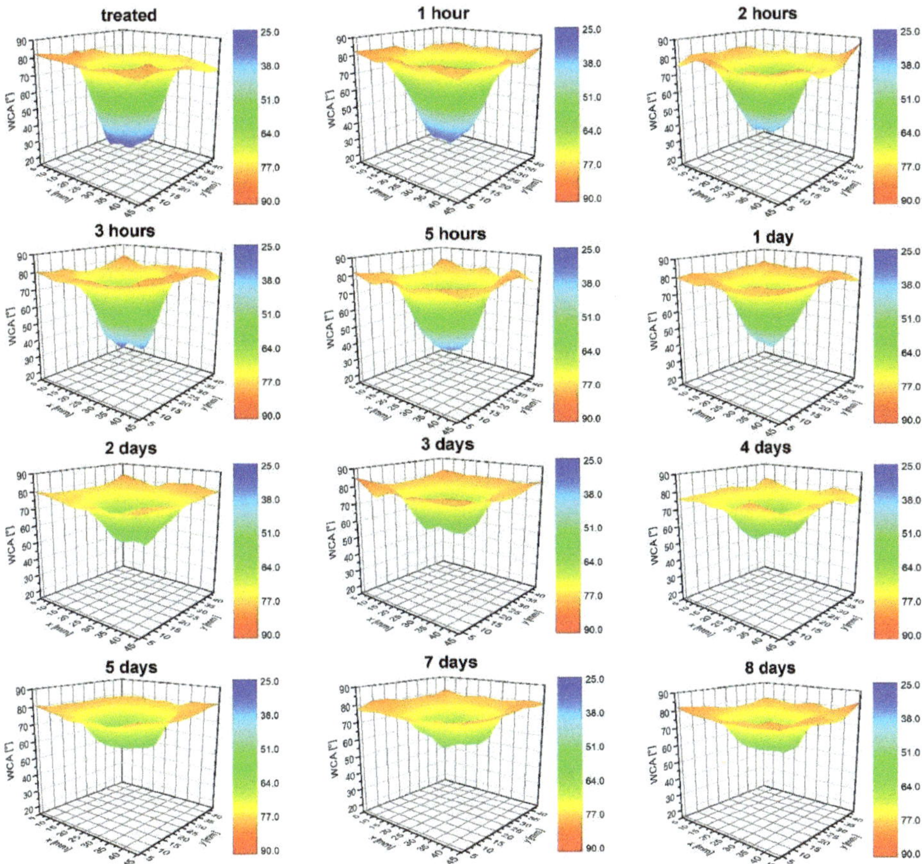

Figure 12. 3D evolution of water contact angles versus aging time.

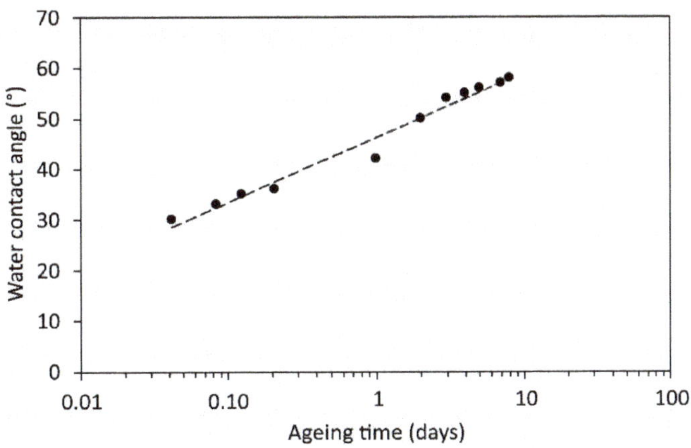

Figure 13. The evolution of the minimum water contact angles versus aging time.

Although the wettability is well known to depend on the surface morphology and concentration of functional groups, we performed 2D XPS mapping to prove the origin of the surface activation. Because the XPS mapping is time consuming, we only performed the measurements for the sample treated 5 mm from the nozzle for 3 min. The corresponding wettability for this particular sample is shown in Figure 7 (marked with 3 min). The ratio between oxygen and carbon as calculated from XPS survey spectra of this sample is shown in Figure 14. Although the spot size as detected by XPS is somehow smaller (because of the limits of the XPS device, a grid of measured points was not as dense as in the case of WCA measurements), the major features are analogous. In both cases, there is a rather large spot of a diameter just below 2 cm, where the surface is saturated; i.e., in Figure 7, the saturation is observed for the case of the water contact angle, whereas, in Figure 14 in terms as O/C ratio. The surface activation as elaborated in Figures 7, 9 and 10, therefore, arise from functionalization of the polymer surface with oxygen rich functional groups. In both cases (surface wettability and O/C ratio), a rather sharp interphase between saturated area and the not affected area is observed. As already discussed above, such a sharp interphase is a consequence of very strong gradients of reactive species at the edge of the plasma jet.

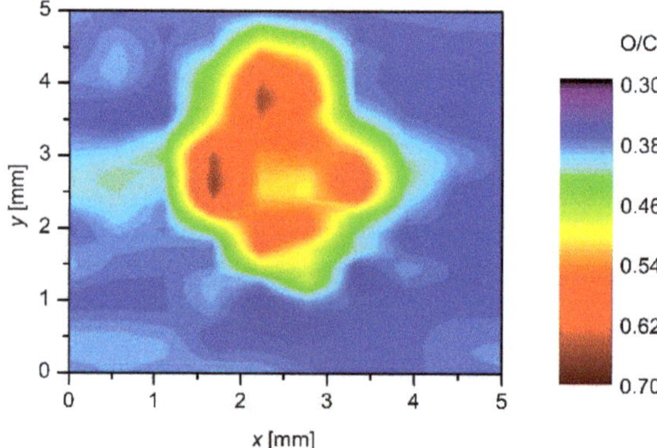

Figure 14. 2D XPS mapping of the O/C ratio of a sample treated for 3 min at a distance of 5 mm from the nozzle.

As already reported, the surface morphology of the sample may change as a result of plasma treatment [16]. Figure 15 shows typical AFM images obtained over a surface area of 2×2 μm^2. The left image is for the untreated sample, and the right one for the sample treated for 10 min at a distance of 5 mm from the nozzle. The plasma-treated sample assumes a morphology resembling nanofeatures.

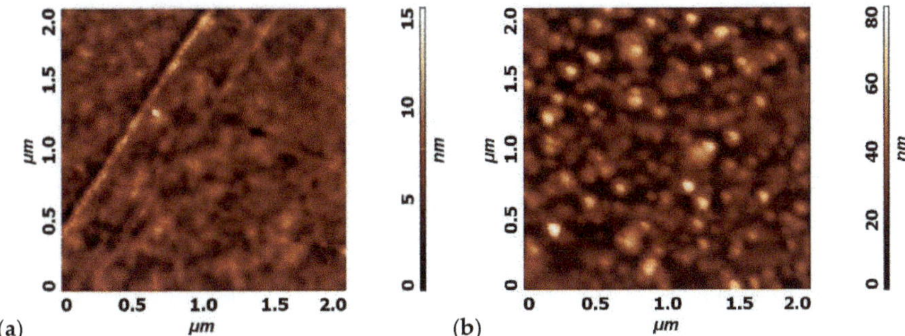

Figure 15. AFM images of: (**a**) The untreated PET foil and (**b**) the PET foil treated for 10 min at a distance of 5 mm.

Finally, it is worth mentioning that there are always exothermic reactions on the surface of a material treated by gaseous plasma despite the fact that plasma jets as adopted in this study are cold. As discussed above, the list of exothermic reactions includes oxidation of the surface because of interaction with OH and some other radicals, as well as absorption of light quanta, neutralization of charged particles and relaxation of metastables. To estimate the heat load, we measured the surface temperature versus treatment time with the IR pyrometer. Although the albedo may change during the measurement because of variation of surface morphology, composition and structure, we took into account a constant emissivity in the range of wavelengths probed with the pyrometer. Figure 16 shows the temperature evolutions versus treatment time for the cases when the PET sample was positioned 5 and 30 mm from the nozzle of the discharge tube. For a rather large distance (i.e., 30 mm), the surface temperature as probed with the pyrometer increased for few degrees centigrade; therefore, the thermal load is regarded marginal. For a short distance (i.e., 5 mm), however, the surface temperature increased to approximately 35 °C in the first 100 s. Thereafter, it does not stabilize, but keeps increasing almost linearly with treatment time. The linear increase is explained by a huge thermal capacity of the sample holder as compared with the sample itself. The heat is used for increasing the temperature of the sample holder; thus, the sample temperature remains reasonable, i.e., 40 °C for 10 min of treatment. Such a low temperature has not been reported by many authors who have used a similar experimental setup. For example, Dowling et al. [21] reported temperatures exceeding 60 °C. The discrepancy is explained by differences in experimental conditions, in particular, the thermal contact between the substrate and the sample holder, as well as thermal properties of the holder. After turning off the discharge, the sample temperature quickly dropped to approximately 35 °C and then it slowly decreased as revealed from Figure 16. The fact that the sample temperature did not drop to initial temperature (i.e., 24 °C) is explained by a huge thermal capacity of the sample holder providing the albedo has not changed.

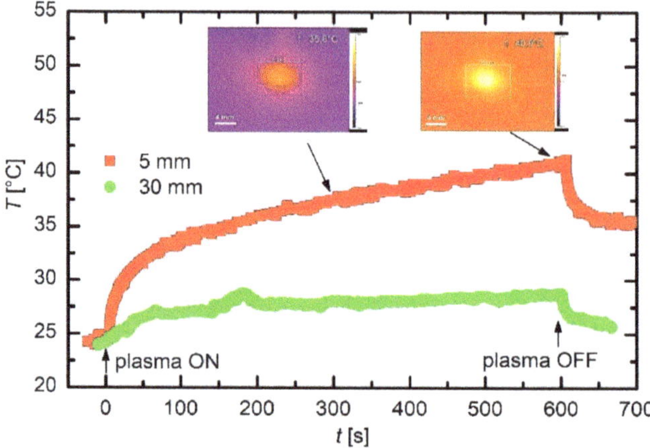

Figure 16. Evolution of sample temperatures as determined with the IR Pyrometer for the case when samples were placed 5 and 30 mm from the nozzle of the discharge tube.

4. Conclusions

Results of systematic 2D mapping of the surface wettability enable an insight in processes responsible for surface activation of PET polymer using a simple atmospheric pressure plasma jet. The discharge tube was only flushed with Ar gas to get rid of permanent gases, such as nitrogen, oxygen and CO_2, but the water remained on any surfaces as a result of the humidity of the laboratory atmosphere. The OES spectra showed significant radiation from the OH radicals. Unlike nitrogen bands, which were observed only when the sample was close to the exhaust of the dielectric discharge tube, OH radiation persisted even for long distances of the sample from the nozzle. This observation indicated that the OH origin was water vapor which was slowly desorbing from the surfaces of the discharge tube and/or electrode when the sample was treated with Ar plasma. The wettability of samples exhibited saturation at the water contact angle of approximately 25° as long as the samples were exposed to the glowing plasma at a short distance from the nozzle. At longer distances of the sample from the nozzle, the minimal contact angle of 25° was achieved at a much longer time which is impractical. A rather sharp interphase between the highly wettable area saturated with oxygen functional groups and the surrounding unaffected area was observed in all cases. The results were explained with both radial and axial gradients of reactive species useful for activation of the polymer materials. Almost perfect overlapping of the area saturated with oxygen functional groups with highly wettable area suggests that the major reactants causing activation of the polymer at these experimental conditions are OH radicals.

Author Contributions: Methodology, R.Z. and G.P.; validation, R.Z. and G.P.; formal analysis, R.Z. and A.V.; investigation, R.Z. and A.V.; data curation, G.P.; writing—original draft preparation, A.V.; writing—review and editing, A.V. and M.M.; supervision, A.V.; project administration, M.M.; funding acquisition, M.M. All authors have read and agreed to the published version of the manuscript.

Funding: This research was funded by the Slovenian Research Agency, grant number L2-8179 (Evaluation of the range of plasma parameters suitable for nanostructuring of polymers on an industrial scale) and P2-0082 (Thin film structures and plasma surface engineering).

Conflicts of Interest: The authors declare no conflict of interest. The funders had no role in the design of the study; in the collection, analyses, or interpretation of data; in the writing of the manuscript, or in the decision to publish the results.

References

1. Swilem, A.E.; Lehocky, M.; Humpolicek, P.; Kucekova, Z.; Novak, I.; Micusik, M.; Abd El-Rehim, H.A.; Hegazy, E.A.; Hamed, A.A.; Kousal, J. Description of D-glucosamine immobilization kinetics onto poly(lactic acid) surface via a multistep physicochemical approach for preparation of novel active biomaterials. *J. Biomed. Mater. Res. A* **2017**, *105*, 3176–3188. [CrossRef] [PubMed]

2. Ozaltin, K.; Lehocky, M.; Humpolicek, P.; Vesela, D.; Mozetic, M.; Novak, I.; Saha, P. Preparation of active antibacterial biomaterials based on sparfloxacin, enrofloxacin, and lomefloxacin deposited on polyethylene. *J. Appl. Phys. Polym. Sci.* **2018**, *135*. [CrossRef]

3. Bilek, F.; Sulovska, K.; Lehocky, M.; Saha, P.; Humpolicek, P.; Mozetic, M.; Junkar, I. Preparation of active antibacterial LDPE surface through multistep physicochemical approach II: Graft type effect on antibacterial properties. *Colloids Surf. B* **2013**, *102*, 842–848. [CrossRef] [PubMed]

4. Asadinezhad, A.; Novak, I.; Lehocky, M.; Sedlarik, V.; Vesel, A.; Junkar, I.; Saha, P.; Chodak, I. A physicochemical approach to render antibacterial surfaces on plasma-treated medical-grade PVC: Irgasan coating. *Plasma Process. Polym.* **2010**, *7*, 504–514. [CrossRef]

5. Ozaltin, K.; Lehocky, M.; Kucekova, Z.; Humpolicek, P.; Saha, P. A novel multistep method for chondroitin sulphate immobilization and its interaction with fibroblast cells. *Mater. Sci. Eng. C* **2017**, *70*, 94–100. [CrossRef]

6. Bilek, F.; Krizova, T.; Lehocky, M. Preparation of active antibacterial LDPE surface through multistep physicochemical approach: I. Allylamine grafting, attachment of antibacterial agent and antibacterial activity assessment. *Colloids Surf. B* **2011**, *88*, 440–447. [CrossRef]

7. Vesel, A.; Mozetic, M. New developments in surface functionalization of polymers using controlled plasma treatments. *J. Phys. D Appl. Phys.* **2017**, *50*. [CrossRef]

8. Boselli, M.; Chiavari, C.; Colombo, V.; Gherardi, M.; Martini, C.; Rotundo, F. Atmospheric pressure non-equilibrium plasma cleaning of 19th century daguerreotypes. *Plasma Process. Polym.* **2017**, *14*. [CrossRef]

9. Schäfer, J.; Fricke, K.; Mika, F.; Pokorná, Z.; Zajíčková, L.; Foest, R. Liquid assisted plasma enhanced chemical vapour deposition with a non-thermal plasma jet at atmospheric pressure. *Thin Solid Films* **2017**, *630*, 71–78. [CrossRef]

10. Reuter, S.; von Woedtke, T.; Weltmann, K.-D. The kINPen—A review on physics and chemistry of the atmospheric pressure plasma jet and its applications. *J. Phys. D Appl. Phys.* **2018**, *51*. [CrossRef]

11. Kredl, J.; Kolb, J.F.; Schnabel, U.; Polak, M.; Weltmann, K.-D.; Fricke, K. Deposition of antimicrobial copper-rich coatings on polymers by atmospheric pressure jet plasmas. *Materials* **2016**, *9*, 274. [CrossRef] [PubMed]

12. Ulrich, C.; Kluschke, F.; Patzelt, A.; Vandersee, S.; Czaika, V.A.; Richter, H.; Bob, A.; von Hutten, J.; Painsi, C.; Huge, R.; et al. Clinical use of cold atmospheric pressure argon plasma in chronic leg ulcers: A pilot study. *J. Wound Care* **2015**, *24*, 196–203. [CrossRef] [PubMed]

13. Daeschlein, G.; von Woedtke, T.; Kindel, E.; Brandenburg, R.; Weltmann, K.-D.; Jünger, M. Antibacterial activity of an atmospheric pressure plasma jet against relevant wound pathogens in vitro on a simulated wound environment. *Plasma Process. Polym.* **2010**, *7*, 224–230. [CrossRef]

14. Daeschlein, G.; Scholz, S.; Arnold, A.; von Podewils, S.; Haase, H.; Emmert, S.; von Woedtke, T.; Weltmann, K.-D.; Jünger, M. In vitro susceptibility of important skin and wound pathogens against low temperature atmospheric pressure plasma jet (APPJ) and dielectric barrier discharge plasma (DBD). *Plasma Process. Polym.* **2012**, *9*, 380–389. [CrossRef]

15. Privat-Maldonado, A.; O'Connell, D.; Welch, E.; Vann, R.; van der Woude, M.W. Spatial dependence of DNA damage in bacteria due to low-temperature plasma application as assessed at the single cell level. *Sci. Rep.* **2016**, *6*. [CrossRef]

16. Birer, Ö. Reactivity zones around an atmospheric pressure plasma jet. *Appl. Surf. Sci.* **2015**, *354*, 420–428. [CrossRef]

17. Kostov, K.G.; Nishime, T.M.C.; Castro, A.H.R.; Toth, A.; Hein, L.R.O. Surface modification of polymeric materials by cold atmospheric plasma jet. *Appl. Surf. Sci.* **2014**, *314*, 367–375. [CrossRef]

18. Jofre-Reche, J.A.; Pulpytel, J.; Fakhouri, H.; Arefi-Khonsari, F.; Martín-Martínez, J.M. Surface treatment of polydimethylsiloxane (PDMS) with atmospheric pressure rotating plasma jet: Modeling and optimization of the surface treatment conditions. *Plasma Process. Polym.* **2016**, *13*, 459–469. [CrossRef]

19. Van Deynse, A.; Cools, P.; Leys, C.; Morent, R.; De Geyter, N. Surface modification of polyethylene in an argon atmospheric pressure plasma jet. *Surf. Coat. Technol.* **2015**, *276*, 384–390. [CrossRef]

20. Shaw, D.R.; Gyuk, P.M.; West, A.T.; Momoh, M.; Wagenaars, E. Surface modification of polymer films using an atmospheric-pressure plasma jet. In Proceedings of the 22nd International Symposium on Plasma Chemistry, Antwerp, Belgium, 5–10 July 2015.

21. Dowling, D.P.; O'Neill, F.T.; Langlais, S.J.; Law, V.J. Influence of DC pulsed atmospheric pressure plasma jet processing conditions on polymer activation. *Plasma Process. Polym.* **2011**, *8*, 718–727. [CrossRef]

22. Lommatzsch, U.; Pasedag, D.; Baalmann, A.; Ellinghorst, G.; Wagner, H.-E. Atmospheric pressure plasma jet treatment of polyethylene surfaces for adhesion improvement. *Plasma Process. Polym.* **2007**, *4*, S1041–S1045. [CrossRef]

23. Liu, X.; Wang, G.; Liu, J.; Zhang, J.; Liu, S.; Wang, C.; Yang, Z.; Sun, J.; Song, J. Influence of water addition on the modification of polyethylene surface by nitrogen atmospheric pressure plasma jet. *J. Appl. Phys. Polym. Sci.* **2019**, *136*. [CrossRef]

24. Luan, P.; Kondeti, V.S.S.K.; Knoll, A.J.; Bruggeman, P.J.; Oehrlein, G.S. Effect of water vapor on plasma processing at atmospheric pressure: Polymer etching and surface modification by an Ar/H_2O plasma jet. *J. Vac. Sci. Technol. A* **2019**, *37*. [CrossRef]

25. Sarani, A.; Nikiforov, A.Y.; De Geyter, N.; Morent, R.; Leys, C. Surface modification of polypropylene with an atmospheric pressure plasma jet sustained in argon and an argon/water vapour mixture. *Appl. Surf. Sci.* **2011**, *257*, 8737–8741. [CrossRef]

26. Foest, R.; Bindemann, T.; Brandenburg, R.; Kindel, E.; Lange, H.; Stieber, M.; Weltmann, K.-D. On the vacuum ultraviolet radiation of a miniaturized non-thermal atmospheric pressure plasma jet. *Plasma Process. Polym.* **2007**, *4*, S460–S464. [CrossRef]

27. Knoll, A.J.; Luan, P.; Bartis, E.A.J.; Kondeti, V.S.S.K.; Bruggeman, P.J.; Oehrlein, G.S. Cold atmospheric pressure plasma VUV interactions with surfaces: Effect of local gas environment and source design. *Plasma Process. Polym.* **2016**, *13*, 1069–1079. [CrossRef]

28. Schneider, S.; Lackmann, J.W.; Narberhaus, F.; Bandow, J.E.; Denis, B.; Benedikt, J. Separation of VUV/UV photons and reactive particles in the effluent of a He/O_2 atmospheric pressure plasma jet. *J. Phys. D Appl. Phys.* **2011**, *44*. [CrossRef]

29. Onyshchenko, I.; Yu Nikiforov, A.; De Geyter, N.; Morent, R. Local analysis of PET surface functionalization by an atmospheric pressure plasma jet. *Plasma Process. Polym.* **2015**, *12*, 466–476. [CrossRef]

30. Onyshchenko, I.; De Geyter, N.; Morent, R. Improvement of the plasma treatment effect on PET with a newly designed atmospheric pressure plasma jet. *Plasma Process. Polym.* **2017**, *14*. [CrossRef]

31. Foest, R.; Kindel, E.; Lange, H.; Ohl, A.; Stieber, M.; Weltmann, K.-D. RF capillary jet—A tool for localized surface treatment. *Contrib Plasma Phys.* **2007**, *47*, 119–128. [CrossRef]

32. Lange, H.; Foest, R.; Schafer, J.; Weltmann, K.D. Vacuum UV radiation of a plasma jet operated with rare gases at atmospheric pressure. *IEEE Trans. Plasma Sci.* **2009**, *37*, 859–865. [CrossRef]

33. Dufour, T.; Hubert, J.; Vandencasteele, N.; Viville, P.; Lazzaroni, R.; Reniers, F. Competitive and synergistic effects between excimer VUV radiation and O radicals on the etching mechanisms of polyethylene and fluoropolymer surfaces treated by an atmospheric $He–O_2$ post-discharge. *J. Phys. D Appl. Phys.* **2013**, *46*. [CrossRef]

34. Cho, K.; Takenaka, K.; Setsuhara, Y.; Shiratani, M.; Sekine, M.; Hori, M. Effects of irradiation with ions and photons in ultraviolet–vacuum ultraviolet regions on nano-surface properties of polymers exposed to plasmas. *Jpn. J. Appl. Phys.* **2012**, *51*. [CrossRef]

35. Vesel, A.; Junkar, I.; Cvelbar, U.; Kovac, J.; Mozetic, M. Surface modification of polyester by oxygen- and nitrogen-plasma treatment. *Surf. Interface Anal.* **2008**, *40*, 1444–1453. [CrossRef]

36. De Geyter, N.; Morent, R.; Leys, C.; Gengembre, L.; Payen, E.; Van Vlierberghe, S.; Schacht, E. DBD treatment of polyethylene terephthalate: Atmospheric versus medium pressure treatment. *Surf. Coat. Technol.* **2008**, *202*, 3000–3010. [CrossRef]

37. Jeong, J.Y.; Park, J.; Henins, I.; Babayan, S.E.; Tu, V.J.; Selwyn, G.S.; Ding, G.; Hicks, R.F. Reaction chemistry in the afterglow of an oxygen-helium, atmospheric-pressure plasma. *J. Phys. Chem. A* **2000**, *104*, 8027–8032. [CrossRef]

38. Schröter, S.; Wijaikhum, A.; Gibson, A.R.; West, A.; Davies, H.L.; Minesi, N.; Dedrick, J.; Wagenaars, E.; de Oliveira, N.; Nahon, L.; et al. Chemical kinetics in an atmospheric pressure helium plasma containing humidity. *Phys. Chem. Chem. Phys.* **2018**, *20*, 24263–24286. [CrossRef]

 *polymers*

Article

Dual Crosslinked Collagen/Chitosan Film for Potential Biomedical Applications

Rushita Shah [1,*], Pavel Stodulka [2], Katerina Skopalova [1] and Petr Saha [1]

[1] Centre of Polymer Systems, University Institute, Tomas Bata University in Zlin, Tř. T. Bati 5678, 760 01 Zlín, Czech Republic; skopalova@utb.cz (K.S.); saha@utb.cz (P.S.)

[2] Gemini Eye Clinic, U Gemini 360, 760 01 Zlín, Czech Republic; stodulka@lasik.cz

* Correspondence: rusheeta.shah@gmail.com; Tel.: +420-57603-1709

Received: 20 October 2019; Accepted: 12 December 2019; Published: 14 December 2019

Abstract: The application of polymeric biomaterial scaffolds utilizing crosslinking strategy has become an effective approach in these days. In the present study, the development and characterization of collagen–chitosan hydrogel film has been reported on using dual crosslinking agent's, i.e., tannic acid and genipin simultaneously. Incorporation of genipin imparts a greenish-blue color to the polymeric film. The effect of dual crosslinking and their successful interaction within the matrix was evaluated by infrared analysis spectroscopy. The porosity of the film was examined using scanning electron microscopy (SEM). Results of TGA determine the intermediate thermal degradation. Further, the crosslinking phenomenon has found primary impact on the strength of the films. Enzymatic degradation for the films was performed with lysozyme and lipase. The cell adhesion and proliferation was also accomplished using mouse embryonic cell lines wherein the cells cultured on the dual crosslinked film. The thriving utilization of such dual crosslinked polymeric film finds their applications in ophthalmology especially as an implant for temporary injured cornea and skin tissue regeneration.

Keywords: chitosan; collagen; dual crosslinking; polymeric biomaterial

---

1. Introduction

Regenerative medicine is a broad field that includes tissue engineering, which has a promising and developing approach with a great potential of replacement and rejuvenate any injured or diseased tissues [1]. It focuses also on preparing bio interactive scaffolds that mimic extracellular matrix of the body as well as treats the complex, often chronic diseases. Hence, it has gained much interest among the researchers from different disciplines such as material science, chemistry and medicine [1]. Tissue engineering materials are destined to react with the living body, so there are many conditions associated with selecting the materials [2]. The matrix for tissue supporting should exhibit high porosity, appropriate mechanical and physicochemical properties [3]. Hence, polymeric materials are preferred by the researchers. Several synthetic polymers like poly(lactic acid), poly(a-hydroxyesters) and poly(l-glycolic acid) are widely utilized in tissue engineering because of their biodegradable nature [2]. However, there are some drawbacks associated with the use of synthetic based polymers, i.e., the intermediate product obtained during non-enzymatic hydrolysis and the decrease of the local pH, which ultimately give rise to inflammatory reactions and harm the cells surrounding the implant site [4]. Further, it also reduces the polymers melting point as well as its degradation rate. To overcome this problem, biomaterial constructed utilizing natural origin based biopolymers like collagen [1], gelatin, chitosan [2], hyaluronic acid [5] and alginate [6] with improved biological activity, cell adhesion and proliferation are utilized as safe materials for tissue engineering.

Collagen and chitosan are the most abundant biopolymers found in nature and exhibit a large spectrum of application in the biomedical field of science [7]. Collagen is a group of naturally occurring protein and is the mostly preferred biomaterial used in tissue engineering due to its excellent biocompatibility and lower antigenicity [8]. It is the main component of the connective tissue and proteins in mammals, overall constituting up to 25%–35% of the whole body protein content [9]. Collagens from bovine and pig skins are the main industrial source and utilized in functional foods, cosmetics and biomedical materials [10]. However, the major drawback dealing with homopolymer, collagen-based biomaterial is the rapid degradation and poor mechanical properties, which creates hindrance in several tissue engineering applications [11]. Thus, it becomes essential for blending of collagen with other polymers. This will enhance superiority, processability and performance of the materials generated thereafter. Chitosan, a polyatomic polysaccharide is present in soft bodied insects, crustaceans, bone plate of cuttlefish and squids [12,13]. It is synthesized through deacetylation reaction using chitin as raw material and it is a co-polymer of D-glucosamine and *N*-acetyl-D-glucosamine [12]. The molecular weight and degree of deacetylation of chitosan can be easily modified [14]. Chitosan is well-known for its biocompatibility allowing its use in topical ocular application, implantation, drug delivery, wound healing, hemodialysis membrane, tissue engineering, etc. [15–17]. It is bio-adhesive in nature because of its positive charge at physiological pH [18]. Pure chitosan possesses properties such as non-toxicity, biodegradable, antifungal, antibacterial as well as biological inertness and stability. Until now, it has been utilized in form of hydrogels, fiber membranes, beads, blends and porous scaffolds for several biological and clinical purposes [14].

In tissue engineering, the crosslinking phenomenon is reflecting more concern due to excellent stability among the polymeric blends and decreasing reactivity [19]. Crosslinking results in elasticity, decreasing solubility and viscosity of the polymer, gives strength and toughness to the biomaterial developed [20]. Crosslinking can be physical for, e.g., UV-radiation, microwave and dihydrothermal treatment [21]. Even though these methods can avoid the introduction of potential toxic residue, they fail to yield an increasing crosslinking degree [11]. Chemical crosslinkers commonly used are aldehydes (for e.g., glutaraldehyde, formaldehyde), carbodimides, polyepoxy compounds, etc. The main limitation concerned with chemical crosslinkers is the unreacted crosslinkers inside the scaffolds, which ultimately gives the risk of the formation of toxic products and also limited mechanical strength [21]. To overcome this issue, natural alternatives like tannic acid, genipin, citric acid, proanthocyanidin and ferulic acid are preferred.

Tannic acid is a natural plant based polyphenol compound that has diverse biological functions such as antiviral, anti-inflammatory, antioxidant and antimicrobial properties [22]. It can interact readily with biopolymers like collagen, chitosan, albumin and gelatin through non-covalent interactions like H_2-bonding and hydrophobic effects [23,24]. Sionkowska A and group for the first time utilized tannic acid as crosslinker to modify the properties of chitosan-collagen mixture [7]. Genipin is an aglycone of geniposide comprising of a dihydropyran ring and an ester group, which is derived from the fruits of Gardenia Jasminoides Ellis. Traditionally, it is used to treat pyrogenic infection, febrile disease, sprain, swelling, etc. [25–27]. Genipin is materializing with a number of polymeric materials comprising primary amino groups for, e.g., chitosan, certain peptides, polypeptides by crosslinked covalent grid and gives blue colored fluorescent [21,27]. Due to its lowered toxicity, genipin has gained increasing interest in the field of biomaterial processing technique [28,29]. With respect to the prospective results, it is possible to develop genipin crosslinked biomaterial for ocular therapeutics, tissue repair and pharmacology [29,30]. There are several examples for the approach to utilize genipin ophthalmology or tissue engineering such as the Mi FL et al. group found out that genipin not only exhibits decreasing cytotoxicity as compared to glutaraldehyde and epoxy compounds but is also able to efficiently crosslink cellular tissues and biomaterials comprising of free amino groups [31]. The genipin cross-linked chitosan thin membrane improved the preservation of corneal endothelial cell density as well as showed anti-inflammatory activity, which was reported by Jui-Yang Lai [29]. Maria Grolik et al. reported genipin cross-linked chitosan-collagen blends for corneal tissue engineering [32]. Long Bi also used genipin cross-linked chitosan-collagen for cartilage regeneration [33].

The present research portrayed for the first time preparation of collagen–chitosan hydrogel film utilizing two different natural crosslinkers, i.e., tannic acid and genipin simultaneously. At the moment, there is also no such information reported hence, emphasis is put on fabrication of collagen/chitosan-based biomaterial that should persist an inert effect on the human body, moreover it is user friendly. The properties of the newly formed film were thoroughly studied in the form of its physical appearance, physico-chemical structure, swelling behavior, thermal analysis, mechanical properties, biodegradation and cell (mouse embryonic cell lines) adhesion studies. This kind of film will find potential application in ophthalmology especially wherein corneal epithelium is injured externally, skin tissue engineering, wound dressing and/or cartilage/tissue regeneration.

2. Materials and Methods

2.1. Materials

Chitosan (from crab shell, highly viscous, viscosity >400 mPa·s, catalog number-9012-76-4), collagen (from bovine, catalog number-48165), phosphate buffer saline (sterile liquid, D8662), tannic acid (catalog number-403040) and genipin (catalog number-G4796) were obtained from Sigma-Aldrich, Prague, Czech Republic. Whereas acetic acid (99.8%, catalog number 19990-11000), sodium hydroxide (catalog number-15740-31000) and dimethyl sulfoxide (DMSO, catalog number 12630-11000) from Penta, Prague, Czech Republic.

2.2. Preparation of Dual Crosslinked Collagen–Chitosan Film

Dual crosslinked collagen/chitosan (Col/Chi) film was prepared through a casting technique, utilizing tannic acid and genipin as natural origin crosslinkers. To the chitosan solution (dissolved in 1% acetic acid), collagen was added in the ratio of 75:25 respectively. Then the primary cross linker, i.e., tannic acid (concentration range: 0.5%–3%, solvent: demineralized water) was slowly added portion wise into the mixture of Col/Chi. The entire mixture was stirred for 30 min at room temperature under magnetic stirring with the rotation speed of 500 rpm to ensure complete homogeneity. Thereafter, the solution is casted onto the polystyrene plates and allowed us to dry at room temperature. A smooth, transparent and flexible tannic acid crosslinked Col/Chi (Col/Chi-Ta) film was obtained. For achieving dual crosslinking in the system, the solution of genipin (concentration: 0.25%, solvent: DMSO and phosphate buffer solution) was prepared. The Col/Chi-Ta film was incubated for 48–72 h in the genipin solution for the crosslinking reaction to be achieved. Finally, the resulted genipin crosslinked Col/Chi-Ta (GpCol/Chi-Ta) film obtained had a thin, smooth texture with a greenish-blue color appearance and is termed as DC-Col/Chi (dual crosslinked collagen/chitosan) film. The entire preparation is represented in Figure 1. The blank sample was also prepared, which is devoid of any cross-linkers and comprises only of Col/Chi.

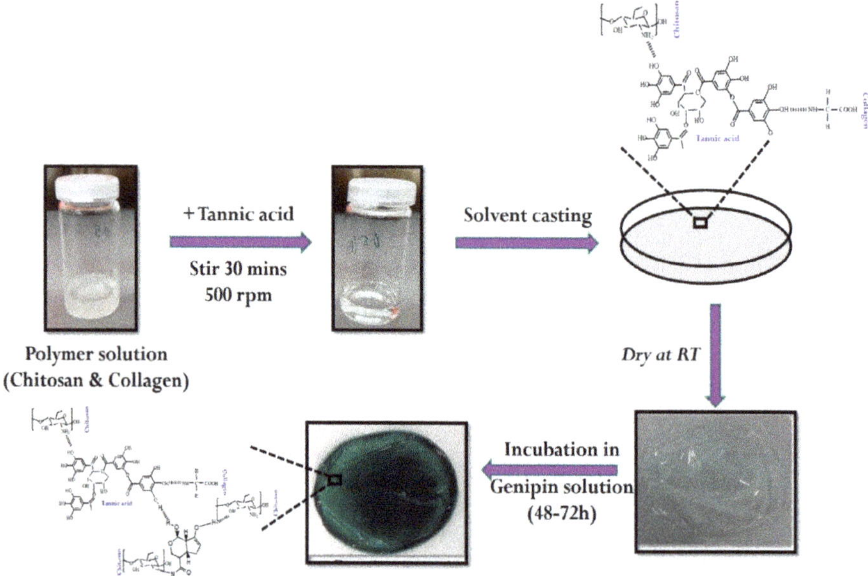

Figure 1. Schematic approach for the preparation of dual crosslinked collagen–chitosan films.

2.3. Characterization of Films

2.3.1. Fourier Transform Infrared Spectroscopy

FTIR spectra of the films Col/Chi, Col/Chi-Ta and DC-Col/Chi were obtained at wave number of 2000–600 cm^{-1} at room temperature with uniform resolution of 4 cm^{-1} and 64 scans. For this, single beam Fourier transform infrared spectroscopy (FTIR) equipped with iD5 attenuated reflectance (ATR) was used. This ATR-FTIR was equipped with the "Omnic" software package. The crystal utilized for detecting the spectra was germanium (iD5-Ge-ATR).

2.3.2. Scanning Electron Microscopy

The morphology and porous structure of films Col/Chi, Col/Chi-Ta and DC-Col/Chi were determined by scanning electron microscopy on VEGA II LMU (TESCAN) operating at high-vacuum with an accelerating voltage 5–20 kV. The images were taken at a magnification of 100–10,000×. All the samples were sputter coated with a thin layer of palladium/gold alloy to improve the surface conductivity and tilted 30° for better observation.

2.3.3. Thermogravimetric Analysis

The TA Q500 apparatus (TA Instruments, New Castle, DE, USA) was used for thermogravimetric analysis (TGA). This analysis was performed at the constant heating rate of 10°C/min from temperature range of 25–700 °C under nitrogen atmosphere. The amount of each selected sample was approximately 10 mg.

2.3.4. Swelling and Invitro Degradation Studies

Water uptake or swelling studies of the films Col/Chi, Col/Chi-Ta and DC-Col/Chi (diameter: 10 mm × 10 mm and thickness: 10 μm) were performed in phosphate buffered saline (pH 7.4; control) and in enzymatic solutions, i.e., lysozyme from chicken white (13 mg/L) and lipase from *Aspergillus oryzae* (110 U/L) at 37 °C for 2 weeks. Thereafter, time to time the swollen samples were removed from the medium and the excess water from the sample surface was removed by the filter paper. The water uptake was determined with respect to swollen samples in the control solution, i.e., PBS as well as eventual or partial degradation in the enzymatic solutions by the following equation, [34–36].

$$\text{Water uptake } (\%) = \left(\frac{Ws - Wf}{Wf}\right) \times 100 \tag{1}$$

where, W_s and W_f are the weight of the swollen and final dry weight of test samples, respectively.

The weight loss was calculated using the equation below:

$$\text{Weight loss } (\%) = \left(\frac{Wi - Wf}{Wf}\right) \times 100 \tag{2}$$

where, W_i and W_f are the initial and final dry weight of test samples respectively.

2.3.5. Cell Adhesion and Proliferation Studies

The cell compatibility was detected according to previously described procedures [37]. The test samples (films Col/Chi, Col/Chi-Ta and DC-Col/Chi) were sterilized by UV radiation prior to testing. Mouse embryonic fibroblast cell line (ATCC CRL-1658 NIH/3T3; Marlboro, MA, USA) was used to test the adhesion and proliferation of cells on the surfaces. ATCC-formulated Dulbecco's modified Eagle's medium (Biosera, Nuaille, France) containing 10% calf serum (Biosera) and 100 U·mL^{-1} penicillin/streptomycin (PAA, Trasadingen, Switzerland) was used as the culture medium.

In the case of adhesion, the cells were seeded on the samples in the concentration of 1.106 cells mL^{-1}. After one hour, the cells were stained with Hoechst 33258 (Molecular Probes, Carlsbad, CA, USA). To determine the ability of cells to proliferate on the surfaces, the cells were seeded at an initial concentration of 1.105 cells mL^{-1} and cultivated for 48 h. After 48 h the cells were fixed and stained with Hoechst 33258 and ActinRed 555 (Thermo Fisher Scientific, Waltham, MA, USA). Micrographs were taken using the fluorescence microscope Olympus IX 81 (Olympus, Tokyo, Japan).

3. Results and Discussion

3.1. FTIR Analysis of Films

The FTIR spectra of Col/Chi, Col/Chi-Ta and DC-Col/Chi films are shown in Figure 2. In Col/Chi film spectra, the absorption bands at 3307 and 2927 represent the stretching of –OH and –CH$_3$ respectively. The band at 1457 cm^{-1} is due to stretching of the pyrrolidine ring [38]. The spectrum shows the presence of collagen through the vibration band at 1245 cm^{-1} and 1552 cm^{-1} arise due to N–H bending coupled with C–N stretching vibrations indicate amide II absorbance. However, for the collagen detection, the strong signal always arises between 1700 and 1600 cm^{-1} where in the present case, it is observed at 1633 cm^{-1}.

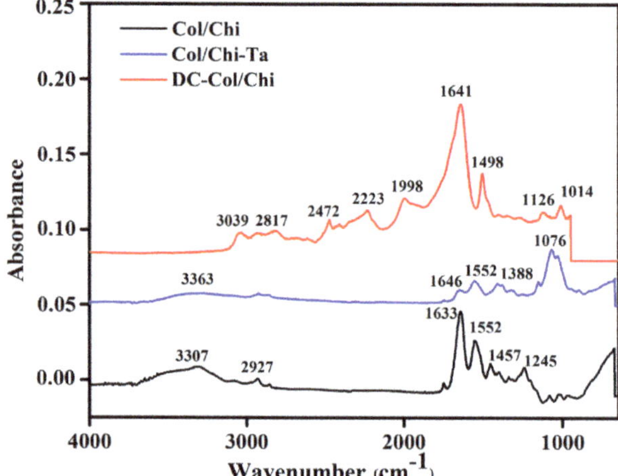

Figure 2. FTIR spectra of collagen/chitosan (Col/Chi), tannic acid crosslinked Col/Chi (Col/Chi-Ta) and dual crosslinked Col/Chi (DC-Col/Chi) films.

After crosslinking Col/Chi film with tannic acid spectra the shifts of bands are noticed. Tannic acid has the ability to form hydrogen bonds with the chemical moieties found in collagen and chitosan type of biopolymers. In the Col/Chi-Ta spectra, the band between around 3363 cm^{-1} represents the aliphatic –OH stretching of chitosan and tannic acid, 1646 cm^{-1} (amide I C=O stretching), 1552 cm^{-1} (amide II N–H bending and C–N stretching) of collagen and 1076 cm^{-1} for v(C–O–C) absorption [7]. Col/Chi-Ta shows new medium intensity peaks at 1388 cm^{-1} may be due to the CN stretching coupled with N–H bending [7,39].

In the crosslinking mechanism of genipin over Col/Chi-Ta, physical interaction takes place. Here there is a nucleophilic attack of the genipin C3 carbon atom with the primary amino group of the biopolymer and then embedding a tertiary N2 in the six-membered ring in place of oxygen atom [28]. The spectra of DC-Col/Chi reveals the peaks at 1641 cm^{-1}, is assigned to the C=C ring stretching. These results are in accordance with the Dimida, S et al. group wherein the interaction of genipin with free amino group of polymers is shown [40]. Further, the band at 1126 cm^{-1} was assigned to the C–N stretch of the tertiary aromatic amine of genipin crosslinked with the Col/Chi-Ta. The amide II at N–H bending and C–N stretching is shifted from 1552 to 1498 cm^{-1}. This is due to reaction between the genipin ester and hydroxyl groups and the amino group of chitosan within the polymeric film [41,42]. There also exists small intensity peaks between 2200 and 3000 cm^{-1}, which are associated with the –OH group [43,44]. The absorption band between 1000 and 1100 cm^{-1} was attributed to C–O and C–N stretching vibrations, and C–C–N bending vibrations [41].

3.2. SEM Micrographs

Figure 3 represents the interior microstructure of crosslinked and uncrosslinked films Col/Chi, Col/Chi-Ta and DC-Col/Chi. In Figure 3a, Col/Chi matrix shows flake like structure, which is irregular in size and shape. This could be because chitosan being semi-crystal polymer tends to form the membrane structure and also there exists physical interactions of the bonds within polymers [45]. Moreover, the matrix exhibits porous nature with lack of proper alignment. After crosslinking with tannic acid and genipin consecutively significant changes were noted. The porosity of the further crosslinked Col/Chi-Ta matrix in Figure 3b was quite high even though it exhibits an irregular structure of interconnected pores. Apart from this, a honeycomb like structure was noticed with dissimilar pore sizes. However, when genipin got crosslinked to Col/Chi-Ta (Figure 3c) film, it should reveal more crosslinked structure and rehydrated, but on the contrary there was a collapse noticed in the porous structure and the fusion of the interconnected pores led to a decrease in the number of pores.

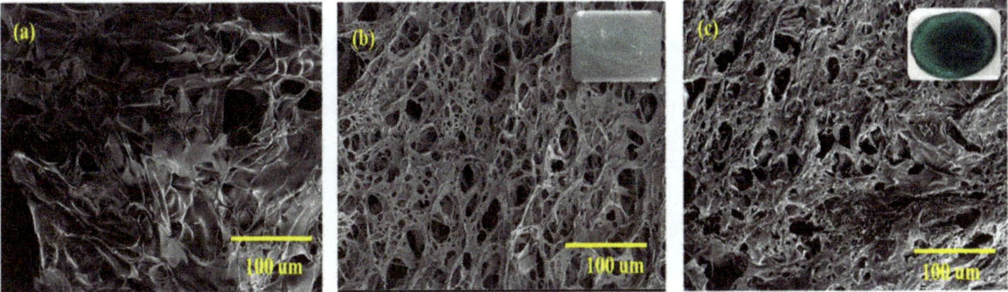

Figure 3. Cross section images of (**a**) Col/Chi, (**b**) Col/Chi-Ta and (**c**) DC-Col/Chi films.

3.3. TGA Analysis

Thermal properties of the Col/Chi, Col/Chi -Ta and DC-Col/Chi films were studied using TGA analysis as depicted in Figure 4. Here, the initial weight loss up to 150 °C was assigned to the loss of structural bound water. The second weight loss between 300 and 340 °C was attributed to chitosan degradation. The third weight loss around 400 °C corresponded to collagen degradation, as identified by Horn et al. [46]. Usually, the weight loss until 400 °C is due to complex processes like the dehydration of the polysaccharide rings, with vaporization and removal of volatile products [41,47].

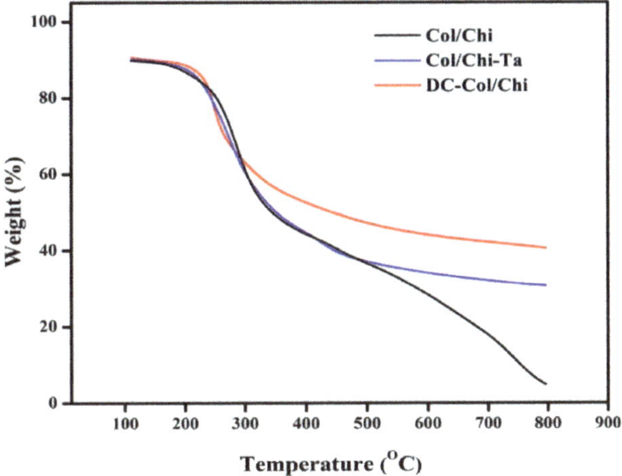

Figure 4. TGA of Col/Chi, Col/Chi-Ta and DC-Col/Chi films.

After crosslinking of Col/Chi with tannic acid and genipin, the polymer film exhibits a different degradation pattern. A crosslinked Col/Chi film shows water loss at the lowest heating temperature. Such findings represent the strong or weak interaction of water molecules with polysaccharides and this is clearly described by Beppu and coworkers [48]. The second stage loss between 290 and 310 °C can be due to partial decomposition of gallic acid, tannic acid or gallic acid dimers as explained by Peña and coworkers [49]. It is also interesting to know that weight loss decreases after cross linking and the stability in degradation was observed after 500 °C in both Col/Chi-Ta and DC-Col/Chi films. In DC-Col/Chi, it is clearly seen there was a significant loss after 200 °C. However, it could be concluded that the water content and thermal stability of the polymers were greatly influenced through crosslinking degree and intermolecular chain interaction.

3.4. Swelling and In Vitro Degradation Studies

The water uptake capacity or swelling studies of any polymeric gel/hydrogel depends on their composition, degree of crosslinking, several external conditions like temperature, pH, salt concentration, etc. The mechanism to absorb any solution (for, e.g., water, body fluids and cell nutrients) by the gel like matrixes is well explained through Donnan equilibrium theory [50]. Here, to broaden the application of the prepared films Col/Chi, Col/Chi-Ta and DC-Col/Chi, the swelling studies were performed in PBS and two different enzymes present in the human blood serum, i.e., lysozyme (also found in eyes) and lipase [36,51]. The swelling studies were carried out in the static conditions. Moreover, the swelling capacity also depends on the hydrophilicity as well as microstructure of the scaffolds. Chitosan and collagen being hydrophilic polymers, has higher water absorbing capacity [52]. So, after immersing Col/Chi film in PBS, it was impossible to assess the swelling, as it readily dissolves in the PBS solution.

From the Figure 5, it is clearly visible that the swelling behavior of the Col/Chi-Ta film in PBS is significantly higher as compared to DC-Col/Chi. This could be because of the porous nature of Col/Chi-Ta, it can entrap and seize more water through capillary action. Thus the crosslinking treatment improves the scaffolds structural stability and allowing more water retention ability. However, when Col/Chi-Ta and DC-Col/Chi film is again crosslinked with genipin, there is a decrease noticed in the swelling behavior as there could be reduction in the hydrophilic groups (for, e.g., amino, hydroxyl or carboxylic groups). Moving further, the same trend (as in PBS) was observed when Col/Chi-Ta and DC-Col/Chi films were swelled in enzymatic solutions. The increasing value in the swelling of Col/Chi-Ta and DC-Col/Chi in the presence of lysozyme and lipase could be due to the degradation of the polymeric films.

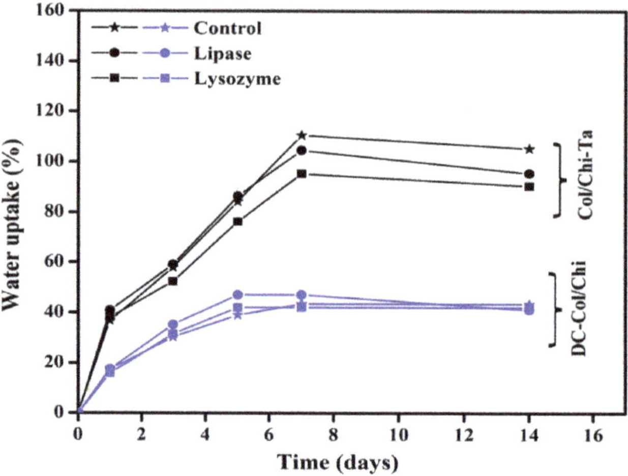

Figure 5. Swelling studies of Col/Chi -Ta and DC-Col/Chi films.

When analyzing the degradation studies of Col/Chi-Ta and DC-Col/Chi films that were supplemented with lysozyme and lipase individually in PBS, the residual weight of Col/Chi-Ta and DC-Col/Chi films after degradation is depicted in Table 1. The Col/Chi-Ta film maintained 86% of its initial weight in lysozyme and 85% in lipase solution after a 14 days incubation period. The DC-Col/Chi film retained 89% and 88% of its initial weight in lysozyme and lipase solution respectively. The hydrolytic nature of the enzymes could be the cause for the degradation of the film. As a whole, both the films depict minimal degradation property reflecting positive evidence about its use as a medical implant (especially in corneal tissue engineering) and also in wound treatment.

Table 1. Degradation of Col/Chi-Ta and DC-Col/Chi films.

Sample Index	Degradation of Film (%)	
	Lysozyme	Lipase
Col/Chi-Ta	14	15
DC-Col/Chi	11	12

3.5. Cell Adhesion and Proliferation Studies

The ability of cells to adhere and proliferate on the tested surfaces, the mouse embryonic fibroblast cell line (NIH/3T3), which is one of the most frequently used lines. The cell adhesion results are shown in the Figure 6. The best cell adhesion was observed on the sample DC-Col/Chi. However, the amount of adherent cells was lower than that of the reference (tissue culture plastic). No cell adhesion was observed on the Col/Chi and Col/Chi-Ta after one hour.

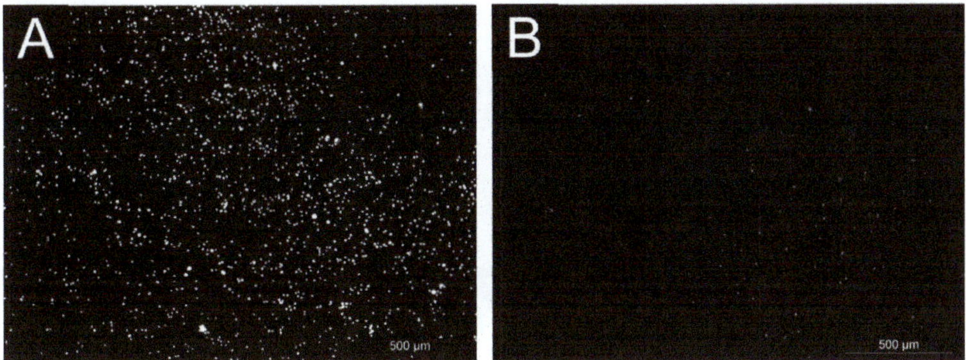

Figure 6. Microphotographs of cell adhesion visualized as number of cell nucleus (DNA dyed by Hoechst 33258) on reference (**A**) and DC-Col/Chi (**B**). No cell adhesion was observed on the Col/Chi and Col/Chi-Ta.

The cell proliferation on tested surfaces is shown in the Figure 7. Due to low cell adhesion, limited cell proliferation was expected on the Col/Chi and Col/Chi-Ta films. There was no cell growth and proliferation observed on Col/Chi film, whereas on the Col/Chi-Ta film the cells were able to grow, however their proliferation was limited. It could be predicted that polycationic nature of chitosan molecules might interact with fibroblasts membrane, thus causing cell death or apoptosis. Hence, lesser the chitosan molecules within the polymeric matrix reduce the cell membrane damage [53]. Remarkable was the proliferation on the DC-Col/Chi were a number of nucleus observed, but no acting fibers were present. This is probably because the cells were actually damaged, and only a residual nucleus was present.

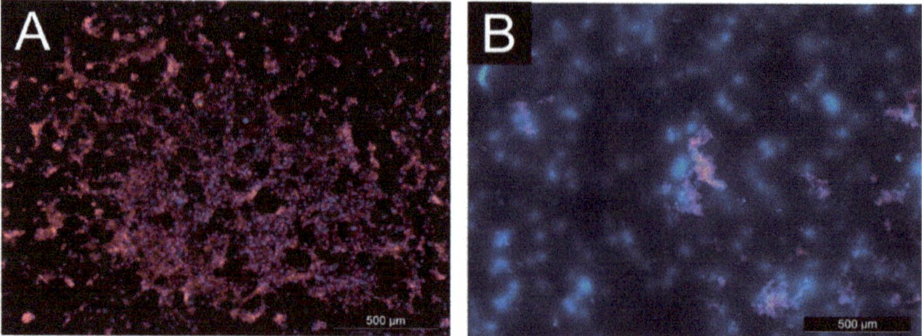

Figure 7. Microphotographs of cell proliferation on reference (**A**) Col/Chi-Ta No cells were observed on Col/Chi. No actin was present within the cells on (**B**) DC-Col/Chi (DNA dyed by Hoechst 33258, actin dyed by ActinRed 555).

4. Conclusions

The objective of this research work was to develop a novel dual crosslinked film that depicts a promising future in ophthalmology, skin tissue engineering and wound dressing. Firstly, Col/Chi film was prepared by a solvent casting technique and utilizing two crosslinking agents together, i.e., tannic acid and genipin. The obtained final dual crosslinked film was a translucent, thin and greenish-blue in color. The distinguishable differences among their physico-chemical properties were recorded through IR spectroscopy. The difference in the internal morphology (porosity) of the crosslinked films was visualized through SEM analysis. The thermal property was also studied using TGA analysis. Further, it was noticed that genipin crosslinked Col/Chi-Ta film exhibited lower swelling capacity. However, the degradation studies show more than 80% of the initial film weight that was retained even after 2 weeks of the incubation within the enzymatic solutions (i.e., lysozyme and lipase). Finally, the mouse fibroblasts cell adhesion and proliferation was performed indicating success in the adhesion of cells onto the genipin crosslinked matrix. However, this study definitely shows that the polymeric film constructed after the crosslinking could serve as a temporary graft in the field of ophthalmology especially for embedding over the cornea of the eyes.

Author Contributions: Conceptualization, R.S. and P.S. (Pavel Stodulka); methodology, R.S.; validation, R.S.; formal analysis, R.S. and K.S.; investigation, R.S.; resources, R.S. and P.S. (Petr Saha); writing-original draft preparation, R.S.; writing-review and editing, R.S. and P.S. (Petr Saha); visualization, R.S. and P.S. (Petr Saha); supervision, P.S. (Petr Saha)

Funding: This work is supported by the Ministry of Education, Youth and Sports of the Czech Republic-Program NPUI (LO1504).

Conflicts of Interest: The authors declare no conflict of interest

References

1. Fiejdasz, S.; Szczubiałka, K.; Lewandowska-Lancucka, J.; Osyczka, A.M.; Nowakowska, M. Biopolymer-based hydrogels as injectable materials for tissue repair scaffolds. *Biomed. Mater.* **2013**, *8*, 035013. [CrossRef] [PubMed]

2. Kuo, Y.C.; Ku, H.F.; Rajesh, R. Chitosan/γ-poly(glutamic acid) scaffolds with surface-modified albumin, elastin and poly-L-lysine for cartilage tissue engineering. *Mater. Sci. Eng. C Mater. Biol. Appl. C* **2017**, *78*, 265–277. [CrossRef] [PubMed]

3. Ma, P.X. Biomimetic materials for tissue engineering. *Adv. Drug Deliv. Rev.* **2008**, *60*, 184–198. [CrossRef] [PubMed]

4. Liu, H.; Slamovich, E.B.; Webster, T.J. Less harmful acidic degradation of poly(lactic-coglycolic acid) bone tissue engineering scaffolds through titania nanoparticle addition. *Int. J. Nanomed.* **2006**, *1*, 541–545. [CrossRef]

5. Yamane, S.; Iwasaki, N.; Majima, T.; Funakoshi, T.; Masuko, T.; Harad, K.; Minami, A.; Monde, K.; Nishimura, S.-h. Feasibility of chitosan-based hyaluronic acid hybrid biomaterial for a novel scaffold in cartilage tissue engineering. *Biomaterials* **2005**, *26*, 611–619. [CrossRef]

6. Han, J.; Zhou, Z.Y.; Yin, R.X.; Yang, D.Z.; Nie, J. Alginate-chitosan/hydroxyapatite polyelectrolyte complex porous scaffolds: Preparation and characterization. *Int. J. Biol. Macromol.* **2010**, *46*, 199–205. [CrossRef]

7. Sionkowska, A.; Kaczmarek, B.; Lewandowska, K. Modification of collagen and chitosan mixtures by the addition of tannic acid. *J. Mol. Liq.* **2014**, *199*, 318–323. [CrossRef]

8. Lee, C.H.; Singla, A.; Lee, Y. Biomedical applications of collagen. *Int. J. Pharm.* **2001**, *221*, 1–22. [CrossRef]

9. Krki, N.; Lazi, V.; Petrovi, L.; Gvozdenovi, J.; Peji, D. Properties of Chitosan-Laminated Collagen Film. Food Technol. *Biotechnology* **2012**, *50*, 483–489.

10. Liu, H.; Zhao, L.; Guo, S.; Xia, Y.; Zhou, P. Modification of fish skin collagen film and absorption property of tannic acid. *J. Food Sci. Technol.* **2014**, *51*, 1102–1109. [CrossRef]

11. Yan, L.; Wang, Y.; Ren, L.; Wu, G.; Caridade, S.G.; Fan, J.; Wang, L.; Ji, P.; Oliveira, J.M.; Oliveira, J.T.; et al. Genipin-cross-linked collagen/chitosan biomimetic scaffolds for articular cartilage tissue engineering applications. *J. Biomed. Mater. Res. A* **2010**, *95A*, 466–475. [CrossRef] [PubMed]

12. Yeh, J.T.; Chen, C.L.; Huang, K.S.; Nien, Y.H.; Chen, J.L.; Huang, P.Z. Synthesis, Characterization, and Application of PVP/Chitosan Blended Polymers. *J. Appl. Polym. Sci.* **2006**, *101*, 885–891. [CrossRef]

13. Lewandowska, K. Miscibility and interactions in chitosan acetate/poly (N-vinylpyrrolidone) blends. *Thermochim. Acta* **2011**, *517*, 90–97. [CrossRef]

14. Nwe, N.; Furuike, T.; Tamura, H. The Mechanical and Biological Properties of Chitosan Scaffolds for Tissue Regeneration Templates Are Significantly Enhanced by Chitosan from Gongronella butleri. *Materials* **2009**, *2*, 374–398. [CrossRef]

15. Chaiyasan, W.; Srinivas, S.P.; Tiyabooncha, W. Crosslinked chitosan-dextran sulfate nanoparticle for improved topical ocular drug delivery. *Mol. Vis.* **2015**, *21*, 1224–1234. [PubMed]

16. Seol, Y.-J.; Lee, J.-Y.; Park, Y.-J.; Lee, Y.-M.; Ku, Y.; Rhyu, I.-C.; Lee, S.-J.; Han, S.-B.; Chung, C.-P. Chitosan sponges as tissue engineering scaffolds for bone formation. *Biotechnol. Lett.* **2004**, *26*, 1037–1041. [CrossRef]

17. Tachibana, M.; Yaita, A.; Taniura, H.; Fukasawa, K.; Nagasue, N.; Nakamura., T. The use of chitin as a new absorbable suture material-an experimental study. *Jpn. J. Surg.* **1988**, *18*, 533–539. [CrossRef]

18. Bergera, J.; Reista, M.; Mayera, J.M.; Feltb, O.; Gurny, R. Structure and interactions in chitosan hydrogels formed by complexation or aggregation for biomedical applications. *Eur. J. Pharm. Biopharm.* **2004**, *57*, 35–52. [CrossRef]

19. Mane, S.; Ponrathnam, S.; Chavan, N. Effect of Chemical Cross-linking on Properties of Polymer Microbeads: A Review canchemtrans. *Can. Chem. Trans.* **2015**, *3*, 473–485.

20. Maitra, J.; Shukla, V.K. Cross-linking in Hydrogels—A Review. *Am. J. Appl. Polym. Sci.* **2014**, *4*, 25–31.

21. Chiono, V.; Pulieri, E.; Vozzi, G.; Ciardelli, G.; Ahluwalia, A.; Giusti, P. Genipin-crosslinked chitosan/gelatin blends for biomedical Applications. *J. Mater. Sci. Mater. Med.* **2008**, *19*, 889–898. [CrossRef] [PubMed]

22. Sahiner, M.; Sagbas, S.; Bitlisli, B.O. p(AAm/TA)-based IPN hydrogel films with antimicrobial and antioxidant properties for biomedical applications. *J. Appl. Polym. Sci.* **2015**, *132*, 41876.

23. Sionkowska, A.; Kaczmarek, B.; Gnatowska, M.; Kowalonek, J. The influence of UV-irradiation on chitosan modified by the tannic acid addition. *J. Photochem. Photobiol. B Biol.* **2015**, *148*, 333–339. [CrossRef] [PubMed]

24. Cheryl, P. Tannic Acid Crosslinked Collagens and Potential for Breast Tissue Engineering. Master's Thesis, Clemson University, Clemson, South Carolina, 2006.

25. Muzzarelli, R.A.A.; Mehtedi, M.E.; Bottegoni, C.; Aquili, A.; Gigante, A. Genipin-Crosslinked Chitosan Gels and Scaffolds for Tissue Engineering and Regeneration of Cartilage and Bone. *Mar. Drugs* **2015**, *13*, 7314–7338. [CrossRef]

26. Matcham, S.; Novakovic, K. Fluorescence Imaging in Genipin Crosslinked Chitosan–Poly (vinyl pyrrolidone) Hydrogels. *Polymers* **2016**, *8*, 385. [CrossRef]

27. Manickam, B.; Sreedharan, R.; Elumalai, M. 'Genipin'—The Natural Water Soluble Cross-linking Agent and Its Importance in the Modified Drug Delivery Systems: An Overview. *Curr. Drug Deliv.* **2014**, *11*, 139–145. [CrossRef]

28. Sundararaghavan, H.G.; Monteiro, G.A.; Lapin, N.A.; Chabal, Y.J.; Miksan, J.R.; Shreiber, D.I. Genipin-induced changes in collagen gels: Correlation of mechanical properties to fluorescence. *J. Biomed. Mater. Res. Part A.* **2008**, *87*, 308–320. [CrossRef]

29. Lai, J.Y. Biocompatibility of Genipin and Glutaraldehyde Cross-Linked Chitosan Materials in the Anterior Chamber of the Eye. *Int. J. Mol. Sci.* **2012**, *13*, 10970–10985. [CrossRef]

30. Lai, J.Y.; Li, Y.T.; Wang, T.P. In vitro response of retinal pigment epithelial cells exposed to chitosan materials prepared with different cross-linkers. *Int. J. Mol. Sci.* **2010**, *11*, 5256–5272. [CrossRef]

31. Mi, F.L.; Tan, Y.C.; Liang, H.F.; Sung, H.W. In vivo biocompatibility and degradability of a novel injectable-chitosan-based implant. *Biomaterials* **2002**, *23*, 181–191. [CrossRef]

32. Grolik, M.; Szczubiałka, K.; Wowra, B.; Dobrowolski, D.; Orzechowska-Wylęgała, B.; Wylęgała, E.; Nowakowska, M. Hydrogel membranes based on genipin-cross-linked chitosan blends for corneal epithelium tissue engineering. *J. Mater. Sci. Mater. Med.* **2012**, *23*, 1991–2000. [CrossRef] [PubMed]

33. Bi, L.; Cao, Z.; Hu, Y.; Song, Y.; Yu, L.; Yang, B.; Mu, J.; Huang, Z.; Han, Y. Effects of different cross-linking conditions on the properties of genipin-cross-linked chitosan/collagen scaffolds for cartilage tissue engineering. *J. Mater. Sci. Mater. Med.* **2011**, *22*, 51–62. [CrossRef] [PubMed]

34. Shah, R.; Saha, N.; Kuceková, Z.; Humpolicek, P.; Saha, P. Properties of biomineralized (CaCO3) PVP-CMC hydrogel with reference to its cytotoxicity. *Int. J. Polym. Mater.* **2016**, *65*, 619–628. [CrossRef]

35. Fathi, M.; Entezami, A.A.; Pashaei-Asl, R. Swelling/deswelling, thermal, and rheological behavior of PVA-g-NIPAAm nanohydrogels prepared by a facile free-radical polymerization method. *J. Polym. Res.* **2013**, *20*, 125. [CrossRef]

36. Costa-Pinto, A.R.; Martins, A.M.; Castelhano-Carlos, M.J.; Correlo, V.M.; Sol, P.; Longatto-Filho, A.; Battacharya, M.; Reis, R.L.; Neves, N.M. In vitro degradation and in vivo biocompatibility of chitosan–poly (butylene succinate) fiber mesh scaffolds. *J. Bioact. Compat. Polym.* **2014**, *29*, 137–151. [CrossRef]

37. Rejmontová, P.; Capáková, Z.; Mikušová, N.; Maráková, N.; Kašpárková, V.; Lehocký, M.; Humpolíček, P. Adhesion, proliferation and migration of NIH/3T3 cells on modified polyaniline surfaces. *Int. J. Mol. Sci.* **2016**, *17*, 1439. [CrossRef]

38. Fernandes, L.L.; Resende, C.X.; Tavares, D.S.; Soares, G.A. Cytocompatibility of Chitosan and Collagen-Chitosan Scaffolds for Tissue Engineering. *Polímeros* **2011**, *21*, 1–6. [CrossRef]

39. Natarajan, V.; Krithica, N.; Madhan, B.; Sehgal, P.K. Preparation and properties of tannic acid cross-linkedcollagen scaffold and its application in wound healing. *J. Biomed. Mater. Res. B Appl. Biomater.* **2013**, *101 Pt B*, 560–567. [CrossRef]

40. Dimida, S.; Demitri, C.; Benedictis, V.M.D.; Scalera, F.; Gervaso, F.; Sannino, A. Genipin-cross-linked chitosan-based hydrogels: Reaction kinetics and structure-related characteristics. *J. Appl. Polym. Sci.* **2015**, *132*, 42256. [CrossRef]

41. Klein, M.P.; Hackenhaar, C.R.; Lorenzoni, A.S.G.; Rodrigues, R.C.; Costa, T.M.H.; Ninow, J.L.; Hertz, P.F. Chitosan crosslinked with genipin as support matrix for application in food process: Support characterization and B-d-galactosidase immobilization. *Carbohydr. Polym.* **2016**, *137*, 184–190. [CrossRef]

42. Dimida, S.; Barca, A.; Cancelli, N.; de Benedictis, V.; Raucci, M.G.; Demitri, C. Effects of Genipin Concentration on Cross-Linked Chitosan Scaffolds for Bone Tissue Engineering: Structural Characterization and Evidence of Biocompatibility Features. *Int. J. Polym. Sci.* **2017**, *2017*, 8410750. [CrossRef]

43. Mayra, A.P.C.; Horacio, G.R. Study by infrared spectroscopy and thermogravimetric analysis of Tannins and Tannic acid. *Lat. Am. J. Chem.* **2011**, *39*, 107–112.

44. Mirzaei, E.; Majidi, R.F.; Shokrgozar, M.A.; Paskiabi, F.A. Genipin cross-linked electrospun chitosan-based nanofibrous mat as tissueengineering scaffold. *Nanomed. J.* **2014**, *1*, 137–146.

45. Ma, L.; Gao, C.Y.; Mao, Z.W.; Shen, J.C.; Hu, X.Q.; Han, C.M. Thermal dehydration treatment and glutaraldehyde cross-linking to increase the biostability of collagen-chitosan porous scaffolds used as dermal equivalent. *J. Biomater. Sci. Polym. Ed.* **2003**, *14*, 861–874.

46. Horn, M.M.; Martins, V.C.A.; Plepis, A.M.G. Interaction of anionic collagen with chitosan: Effect on thermal and morphological characteristics. *Carbohydr. Poly.* **2009**, *77*, 239–243. [CrossRef]

47. Rivero, S.; García, M.A.; Pinotti, A. Physical and Chemical Treatments on Chitosan Matrix to Modify Film Properties and Kinetics of Biodegradation. *Mater. Chem. Phys.* **2013**, *1*, 51–57.

48. Beppu, M.M.; Vieira, R.S.; Aimoli, C.G.; Santana, C.C. Crosslinking of chitosan membranes using glutaraldehyde: Effect on ion permeability and water absorption. *J. Memb. Sci.* **2007**, *301*, 126–130. [CrossRef]

49. Peña, C.; Caba, K.; Eceiza, A.; Ruseckaite, R.; Mondragon, I. Enhancing water repellence and mechanical properties of gelatin films by tannin addition. *Bioresour. Technol.* **2010**, *101*, 6836–6842. [CrossRef]

50. Sadeghi, M.; Hosseinzadeh, H. Synthesis and super-swelling behavior of a novel low salt-sensitive protein-based superabsorbent hydrogel: Collagen-g-poly(AMPS)H. *Turk. J. Chem.* **2010**, *34*, 739–752.

51. Hankiewicz, J.; Swierczek, E. Lysozyme in human body fluids. *Clin. Chim. Acta* **1974**, *57*, 205–209. [CrossRef]

52. Ahmadi, F.; Oveisi, Z.; Mohammadi Samani, S.; Amoozgar, Z. Chitosan based hydrogels: Characteristics and pharmaceutical applications. *Res. Pharm. Sci.* **2015**, *10*, 1–16.

53. Gao, L.; Gan, H.; Meng, Z.; Gu, R.; Wu, Z.; Zhang, L.; Zhu, X.; Sun, W.; Li, J.; Zheng, Y.; et al. Effects of genipin cross-linking of chitosan hydrogels on cellularadhesion and viability. *Colloids Surf. B Biointerfaces* **2014**, *117*, 398–405. [CrossRef]

 *polymers*

Article

Atmospheric Pressure Plasma Polymerized Oxazoline-Based Thin Films—Antibacterial Properties and Cytocompatibility Performance

Pavel Sťahel [1], Věra Mazánková [2,3], Klára Tomečková [2], Petra Matoušková [4], Antonín Brablec [1], Lubomír Prokeš [1], Jana Jurmanová [1], Vilma Buršíková [1], Roman Přibyl [1], Marián Lehocký [5], Petr Humpolíček [5], Kadir Ozaltin [5] and David Trunec [1,*]

[1] Department of Physical Electronics, Faculty of Science, Masaryk University, Kotlářská 2, 611 37 Brno, Czech Republic; pstahel@physics.muni.cz (P.S.); abr92@sci.muni.cz (A.B.); luboprok@gmail.com (L.P.); janar@physics.muni.cz (J.J.); vilmab@physics.muni.cz (V.B.); 451677@mail.muni.cz (R.P.)
[2] Faculty of Chemistry, Institute of Physical and Applied Chemistry, Brno University of Technology, Purkyňova 118, 612 00 Brno, Czech Republic; mazankova@fch.vut.cz (V.M.); ktomeckova11@gmail.com (K.T.)
[3] Department of Mathematics and Physics, Faculty of Military Technology, University of Defence in Brno, Kounicova 65, 662 10 Brno, Czech Republic
[4] Faculty of Chemistry, Institute of Food Science and Biotechnology, Brno University of Technology, Purkyňova 118, 612 00 Brno, Czech Republic; matouskova@fch.vut.cz
[5] Centre of Polymer Systems, Tomas Bata University in Zlín, Trida Tomase Bati 5678, 760 01 Zlín, Czech Republic; lehocky@post.cz (M.L.); humpolicek@utb.cz (P.H.); kadirozaltin@hotmail.com (K.O.)
* Correspondence: trunec@physics.muni.cz; Tel.: +420-549-497-763

Received: 1 November 2019; Accepted: 6 December 2019; Published: 12 December 2019

Abstract: Polyoxazolines are a new promising class of polymers for biomedical applications. Antibiofouling polyoxazoline coatings can suppress bacterial colonization of medical devices, which can cause infections to patients. However, the creation of oxazoline-based films using conventional methods is difficult. This study presents a new way to produce plasma polymerized oxazoline-based films with antibiofouling properties and good biocompatibility. The films were created via plasma deposition from 2-methyl-2-oxazoline vapors in nitrogen atmospheric pressure dielectric barrier discharge. Diverse film properties were achieved by increasing the substrate temperature at the deposition. The physical and chemical properties of plasma polymerized polyoxazoline films were studied by SEM, EDX, FTIR, AFM, depth-sensing indentation technique, and surface energy measurement. After tuning of the deposition parameters, films with a capacity to resist bacterial biofilm formation were achieved. Deposited films also promote cell viability.

Keywords: antibiofouling; plasma polymer; oxazoline

---

1. Introduction

Polyoxazolines (POx) are a promising and important class of polymers that have attracted substantial attention recently due to their antibiofouling properties [1,2] and good biocompatibility [3]. Usually, POx are prepared by living-cationic ring-opening polymerization, which is a lengthy wet process conducted in organic solvents. POx thin films are created in a subsequent step, which needs to be tailored for any particular type of substrates [4,5]. Other possible polymerization techniques are photocoupling [6] and grafting [7], both of which require the premodification of substrates. So, the formation of polyoxazoline coatings using conventional methods is a slow and complex multistep procedure, which can be conducted only on a limited range of substrates. The difficulties of these conventional methods can be overcome via plasma polymerization. Plasma polymerization is known to be a suitable method for the deposition of many biomaterial coatings [8,9]. Plasma polymerization is in the class of plasma-enhanced chemical vapor deposition (PECVD) methods, which are successfully used for example for thin film deposition [10], surface modification [11,12], or growing of nanomaterials [13–15]. Plasma deposition of 2-methyl-2-oxazoline and 2-ethyl-2-oxazoline has already been performed in low-pressure radio frequency (RF) discharge [16–18]. However, the necessity to use expensive vacuum pumping systems is the disadvantage of low-pressure plasma deposition techniques. Recently, plasma deposition in atmospheric pressure discharges has become a new promising technology due to its economic and ecological advantages. A suitable discharge type for plasma deposition at atmospheric pressure is a homogeneous dielectric barrier discharge (DBD), which can be obtained in nitrogen. This discharge type is called atmospheric pressure Townsend-like discharge (APTD) [19]. APTD in the mixture N_2–SiH_4–N_2O was used for SiO_2 thin film deposition [20], APTD in the mixture of N_2 with hexamethyldisiloxane (HMDSO) was used for deposition of organosilicon polymer films [21]. The properties of films deposited in APTD are different from properties of films deposited in low-pressure discharges due to different plasma and discharge parameters, e.g., ion energies in APTD are lower than those in low pressure discharges. The film properties can be changed by increasing the substrate temperature at deposition [22]. Recently, POx coatings were deposited using discharges at atmospheric pressure. Atmospheric pressure helium plasma jet was used for plasma polymerization of 2-methyl-2-oxazoline on heated silicon substrates [23]. The film stability in buffer solution was substantially improved when the films were deposited at substrate temperatures above 50 °C in these experiments. Near atmospheric pressure (0.5 bar) plasma polymerization of 2-alkyl-2-oxazolines in argon DBD was used for the study of the influence of the aliphatic side-chain length on the plasma polymerization process conditions as well as on the properties of the deposited coatings [24].

In the current study, POx thin films were deposited in nitrogen APTD using 2-methyl-2-oxazoline as monomer. The substrate temperature was changed from 20 °C to 150 °C at the film deposition. This temperature change leads to different film properties. The films with the highest biocompatibility and the best antibiofoulding properties were obtained at the deposition with a substrate temperature of 150 °C.

2. Materials and Methods

2.1. Materials

Glass plates (soda-lime glass, 150 × 100 mm, thickness 1.1 mm) were used as substrates for deposition. 2-Methyl-2-oxazoline (98%, Sigma-Aldrich, Munich, Germany) was used as a monomer for plasma deposition. Preliminary antibacterial tests were done with *Staphyloccocus epidermidis* (CCM 4418), supplied by the Czech Collection of Microorganisms in Brno. Bacterial culture was grown into commercial BHI medium (Brain Heart Infusion Broth, HiMedia, Mumbai, India). Basic Red 2 (Safranin O, Sigma-Aldrich, Munich, Germany) stain was used for biofilms visualization. Other antibacterial tests were done with *Staphylococcus aureus* (CCM 4516) and *Escherichia coli* (CCM 4517), both supplied by the Czech Collection of Microorganisms in Brno.

2.2. Plasma Deposition

Plasma polymerization was performed in a custom-built reactor (metallic chamber with dimensions 500 mm × 500 mm × 500 mm) with dielectric barrier discharge [22]. The discharge was generated between two planar metal electrodes. The bottom electrode with dimensions 150 mm × 55 mm could be heated using a heating spiral, and the electrode temperature was measured with a thermocouple. The upper electrode with dimensions 55 mm × 40 mm was covered with glass,

1.5 mm in thickness. A slit 2 mm wide in the center of the upper electrode was used for the supply of working gas with the monomer to the discharge. The electrode system is shown in Figure 1.

Figure 1. The image of electrode system. 1—gas inlet; 2—upper electrode; 3—glass; 4—lower heated electrode (without substrate).

The scheme of the electrode system can also be found in a previous study [25], where gas flow in the discharge gap was also modeled. The glass substrates were cleaned in a mixture of cyclohexane and isopropyl alcohol (1:1) and dried in airflow. Clean substrates were put into the reactor on the bottom electrode, which was then entirely covered by the substrate. The discharge gap between the substrate and the upper electrode was set to be 1.0 mm. Before starting the depositions, the discharge chamber was pumped down to a pressure of 100 Pa and then filled with nitrogen to a pressure of 101 kPa. Atmospheric pressure during the deposition was maintained by slight pumping.

Nitrogen flow with flow rates from 50 sccm to 400 sccm bubbled through the liquid 2-methyl-2-oxazoline monomer in a glass bottle container. This flow was then mixed with the main nitrogen flow with the flow rate of 500 sccm. The temperature of the monomer was kept constant and set to 20 °C. The monomer flow rate was determined by weighing the monomer before and after the deposition, the flow rate of the monomer was 20 mg s^{-1} for the nitrogen flow of 100 sccm through the liquid monomer. A high voltage with a frequency 6 kHz was used for the generation of a dielectric barrier discharge in the discharge gap. The discharge was burning in homogeneous APTD mode [21], which is suitable for the deposition of homogeneous thin films. The discharge mode was checked by current–voltage measurements using an oscilloscope. The input power to a high voltage source was set to 55 W and kept constant in all experiments presented in this paper. The temperature of the bottom electrode was increased to a given value before the deposition. The upper electrode was periodically moving with a speed of 0.6 cm s^{-1} above the substrate during the deposition in order to ensure an even greater homogeneity of the deposited film. The deposition time was 23 min.

2.3. Surface Characterization

Deposited films were imaged with scanning electron microscope (SEM) MIRA3 (TESCAN, Brno, Czech Republic) with a Schottky field emission electron gun equipped with secondary electron and back-scattered electron detectors as well as a characteristic X-ray detector (EDX) analyzer (Oxford Instruments, High Wycombe, UK). The IR spectra of deposited films were measured by FTIR spectrometer Alpha (Bruker, Billerica, MA, USA) using a single reflection ATR module Platinum. The total surface free energy of the films was determined from measurements of contact angles between testing liquids and the film surfaces using a sessile-drop technique. Acid-base theory was used for the calculation of total surface free energy. Atomic Force Microscope (AFM) Ntegra Prima (NT-MDT, Apeldoorn, The Netherlands) was used to study the surface topography of the POx films. The measurements were performed in semicontact mode on 10×10 μm^2 and 5×5 μm^2 areas of each coating with a scanning rate of 0.5 Hz. The 3D roughness parameters of the thin films were evaluated according to the ASME B46 standard using the NovaPx software (NT-MDT). The film thickness was measured using a Dektak XT (Bruker, Tucson, AZ, USA) mechanical profilometer.

2.4. Characterization of Mechanical Properties

Hardness (H) and effective elastic modulus (E_{eff}) of POx films were assessed by a nanoindentation tester Hysitron TI 950 TriboIndenter (Bruker, Minneapolis, MN, USA) with a load resolution of 1 nN. The effective elastic modulus could be expressed according to $E_{eff} = E/(1 - v^2)$, where v and E are the Poisson's ratio and Young's modulus of the material, respectively. The standard Oliver and Pharr method [26] was used to calculate the above listed parameters. For a reliable characterization of mechanical properties of POx films several indentation modes were used including basic quasistatic indentation tests with trapezoid load function (5 s loading, 2 s creep, 5 s unloading), quasistatic nanoindentation tests with 33 partial unloading segments, and nanodynamic indentation (nanoDMA) in constant strain-rate measuring mode. The nanoDMA indentation tests were carried out by superimposing a sinusoidal load with a small amplitude (30 μN to 0.2 mN) and a frequency of 220 Hz on the quasistatic loading curve. From the sample response to this dynamic loading force, the dynamic displacement amplitude and the phase shift ϕ between this dynamic displacement and the input sinusiodal loading force were acquired. In order to decouple the indenter and the sample contribution from the the total spring stiffness, k, and the total damping coefficient, C, the Kelvin–Voigt mechanical equivalents model was used [27]. In this model, the indenter and the sample are described by a viscous damper and a purely elastic spring, which are connected in parallel. On the basis of the above described method, the storage modulus E', loss modulus E'', and loss factor $\tan \phi$ were calculated. Equation (1) shows the relationship between E', E'' and complex modulus (denoted as E^*):

$$E' = \frac{k_s \sqrt{\pi}}{2\sqrt{A_c}}; \quad E'' = \frac{\omega C_s \sqrt{\pi}}{2\sqrt{A_c}}; \quad E^* = E' + iE'', \tag{1}$$

where i is the imaginary unit, A_c is the projected contact area between the tip and the sample, ω is the angular frequency of the oscillating load, k_s is the sample spring stiffness, and C_s is the sample damping coefficient.

The maximum indentation load was varied between 100 μN and 11 mN. The mechanical parameters were determined at indentation depths <1/10 of film thickness to avoid any substrate effect.

2.5. Antibacterial Tests

Bacterial culture *Staphyloccocus epidermidis* was grown into brain heart infusion (BHI) medium, and the temperature for microorganism cultivation was 37 °C. After 24 h incubation, *S. epidermidis* was diluted with new sterile medium to 1×10^8 CFU per mL based on turbidity (NanoPhotometer™ P300, Implen, Munich, Germany). Then, *S. epidermidis* suspension (500 μL) was added on each sample. The bacteria on the surface of the samples were incubated for 24 h to allow the formation of biofilms.

After incubation, all samples were washed twice with Milli-Q water to remove any loosely bound biofilm. For biofilms visualization, 200 µL of Basic Red 2 stain was used. The excess of Basic Red 2 stain was then washed off and the results were evaluated using optical microscopy (Optical microscope Intraco Micro LM 666 PC/∞ with Dino-Capture 2.0 software, Tachlovice, Czech Republic). Samples were imaged at least fifteen times each at random points, and the surface area covered by bacteria was quantified using images.

Other antibacterial tests were performed according to the ISO 22196 procedure with modifications. Before antibacterial testing, samples were sterilized by UV–radiation (wavelength of 258 nm) for 30 min. For the determination of antibacterial performance, gram-positive *Staphylococcus aureus* and gram-negative *Escherichia coli* were used. Bacterial suspensions (*E. coli* 3.4×10^6 CFU mL^{-1}; *S. aureus* 8.9×10^5 CFU mL^{-1}) were prepared in 1/500 Nutrient broth (HiMedia laboratories, Mumbai, India). The bacterial suspension was dispensed on the sample surface (dimensions 25 mm $\times$ 25 mm) in the volume of 100 µL and the sample was covered with the polypropylene foil (dimensions 20 mm $\times$ 20 mm). Samples with foils were cultivated at 35 °C and 100% relative humidity for 24 h. After the incubation time, polypropylene foil was removed and each sample was completely washed by SCDLP (Soybean, Casein, Digest, Lecithin, Polysorbate) broth (HiMedia laboratories, India), which was subsequently collected. The viable bacteria count was determined by the pour plate culture method (PCA, HiMedia laboratories, India).

2.6. Cytocompatibility Test

The mouse embryonic fibroblast continuous cell line (NIH/3T3, ATCC® CRL-1658™, Teddington, UK) was used for cytocompatibility test, according to the EN ISO 10993-5 standard, with modification. As a culture medium, the ATCC-formulated Dulbecco's Modified Eagle's Medium (BioSera, Nuaille, France), containing 10% calf serum (BioSera, France) and Penicillin/Streptomycin at 100 U mL^{-1} (PAA Laboratories GmbH, Pasching, Austria) was used. The tested samples were prepared with a dimension of 10 mm $\times$ 10 mm and sterilized by UV–radiation (wavelength of 258 nm) for 30 min and placed into the 24 well-plate. The cells were seeded onto the samples in the concentration of 1×10^4 for an hour for adhesion of the cells. After the precultivation, a sufficient amount of the medium was added and incubated for 72 h at 37 °C. The changes in cell morphology were observed with an inverted fluorescent microscope (Olympus, IX 81). In order to assess the cytotoxic effect, an MTT assay (Duchefa, Biochemie, Haarlem, The Netherlands) was performed. The absorbance was measured by an Infinite M200 Pro NanoQuant absorbance reader (Tecan, Männedorf, Switzerland). All tests were performed three times.

3. Results

The POx films were deposited at substrate temperatures 60 °C, 90 °C, 120 °C, 150 °C, and at nitrogen flow of 100 sccm through the monomer. It was found that the POx films deposited at substrate temperatures up to 90 °C can be washed by water from the substrate. Similar results were also observed in a study by other authors [18], where the POx films deposited at low RF power supplied to RF discharge were also soluble in water. So, films deposited at the substrate temperature of 150 °C will be characterized in the following subsections. The changes of nitrogen flow rate through the monomer influenced the film thickness only, so the depositions were performed with a nitrogen flow of 100 sccm.

3.1. Surface Characterization

The image from SEM of films deposited at 150 °C is shown in Figure 2. It can be seen from this image (and other images) that deposited films are smooth and without any pinholes. The elemental composition of POx films deposited at different substrate temperatures determined by EDX is shown in Table 1.

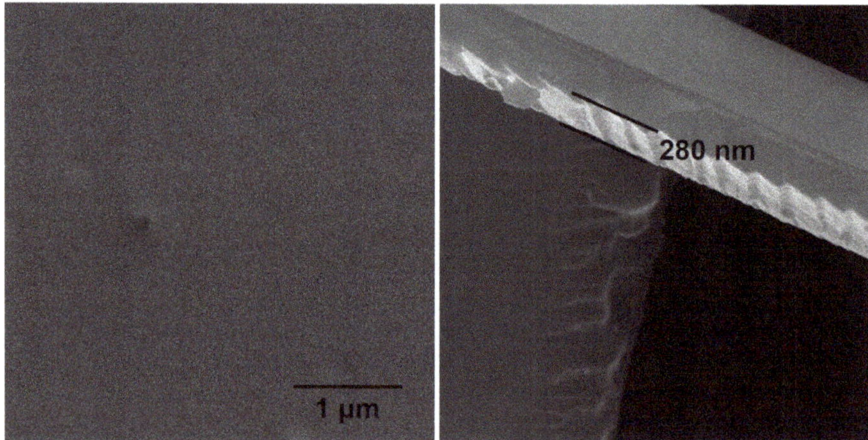

Figure 2. SEM images of polyoxazoline (POx) film deposited at 150 °C. Left images—surface of deposited film, magnification 10k×; right image—deposited film removed from the substrate by scratch, magnification 50k×.

Table 1. The elemental composition of films deposited at different substrate temperatures. The elemental composition is given in atomic %.

Element	60 °C	90 °C	120 °C	150 °C
C	43	43	47	48
N	40	41	42	39
O	17	16	11	13

The nitrogen content in the films was increased by up to 40% in comparison with nitrogen content in 2-methyl-2-oxazoline. The oxygen content corresponds to oxygen content in 2-methyl-2-oxazoline at substrate temperatures of 60 °C and 90 °C and it decreases to lower values at substrate temperatures of 120 °C and 150 °C. Additionally, carbon content increases at higher substrate temperatures.

The contact angles between the test liquids and POx films were measured in order to determine the total surface free energy using the sessile drop technique. Three test liquids were used—distilled water, glycerol, and diiodomethane (CH_2I_2). The acid–base theory with multiple regression [28] was used to calculate the total surface free energy and its components—the Lifshitz–van der Walls (LW) interaction component and the acid–base (AB) interaction component. The surface free energy and its abovementioned components of POx films are given in Table 2.

Table 2. The contact angles for different liquids and surface free energy and its components of POx films deposited at different substrate temperatures.

Sample	Contact Angle (°)			Surface Free Energy (mJ/m^2)		
	CH_2I_2	Glycerol	Water	Total	LW	AB
substrate	59.8 ± 1.2	35.5 ± 2.0	33.4 ± 2.3	52.6 ± 1.0	28.7 ± 0.7	23.9 ± 2.0
60 °C	40.5 ± 1.0	28.9 ± 0.8	10.0 ± 2.0	56.6 ± 0.7	39.4 ± 0.6	17.3 ± 1.2
90 °C	40.8 ± 1.4	38.3 ± 0.5	16.1 ± 0.9	50.3 ± 0.9	39.2 ± 0.6	11.2 ± 1.4
120 °C	60.8 ± 1.7	50.5 ± 1.8	40.0 ± 2.8	42.4 ± 1.9	28.1 ± 1.9	14.3 ± 2.8
150 °C	60.6 ± 4.0	46.6 ± 1.7	21.9 ± 2.7	43.3 ± 1.8	27.8 ± 2.8	15.5 ± 4.0

All POx films were hydrophilic, their surface energy is in the range from 42.4 to 56.6 mJ m^{-2}. The thickness, hardness, and effective elastic modulus of POx films are given in Table 3.

Table 3. The film thickness, hardness, and effective elastic modulus of POx films deposited at different substrate temperatures.

Sample	Thickness (μm)	Hardness (GPa)	E_{eff} (GPa)
60 °C	1.5 ± 0.2	0.70 ± 0.10	15 ± 1
90 °C	1.7 ± 0.1	0.55 ± 0.05	11 ± 1
120 °C	1.1 ± 0.1	0.55 ± 0.05	11 ± 1
150 °C	0.6 ± 0.1	0.60 ± 0.05	15 ± 1

The film thickness decreases with increasing substrate temperature at the deposition. This dependence is in agreement with the findings in previous experiments [22,23]. From the point of view of mechanical properties, the films showed polymer-like viscoelastic character. However, compared to common polymer materials (e.g., polycarbonate, polyethylene, polypropylene), they exhibited significantly higher hardness and effective elastic modulus (for example the hardness of polycarbonate is 0.18 GPa and its elastic modulus is around 3 GPa). The time dependent mechanical response of polymer materials may be characterized using the storage modulus E', loss modulus E'', and loss factor $\tan\phi$. The storage modulus E' is the measure of the energy stored and recovered during the loading period, and the loss modulus E'' is the measure of the energy dissipated in the studied material during the loading period. The loss factor $\tan\phi$ is the measure of viscoelastic behavior of materials [27]. The studied POx films exhibited relatively low $\tan\phi$ values indicating a predominantly elastic behavior (low damping) of the film material, see Table 4.

The AFM images of POx films are shown in Figure 3.

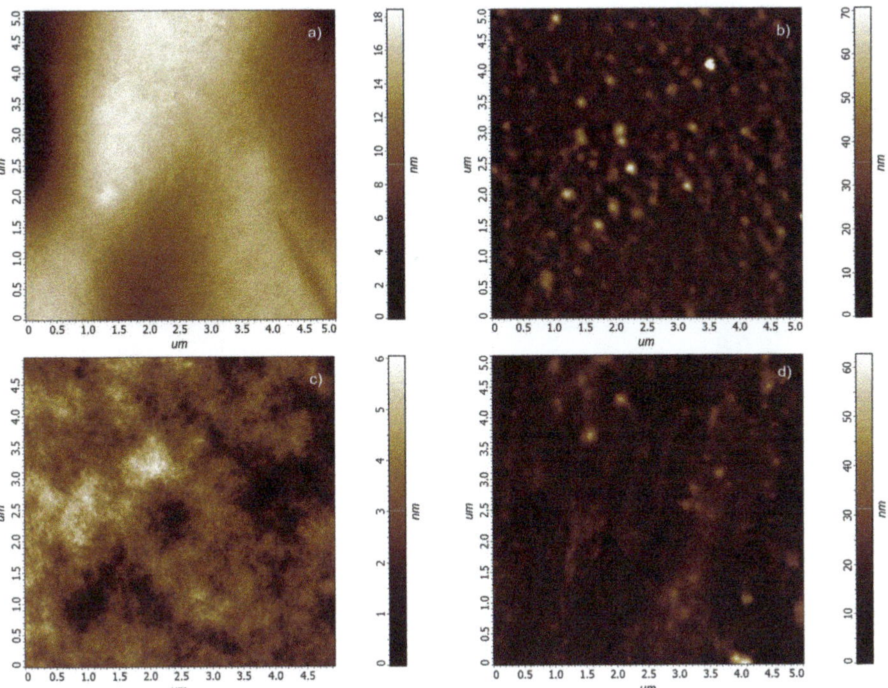

Figure 3. Atomic Force Microscope (AFM) images of POx films deposited at different substrate temperatures. (**a**) 60 °C; (**b**) 90 °C; (**c**) 120 °C; (**d**) 150 °C.

Table 4. The results of the nanodynamic indentation (nanoDMA) measurements: storage modulus E', loss modulus E'', and loss factor $\tan \phi$ of POx films deposited at different substrate temperatures.

Sample	E' (GPa)	E'' (GPa)	$\tan \phi$
60 °C	16 ± 2	0.50 ± 0.05	0.031 ± 0.005
90 °C	11 ± 1	0.30 ± 0.04	0.027 ± 0.006
120 °C	12 ± 1	0.37 ± 0.05	0.031 ± 0.005
150 °C	14 ± 1	0.48 ± 0.05	0.034 ± 0.005

The film roughness values determined from AFM measurements are given in Table 5. Instead of the commonly used root mean square (RMS) roughness and average roughness values, the area peak-to-valley height S_t and area peak density Sd_s values [29] were added in order to illustrate the change of the film surface structure with the preparation temperature.

Table 5. The film roughness of POx films deposited at different substrate temperatures. S_q—root mean square (RMS) roughness; S_a—average roughness; S_t—RMS roughness, area peak-to-valley height; Sd_s—area peak density.

Sample	S_q (nm)	S_a (nm)	S_t (nm)	Sd_s (μm^{-2})
60 °C	3.3	2.7	86.0	170
90 °C	6.2	4.2	70.6	140
120 °C	7.3	6.8	48.9	88
150 °C	4.8	3.7	62.6	278

3.2. FTIR Analysis

The FTIR spectra of deposited POx thin films are shown in Figure 4.

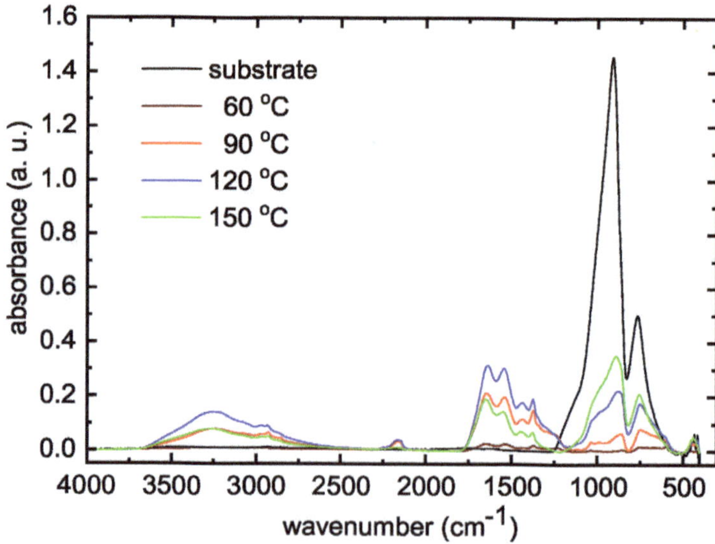

Figure 4. FTIR spectra of thin films deposited at different substrate temperatures.

Very low absorbance for the film deposited at substrate temperature of 60 °C was caused by small film thickness. Broad absorption band in the range 3000–3600 cm^{-1} consists of several peaks belonging to OH, NH, and NH$_2$ groups. The bands at 2950 cm^{-1}, 1450 cm^{-1}, and 1370 cm^{-1} are characteristic for vibrations of CH$_3$ and CH$_2$ groups. The band at 2170 cm^{-1} can be attributed to alkyne

C≡C and/or isocyanate O=C=N and nitrile C≡N. Such chemical bonds are not present at traditional polymerization of oxazolines and they can be attributed to fragmentation and recombination of the oxazoline monomer during plasma polymerization. The band between 1790 cm^{-1} and 1590 cm^{-1} is characteristic for stretching vibration C=N bond constituting the oxazoline ring. Its presence in the IR spectrum indicates the presence of oxazoline rings in deposited films. The band around 1550 cm^{-1} belongs to N-H bonds. Finally, the bands below 1000 cm^{-1} belong to Si-O-Si or Si-O bonds from substrate glass.

3.3. Antibacterial Properties

Firstly, the antibacterial test was done using *S. epidermidis*, which was also used in previous studies [16,18]. Experiments were repeated twice using two identical series. It can be concluded that the results showed significant differences between the uncoated sample and samples with deposited films, see Figure 5. Bacteria on POx films were not able to form a biofilm. It can be observed in the microscopy image where only individual or small colonies are observed. In comparison, the bacteria growing on the untreated surface formed a biofilm. The possible explanations of antibiofouling properties can be found in a recent review [30].

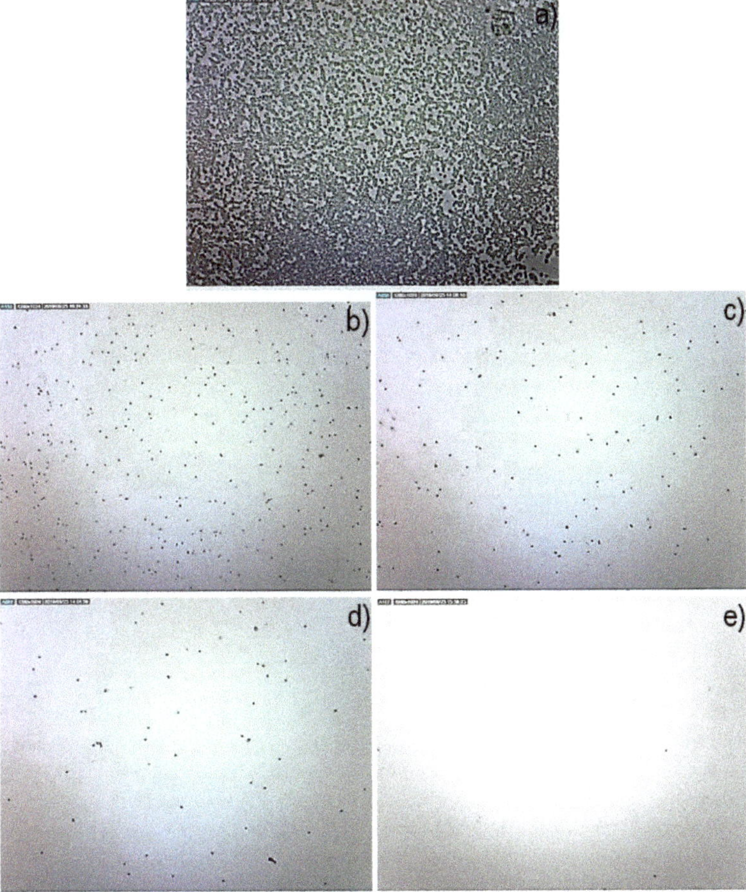

Figure 5. Images of films deposited at different substrate temperatures with bacteria *S. epidermidis*. (**a**) blank substrate; (**b**) 60 °C; (**c**) 90 °C; (**d**) 120 °C; (**e**) 150 °C.

Before adding the micro-organism and incubation, some samples were washed once or twice with sterile Milli-Q water in order to test the stability of prepared POx films. The results are shown in Figure 6.

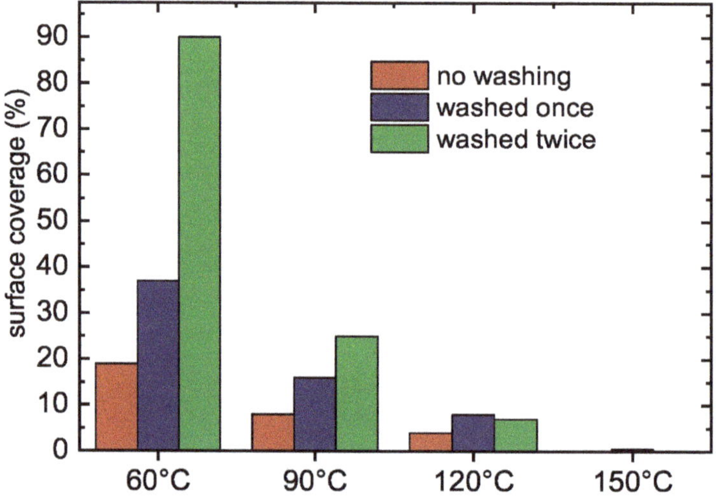

Figure 6. Bacteria *S. epidermidis* surface coverage area (percent) formed on POx films deposted at different substrate temperatures.

The films deposited at substrate temperatures of 60 °C and 90 °C are soluble in water and their antibiofouling properties decrease after their washing by water. On the other hand, the films deposited at substrate temperatures of 120 °C and 150 °C are not soluble in water and the washing does not decrease their antibiofouling properties significantly.

Secondly, the other antibacterial tests were done using *S. aureus* and *E. coli* according to the ISO 22196 procedure. Antibacterial activity against *S. aureus* and *E. coli* strains after 72 h of incubation time for the samples is listed in Table 6.

Table 6. Antibacterial activity results of studied POx films.

Sample	*S. aureus* (CFU/cm^2)	*E. coli* (CFU/cm^2)
substrate	1.3×10^6	2.0×10^5
60 °C	<1	4.4
90 °C	<1	1.1
120 °C	<1	4.4
150 °C	1.6	5.4

The reference substrate glass was open to both gram-positive and -negative bacterial contamination and did not perform any antibacterial effect, as expected. Nevertheless, counted viable gram-positive *S. aureus* level was found almost five times higher than the gram-negative *E. coli* strain. The oxazoline-based thin film deposited samples were highly active against both *S. aureus* and *E. coli* strains. The effect of the oxazoline-based thin film against *S. aureus* was slightly higher compared to *E. coli*, but the differences were negligible. Among them, as it can be seen in Table 6, the sample deposited at 150 °C performed slightly lower antibacterial activity against both strains compared to other oxazoline-based thin film deposited samples, probably due to its higher bonding performance. However, it is also negligible, since the difference is extremely low. The antibacterial activity against

S. aureus and *E. coli* depends on the cell-wall compositions of the bacterial strains and physicochemical characteristics with an efficiency and the releasing performance of the deposited oxazoline-based thin films. Therefore, the level of successfully deposited oxazoline-based thin films onto the glass substrate plays a critical role. Moreover, it strictly depends on the surface characteristics, such as wettability, roughness, charge density, and functionality.

3.4. Cytocompatibility Results

An in vitro cytocompatibility test was performed by mouse embryonic fibroblast cells (NIH/3T3) on the samples for 72 h and results are shown in Figure 7.

As can be seen from Figure 7, all types of modifications promote cell viability compared to blank glass substrate which was taken as reference. Films deposited at 60 °C and 90 °C exhibit almost identical cell viability as the reference. However, in the case of the film deposited at 120 °C, the viability increased significantly when reaching the value of almost 180%. The highest fibroblasts cell viability was observed for the film deposited at 150 °C when the value was more than 270% which signifies an excellent compatibility to the used cell substrate. The abovementioned results suggest that oxazoline-based thin films are affecting cell behavior, especially cell attachment. Moreover, with rising deposition temperature, the cell viability value is increasing.

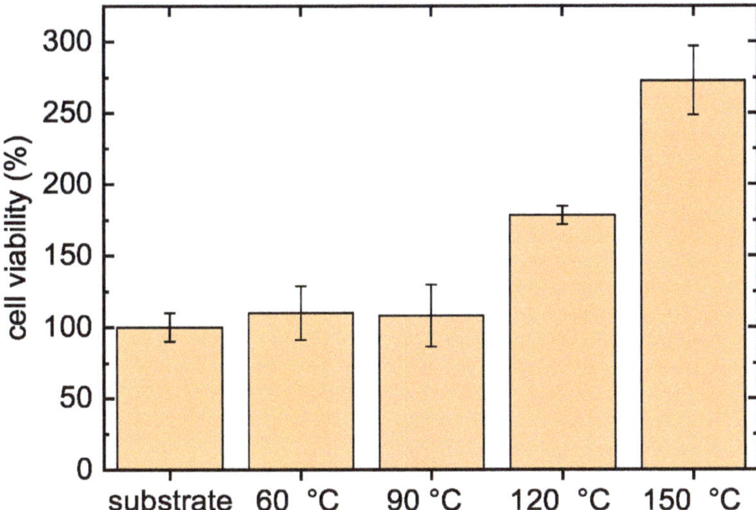

Figure 7. In vitro cytocompatibility results of tested POx films deposited at different substrate temperatures.

4. Discussion

The oxazoline-based thin films were successfully deposited on glass substrates in atmospheric pressure dielectric barrier discharge. Nitrogen was used as the working gas for the discharge, 2-methyl-2-oxazoline vapors were admixed to the nitrogen flow and used as the monomer. This gas composition made it possible to obtain a homogeneous discharge, which led to the deposition of homogeneous thin films. To improve the film properties, it was necessary to increase the substrate temperature during the deposition. The deposited films were smooth and without any pinholes. The films deposited at substrate temperatures of 60 °C and 90 °C were soluble in water and they also did not exhibit increased cell viability compared to the blank glass substrate. It was found from FTIR spectra that some oxazoline rings are still present in the plasma polymer in contrast to classical oxazoline polymerization. The retention of the oxazoline ring is assumed to be highly beneficial for selected biomedical applications. On the other hand, some fragmentation and recombination of the oxazoline monomer was also observed in FTIR spectra. These findings are in agreement with the results of oxazoline plasma polymerization in low-pressure RF discharge [16]. However, higher N:C elemental ratio (0.93–0.81) was observed in films deposited in nitrogen APTD in this study, whereas lower N:C ratio (0.31–0.24) was observed at films deposited in RF discharge [16]. Nitrogen-rich surfaces are known for their excellent biocompatibility [31]. The best cell viability results were found at films deposited at substrate temperatures of 120 °C and 150 °C. Lower cell viability on films deposited at substrate temperatures of 60 °C and 90 °C can be explained by their dissolvability in aqueous solvents. All deposited films exhibited excellent antibacterial properties against all bacterial strains used for antibacterial tests. Deposited films could be used as coatings with antibacterial and antibiofouling properties for biomedical applications.

Author Contributions: Conceptualization, D.T. and M.L.; methodology, P.M. and P.H.; validation, P.S. and M.L.; formal analysis, D.T. and L.P.; investigation, P.S., K.T., P.M., A.B., J.J., L.P., V.B., R.P., P.H., and K.O.; resources, P.M.; writing—original draft preparation, D.T., M.L., and P.M.; writing—review and editing, D.T. and M.L.; visualization, V.M.; supervision, D.T.

Funding: This research was funded by the Ministry of Education, Youth and Sports of the Czech Republic, project LO1411 (NPU I), and by Czech Science Foundation under project GACR 19-15240S. V.M. was supported by institutional support for the research organization development awarded by the Ministry of Defence of the Czech Republic.

Conflicts of Interest: The authors declare no conflict of interest.

References

1. Woodle, M.C.; Engbers, C.M.; Zalipsky, S. New Amphipatic Polymer Lipid Conjugates Forming Long-Circulating Reticuloendothelial System-Evading Liposomes. *Bioconj. Chem.* **1994**, *5*, 493–496. [CrossRef] [PubMed]

2. Zalipsky, S.; Hansen, C.B.; Oaks, J.M.; Allen, T.M. Evaluation of Blood Clearance Rates and Biodistribution of Poly(2-oxazoline)-grafted Liposomes. *J. Pharm. Sci.* **1996**, *85*, 133–137. [CrossRef] [PubMed]

3. Goddard, P.; Hutchinson, L.E.; Brown, J.; Brookman, L.J. Soluble Polymeric Carriers for Drug Delivery. Part 2. Preparation and in Vivo Behaviour of N-acylethylenimine Copolymers. *J. Control. Release* **1989**, *10*, 5–16. [CrossRef]

4. Jordan, R.; Ulman, A. Surface Initiated Living Cationic Polymerization of 2-Oxazolines. *J. Am. Chem. Soc.* **1998**, *120*, 243–247. [CrossRef]

5. Bouten, P.J.M.; Hertsen, D.; Vergaelen, M.; Monnery, B.D.; Boerman, M.A.; Goossens, H.; Catak, S.; van Hest, J.C.M.; Van Speybroeck, V.; Hoogenboom, R. Accelerated Living Cationic Ringopening Polymerization of a Methyl Ester Functionalized 2-Oxazoline Monomer. *Polym. Chem.* **2015**, *6*, 514–518. [CrossRef]

6. Wang, H.; Li, L.; Tong, Q.; Yan, M. Evaluation of Photochemically Immobilized Poly(2-ethyl-2-oxazoline) Thin Films as Protein-Resistant Surfaces. *ACS Appl. Mater. Interfaces* **2011**, *3*, 3463–3471. [CrossRef]

7. Pidhatika, B.; Rodenstein, M.; Chen, Y.; Rakhmatullina, E.; Mühlebach, A.; Acikgöz, C.; Textor, M.; Konradi, R. Comparative Stability Studies of Poly(2-methyl-2-oxazoline) and Poly(ethyleneglycol) Brush Coatings. *Biointerphases* **2012**, *7*. [CrossRef]

8. Vasilev, K. Nanoengineered Plasma Polymer Films for Biomaterial Applications. *Plasma Chem. Plasma Process.* **2014**, *34*, 545–558. [CrossRef]

9. Siow, K.S.; Britcher, L.; Kumar S.; Griesser, H.J. Plasma Methods for the Generation of Chemically Reactive Surfaces for Biomolecule Immobilization and Cell Colonization—A Review. *Plasma Processes Polym.* **2006**, *3*, 392–418. [CrossRef]

10. Lieberman, M.A.; Lichtenberg, A.J. *Principles of Plasma Discharges and Materials Processing*, 2nd ed.; John Wiley & Sons: Hoboken, NJ, USA, 2005.

11. Duan, S.; Liu, X.; Wang, Y.; Meng, Y.; Alsaedi, A.; Hayat, T.; Li, J. Plasma Surface Modification of Materials and Their Entrapment of Water Contaminant: A Review. *Plasma Process Polym.* **2017**, *14*, e1600218. [CrossRef]

12. Yue, M.; Zhou, B.; Jiao, K.; Qian, X.; Xu, Z.; Teng, K.; Zhao, L.; Wang, J.; Jiao, Y. Switchable Hydrophobic/Hydrophilic Surface of Electrospun Poly(l-lactide) Membranes Obtained by CF4 Microwave Plasma Treatment. *Appl. Surf. Sci.* **2015**, *327*, 93–99. [CrossRef]

13. Wu, H.; Zhou, Z.; Chen, L.; Li, W.; Han, Q.; Li, C.; Xu, Z.; Qian, X. PECVD-Induced Growing of Diverse Nanomaterials on Carbon Nanofibers under Various Conditions. *Mater. Lett.* **2018**, *216*, 291–294. [CrossRef]

14. Zhang, C.; Liu, L.; Xu, Z.; Lv, H.; Wu, N.; Zhou, B.; Mai, W.; Zhao, L.; Tian, X.; Guo, X. Improvement for Interface Adhesion of Epoxy/Carbon Fibers Endowed With Carbon Nanotubes via Microwave Plasma-Enhanced Chemical Vapor Deposition. *Polym. Compos.* **2018**, *39*, E1262–E1268. [CrossRef]

15. Elias, M.; Kloc, P.; Jasek, O.; Mazankova, V.; Trunec, D.; Hrdy, R.; Zajickova, L.; Atmospheric Pressure Barrier Discharge at High Temperature: Diagnostics and Carbon Nanotubes Deposition. *J. Appl. Phys.* **2015**, *117*, 103301. [CrossRef]

16. Ramiasa, M.; Cavallaro, A.; Mierczynska, A.; Christo, S.; Gleadle, J.; Hayball, J.D.; Vasilev, K. Plasma Polymerised PolyOxazoline Thin Films for Biomedical Applications. *Chem. Commun.* **2015**, *51*, 4279–4282. [CrossRef] [PubMed]

17. Macgregor-Ramiasa, M.N.; Cavallaro, A.A.; Vasilev, K. Properties and Reactivity of Polyoxazoline Plasma Polymer Films. *J. Mater. Chem. B* **2015**, *3*, 6327–6337. [CrossRef]

18. Cavallaro, A.A.; Macgregor-Ramiasa, M.N.; Vasilev, K. Antibiofouling Properties of Plasma-Deposited Oxazoline-Based Thin Films. *ACS Appl. Mater. Interfaces* **2016**, *8*, 6354–6362. [CrossRef]

19. Gherardi, N.; Gouda, G.; Gat, E.; Ricard, A.; Massines, F. Transition from Glow Silent Discharge to Micro-discharges in Nitrogen Gas. *Plasma Sources Sci. Technol.* **2000**, *9*, 340–346. [CrossRef]

20. Gherardi, N.; Martin, S.; Massines, F. A New Approach to SiO_2 Deposit using a N_2–SiH_4–N_2O Glow Dielectric Barrier-Controlled Discharge at Atmospheric Pressure. *J. Phys. D: Appl. Phys.* **2000**, *33*, L104–L108. [CrossRef]

21. Trunec, D.; Navratil, Z.; Stahel, P.; Zajickova, L.; Bursikova, V.; Cech, J. Deposition of Thin Organosilicon Polymer Films in Atmospheric Pressure Glow Discharge. *J. Phys. D Appl. Phys.* **2004**, *37*, 2112–2120. [CrossRef]

22. Trunec, D.; Zajickova, L.; Bursikova, V.; Studnicka, F.; Stahel, P.; Prysiazhnyi, V.; Perina, V.; Houdkova, J.; Navratil, Z.; Franta, D. Deposition of Hard Thin Films from HMDSO in Atmospheric Pressure Dielectric Barrier Discharge. *J. Phys. D Appl. Phys.* **2010**, *43*, 225403. [CrossRef]

23. Al-Bataineh, S.A.; Cavallaro, A.A.; Michelmore, A.; Macgregor, M.N.; Whittle, J.D.; Vasilev, K., Deposition of 2-oxazoline- based Plasma Polymer Coatings using Atmospheric Pressure Helium Plasma Jet. *Plasma Process Polym.* **2019**, *16*, e1900104. [CrossRef]

24. Van Guyse, J.F.R.; Cools, P.; Egghe, T.; Asadian, M.; Vergaelen, M.; Rigole, P.; Yan, W.; Benetti, E.M.; Jerca, V.; Declercq, H.; et al. Influence of the Aliphatic Side Chain on the Near Atmospheric Pressure Plasma Polymerization of 2-Alkyl-2-oxazolines for Biomedical Applications. *ACS Appl. Mater. Interfaces* **2019**, *11*, 31356–31366. [CrossRef] [PubMed]

25. Obrusnik, A.; Jelinek, P.; Zajickova, L. Modelling of the Gas Flow and Plasma Co-polymerization of Two Monomers in an Atmospheric-Pressure Dielectric Barrier Discharge. *Surf. Coat. Technol.* **2017**, *314*, 139–147. [CrossRef]

26. Oliver, W.C.; Pharr, G.M. An Improved Technique for Determining Hardness and Elastic Modulus using Load and Displacement Sensing Indentation Experiments. *J. Mater. Res.* **1992**, *7*, 1564–1583. [CrossRef]

27. Zlotnikov, I.; Zolotoyabko, E.; Fratzl, P. Nano-scale Modulus Mapping of Biological Composite Materials: Theory and Practice *Prog. Mater. Sci.* **2017**, *87*, 292–320. [CrossRef]

28. Navratil, Z.; Bursikova, V.; Stahel, P.; Sira, M.; Zverina, P. On the Analysis of Surface Free Energy of DLC Coatings Deposited in Low Pressure RF Discharge. *Czech. J. Phys. (Suppl. C)* **2004**, *54*, C877–C882. [CrossRef]

29. Deltombe R.; Kubiak K.J.; Bigerelle M. How to Select the Most Relevant 3D Roughness Parameters of a Surface. *Scanning* **2014**, *36*, 150–160. [CrossRef]

30. Katsikogianni, M.; Missirlis, Y.F. Concise Review of Mechanisms of Bacterial Adhesion to Biomaterials and of Techniques Used in Estimating Bacteria–Material Interactions. *Eur. Cells Mater.* **2004**, *8*, 37–57. [CrossRef]
31. Lerouge, S.; Major, A.; Girault-Lauriault, P.-L.; Raymond, M.-A.; Laplante, P.; Soulez, G.; Mwale, F.; Wertheimer, M.R.; Hébert, M.-J. Nitrogen-Rich Coatings for Promoting Healing around Stent-Grafts after Endovascular Aneurysm Repair. *Biomaterials* **2007**, *28*, 1209–1217. [CrossRef]

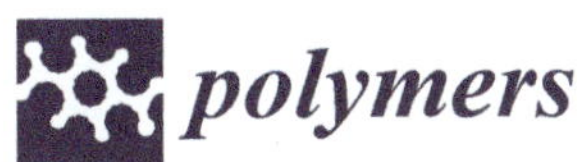 *polymers*

Article

Caseinate-Stabilized Emulsions of Black Cumin and Tamanu Oils: Preparation, Characterization and Antibacterial Activity

Lucie Urbánková [1],*, Věra Kašpárková [1,2], Pavlína Egner [1], Ondřej Rudolf [1] and Eva Korábková [2]

[1] Department of Fat, Surfactant and Cosmetics Technology, Faculty of Technology, Tomas Bata University in Zlín, nám. T. G. Masaryka 5555, 760 01 Zlín, Czech Republic; vkasparkova@utb.cz (V.K.); egner@utb.cz (P.E.); rudolf@utb.cz (O.R.)

[2] Centre of Polymer Systems, Tomas Bata University in Zlín, nám. T. G. Masaryka 5555, 760 01 Zlín, Czech Republic; e_korabkova@utb.cz

* Correspondence: urbankova@utb.cz; Tel.: +420-576-031-235

Received: 3 November 2019; Accepted: 25 November 2019; Published: 27 November 2019

Abstract: Caseinate-stabilized emulsions of black cumin (*Nigella sativa*) and tamanu (*Calophyllum inophyllum*) oils were studied in terms of preparation, characterization, and antibacterial properties. The oils were described while using their basic characteristics, including fatty acid composition and scavenging activity. The oil-in-water (o/w) emulsions containing the studied oils were formulated, and the influence of protein stabilizer (sodium caseinate (CAS), 1–12 wt %), oil contents (5–30 wt %), and emulsification methods (high-shear homogenization vs. sonication) on the emulsion properties were investigated. It was observed that, under both preparation methods, emulsions of small, initial droplet sizes were predominantly formed with CAS content that was higher than 7.5 wt %. Sonication was a more efficient emulsification procedure and was afforded emulsions with smaller droplet size throughout the entire used concentration ranges of oils and CAS when compared to high-shear homogenization. At native pH of ~ 6.5, all of the emulsions exhibited negative zeta potential that originated from the presence of caseinate. The antibacterial activities of both oils and their emulsions were investigated with respect to the growth suppression of common spoilage bacteria while using the disk diffusion method. The oils and selected emulsions were proven to act against gram positive strains, mainly against *Staphylococcus aureus* (*S. aureus*) and *Bacillus cereus* (*B. cereus*); regrettably, the gram negative species were fully resistant against their action.

Keywords: black cumin (*Nigella sativa*) seed oil; tamanu (*Calophyllum inophyllum*) seed oil; emulsion; formulation; antibacterial activity

---

1. Introduction

In recent years, there has been increasing demand for the use of biopolymers as emulsion stabilizers in the cosmetic, pharmaceutic, and food industry. Additionally, alternative sources of bioactive substances, such as essential fatty acids, are in focus. Many of the natural products that were extracted from plants demonstrate biological activities, and they receive particular attention as a source of valuable fatty acids, antimicrobials, antioxidants, or wound healing agents [1,2].

Nigella sativa, which is commonly called black cumin (BC), and *Calophyllum inophyllum* (tamanu, TA), have been used for its nutritional and therapeutic value for many years. Black cumin and its seeds with a characteristic strong taste have been exploited for both culinary and medicinal purposes. The reason for the biological activity of cumin seeds (antioxidant, anticancer, anti-inflammatory, as well as antibacterial) can be found in their complex composition, as they contain over 100 different constituents, including all essential fatty acids. In general, BC consists of oil, proteins, and carbohydrates [2,3]; oil from black cumin is rich in linoleic and oleic acids, as well as bioactive phytosterols and tocopherols [4]. In addition, the seeds contain active phytochemicals, such as thymoquinone, thymohydroquinone, q-cymene, carvacrol, and 4-terpineol [5]. This oil can be used as a natural resource for the production of pharmaceuticals and novel functional foods thanks to its composition and effects [2,6–8].

Tamanu oil expelled from the seeds of *Calophyllum inophyllum* is mainly beneficial in the treatment of dermal problems. The antibiotic and anti-inflammatory properties make this oil an excellent raw material for regenerating and protective formulations. Calophyllic acid and a lactone with antibiotic properties are the two main bioactive substances in the tamanu oil [9]. *Calophyllum inophyllum* is also known as a rich source of secondary metabolites (coumarins, xanthones, flavonoids, and triterpenes) [10], with some of them having anti-inflammatory, antibacterial, and antiviral properties [11,12]. Additionally, coumarin derivatives (calaustralin, calophyllolide, inophyllum, inophyllum E) that were obtained from a crude extract of the nuts were reported to have activity against *Staphylococcus aureus* [1] and antioxidant properties [13].

Nevertheless, the presence of unsaturated fatty acids in both oils causes their sensitivity to oxidation. This makes encapsulation a suitable strategy for protecting the oils, thus allowing for the delivery of lipophilic compounds into aqueous-based systems. For the efficient encapsulation, the production of stable emulsions is crucial and the properties of stabilizing surface layer, oil, as well as the emulsion characteristics are the central factors that can affect the performance of final product; emulsification also plays a key role in optimizing the encapsulation efficiency of oils [14,15].

Emulsification of triacylglycerol-based oils (fish, olive, or sunflower oil) is frequently conducted with milk proteins [16–20]. Particularly, sodium caseinate (CAS) is a common emulsifier and it can be the integral part of formulations due to its amphiphilic structure and surface activity. In o/w emulsions, CAS has been used as emulsifier and encapsulating agent [21], as it forms a barrier by adsorbing to the oil-water interface. This barrier is essential for protecting the bioactive substances against oxidation and the molecules that are adsorbed at oil-water interface also provide an effective shield against flocculation and coalescence due to a combination of electrostatic and steric repulsion. For example, CAS provided good protection of fish oil against oxidation by forming a physical barrier effect [16,22–24]. Casein is also able to protect other oils against oxidation. As published by Hu, et al., corn o/w emulsions that were stabilized with CAS showed both high physical and oxidative stability, which was probably due to ability of CAS to produce a thick layer on emulsion droplet interface and its unique chelating properties [25].

The above summary clearly shows that the incorporation of rarely studied tamanu and black cumin oils in emulsions can be of benefit, as they contain active ingredients with biological and pharmaceutical effects (treatment of asthma, bronchitis, skin diseases; antimicrobial, anti-inflammatory effects; and, gastro-protective properties) [8,9,26,27]. Moreover, the stabilizing of emulsions containing these oils with biocompatible and biodegradable protein CAS can facilitate the delivery of oils to hydrophilic systems while also protecting them against degradation. Therefore, the aim of the work was to prepare o/w emulsions of tamanu and black cumin oil stabilized with CAS, which can serve as carrier systems with therapeutic and physiological benefits for humans. The study also aimed at finding the optimum emulsion formulation with respect to the concentrations of both CAS and oils, and at establishing a suitable emulsification procedure when using two commonly available methods, high-shear homogenization and sonication. The novelty of the study lies in the determination of antimicrobial properties of the oils and their emulsions against both gram positive and gram negative strains, as contamination with spoilage microorganisms is one of the crucial problems encountered in cosmetic and food industry. In the view of the fact that the studies that deal with the emulsification of black cumin and tamanu oils are relatively scarce and only limited data on this topic is found in the scientific literature, the presented study is original and it brings about topical information in this area.

2. Materials and Methods

2.1. Materials

Casein sodium salt (CAS), DPPH (1,1-diphenyl-2-picrylhydrazyl), toluene, and sodium thiosulfate were provided from Sigma-Aldrich (Darmstadt, Germany). Non-traditional vegetable oils from *Calophyllum inophyllum* (Tamanu oil, TA) and *Nigella sativa* (Black cumin oil, BC) were obtained from Nobilis Tilia (Krásná Lípa, Czech Republic). All of the chemicals and reagents were of analytical grade and used without further purification.

2.2. Characterization of Oils

2.2.1. Basic Characteristics

Iodine value (IV; Hanus) expressing the amount of unsaturation in fatty acids present in the oil, saponification value (SV), which is a measure of average molecular mass of all fatty acids, acid value (AV) corresponding to amount of free fatty acid in the oil, and peroxide value (oxidation stability) were determined while using volumetric analysis. The standard methods that were performed with slight modifications were followed [28].

2.2.2. Fatty Acid Composition by Gas Chromatography-Flame Ionization Detector (GC-FID)

The composition of fatty acids in oils was determined by GC-FID after their conversion to respective methyl esters (FAME). In the case of tamanu oil, 30 mL methanol, 2 mL methanolic phenolphthalein solution, and 5 mL hexane were used to dissolve 2 g oil. The solution was heated for approx. 15 min. and the hot sample was then titrated with 0.5M methanolic KOH to neutralize the sample. The sample was added 1 mL 1M methanolic potassium hydroxide and then heated again for 30 min. After the reaction was completed, the FAMEs were twice extracted with 2 × 10 mL hexane. FAME of black cumin oil were prepared via mixing of 2 g oil with 20 mL methanol and 0.5 mL methanolic KOH (1 M). The solution was heated for 30 min. and the resulting products were twice extracted with 2 × 10 mL hexane. The samples of methyl esters were diluted with hexane for analysis.

GC analyses were conducted while using a DANI Master GC Fast Gas Chromatograph (DANI Instruments, Cologno Monzese, Italy) that was equipped with a flame ionization detector (FID) (DANI Instruments, Cologno Monzese, Italy) and capillary column Phenomenex Zebron™ ZB-5MS (30 m × 0.25 mm × 0.5 μm, Phenomenex Inc., Torrance, CA, USA). The injection volume was of 1 μL and flow of a nitrogen carrier gas of 1 mL·min^{-1}. The temperature gradient started at 110 °C, followed by increase of 5 °C·min^{-1} to 240 °C and then to 280 °C at the rate of 2 °C·min^{-1}. The sample was held at the temperature for 20 min. The temperatures of injector and detector were set to 230 and 300 °C, respectively. The FAMEs were identified by comparing their retention times with those of *SUPELCO*™ *37 Component FAME Mix* standard (Sigma Aldrich, Darmstadt, Germany). The quantification of FAME was performed according to internal normalization method. The results are reported in percentage (wt %) of respective fatty acid/100 g total fatty acid content.

2.2.3. Antioxidant Activity

Free radical scavenging activity of BC and TA oils was assayed according to [4]. The DPPH radicals were dissolved in toluene at concentration of 10^{-4} M. For evaluation, 10 mg of oil (in 100 μL toluene) was mixed with 390 μL toluenic DPPH• solution and then vortexed for 20 s. The decrease in absorbance at 515 nm was measured after 1, 15, 30, 45, and 60 min. of mixing while using a Photolab 6600 UV-VIS photometer (Xylem Analytics Germany Sales GmbH & Co. KG, WTW, Weilheim, Germany) with toluene as the blank. The scavenging activity of the samples was calculated while using the equation $I = (A_0 - A_1)/A_0 \times 100$ (%), where I is inhibition activity, A_0 is the absorbance of the blank and A_1 is the absorbance of the mixture.

2.2.4. Interfacial Tension

The interfacial tension of oil-water systems was determined by the pendant drop technique while using an Attension Theta optical tensiometer (Biolin Scientific, Gothenburg, Sweden). The images of the droplets were recorded with a black and white digital camera and the surface tension was obtained by iterative fitting of the shape of the droplet with the Young-Laplace Equation. The droplet was formed while using a 0.718 mm (22 gauge) stainless steel needle.

2.3. Preparation of Emulsions

The o/w emulsions were formulated with 1, 2, 5, 7.5, 10, or 12 wt % CAS. The CAS solutions were prepared by dispersing the powder in deionized water under gentle stirring at room temperature for 4 h. The sample was left to stand at 4 °C overnight to allow for complete hydration and, prior to emulsification, the solution was allowed to equilibrate to room temperature under stirring. Emulsifications were carried out by mixing each of the above CAS solutions and oil phase (5, 10, 20, and 30 wt %; TA or BC oil) at 25 °C. The pH of CAS solutions was not adjusted prior to emulsification.

Two different methods were carried out to prepare the emulsions: a) a high-shear homogenization with an rotor-stator device Ultra-Turrax T25 (IKA, Staufen, Germany) for 12 min. at 13, 400 rpm and b) an ultrasonic homogenization (UP400S, Hielscher Ultrasonics, Teltow, Germany) with 400 W, 24 KHz, for 1 min., operated with 100% amplitude. The total sample volume was of 20 mL. Emulsions were kept in the ice bath during sonication. After emulsification, the pH of prepared emulsions was measured while using pH meter (CPH 51, Elteca, Turnov, Czech Republic) and values of 6.44 ± 0.03 and 6.45 ± 0.04 were recorded for TA and BC oil, respectively.

2.4. Characterization of Emulsions

2.4.1. Size of Emulsion Droplets and Zeta Potential

The size and distribution of emulsion droplets were measured while using the laser diffraction instrument Mastersizer 3000 (Malvern Instruments, Malvern, UK) that was capable of measuring within the size range of 0.01 to 3500 µm. The emulsions were suspended in recirculating milliQ-water flowing through the Hydro SM measuring cell of the instrument, at a pump velocity of 2, 400 rpm. The volume moment mean diameter $D_{(4,3)}$ was determined by instrument software according to equation $D_{(4,3)} = \Sigma ni \cdot d^4 / \Sigma\, ni \cdot d^3$, where n_i is the number of droplets and d represents the droplet diameter. The value of $D_{(4,3)}$ corresponds to a mean diameter of spheres with the same volume as the analysed droplets. The emulsions were also observed with optical microscope Zeiss AxioCam MR 5 (Carl Zeiss MicroImaging GmbH, Oberkochen, Germany). Prior to observation, a droplet of emulsion (10 µL) was placed onto a glass microscope slide and then viewed under 10–100× magnification.

The zeta potential of the samples was measured while using a Zetasizer Nano (Malvern Instruments, Malvern, UK) and calculated using the Smoluchowski model. For measurements, 5 µL of the sample were diluted with 1 mL of twice filtered (Millipore, 0.22 µm) deionized water. The average of three records on the freshly prepared samples is reported.

2.4.2. Creaming Index

The stability of emulsions was assessed by visual observation, daily during the first week of preparation and then at seven-day intervals. The emulsions were stored at ambient temperature. The stability was expressed as the height of emulsion layer (H_{emul}), relative to the total height of the emulsion (H_{total}), which is referred to as the creaming index (CI); ($CI = H_{emul}/H_{total} \times 100$ (%)).

2.5. Antimicrobial Activity

The antimicrobial activities of oils and emulsions were evaluated while using eight bacterial strains that were obtained from the Czech Collection of Microorganisms (CCM, Brno, Czech Republic). The bacteria were selected to represent major food-borne classes. The gram positive (G$^+$; *Micrococcus luteus* CCM 732, *Staphylococcus aureus* subps. *aureus* CCM 3953, *Bacillus cereus* CCM 2010, *Enterococcus faecalis* CCM 2665) and gram negative (G$^-$; *Escherichia coli* CCM 3954, *Pseudomonas aeruginosa* CCM 3955, and *Salmonella enterica* subsp. *enterica ser. Enteritidis* CCM 4420, *Serratia marcescens* subsp. *marcescence* CCM 303) strains were employed in the test. All of the microorganisms were maintained on nutrient agar and sub-cultured onto fresh media every two weeks. Suspensions of bacterial strains were prepared by inoculation from pure culture on a Petri plate into sterile tube with nutrient broth and incubation at 30 °C for 24 h (*Pseudomonas aeruginosa* and *Bacillus cereus*). Other bacterial strains were incubated at 37 °C for 24 h.

The antimicrobial activities of oils and emulsions were determined while using the disc diffusion assay. An overnight culture of each of the bacterial strains in nutrient broth was adjusted to 10^6 CFU/mL and 100 µL of inoculum was spread onto Mueller-Hinton Agar (MHA) sterile agar plates (Hi-Media Laboratories, Mumbai, India). Sterile paper discs (5 mm in diameter) were soaked with 7 µL of oil or with 5 µL of emulsion and then placed on the inoculated agar. As references, discs that were soaked with sterile deionized water and 7.5 wt % aqueous CAS dispersions were used. All of the inoculated plates were incubated at 37 °C for 24 h for mesophilic bacterial strains and at 30 °C for 24 h for *Pseudomonas aeruginosa* and *Bacillus cereus*. After incubation, the antimicrobial activity of the tested samples was evaluated by measuring the diameter of inhibition zone (mm). The diameter of the disc was subtracted from the diameter of inhibition zone.

2.6. Statistical Analysis

All the analyses were conducted at least in triplicates, with the Dean-Dixon method being utilized to calculate the means and standard deviations. The Student T-test was applied to detect any statistical differences between the samples (Statistica, StatSoft, Inc., Palo Alto, CA, USA). The P (probability) value of ≤0.05 was considered to be statistically significant.

3. Results and Discussion

3.1. Properties and Antioxidant Activity of Oils

3.1.1. Basic Characteristics and Fatty Acid Composition

To evaluate differences between the oils, their properties were determined by measuring their basic characteristics, namely iodine (IV), saponification (SV), acid (AV), and peroxide (PV) values, and via determining their fatty acid composition (Tables 1 and 2). The SV and PV were similar for both oils and they are in agreement with the previously published values. The AVs of TA oil were significantly higher (43.6 ± 0.6 mg KOH·g^{-1}) in comparison with those of BC oil (8.4 ± 0.7 mg KOH·g^{-1}) and they roughly corresponded to the values that were reported in literature [29,30]. In this respect, it can be also stressed that the acid value of TA oil was notably higher than that of the common vegetable oils, thus giving evidence of an increased amount of free fatty acids in this oil. Both of the oils showed high SV, which is a measure of the average molecular mass of fatty acid in the oil. The IV values indicated a higher degree of unsaturation for cumin oil, which was also confirmed by GC analyse, showing a higher content of linoleic acid C18:2 in the sample (Table 1).

Table 1. Basic characteristics of tamanu (TA) and black cumin (BC) oils determined as iodine (IV), saponification (SV), acid (AV), and peroxide (PV) values.

Oil	IV (g I$_2$/100 g)	SV (mg KOH·g^{-1})	AV (mg KOH·g^{-1})	PV (µval·g^{-1})
TA	83.3 ± 0.5	200.8 ± 1.9	43.6 ± 0.6	1.23 ± 0.0
BC	116.0 ± 0.9	197.2 ± 0.5	8.4 ± 0.7	1.05 ± 0.0

The composition of fatty acids in TA oil proved the presence of oleic acid C18:1 (41 wt %) as the dominating fatty acid, followed by linoleic acid C18:2 (28 wt %); stearic C18:0 (16 wt %) and palmitic C16:0 (14 wt %) acids were the major saturates. The minor amounts of C16:1, C17:0, and C20:0 were also detected. In BC oil, linoleic acid C18:2 (57 wt %) was the prevailing fatty acid, followed by oleic C18:1 (25 wt %), palmitic C16:0 (13 wt %), and stearic C18:0 (3.4 wt %) acids (Table 2). The results complied well with data regarding fatty acid composition of the oils reported in literature [29,31,32].

Table 2. Composition of fatty acids (wt %) in tamanu (TA) and black cumin (BC) oils determined by Gas Chromatography (GC).

Oil	Fatty Acid (% of Total Content)							
	C16:0	C16:1	C17:0	C18:0	C18:1	C18:2	C20:0	C22:0
TA	14.1 ± 0.3	0.3 ± 0.1	0.1 ± 0	15.7 ± 0.5	41.2 ± 0.2	27.6 ± 0.5	0.2 ± 0	0.2 ± 0
BC	12.6 ± 0.0	n.d.	n.d.	3.4 ± 0.6	24.9 ± 1.9	57.1 ± 2.6	n.d.	n.d.

n.d. not determined.

3.1.2. Radical Scavenging Effect of Oils

The DPPH free radical scavenging assay is frequently used to estimate the antioxidant capacity of the substances. In current work, the time development of scavenging effects of BC and TA oils on DPPH• was assayed in toluene and illustrates a slightly higher activity of cumin oil than that of TA oil (Figure 1). After 60 min. of incubation with radicals, 96% of DPPH• were quenched by cumin oil, while TA oil was able to inhibit 94% radicals. The oils were already efficient after 1 min. of incubation time, with 89 and 78% activity for TA and black cumin oil, respectively. The current results are somewhat higher than those that were reported by [4], who used the same procedure and found out that after 60 min. of incubation, 60% of radicals were inhibited by BC seed oil. The scavenging action of both oils can be mainly ascribed to (1) the content and composition of unsaponifiables, (2) the diversity in structural characteristics of phenolic antioxidants present, (3) the synergism of the phenolic antioxidants with other bioactive components, and (4) the different kinetic behaviours of potential antioxidants [4].

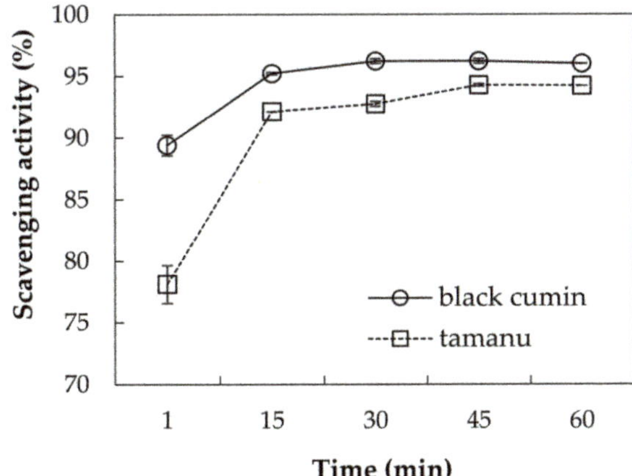

Figure 1. 1,1-diphenyl-2-picrylhydrazyl (DPPH) free radical-scavenging activities and their development in time determined for BC and TA oils after dissolving in toluene.

3.2. Emulsion Properties, Influence of Processing Conditions, and Composition

3.2.1. Droplet Size and Distribution

The size of emulsion droplets is an important parameter with a crucial effect on emulsion stability. Light diffraction analyses showed that the emulsion droplet size ($D_{(4,3)}$) was influenced by all of the studied variables, especially by the method of preparation and concentration of stabilizing CAS; and, to lesser extent, by the type and content of both oils.

In this study, the emulsions were prepared by two different emulsification procedures, which involved high-shear homogenization (Ultra-Turrax, UT) or sonicator (US), which obviously controlled the size of emulsion droplets. Emulsification with UT led to coarse emulsions with bigger droplets that ranged from 0.3 to 13 µm, with their properties being notably affected by composition, such as o/w ratio and CAS content (Figure 2A,C). On the other side, the homogenization with US yielded fine emulsions, with $D_{(4,3)}$ varying from 0.3 to 0.9 µm and 0.4 to 1.5 µm for BC and TA emulsions, respectively; hence, with a sufficiently small size that is a prerequisite for production of stable systems (Figure 2B,D) [33]. Although droplets after sonication were notably smaller than droplets that were treated with Ultra-Turrax, some differences between emulsions containing BC and TA oils were observed. These can be explained by the character, composition, and physical properties of the oils, which affect the emulsifying efficiency and hence the droplet size of emulsions. Here, an important role can play two crucial characteristics, namely viscosity of the oil and interfacial tension at oil-water interface. The viscosities of the used oils are notably different, TA oil is viscous (20–26 mPa·s) [34], whilst the viscosity of BC oil is lower (6.3 mPa·s) [35]. On the other hand, both of the oils have low and rather similar interfacial tensions at the oil-water interface, and values of 8.4 ± 1.0 mN·m^{-1} and 4.3 ± 0.5 mN·m^{-1} were measured for BC and TA oil, respectively. Therefore, the larger droplets of TA emulsions are likely due to the higher viscosity rather than the differences in the interfacial tension between oils. Interestingly, Wooster et al. also referred that high-viscosity oils formed emulsions with larger droplets than oils with low viscosity [36].

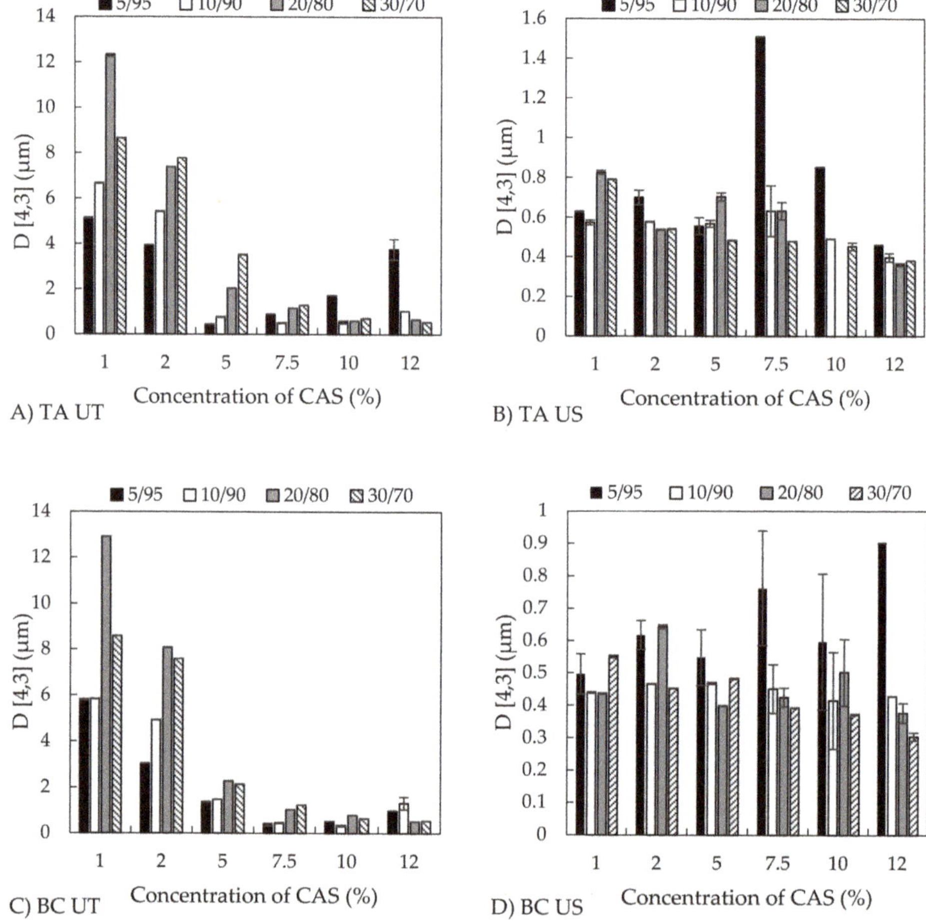

Figure 2. Influence of casein sodium salt (CAS) concentration and o/w ratio on volume weighed diameter of emulsion droplets ($D_{(3,4)}$) of (**A**) TA oil emulsions prepared by Ultra-Turrax (UT) (**B**) TA oil emulsions prepared by sonication (US) (**C**) BC oil emulsions prepared by UT, and (**D**) BC oil emulsion prepared by US.

In addition, the droplet distributions in US emulsions were narrower when compared to those that were prepared with UT (Figure 3). This finding is in agreement with studies that were performed on emulsions stabilized by other types of biopolymers, such as CAS/polysaccharides [17,37], whey protein concentrate [38], or on emulsions containing modified starch and maltodextrin [39].

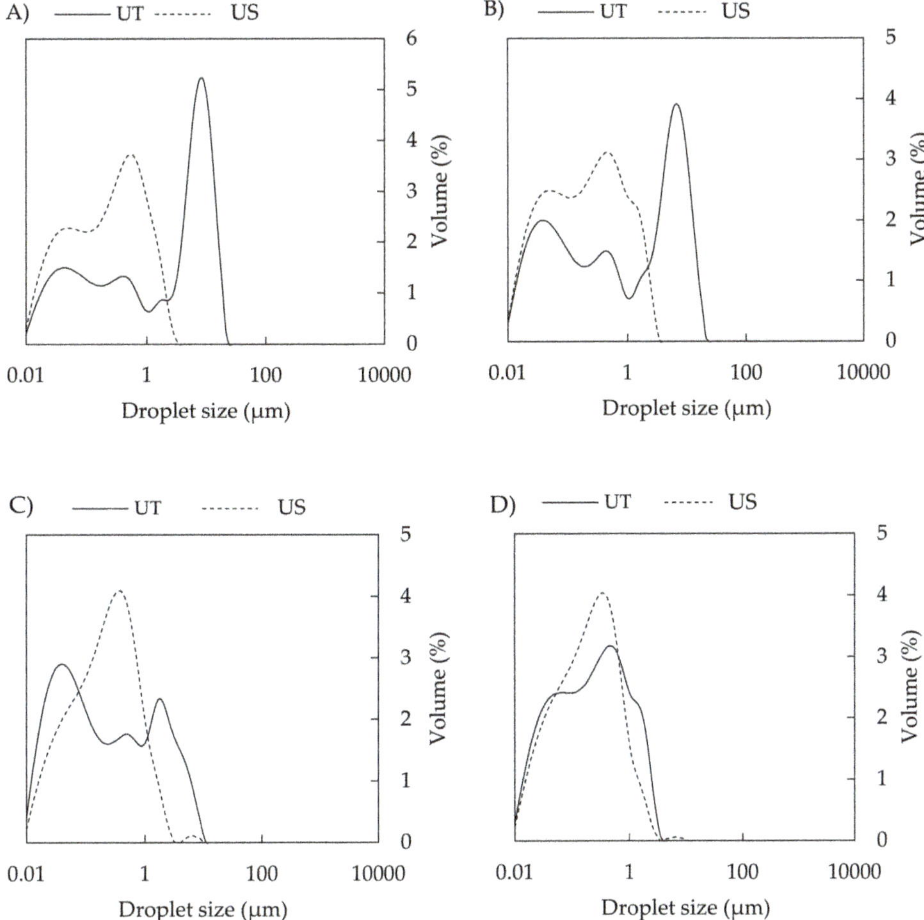

Figure 3. Influence of homogenization method on distribution curves of CAS stabilized emulsions. Ultra-Turraxed (UT) and sonicated (US) emulsions are compared across oil type, o/w ratio and CAS concentrations: (**A**) BC emulsions with 2 wt % of CAS (10/90); (**B**) TA emulsions with 5 wt % of CAS (30/70); (**C**) BC emulsions with 7.5 wt % of CAS (20/80); and, (**D**) BC emulsions with 12 wt % of CAS (20/80).

The influence of the processing method on droplet size must always be considered in combination with a composition of emulsions. Regarding the concentration of CAS, the droplet size of UT-prepared emulsions gradually decreased with increasing protein concentration (1 to 5 wt %); however, $D_{(4,3)}$ was significantly reduced as CAS concentrations increased to 7.5–12 wt % (Figure 2A,C). Full coverage of droplets with protein occurs at a CAS content of approximately 3 wt % (emulsions with 35 % tetradecane oil phase) [40]. At a higher protein concentration (4 wt %), a higher amount of CAS than required for saturation coverage of the oil droplets is presented; hence, a range of CAS concentrations used in current work is sufficient for completely covering the arising droplets. Therefore, the bigger droplets that were prepared by UT (in the case of sufficient CAS available in the system) can be assigned to the discrepancy between the higher rate of coalescence of oil droplets being formed by a given energy input and the lower rate of CAS adsorption at the oil-water interface during homogenization. Under these conditions, the uncovered droplets tend to come together and form larger droplets again.

On the other hand, when the CAS content in aqueous phase is insufficient, flocculation might occur due to casein bridging between droplets, also resulting in rapid creaming [17,20,41,42]. The results from particle sizing were also supported by the distribution curves of UT-emulsions, where a shift in droplet diameter toward smaller sizes was observed as the protein concentration increased, and distribution curves changed from multimodal (1 wt % CAS) to monomodal (10 wt % CAS), which indicated the formation of a stable emulsions (Figure 4A) [43].

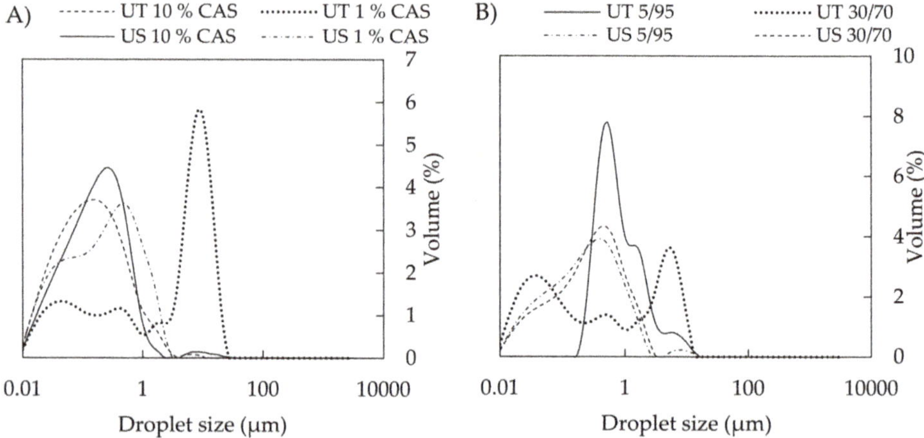

Figure 4. (**A**) Effect of CAS concentration on distribution curves of BC emulsions with o/w ratio 10/90 prepared by UT and US; (**B**) effect of o/w ratio on distribution curves of BC emulsions with 5 wt % CAS. For tamanu oil, the trend is similar (data not shown).

On the contrary, emulsions that were prepared by US contained much smaller droplets, and only minor changes in the droplet size alongside change in CAS concentration were observed (Figure 2B,D). For emulsions that were prepared with the aid of US, the distribution curves were mostly monomodal with the simultaneous shift in direction to lower droplet sizes (Figure 4A). Figure 5 visualizes an example of light microscopy figure captured on the emulsion prepared with US.

The volume fraction of oil is another parameter with an impact on emulsion droplet size. In the current study, the increasing oil content influenced ($D_{(4,3)}$); however, only at lower concentrations of CAS (1–7.5 wt %) and mainly in the case of UT-prepared emulsions. Here, the droplet size increased with an increasing oil fraction. The effect is documented in Figure 4B, showing the broad distribution curve with three droplet populations being recorded for UT emulsions containing 30 wt % BC oil and 5 wt % CAS; on the other hand, by lowering the oil content to 5 wt %, monomodal distributions were obtained. CAS contents of 10 and 12 wt % afforded droplets that were notably smaller and influence of oil fraction was only marginal. In contrast, the situation was different when it came to emulsions that were prepared by sonication (Figure 4B) and only minor changes in $D_{(4,3)}$ and distributions with the increasing oil content were observed at the entire concentration range of CAS used. In systems containing high concentrations of both protein and oil, the emulsions can be stabilized through the formation of a weak particle-based gel network, which slowly reorganizes under the influence of gravity and internal osmotic stresses [40].

Alongside with other variables, the type of oil plays also a role with respect to the droplet size in emulsions, and the droplets of the BC emulsions that were prepared by UT were smaller as compared to droplets containing the TA oil. As already discussed above, the differences can be explained by composition and physical properties of the oils, mainly viscosity and interfacial tension. This difference was again wiped off in emulsions that were prepared using US. The fact that TA oil also contains, in addition to neutral lipids and glycolipids, phospholipids (1.6%), small amounts of sterols, and mono-

and diacylglycerols, which all have emulsifying properties, can also contribute to easier formation of smaller droplets [9].

Figure 5. Optical microscopy (Magnification 20x) of BC emulsions with o/w 20/80 prepared with 1 wt % CAS while using sonication. Scale bar is 10 μm.

3.2.2. Zeta Potential

Zeta (ξ) potential is commonly considered as a stability indicating parameter of dispersion systems. However, the values of ξ potential alone are not always capable of predicting the stability of emulsions, mainly those that are sterically stabilized. Immediately after preparation, the ξ potential ranged from −53 to −41 mV for emulsions with BC oil (prepared with US) and from −60 to −46 mV for those that were prepared with UT. TA emulsions behaved similarly with the potential values lying from −55 to −41 mV and −59 to −44 mV for US and UT emulsions, respectively. In general, UT treatment seemed to deliver slightly lower ξ potentials and, obviously, no or minor systematic influence of oil content on the potential was observed. However, CAS concentrations above 5 wt % might have a positive influence in potential lowering, mainly for emulsions with TA oil (Figure 6). The negative charges of emulsion droplets are, therefore, due to the presence of CAS stabilizer. As such, this protein is negatively charged at a natural pH and its aqueous dispersion in the absence of oil showed a potential of −18.7 mV.

In current study, ξ potential measurements were conducted on emulsions with non-adjusted pH (TA oil emulsions pH = 6.44 ± 0.03, BC oil emulsions pH = 6.45 ± 0.04). At a pH close to caseinate isoelectric point (pH 4.6), the CAS emulsions are unstable [44], thanks to a reduction in electrostatic repulsion between the droplets. However, at pH 6, the negative ξ potential of CAS assured emulsion stability due to electrostatic and steric stabilization mechanisms.

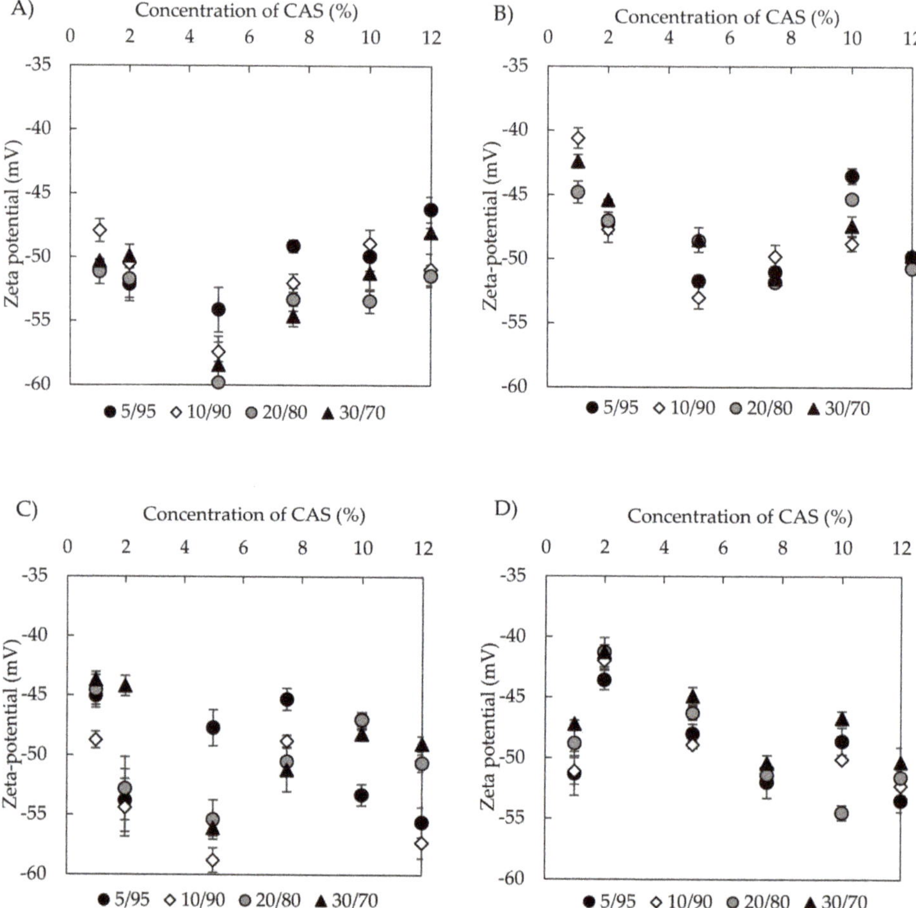

Figure 6. Zeta potential of freshly prepared emulsions at native pH of ~6,5 as a function of o/w ratio and CAS content (**A**) BC emulsions prepared with Ultra-Turrax (UT), (**B**) BC emulsions prepared with sonication (US), (**C**) TA emulsions prepared with UT, and (**D**) TA emulsions with US.

Differences in the ξ potentials between emulsions that were prepared by UT and US can be assigned to combined effect of sonication treatment and oil properties. When compared with many other proteins, caseins are particularly disordered and substantially hydrophobic, which assist in their rapid adsorption to oil droplet surfaces during emulsification. In aqueous solutions, CAS forms a random coil with only small amount of secondary structure due to the relatively high content of hydrophobic amino acids (proline residues) [45]. During sonication, the hydrophobicity of CAS changes, thanks to the different orientation of proteins at the oil-water interface, which can impact the ξ potential of emulsions formed by US treatment [46]. The impact of oil type on ξ potential, which is observed in our work, was also reported for soy bean protein isolate emulsions containing medium chain triacylglycerols, palm, soybean, and rapeseed oils after US treatment [47]. The lowering of ξ potential was ascribed to the formation of protein aggregates and different amounts of protein adsorbed onto emulsions containing studied oils of different compositions.

3.2.3. Phase Studies and Emulsion Stability

When proteins are used to stabilize emulsions, the main mechanism governing their creaming in the presence of non-adsorbed stabilizer is related to depletion flocculation [48]. Kinetic stability of emulsions, as described by creaming index (CI), provides indirect information on droplet flocculation and destabilization occurring in an emulsion.

Not all of the emulsions prepared in the current study demonstrated kinetic stability, even immediately after preparation. When observed by naked eye, some of them formed a bottom serum layer and a small cream phase on the top. On the other hand samples with specific formulations, mainly with high CAS content, exhibited good resistance to creaming. Figure 7 shows the CI of the freshly prepared emulsions. Stable emulsions (CI = 100%) were observed at CAS concentrations of 10–12 wt % (UT emulsion) and 7.5–12 wt % CAS (US emulsions). However, samples containing lower concentrations of CAS destabilised readily and the extent of destabilization, expressed as CI, was obviously dependent on oil content. More specifically, the CI increased with an increasing amount of oil, which was more notable on UT emulsions in comparison with emulsions that were prepared by US. Therefore, the formation of stable emulsions was considered to be a synergic effect of the US processing, resulting in small droplets with narrower distribution and the sufficient amount of CAS. Only a minor difference in CI was observed between emulsions with TA or BC oils.

The long-term stability of the emulsions was evaluated by the recording of changes in their visual appearance during storage until the point of emulsion breaking appeared. Not surprisingly, the most stable emulsions were those that contained the highest protein content (12 wt % CAS). For example, BC emulsions with 30 wt % oil prepared by both procedures remained unchanged for seven days of storage with CI being of 100%. Stability studies also corroborated the fact that US treatment assured the production of emulsions with limited creaming. At a CAS concentration of 10 wt %, the emulsions that were manufactured with US were stable towards creaming for seven days, whilst in emulsions that were prepared with UT, the creaming already started three days after preparation. Moreover, UT manufactured emulsions with even lower CAS content were prone to breakdown and underwent phase separation approximately after a week of storage. Interestingly, in most emulsions, although the CI value was lower immediately after emulsification, it did not notably change during their storage. From the CI data, it was also apparent that the creaming process took about a week to be completed, thereafter compaction of the cream layer started, which stopped when no more oil droplets and proteins could be packed in the top cream layer. The compaction/compression of the cream layer was also reported by [49]. These results can be generalized for both BC and TA emulsions. The above observations demonstrate that an important role in stability of CAS-containing emulsions is related to concentration of this biopolymer in the aqueous phase, which closely correlates with emulsion viscosity. The viscosity of CAS emulsions is controlled by the interactions between the droplets and mainly by the nature and strength of interparticle attractive forces, which depends on the structure of the CAS layer that was adsorbed at the oil-water interface. Additionally, the CAS self-assembly and aggregation of non-adsorbed CAS in the aqueous phase play an important role with respect to the viscosity and stability behaviour of CAS stabilized emulsions [50].

The data from the observation of creaming were in reasonably good correlation with particle sizing measurements, which were also conducted during emulsion storage. The results revealed that the $D_{(4,3)}$ increased during a month-long storage at an ambient temperature. The US emulsions changed their droplet sizes to higher extent than the UT systems; however, emulsions that were prepared with the aid of UT were more prone to phase separation and their breakdown occurred earlier. Here again, the key role played the content of CAS in emulsions. Emulsions with 1 or 2 wt % CAS were stable in terms of size only during the first days, emulsion with higher CAS content for a week and emulsions with 10 and 12 wt % CAS remained unchanged for a longer period of time (Figure 8). This is in agreement with findings in [51]. However, when considering the practical application of TA and BC emulsions, their long-term stability is not satisfactorily. In this respect, the prolongation of sonication time or two-stage homogenization with an additional treatment of coarse emulsions with high-pressure homogenizer can be used for the improvement of stability. The effect of high-pressure homogenization on properties of CAS emulsions studied authors [33]. They compared properties of thus prepared emulsions with emulsions that were prepared by Ultra-Turrax and concluded that UT produced larger droplets and less stable systems than the homogenizer did. However, during homogenization, the choice of correct pressure was essential, as the over-processing of emulsions and conformation changes of proteins can occur at higher pressures [52].

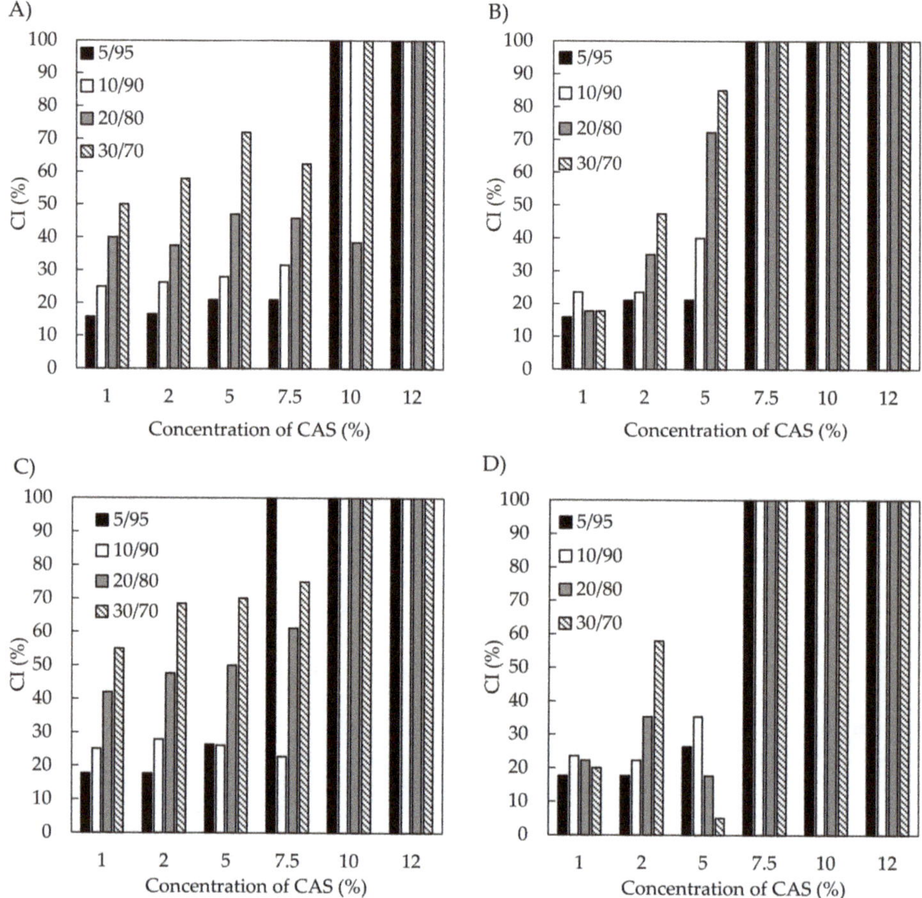

Figure 7. Comparison of creaming index (CI; CI—100% is for stable emulsion) determined on freshly prepared emulsions, as affected by processing method and composition of emulsions (**A**) BC emulsions prepared with Ultra-Turrax (UT), (**B**) BC emulsions prepared with sonication (US), (**C**) TA emulsions prepared with UT, and (**D**) TA emulsions with US.

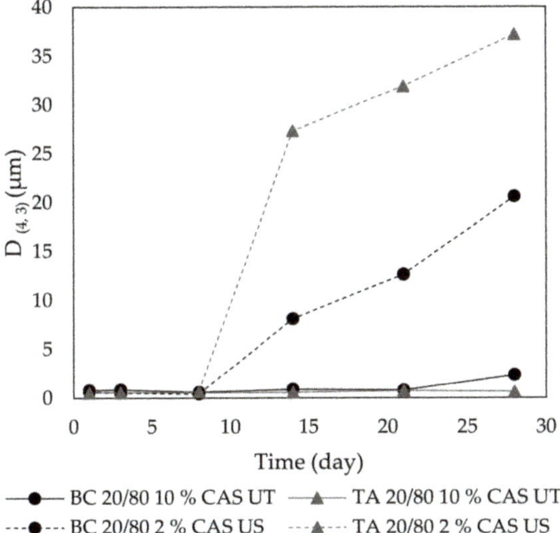

Figure 8. Long term stability of TA and BC emulsions stabilized with CAS determined at room temperature: development of volume weighed diameter of emulsion droplets ($D_{(4,3)}$) in time.

3.3. Antimicrobial Activity

Illnesses that are caused by pathogenic bacteria and/or their toxins are of great concern to public health. For the suppression of microorganism growth, plant extracts with antimicrobial activity can also be used [4].

Antibacterial activity of oils and emulsions (o/w 30/70) prepared thereof with 2 and 7.5 wt % CAS was determined while using the disc diffusion method and expressed in terms of the size of the inhibition and halo zones (mm).

Disc diffusion assay proved the inhibitory activity of both oils and selected emulsions against the gram positive bacteria, which was controlled by the type of the oil and the formulation of the emulsion. Regrettably, none of the samples were capable of suppressing the growth of the tested gram-negative species (*Escherichia coli, Pseudomonas aeruginosa, Salmonella enterica*, and *Serratia marcescens*). Table 3 shows that the effect of TA oil was significantly higher ($p \geq 0.5$) than that of BC oil, both for *B. cereus* and *S. aureus*. It can be seen from the bigger inhibition zones with the sizes ranging from 8.2 ± 0.4 mm (*B. cereus*) to 6.0 ± 1.2 mm (*S. aureus*). This oil also caused lower bacterial growth (14.8 ± 1.3 to 7.7 ± 0.7 mm) around the inhibition zones (halo zone) and it was able to act, at least partially, against *M. luteus* and *E. faecalis* (halo zone only). The effects of TA oil encapsulated in emulsions with 2 and 7.5 wt % of CAS were rather similar, although the activity of emulsion stabilized with 2 wt % CAS was comparable with that of non-encapsulated TA oil ($p \geq 0.5$). The antibacterial activity of the BC oil was, in comparison with TA oil, weaker, and this oil only showed antibacterial activity against *B. cereus* (5.7 ± 1.5 mm) and *S. aureus* (4.2 ± 0.7 mm). Emulsion that was stabilized with 2 wt % CAS containing BC oil exerted the comparable effect, which was weaker than that of BC oil alone. Neither BC oil nor its emulsions were efficient against *M. luteus* and *E. faecalis*. The example of growth media images with inhibition zones observed after application of TA and BC emulsions are shown in Figure 9. The lower activity of oils and emulsions against gram negative bacteria is a result of composition in their cell walls, as they contain an outer lipopolysaccharide membrane, which more efficiently protects the bacteria from the disruption caused by oils. As reported in literature, the antimicrobial activity of BC oil mainly results from the presence of thymoquinone, p-cymene, longifolene, and thymohydroquinone. Moreover, unsaturated, long chain fatty acids, such as linoleic and oleic acids present in the oil, were also reported to possess antibacterial and antifungal activity [8]. The study conducted by [4] supported the above results regarding the bioactivity of BC oil, also extending the range of its inhibitory activity to action against yeasts, namely *Saccharomyces cerevisiae* and *Candida albicans*. Regarding triacylglycerol-based TA oils, their bioactivity was thoroughly investigated by [1], who reported that TA oil exhibited two distinct antibacterial effects: the first against gram positive bacteria acting by direct inhibition of mitotic growth and the second potent effect against gram negative strains due to an increased release of β-defensin 2 peptide by macrophages. However, the results of our study have not confirmed the effect against gram negative strains.

Table 3. Antibacterial activity of BC and TA oils and their 30/70 o/w emulsions against Gram positive strains; No inhibition zone towards *Micrococcus luteus* (*M. luteus*) and *Enterococcus faecalis* (*E. faecalis*) were observed.

	Size of Zone ± SD (mm)					
	B. cereus		S. aureus		M. luteus	E. faecalis
	Inhibition	Halo	Inhibition	Halo	Halo	Halo
			Tamanu			
Oil	8.2 ± 0.4	14.8 ± 1.3	6.0 ± 1.2	9.5 ± 1.3	9.3 ± 1.2	7.7 ± 0.7
Emulsion 2 % CAS	7.8 ± 1.6	14.3 ± 0.7	6.0 ± 0.6	8.8 ± 0.4	14.3 ± 3.7	8.3 ± 1.4
Emulsion 7.5 % CAS	6.3 ± 0.7	11.2 ± 1.3	4.7 ± 0.5	7.3 ± 0.5	9.0 ± 0.8	11.0 ± 3.1
			Black Cumin			
Oil	5.7 ± 1.5	8.7 ± 0.9	4.2 ± 0.7	6.7 ± 0.9	*n.d*	*n.d*
Emulsion 2 % CAS	2.0 ± 0.8	*n.d.*	3.3 ± 0.9	*n.d*	*n.d*	*n.d*
Emulsion 7.5 % CAS	1.7 ± 0.5	*n.d.*	2.0 ± 0.6	*n.d*	*n.d*	*n.d*

n.d. not determined.

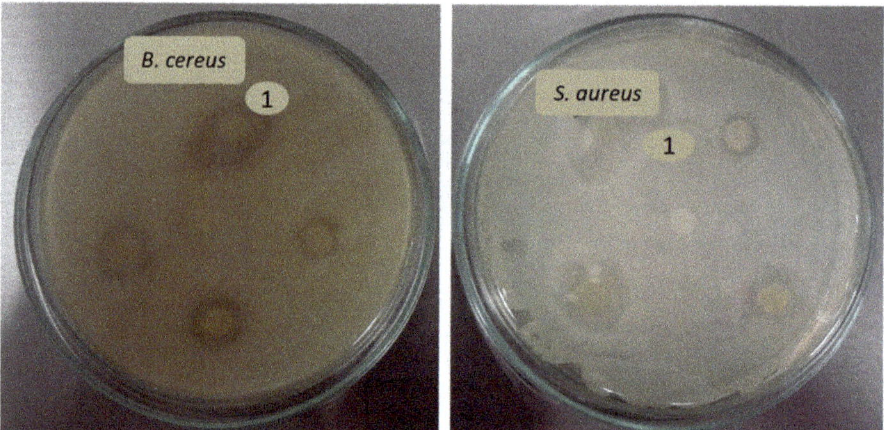

Figure 9. Images of growth media with inhibition zones observed after application of TA and BC emulsions: from "1" clockwise: TA oil with 2 wt % CAS, BC oil with 2 wt % CAS, TA oil with 7.5 wt % CAS, BC oil with 7.5 wt % CAS; in the middle is negative control.

4. Conclusions

In this work, sodium caseinate was used to stabilize emulsions containing bioactive tamanu and black cumin oils. The emulsions were prepared by ultrasound treatment or high-shear homogenization with Ultra-Turrax. The analysis of fatty acid composition in the oils revealed a higher degree of unsaturation for cumin oil with higher content of linoleic acid C18:2, which corresponds to the higher iodine value determined for this oil. Both of the oils effectively scavenged DPPH radicals, thus showing antioxidant activity. The results revealed that, under emulsification, sonication was the more efficient procedure and it afforded emulsions with a small particle size (0.3 to 1.5 μm) throughout the entire used concentration ranges of oils and caseinate. In addition to the emulsification technique, the ability to form stable emulsions of small, initial droplet sizes was mainly controlled by concentration of stabilizing caseinate and, to lesser extent, by the type and amount of used oils. In comparison with sonicated emulsions, the properties of emulsions that were prepared with Ultra-Turrax depended, to a higher extent, on their composition and the emulsions were more prone to destabilization. The oils and their selected emulsions exhibited antibacterial activity against gram positive strains (*S. aureus* and *B. cereus*); regrettably, the gram negative species were fully resistant against their action. Of the studied oils, the tamanu oil and its emulsions were more efficient. The antibacterial and antioxidant properties of the oils, together with their beneficial fatty acid composition, make them suitable for use as carriers for various lipophilic bioactive substances; the incorporation in caseinate emulsions further increases their applicability in hydrophilic systems.

Author Contributions: Conceptualization, L.U., V.K. and P.E.; methodology, L.U., V.K., P.E., O.R. and E.K.; validation, L.U. and V.K.; formal analysis, L.U., O.R. and E.K.; investigation, L.U. and V.K.; resources, L.U. and V.K.; writing—original draft preparation, L.U.; writing—review and editing, L.U. and V.K.; visualization, L.U. and V.K.; supervision, V.K.

Funding: This work was supported by the Czech Science Foundation (19-16861S) and by the Ministry of Education, Youth and Sports of the Czech Republic – Program NPU I (LO1504). One of us (LU) also appreciates support of the internal grants of TBU in Zlín IGA/CPS/2019/004 funded from the resources of specific academic research.

Conflicts of Interest: The authors declare no conflict of interest.

References

1. Léguillier, T.; Lecsö-Bornet, M.; Lémus, C.; Rousseau-Ralliard, D.; Lebouvier, N.; Hnawia, E.; Nour, M.; Aalbersberg, W.; Ghazi, K.; Raharivelomanana, P.; et al. The Wound Healing and Antibacterial Activity of Five Ethnomedical Calophyllum Inophyllum Oils: An Alternative Therapeutic Strategy to Treat Infected Wounds. *PLoS ONE* **2015**, *10*, e0138602. [CrossRef]

2. Piras, A.; Rosa, A.; Marongiu, B.; Porcedda, S.; Falconieri, D.; Dessì, M.A.; Ozcelik, B.; Koca, U. Chemical Composition and in Vitro Bioactivity of the Volatile and Fixed Oils of *Nigella Sativa*, L. Extracted by Supercritical Carbon Dioxide. *Ind. Crop. Prod.* **2013**, *46*, 317–323. [CrossRef]

3. Amin, B.; Hosseinzadeh, H. Black Cumin (*Nigella Sativa*) and its Active Constituent, Thymoquinone: An Overview on the Analgesic and Anti-Inflammatory Effects. *Planta Med.* **2016**, *82*, 8–16. [CrossRef] [PubMed]

4. Ramadan, M.; Asker, M.; Tadros, M. Antiradical and Antimicrobial Properties of Cold-Pressed Black Cumin and Cumin Oils. *Eur. Food Res. Technol.* **2012**, *234*, 833–844. [CrossRef]

5. Hassanien, M.F.R.; Assiri, A.M.A.; Alzohairy, A.M.; Oraby, H.F. Health-Promoting Value and Food Applications of Black Cumin Essential Oil: An Overview. *J. Food Sci. Technol.* **2015**, *52*, 6136–6142. [CrossRef] [PubMed]

6. Bourgou, S.; Pichette, A.; Marzouk, B.; Legault, J. Bioactivities of Black Cumin Essential Oil and its Main Terpenes from Tunisia. *S. Afr. J. Bot.* **2010**, *76*, 210–216. [CrossRef]

7. Mohammed, N.K.; Tan, C.P.; Abd Manap, M.Y.; Muhialdin, B.J.; Hussin, A.S.M. Production of Functional Non-Dairy Creamer using Nigella Sativa Oil Via Fluidized Bed Coating Technology. *Food Bioprocess Technol.* **2019**, *12*, 1352–1365. [CrossRef]

8. Mazaheri, Y.; Torbati, M.; Azadmard-Damirchi, S.; Savage, G.P. A Comprehensive Review of the Physicochemical, Quality and Nutritional Properties of Nigella Sativa Oil. *Food Rev. Int.* **2019**, *35*, 342–362. [CrossRef]

9. Raharivelomanana, P.; Ansel, J.; Lupo, E.; Mijouin, L.; Guillot, S.; Butaud, J.; Ho, R.; Lecellier, G.; Pichon, C. Tamanu Oil and Skin Active Properties: From Traditional to Modern Cosmetic Uses. *OCL* **2018**, *25*, D504. [CrossRef]

10. Shen, Y.C.; Hung, M.C.; Wang, L.T.; Chen, C.Y. Inocalophyllins A, B and their Methyl Esters from the Seeds of Calophyllum Inophyllum. *Chem. Pharm. Bull.* **2003**, *51*, 802–806. [CrossRef]

11. Itoigawa, M.; Ito, C.; Tan, H.T.-W.; Kuchide, M.; Tokuda, H.; Nishino, H.; Furukawa, H. Cancer Chemopreventive Agents, 4-Phenylcoumarins from Calophyllum Inophyllum. *Cancer Lett.* **2001**, *169*, 15–19. [CrossRef]

12. Bui, C.; Nguyen, B.; Trinh, D.; Vo, N. 1077 Anti-Inflammatory and Wound Healing Activities of Calophyllolide Isolated from Calophyllum Inophyllum Linn. *J. Investig. Dermatol.* **2018**, *138*, S183. [CrossRef]

13. Said, T.; Dutot, M.; Martin, C.; Beaudeux, J.-L.; Boucher, C.; Enee, E.; Baudouin, C.; Warnet, J.-M.; Rat, P. Cytoprotective Effect Against UV-Induced DNA Damage and Oxidative Stress: Role of New Biological UV Filter. *Eur. J. Pharm. Sci.* **2007**, *30*, 203–210. [CrossRef] [PubMed]

14. Ixtaina, V.Y.; Julio, L.M.; Wagner, J.R.; Nolasco, S.M.; Tomás, M.C. Physicochemical Characterization and Stability of Chia Oil Rnicroencapsulated with Sodium Caseinate and Lactose by Spray-Drying. *Powder Technol.* **2015**, *271*, 26–34. [CrossRef]

15. Komaiko, J.; Sastrosubroto, A.; McClements, D.J. Encapsulation of Omega-3 Fatty Acids in Nanoemulsion-Based Delivery Systems Fabricated from Natural Emulsifiers: Sunflower Phospholipids. *Food Chem.* **2016**, *203*, 331–339. [CrossRef]

16. Day, L.; Xu, M.; Hoobin, P.; Burgar, I.; Augustin, M.A. Characterisation of Fish Oil Emulsions Stabilised by Sodium Caseinate. *Food Chem.* **2007**, *105*, 469–479. [CrossRef]

17. Huck-Iriart, C.; Pizones Ruiz-Henestrosa, V.; Candal, R.; Herrera, M. Effect of Aqueous Phase Composition on Stability of Sodium Caseinate/Sunflower Oil Emulsions. *Food Bioprocess Technol.* **2013**, *6*, 2406–2418. [CrossRef]

18. Keogh, M.K.; O'Kennedy, B.T.; Kelly, J.; Auty, M.A.; Kelly, P.M.; Fureby, A.; Haahr, A.-M. Stability to Oxidation of Spray-Dried Fish Oil Powder Microencapsulated using Milk Ingredients. *J. Food Sci.* **2001**, *66*, 217–224. [CrossRef]

19. Villiere, A.; Viau, M.; Bronnec, I.; Moreau, N.; Genot, C. Oxidative Stability of Bovine Serum Albumin- and Sodium Caseinate-Stabilized Emulsions Depends on Metal Availability. *J. Agric. Food Chem.* **2005**, *53*, 1514–1520. [CrossRef]

20. Hebishy, E.; Buffa, M.; Juan, B.; Blasco-Moreno, A.; Trujillo, A. Ultra High-Pressure Homogenized Emulsions Stabilized by Sodium Caseinate: Effects of Protein Concentration and Pressure on Emulsions Structure and Stability. *LWT-Food Sci. Technol.* **2017**, *76*, 57–66. [CrossRef]

21. Drusch, S.; Serfert, Y.; Berger, A.; Shaikh, M.Q.; Raetzke, K.; Zaporojtchenko, V.; Schwarz, K. New Insights into the Microencapsulation Properties of Sodium Caseinate and Hydrolyzed Casein. *Food Hydrocoll.* **2012**, *27*, 332–338. [CrossRef]

22. Amine, C.; Dreher, J.; Helgason, T.; Tadros, T. Investigation of Emulsifying Properties and Emulsion Stability of Plant and Milk Proteins using Interfacial Tension and Interfacial Elasticity. *Food Hydrocoll.* **2014**, *39*, 180–186. [CrossRef]

23. Livney, Y.D. Milk Proteins as Vehicles for Bioactives. *Curr. Opin. Colloid Interface Sci.* **2010**, *15*, 73–83. [CrossRef]

24. Nielsen, N.S.; Jacobsen, C. Methods for Reducing Lipid Oxidation in Fish-Oil-Enriched Energy Bars. *Int. J. Food Sci. Technol.* **2009**, *44*, 1536–1546. [CrossRef]

25. Hu, M.; McClements, D.J.; Decker, E.A. Lipid Oxidation in Corn Oil-in-Water Emulsions Stabilized by Casein, Whey Protein Isolate, and Soy Protein Isolate. *J. Agric. Food Chem.* **2003**, *51*, 1696–1700. [CrossRef] [PubMed]

26. Ansel, J.; Lupo, E.; Mijouin, L.; Guillot, S.; Butaud, J.; Ho, R.; Lecellier, G.; Raharivelomanana, P.; Pichon, C. Biological Activity of Polynesian Calophyllum Inophyllum Oil Extract on Human Skin Cells. *Planta Med.* **2016**, *82*, 961–966. [CrossRef]

27. Mukhtar, H.; Qureshi, A.S.; Anwar, F.; Mumtaz, M.W.; Marcu, M. *Nigella Sativa*, L. Seed and Seed Oil: Potential Sources of High-Value Components for Development of Functional Foods and Nutraceuticals/Pharmaceuticals. *J. Essent. Oil Res.* **2019**, *31*, 171–183. [CrossRef]

28. Gunstone, F.D.; Harwood, J.L.; Dijkstra, A.J. (Eds.) *The Lipid Handbook*, 3rd ed.; CRC Press: Boca Raton, FL, USA, 2007.

29. Crane, S.; Aurore, G.; Joseph, H.; Mouloungui, Z.; Bourgeois, P. Composition of Fatty Acids Triacylglycerols and Unsaponifiable Matter in *Calophyllum Calaba*, L. *Oil Guadeloupe Phytochem.* **2005**, *66*, 1825–1831. [CrossRef]

30. Ramadan, M.F.; Mörsel, J. Characterization of Phospholipid Composition of Black Cumin (*Nigella Sativa*, L.) Seed Oil. *Food Nahr.* **2002**, *46*, 240–244. [CrossRef]

31. Cheikh-Rouhou, S.; Besbes, S.; Hentati, B.; Blecker, C.; Deroanne, C.; Attia, H. *Nigella Sativa*, L.: Chemical Composition and Physicochemical Characteristics of Lipid Fraction. *Food Chem.* **2007**, *101*, 673–681. [CrossRef]

32. Singh, S.; Das, S.S.; Singh, G.; Schuff, C.; de Lampasona, M.P.; Catalán, C.A.N. Composition, in Vitro Antioxidant and Antimicrobial Activities of Essential Oil and Oleoresins obtained from Black Cumin Seeds (*Nigella Sativa*, L.). *BioMed Res. Int.* **2014**, *2014*, 918209. [CrossRef] [PubMed]

33. Perrechil, F.A.; Cunha, R.L. Oil-in-Water Emulsions Stabilized by Sodium Caseinate: Influence of pH, High-Pressure Homogenization and Locust Bean Gum Addition. *J. Food Eng.* **2010**, *97*, 441–448. [CrossRef]

34. Amalia Kartika, I.; Cerny, M.; Vandenbossche, V.; Rigal, L.; Sablayrolles, C.; Vialle, C.; Suparno, O.; Ariono, D.; Evon, P. Direct Calophyllum Oil Extraction and Resin Separation with a Binary Solvent of N-Hexane and Methanol Mixture. *Fuel* **2018**, *221*, 159–164. [CrossRef]

35. Mohammed, N.K.; Abd Manap, M.Y.; Tan, C.P.; Muhialdin, B.J.; Alhelli, A.M.; Meor Hussin, A.S. The Effects of Different Extraction Methods on Antioxidant Properties, Chemical Composition, and Thermal Behavior of Black Seed (*Nigella sativa*, L.) Oil. *Evid. Based Complement. Altern. Med. eCAM* **2016**, *2016*, 6273817. [CrossRef] [PubMed]

36. Wooster, T.J.; Golding, M.; Sanguansri, P. Impact of Oil Type on Nanoemulsion Formation and Ostwald Ripening Stability. *Langmuir* **2008**, *24*, 12758–12765. [CrossRef] [PubMed]

37. Álvarez Cerimedo, M.S.; Iriart, C.H.; Candal, R.J.; Herrera, M.L. Stability of Emulsions Formulated with High Concentrations of Sodium Caseinate and Trehalose. *Food Res. Int.* **2010**, *43*, 1482–1493. [CrossRef]

38. Lizarraga, M.S.; Pan, L.G.; Añon, M.C.; Santiago, L.G. Stability of Concentrated Emulsions Measured by Optical and Rheological Methods. Effect of Processing Conditions-I. Whey Protein Concentrate. *Food Hydrocoll.* **2008**, *22*, 868–878. [CrossRef]

39. Jafari, S.M.; He, Y.; Bhandari, B. Production of Sub-Micron Emulsions by Ultrasound and Microfluidization Techniques. *J. Food Eng.* **2007**, *82*, 478–488. [CrossRef]

40. Dickinson, E.; Golding, M. Depletion Flocculation of Emulsions Containing Unadsorbed Sodium Caseinate. *Food Hydrocoll.* **1997**, *11*, 13–18. [CrossRef]

41. Montes de Oca-Ávalos, J.M.; Candal, R.J.; Herrera, M.L. Colloidal Properties of Sodium Caseinate-Stabilized Nanoemulsions Prepared by a Combination of a High-Energy Homogenization and Evaporative Ripening Methods. *Food Res. Int.* **2017**, *100*, 143–150. [CrossRef]

42. Srinivasan, M.; Singh, H.; Munro, P.A. Formation and Stability of Sodium Caseinate Emulsions: Influence of Retorting (121 °C for 15 Min) before or After Emulsification. *Food Hydrocoll.* **2002**, *16*, 153–160. [CrossRef]

43. Dickinson, E.; Radford, S.J.; Golding, M. Stability and Rheology of Emulsions Containing Sodium Caseinate: Combined Effects of Ionic Calcium and Non-Ionic Surfactant. *Food Hydrocoll.* **2003**, *17*, 211–220. [CrossRef]

44. Allen, K.E.; Dickinson, E.; Murray, B. Acidified Sodium Caseinate Emulsion Foams Containing Liquid Fat: A Comparison with Whipped Cream. *LWT-Food Sci. Technol.* **2006**, *39*, 225–234. [CrossRef]

45. Barreto, P.; Roeder, J.; Crespo, J.S.; Maciel, G.R.; Terenzi, H.; Pires, A.; Soldi, V. Effect of Concentration, Temperature and Plasticizer Content on Rheological Properties of Sodium Caseinate and Sodium Caseinate/Sorbitol Solutions and Glass Transition of their Films. *Food Chem.* **2003**, *82*, 425–431. [CrossRef]

46. De Figueiredo Furtado, G.; Mantovani, R.A.; Consoli, L.; Hubinger, M.D.; da Cunha, R.L. Structural and Emulsifying Properties of Sodium Caseinate and Lactoferrin Influenced by Ultrasound Process. *Food Hydrocoll.* **2017**, *63*, 178–188. [CrossRef]

47. Taha, A.; Hu, T.; Hu, H.; Zhang, Z.; Bakry, A.M.; Khalifa, I.; Pan, S. Effect of Different Oils and Ultrasound Emulsification Conditions on the Physicochemical Properties of Emulsions Stabilized by Soy Protein Isolate. *Ultrason. Sonochem.* **2018**, *49*, 283–293. [CrossRef]

48. Liang, Y.; Gillies, G.; Patel, H.; Matia-Merino, L.; Ye, A.; Golding, M. Physical Stability, Microstructure and Rheology of Sodium-Caseinate-Stabilized Emulsions as Influenced by Protein Concentration and Non-Adsorbing Polysaccharides. *Food Hydrocoll.* **2014**, *36*, 245–255. [CrossRef]

49. Yerramilli, M.; Ghosh, S. Long-Term Stability of Sodium Caseinate-Stabilized Nanoemulsions. *J. Food Sci. Technol.* **2017**, *54*, 82–92. [CrossRef]

50. Dickinson, E. Structure Formation in Casein-Based Gels, Foams, and Emulsions. *Colloids Surf. A Physicochem. Eng. Asp.* **2006**, *288*, 3–11. [CrossRef]
51. Silva, E.K.; Gomes, M.T.M.S.; Hubinger, M.D.; Cunha, R.L.; Meireles, M.A.A. Ultrasound-Assisted Formation of Annatto Seed Oil Emulsions Stabilized by Biopolymers. *Food Hydrocoll.* **2015**, *47*, 1–13. [CrossRef]
52. Desrumaux, A.; Marcand, J. Formation of Sunflower Oil Emulsions Stabilized by Whey Proteins with High-Pressure Homogenization (Up to 350 MPa): Effect of Pressure on Emulsion Characteristics. *Int. J. Food Sci. Technol.* **2002**, *37*, 263–269. [CrossRef]

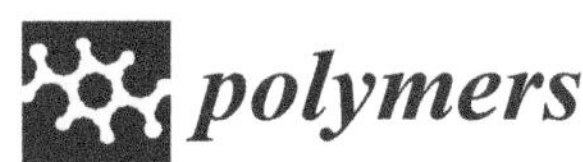 

Communication

In-Vitro Hemocompatibility of Polyaniline Functionalized by Bioactive Molecules

Kateřina Skopalová [1], Zdenka Capáková [1], Patrycja Bober [2], Jana Pelková [3,4], Jaroslav Stejskal [2], Věra Kašpárková [1,5], Marián Lehocký [1,5], Ita Junkar [6], Miran Mozetič [6] and Petr Humpolíček [1,5,*

[1]   Centre of Polymer Systems, Tomas Bata University in Zlín, 760 01 Zlín, Czech Republic;
    skopalova@utb.cz (K.S.); capakova@utb.cz (Z.C.); vkasparkova@utb.cz (V.K.); lehocky@utb.cz (M.L.)
[2]   Institute of Macromolecular Chemistry, Academy of Sciences of the Czech Republic, 162 06 Prague 6,
    Czech Republic; bober@imc.cas.cz (P.B.); stejskal@imc.cas.cz (J.S.)
[3]   Department of Hematology, Tomas Bata Regional Hospital in Zlín, 762 75 Zlín, Czech Republic;
    pelkova@fhs.utb.cz
[4]   Faculty of Humanities, Tomas Bata University in Zlín, 760 01 Zlín, Czech Republic
[5]   Faculty of Technology, Tomas Bata University in Zlín, 760 01 Zlín, Czech Republic
[6]   Jožef Stefan Institute, Jamova 39, Ljubljana 1000, Slovenia; ita.junkar@ijs.si (I.J.); miran.mozetic@ijs.si (M.M.)
*   Correspondence: humpolicek@utb.cz

Received: 6 September 2019; Accepted: 1 November 2019; Published: 11 November 2019

Abstract: Hemocompatibility is an essential prerequisite for the application of materials in the field of biomedicine and biosensing. In addition, mixed ionic and electronic conductivity of conducting polymers is an advantageous property for these applications. Heparin-like materials containing sulfate, sulfamic, and carboxylic groups may have an anticoagulation effect. Therefore, sodium dodecylbenzenesulfonate, 2-aminoethane-1-sulfonic acid and N-(2-acetamido)-2-aminoethanesulfonic acid were used for modification of the representative of conducting polymers, polyaniline, and the resulting products were studied in the context of interactions with human blood. The anticoagulation activity was then correlated to surface energy and conductivity of the materials. Results show that anticoagulation activity is highly affected by the presence of suitable functional groups originating from the used heparin-like substances, and by the properties of polyaniline polymer itself.

Keywords: polymer conductivity; conducting polymer; polyaniline; hemocompatibility

---

1. Introduction

The hemocompatibility of any material is a crucial factor in its application in biomedicine, regenerative medicine, or biosensors. The phenomenon of hemocompatibility is very complex and comprises various processes of which blood coagulation remains at the center of attention when tissue engineering and biosensing applications are considered. Any material that comes in contact with blood should not induce coagulation either by disrupting physiological blood flow or by interactions of the blood with the surface of a material on a molecular level.

In some special applications (e.g., biosensing or tissue engineering of electrosensitive tissues) the conductivity of materials can be an advantage. However, common conducting materials (such as metals or metal oxides) can induce an undesired reaction on contact with tissues (e.g., because of their different elasticity compared to native tissue). Therefore, conducting polymers can be more appropriate. Moreover, conducting polymers have a huge potential for the mentioned applications as they show mixed ionic and electronic conductivity, which is preferred for electrically active biointerfaces. The effect of conductivity on the biological systems is a comprehensive issue. We, therefore, refer to the publication of Rivnay et al. [1], who discuss the topic of ionic and electronic conductivity in depth.

Polypyrrole (PPy), poly(3,4-ethylenedioxythiophene) (PEDOT) and polyaniline (PANI) can be considered as the most frequently studied conducting polymers. The biocompatibility of these polymers was recently studied in many research works. For example, the Ramanaviciene et al. [2] tested the effect of PPy particles on mouse peritoneum cells. The action of injected PPy particles was dependent on their concentration and duration of treatment. Nevertheless, the amount of neutrophils did not exceed the limit values at any of the particle concentrations used. Vaitkuviene et al. [3] also investigated the effect of PPy nanoparticle concentration on living systems. It has been demonstrated that PPy nanoparticles exhibit a cytotoxic effect on murine embryonic stem cells, murine hepatoma cells, and human Jurkat T lymphocytes at concentrations higher than 19.4 μg·mL^{-1}. From a practical point of view, the polymer films are more important as regards bio applications. Vaitkuviene et al. [4] coated gold-plated glass slides by PPy. Proliferation of mouse stem cells on the treated surfaces was comparable to reference (tissue plastic). The PANI, which is the object of this work, was also previously studied in the context of biocompatibility. Its biocompatibility and especially the cytotoxicity were tested on pristine PANI salt and base [5], as well as on its globular and nanotubular forms [6]. In the scientific literature, it is often mentioned that PPy exhibits favorable properties compared to PANI. However, the study of Humpolíček et al. [7] dealing with the comparison of the biocompatibility of PANI and PPy in terms of cytotoxicity and embryotoxicity showed that the form of polymer (salt vs. base) is more important than its type (PPy vs. PANI). Likewise, the present study compares, inter alia, two forms of PANI, salt and base.

The amounts of information found in the scientific literature about the hemocompatibility of conducting polymer differ. Whilst PPy has been intensely studied, predominantly in functionalized forms or as a part of composites, PEDOT or polyaniline has been investigated as possible biomaterials but their hemocompatibility has not been studied. Li et al. [8] dealt with the blood compatibility of PPy films functionalized by heparin. The compatibility was evaluated using plasma re-calcification time and platelet adhesion. The results showed that, with immobilized heparin, platelet adhesion and platelet activation on PPy films were significantly suppressed and, moreover, the plasma re-calcification time was considerably prolonged.

Ferraz et al. [9] tested the hemocompatibility of nanocellulose/PPy membranes modified by a stable heparin coating. The results demonstrated that the heparinized composites were comparable with hemocompatible polysulfone regarding platelet adhesion and thrombin generation, whereas, in terms of complement activation, they were more biocompatible than commercially available membranes. Mao et al. [10] investigated *O*-butyryl chitosan-grafted PPy film for blood compatibility using platelet-rich plasma. The surface with immobilized *O*-butyryl chitosan exhibited lower platelet adhesion and fibrinogen adsorption compared to unmodified PPy. The films showed good blood compatibility and high electrical conductivity.

Previously it was mentioned that the functionalization of neat PANI films with poly(2-acrylamido-2-methyl-1-propanesulfonic acid) (PAMPSA) results in the hindering of blood coagulation [11]. Humpolíček et al. [11] followed up the studies of Paneva et al. [12] and Yancheva et al. [13] who have shown that PAMPSA can act against blood clotting in a similar way as heparin, either alone or incorporated in copolymers. The biological activity of PAMPSA was studied by Šorm et al. [14] who found out that some methacrylic copolymers that contain (similarly to heparin) sulfate, sulfamic, and carboxylic groups have an anticoagulation effect. Therefore, the compounds holding some of these functional groups were chosen for the functionalization of neat PANI films in this work. Dodecylbenzenesulfonic acid sodium salt (SDBS), 2-aminoethane-1-sulfonic acid (taurine), and *N*-(2-acetamido)-2-aminoethanesulfonic acid (ACES) were, therefore, chosen as modifiers with a potential anticoagulation effect.

2. Materials and Methods

2.1. Sample Preparation

PANI films containing the above-given compounds (Figure 1) were prepared using two different procedures: (1) By modification of the surfaces of neat PANI films, PANI salt, or PANI base with SDBS, taurine or ACES, and (2) by adding the respective substance directly into a reaction mixture of aniline hydrochloride and ammonium peroxydisulfate used for the preparation of PANI.

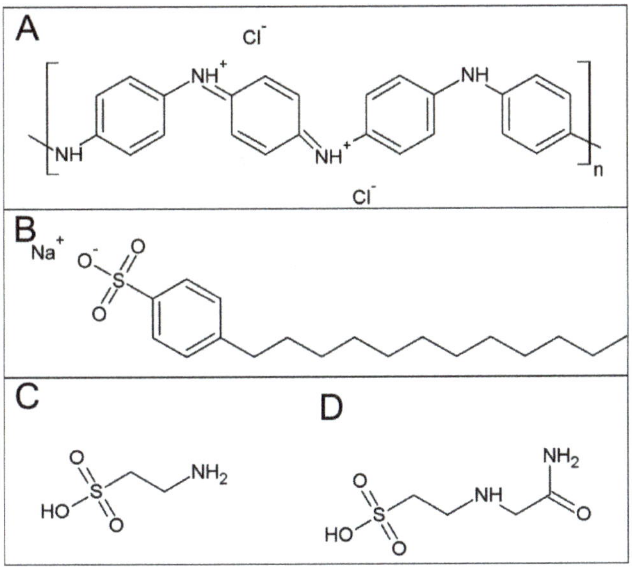

Figure 1. The formula of (**A**) PANI hydrochloride, (**B**) SDBS, (**C**), taurine, and (**D**) ACES.

Neat PANI films were prepared according to a standard procedure described by Stejskal and Gilbert [15]. Aniline hydrochloride (2.59 g, Penta, Czech Republic) was dissolved in water to a 50 mL solution, and ammonium peroxydisulfate (5.71 g, Lach-Ner, Czech Republic) was similarly dissolved to a 50 mL solution. Both solutions were mixed and immediately poured onto the polypropylene foil and into polyethylene terephthalate (PET) blood-collection tubes. After 1 h, the reaction mixture was removed and the resulting films of green conducting salt (PANI-S) deposited on the foils/walls were rinsed with 0.2 M hydrochloric acid, followed by methanol, and were left to dry in air. Some of the films were deprotonated by immersion in 1 M ammonium hydroxide for 12 h and thus converted to blue, non-conducting films of PANI base (PANI-B). In order to prepare reprotonated films, the samples of neat PANI-B were exposed to 2 wt % solutions of SDBS, taurine or ACES (all from Sigma-Aldrich, St. Louis, MO, USA). Please note that the reprotonation of PANI-B with the individual compounds manifests itself by the reverse change in color from blue to green. The process was slow and took several weeks as a rule (for reprotonation with ACES it took three months). Then, the solutions were removed, and the films were rinsed with methanol and left to dry in air. Modified PANI-S films were prepared by simple pouring each of the 2% solutions of SDBS, taurine or ACES onto the neat film. After 24 h, the solutions of bioactive substances were removed, and the films rinsed with methanol and dried in air. The samples were labelled as PANI-BSDBS, PANI-BTaurine, PANI-BACES, PANI-SSDBS, PANI-STaurine, and PANI-SACES.

The second type of PANI film was prepared by adding the respective substance into a mixture of aniline hydrochloride and ammonium peroxydisulfate. Aniline hydrochloride (2.59 g) was again dissolved in 50 mL of an aqueous solution of SDBS or taurine (40 g·L^{-1}). Then, the aqueous solution

(50 mL) of ammonium peroxydisulfate (5.71 g) was added, and the reaction mixture was stirred and poured over the substrates (polypropylene foil and blood-collection tubes). The reaction was left to continue for 1 h. The films of PANI salts formed on the substrates containing respective dopants were then rinsed with 0.2 M HCl and were left to dry in the air. The samples were labelled as PANI-MSDBS and PANI-MTaurine. Polyaniline powders obtained after polymerization along with the films were treated similarly and used for conductivity determination.

2.2. Surface Energy Measurement

Contact angle data was obtained with a Surface Energy Evaluation System (Advex Instruments, Brno, Czech Republic). Deionized water, ethylene glycol, and diiodomethane were utilized as test liquids. The volume of droplets was set to 5 µL for all experiments to avoid errors connected with gravity acting on the sessile drop. Five contact angle readings were averaged to obtain one representative value. The free energy of the film surfaces was determined by the Lifshitz–van der Waals "acid–base" model. Total surface energy (γ^{tot}) was calculated and reported.

2.3. Conductivity

The conductivity of the polyaniline powders collected after polymerization and compressed to pellets was measured by the four-point van der Pauw method. A programmable electrometer with an SMU Keithley 237 current source and a Multimeter Keithley 2010 voltmeter (Keithley instrument, Cleveland, OH, USA) with a 2000 SCAN 10-channel scanner card were employed. Measurements were carried out at ambient temperature.

2.4. Anticoagulation Test

In all tests, venous blood was collected from healthy donors by venipuncture using the vacuum blood collection system into the 5 mL collecting tubes (VACUETTE, Greiner Bio-One, Kremsmünster, Austria) after obtaining informed consent. All tests were conducted in accordance with the Helsinki Declaration. Plasma was prepared from the venous blood by centrifugation (15 min, 3000× g). Plasma was subsequently transferred to the blood-collection tubes coated with studied PANI films. The measurement of coagulation was performed using an instrument commercially used in hospitals, the SYSMEX CA-1500 (Siemens, Erlangen, Germany). The principle of measurement is based on the change of turbidity (measured at a wavelength of 660 nm) as a consequence of adding coagulation reagent which induces the formation of fibrin clothes. In the present study, the following coagulation parameters in human blood plasma treated with 0.109 M citric were studied: (1) Thrombin clotting time (TCT), (2) activated partial thromboplastin time (aPTT), and (3) prothrombin time (PT).

3. Results and Discussion

Hemocompatibility is a complex process and its testing by common in-vitro methods is not able to comprehensively cover all its aspects. Blood coagulation as a whole includes not only the coagulation cascade but also the platelet adhesion and can be, for example, triggered by various biochemical cues or blood rheology (flow velocity, flow turbulence, shear stress) [16]. Especially in the case of platelet adhesion, the shear rates of the fluid are important. In this work, we focused on one of the important aspects of blood coagulation, namely on the detection of coagulation induced by the surface properties of modified PANI films. Although the biological properties of PANI in its native or modified form have already been studied, and cytotoxicity [5], interaction with stem cells [17], or interaction with tissues have already been reported, another important parameter of biocompatibility, namely interaction with blood has only been the subject of a few studies. The effect of PANI itself on platelet adhesion, hemolysis and plasma recalcification time has been studied by Li et al. [18]. Here, PANI-coated polyurethane (PU) fibers were tested. Platelets on PANI-coated PU demonstrated lower aggregation (6.87×10^5 cm^{-2}) than platelets on PU fibers without modification (15.63×10^5 cm^{-2}). As regards hemolysis, according to requirements of ISO 10993-4, materials with hemolysis values <5% are considered safe in contact with blood. In the study, PANI-PU fibers exhibited hemolysis values of 0.14% and non-coated PU fibers 0.21%. The tests also showed that PANI prolonged plasma recalcification time by 13 s, improving thus anticoagulation. The impact of PANI on the blood coagulation can be also improved by modification of its surface by bioactive substances. For example, a work of Zhang et al. [19] studied modifications of PANI films with poly(ethylene glycol) (PEG) to reduce the adhesion of proteins and platelets on its surface. In this case, the PEG chain was covalently bound to PANI film. Adhesion of bovine serum albumin (BSA) and γ-globulin decreased with increasing concentration of surface-bound PEG. The same results were obtained for platelet adhesion. Li and Ruckenstein [20] tested BSA and platelet adhesions on native PANI and poly(ethylene oxide) (PEO) grafted films. Correspondingly to the previous study, modification of the PANI film with hydrophilic PEO also reduced protein adhesion, in this case by 80%. Platelet adhesion also decreased significantly. In the case of the unmodified film, there was adhesion of 18×10^3 mm^{-2}. For the modified film, the adhesion value was 2.1×10^3 mm^{-2}. Similar results were achieved in work of Humpolicek et al. [11]. This report investigated the impact of PANI functionalized with PAMPSA on blood coagulation and found out that PANI films reprotonated with this high-molecular-weight acid hindered blood coagulation by interaction with three of the coagulation factors: Xa, Va, and IIa. Moreover, PAMPSA-modified PANI also reduced platelet adhesion. It was generally expected that methacrylic copolymers containing similarly to heparin the sulfate, sulfamic, and carboxylic groups, may have an anticoagulation effect. With respect to this assumption, SDBS, taurine, and ACES might have the potential to influence blood compatibility as well. As PANI is a versatile material that can easily be modified by a variety of methods, the PANI films were prepared using the above-mentioned substances either by their direct addition into the reaction mixture with subsequent preparation of films, or by modification of neat PANI-S and PANI-B films via pouring over the solutions of each of these substances.

The standard polyaniline salt, PANI-S, contains chloride counter-ions (Figure 1A), as well as sulfate ions originating from the reduction of peroxydisulfate. These counter-ions are removed after the conversion to PANI-B. After immersion of PANI-B into the solution of, e.g., taurine, immobilization of this substance onto PANI film takes place resulting in their mutual salt (PANI-BTaurine) containing corresponding taurine counter-ions. As regards PANI-S films modified with studied heparin-like substances, it can be assumed that a layer of the respective substance remains on the surface of neat PANI. When PANI is prepared in the presence of taurine in reaction mixture, PANI-MTaurine, the films contain chloride, sulfate, and taurine counter-ions. The PANI films prepared in various ways thus differ in the type and occurrence of counter-ions, or more general in surface chemistry.

Blood coagulation can be triggered by contact with interfaces due to the following three crucial characteristics/properties of a surface: (1) The surface energy, (2) the acidity (pH), and (3) the interaction of surfaces with coagulation factors of blood. The adherence of plasma proteins and the subsequent adherence of platelets and leukocytes are also of importance regarding the plasmatic coagulation, which again depends on the surface characteristics of the material. The surface energies measured on modified PANI films were similar for almost all samples (Table 1) and did not differ from γ^{tot} determined on neat PANI-S and PANI-B films [21]. The exception here was PANI-BACES with γ^{tot} equal to 41 ± 1 mN cm^{-1}. For practical application, the stability of the samples is crucial, and the effect of aging on the surface energy was also determined. The surface energy of all samples was stable for at least two weeks, except for the PANI-BTaurine and PANI-STaurine films (Table 1).

It is known that the physiological pH of blood lies within a tight range from 7.31 to 7.42 [22], and several non-physiological reactions can occur as a consequence of the pH changes, including inhibition and prolongation of coagulation processes. The pH of blood after being in contact with the samples was therefore determined, and it was unambiguously confirmed that the pH was not notably influenced by any of the tested samples.

Table 1. The surface energy of modified PANI surfaces, γ^{tot}, determined immediately after preparation, and after 7 and 14 days on the stored under laboratory conditions. The values are compared with γ^{tot} of neat PANI-S and PANI-B films.

	$\gamma^{tot} \pm$ SD (mN·m^{-1})		
	Day 0	**Day 7**	**Day 14**
PANI-BSDBS	51 ± 3	51 ± 2	50 ± 7
PANI-SSDBS	51 ± 3	47 ± 4	44 ± 4
PANI-BTaurine	51 ± 7	53 ± 2	38 ± 12
PANI-STaurine	55 ± 8	45 ± 7	47 ± 2
PANI-BACES	41 ± 1	37 ± 1	45 ± 0
PANI-MSDBS	46 ± 5	47 ± 4	46 ± 4
PANI-MTaurine	53 ± 4	44 ± 10	46 ± 4
PANI-S [a]	52.54	n.d.	n.d.
PANI-B [a]	50.88	n.d.	n.d.

[a] Reproduced from [21]; n.d. not determined.

Conductivity is another important characteristic of conducting polymers. The standard films of PANI-S commonly show conductivity within units of S·cm^{-1} [23]. The conductivity measurements on fresh samples, and after 7 and 14 days of their storage at room temperature are summarized in Table 2. A certain conductivity drift is likely to be associated with still decreasing sample humidity. Measurements revealed that only the samples synthesized with SDBS or taurine present in the reaction mixture showed reasonably high conductivity and all other films exhibited conductivity lower than 10^{-4} S·cm^{-1}. It is beyond the scope of this manuscript to define the relationship between the conductivity of materials and their applicability in biomedicine. This is a comprehensive issue and the situation is complicated not only by the fact that conducting polymers combine ionic and electronic conductivity but also by the different resistivity of various tissues (e.g., from 100 Ω·cm for blood to 1000 Ω·cm for some tissues) [24] which corresponds to the conductivity of 0.01 to 0.001 S·cm^{-1}. In this context, the conductivity of here-tested PANI samples, especially these where taurine was present in the reaction mixture, is significantly higher reaching ~18 S·cm^{-1}.

Table 2. The conductivity of bulk PANI (S cm^{-1}) containing the studied substances.

Sample	Day 0	Day 7	Day 14
PANI-MSDBS	7.0 ± 0.03	5.7 ± 0.002	5.2 ± 0.004
PANI-MTaurine	17.9 ± 0.01	16.5 ± 0.004	15.8 ± 0.003

The coagulation parameters of the blood after contact with the studied surfaces are presented in Table 3. The thrombin clotting time (TCT), activated partial thromboplastin time (aPTT), and prothrombin time (PT) were studied as clinically relevant parameters using the methods described earlier [11]. Similarly to neat PANI-S and PANI-B [11], none of the PANI modified with SDBS, taurine or ACES significantly impacted blood coagulation (Table 3), and all the coagulation parameters were within their respective physiological ranges. The expected impact of taurine, SDBS, or ACES on the hemocompatibility was not hence confirmed. So far, the only compound suitable for influencing the hemocompatibility of PANI is, as previously described [11], PAMPSA. Here, an interesting question arises with regard to the reason why SDBS, taurine, or ACES were not able to modify the hemocompatibility of PANI even though they contain functional groups similar to PAMPSA.

Table 3. Impact of PANI with modified surfaces on selected coagulation parameters, prothrombin time (PT), activated partial thromboplastin time (aPPT), thrombin clotting time (TCT), expressed as times to the coagulation start.

	PT [s]	aPPT [s]	TCT [s]
Reference	12.1 ± 0.1	25.2 ± 0.5	16.7 ± 0.3
PANI-BSDBS	12.2 ± 0.6	29.5 ± 4.7	20.0 ± 2.2
PANI-SSDBS	11.8 ± 0.2	25.6 ± 0.3	18.3 ± 0.3
PANI-BTaurin	11.8 ± 0.0	25.4 ± 0.5	17.7 ± 0.5
PANI-STaurin	11.8 ± 0.2	26.9 ± 0.6	19.2 ± 0.4
PANI-BACES	12.3 ± 0.0	25.4 ± 0.3	15.4 ± 0.4
PANI-MSDBS	12.0 ± 0.1	25.8 ± 0.6	18.2 ± 0.5
PANI-MTaurin	12.1 ± 0.3	30.0 ± 2.5	19.7 ± 0.6

Note: As reference coagulation parameters of donor blood were used. Normal ranges for coagulation parameters of a healthy person: PT 11.0–13.5, aPTT 25–32, TCT below 20 s. The values are expressed as mean value ± standard deviation of three tests.

Based on the results presented in Tables 1–3 and on previously published findings, the reason for this can be found in the low-molecular-weight character of these substances, which in all three cases lies below 350 g·mol^{-1}. On the other hand, in the case of hemocompatible PANI-PAMPSA, its anticoagulation activity can be also assigned to the presence of the high-molecular-weight PAMPSA polyanion (molar mass $\approx 10^6$ g·mol^{-1}) bound to polycationic PANI surface. During reprotonation, a fraction of anions on a PAMPSA chain is bound as counter-ions to the PANI backbone. However, a substantial fraction of the functional sulfo groups still remains free to interact with blood and can act as an anticoagulant agent. In this context, it can be mentioned that also the best-known anticoagulation agent, heparin, is a polymer. This polysaccharide consists of chains containing 1,4-bonded residues of uronic acid and D-glucosamine, and its molar mass ranges from 3000 to 30,000 g·mol^{-1}, with an average of 15,000 g·mol^{-1} [25]. However, its anticoagulant activity is closely related to a unique pentasaccharide sequence necessary for linking to antithrombin [26]. In the heparin chain without this pentasaccharide sequence, the anticoagulation activity is absent [27].

In summary, the present pilot study covers only a limited part of the complex hemocompatibility process. The impact of above-mentioned parameters, such as blood flow velocity, flow turbulence, shear stress, or tension on the hemocompatibility of PANI-based surfaces has not been studied so far. It, therefore, opens the possibility for further testing of the effect of the low-molecular-weight substances with sulfo groups on the hemocompatibility, which can be conducted under physiological flow conditions and using blood flow models.

4. Conclusions

The PANI films modified by substances with anticipated anticoagulant activity, sodium dodecylbenzenesulfonate (SDBS), 2-aminoethane-1-sulfonic acid (taurine) and *N*-(2-acetamido)-2-aminoethanesulfonic acid (ACES) have been studied. The hemocompatibility tests conducted on these PANI films confirmed that none of them showed anticipated anticoagulation activity, though the functional groups typical for anticoagulation substances were present. The absence of the activity can be ascribed to the low molecular weight of these compounds used for PANI modification. It can be, therefore, concluded that, in addition to the presence of suitable functional groups in the molecule, the dopants introducing anticoagulation activity should exhibit polymer-like character with sufficiently high molecular weight.

Author Contributions: Conceptualization, P.H., M.L., V.K., J.S., M.M. and P.B.; methodology, Z.C., K.S., I.J.; formal analysis, Z.C.; investigation, J.P. and K.S.; resources, P.H. and M.L.

Funding: This work was funded by the Czech Science Foundation (19-16861S) and the Ministry of Education, Youth and Sports of the Czech Republic (NPU I, LO1504). One of us, K.S., acknowledges the support of an internal grant from TBU in Zlín (IGA/CPS/2019/004) financed from funds of specific academic research. Authors M.L., I.J., and M.M. would like to express their gratitude to the Slovenian Research Agency (P2-0082).

Acknowledgments: Authors thank Lada Utěkalová and Petra Rejmontová for technical support.

Conflicts of Interest: The authors declare no conflict of interest.

References

1. Rivnay, J.; Inal, S.; Collins, B.; Sessolo, M.; Stavrinidou, E.; Strakosas, X.; Tassone, C.; Delongchamp, D.; Malliaras, G. Structural control of mixed ionic and electronic transport in conducting polymers. *Nat. Commun.* **2016**, *7*, 11287. [CrossRef] [PubMed]
2. Ramanaviciene, A.; Kausaite-Minkstimiene, A.; Tautkus, S.; Ramanavicius, A. Biocompatibility of polypyrrole particles: An in-vivo study in mice. *J. Pharm. Pharm.* **2007**, *59*, 311–315. [CrossRef] [PubMed]
3. Vaitkuviene, A.; Kaseta, V.; Voronovic, J.; Ramanauskaite, G.; Biziuleviciene, G.; Ramanaviciene, A.; Ramanavicius, A. Evaluation of cytotoxicity of polypyrrole nanoparticles synthesized by oxidative polymerization. *J. Hazard. Mater.* **2013**, *250*, 167–174. [CrossRef] [PubMed]
4. Vaitkuviene, A.; Ratautaite, V.; Mikoliunaite, L.; Kaseta, V.; Ramanauskaite, G.; Biziuleviciene, G.; Ramanaviciene, A.; Ramanavicius, A. Some biocompatibility aspects of conducting polymer polypyrrole evaluated with bone marrow-derived stem cells. *Colloid Surf. A* **2014**, *442*, 152–156. [CrossRef]
5. Humpolicek, P.; Kasparkova, V.; Saha, P.; Stejskal, J. Biocompatibility of polyaniline. *Synth. Met.* **2012**, *162*, 722–727. [CrossRef]
6. Stejskal, J.; Hajná, M.; Kašpárková, V.; Humpolíček, P.; Zhigunov, A.; Trchová, M. Purification of a conducting polymer, polyaniline, for biomedical applications. *Synth. Met.* **2014**, *195*, 286–293. [CrossRef]
7. Humpolíček, P.; Kašpárková, V.; Pacherník, J.; Stejskal, J.; Bober, P.; Capáková, Z.; Radaszkiewicz, K.A.; Junkar, I.; Lehocký, M. The biocompatibility of polyaniline and polypyrrole: A comparative study of their cytotoxicity, embryotoxicity and impurity profile. *Mater. Sci. Eng. C* **2018**, *91*, 303–310. [CrossRef] [PubMed]
8. Li, Y.L.; Neoh, K.G.; Cen, L.; Kang, E.T. Physicochemical and blood compatibility characterization of polypyrrole surface functionalized with heparin. *Biotechnol. Bioeng.* **2003**, *84*, 305–313. [CrossRef] [PubMed]
9. Ferraz, N.; Carlsson, D.O.; Hong, J.; Larsson, R.; Fellstrom, B.; Nyholm, L.; Stromme, M.; Mihranyan, A. Haemocompatibility and ion exchange capability of nanocellulose polypyrrole membranes intended for blood purification. *J. R. Soc. Interface* **2012**, *9*, 1943–1955. [CrossRef] [PubMed]
10. Mao, C.; Zhu, A.P.; Wu, Q.; Chen, X.B.; Kim, J.H.; Shen, J. New biocompatible polypyrrole-based films with good blood compatibility and high electrical conductivity. *Colloid Surf. B* **2008**, *67*, 41–45. [CrossRef] [PubMed]
11. Humpolíček, P.; Kuceková, Z.; Kašpárková, V.; Pelková, J.; Modic, M.; Junkar, I.; Trchová, M.; Bober, P.; Stejskal, J.; Lehocký, M. Blood coagulation and platelet adhesion on polyaniline films. *Colloid Surf. B* **2015**, *133*, 278–285. [CrossRef] [PubMed]
12. Paneva, D.; Stoilova, O.; Manolova, N.; Danchev, D.; Lazarov, Z.; Rashkov, I. Copolymers of 2-acryloylamido-2-methylpropanesulfonic acid and acrylic acid with anticoagulant activity. *E-Polymers* **2003**, *3*, 11. [CrossRef]
13. Yancheva, E.; Paneva, D.; Danchev, D.; Mespouille, L.; Dubois, P.; Manolova, N.; Rashkov, I. Polyelectrolyte complexes based on (quaternized) poly (2-dimethylamino)ethyl methacrylate: Behavior in contact with blood. *Macromol. Biosci.* **2007**, *7*, 940–954. [CrossRef] [PubMed]
14. Šorm, M.; Nešpůrek, S.; Mrkvičková, L.; Kálal, J.; Vorlová, Z. Anticoagulation activity of some sulfate-containing polymers of the methacrylate type. *J. Polym. Sci. C* **1979**, *66*, 349–356. [CrossRef]
15. Stejskal, J.; Gilbert, R.G. Polyaniline. Preparation of a conducting polymer (IUPAC technical report). *Pure Appl. Chem.* **2002**, *74*, 857–867. [CrossRef]
16. Secomb, T.W. Hemodynamics. *Compr. Physiol.* **2016**, *6*, 975–1003. [PubMed]
17. Bober, P.; Humpolíček, P.; Pacherník, J.; Stejskal, J.; Lindfors, T. Conducting polyaniline based cell culture substrate for embryonic stem cells and embryoid bodies. *RSC Adv.* **2015**, *5*, 50328–50335. [CrossRef]
18. Li, Y.M.; Zhao, R.; Li, X.; Wang, C.Y.; Bao, H.W.; Wang, S.D.; Fang, J.; Huang, J.Q.; Wang, C. Blood-compatible polyaniline coated electrospun polyurethane fiber scaffolds for enhanced adhesion and proliferation of human umbilical vein endothelial cells. *Fiber. Polym.* **2019**, *20*, 250–260. [CrossRef]

19. Zhang, F.; Kang, E.T.; Neoh, K.G.; Wang, P.; Tan, K.L. Reactive coupling of poly(ethylene glycol) on electroactive polyaniline films for reduction in protein adsorption and platelet adhesion. *Biomaterials* **2002**, *23*, 787–795. [CrossRef]

20. Li, Z.F.; Ruckenstein, E. Grafting of poly(ethylene oxide) to the surface of polyaniline films through a chlorosulfonation method and the biocompatibility of the modified films. *J. Colloid Interface Sci.* **2004**, *269*, 62–71. [CrossRef]

21. Humpolíček, P.; Radaszkiewicz, K.; Kašpárková, V.; Stejskal, J.; Trchová, M.; Kuceková, Z.; Vičarová, H.; Pachernik, J.; Lehocký, M.; Minarik, A. Stem cell differentiation on conducting polyaniline. *RSC Adv.* **2015**, *5*, 68796–68805. [CrossRef]

22. Burtis, C.A.; Ashwood, E.R.; Bruns, D.E. *Tietz Textbook of Clinical Chemistry and Molecular Diagnostics*, 4th ed.; Elsevier: St. Louis, MO, USA, 2006; p. 2412.

23. Stejskal, J.; Sapurina, I. Polyaniline: Thin films and colloidal dispersions - (IUPAC technical report). *Pure Appl. Chem.* **2005**, *77*, 815–826. [CrossRef]

24. Kay, C.; Schwan, H. Specific resistance of body tissues. *Circul. Res.* **1956**, *4*, 664–670.

25. Kaushansky, K.; Lichtman, M.; Beutler, E.; Kipps, T.; Prchal, J.; Seligsohn, U. Principles of antithrombotic therapy. In *Williams Hematology*, 7th ed.; McGraw-Hill Professional: New York, NY, USA, 2006.

26. Lam, L.H.; Silbert, J.E.; Rosenberg, R.D. Separation of active and inactive forms of heparin. *Biochem. Biophys. Res. Commun.* **1976**, *69*, 570–577. [CrossRef]

27. Mosier, P.; Krishnasamy, C.; Kellogg, G.; Desai, U. On the specificity of heparin/heparan sulfate binding to proteins. Anion-binding sites on antithrombin and thrombin are fundamentally different. *PLoS ONE* **2012**, *7*, e48632. [CrossRef] [PubMed]

Article

Preparation of Progressive Antibacterial LDPE Surface via Active Biomolecule Deposition Approach

Salma Habib [1], **Marian Lehocky** [2,3], **Daniela Vesela** [2], **Petr Humpolíček** [2,3], **Igor Krupa** [1] and **Anton Popelka** [1,*]

[1] Center for Advanced Materials, Qatar University, P.O. Box 2713, Doha, Qatar; salma.m.habib@hotmail.com (S.H.); igor.krupa@qu.edu.qa (I.K.)

[2] Centre of Polymer Systems, Tomas Bata University in Zlin, Trida Tomase Bati 5678, 760 01 Zlin, Czech Republic; lehocky@post.cz (M.L.); dvesela@utb.cz (D.V.); humpolicek@utb.cz (P.H.)

[3] Faculty of Technology, Tomas Bata University in Zlin, Vavreckova 275, 760 01 Zlin, Czech Republic

* Correspondence: anton.popelka@qu.edu.qa; Tel.: +974-4403-5676

Received: 9 September 2019; Accepted: 15 October 2019; Published: 17 October 2019

Abstract: The use of polymers in all aspects of daily life is increasing considerably, so there is high demand for polymers with specific properties. Polymers with antibacterial properties are highly needed in the food and medical industries. Low-density polyethylene (LDPE) is widely used in various industries, especially in food packaging, because it has suitable mechanical and safety properties. Nevertheless, the hydrophobicity of its surface makes it vulnerable to microbial attack and culturing. To enhance antimicrobial activity, a progressive surface modification of LDPE using the antimicrobial agent grafting process was applied. LDPE was first exposed to nonthermal radio-frequency (RF) plasma treatment to activate its surface. This led to the creation of reactive species on the LDPE surface, resulting in the ability to graft antibacterial agents, such as ascorbic acid (ASA), commonly known as vitamin C. ASA is a well-known antioxidant that is used as a food preservative, is essential to biological systems, and is found to be reactive against a number of microorganisms and bacteria. The antimicrobial effect of grafted LDPE with ASA was tested against two strong kinds of bacteria, namely, *Staphylococcus aureus* (*S. aureus*) and *Escherichia coli* (*E. coli*), with positive results. Surface analyses were performed thoroughly using contact angle measurements and peel tests to measure the wettability or surface free energy and adhesion properties after each modification step. Scanning electron microscopy (SEM) and atomic force microscopy (AFM) were used to analyze the surface morphology or topography changes of LDPE caused by plasma treatment and ASA grafting. Surface chemistry was studied by measuring the functional groups and elements introduced to the surface after plasma treatment and ASA grafting, using Fourier transform infrared (FTIR) spectroscopy and X-ray photoelectron spectroscopy (XPS). These results showed wettability, adhesion, and roughness changes in the LDPE surface after plasma treatment, as well as after ASA grafting. This is a positive indicator of the ability of ASA to be grafted onto polymeric materials using plasma pretreatment, resulting in enhanced antibacterial activity.

Keywords: biointerface; polyethylene; plasma treatment; antibacterial; grafting modification

1. Introduction

Low-density polyethylene (LDPE) is the most common industrial polymer and is mainly used in food packaging because of its useful properties, such as its ease of shaping, handling, and recycling, and its high cost efficiency [1–4]. Nevertheless, LDPE lacks proficiency in traits such as printability, adhesion, and some other surface properties, as it has an inert surface with a very low surface free energy (wettability). Changing the bulk properties of LDPE by mixing additives is not recommended for food packaging applications; therefore, surface modification is a safe, easy, and cost-effective option [5,6].

Many studies have enhanced the surface properties of LDPE using physical or chemical methods together with conventional methods such as flame treatment, chemical grafting, irradiation, free-radical mechanism, and corona and radio-frequency (RF) plasma treatments [7–11]. These plasma treatments have been found to be more effective techniques for the modification of the surface properties of polymers without any disruption of bulk polymer properties [12,13]. Plasma, as the fourth state of matter, is able to initiate the ionization of air species and surface components, converting them into electrons and negative, positive, and neutral ions, in addition to metastable and free radicals [13–15]. The free radicals that are generated are readily reactive, with mainly oxygen-containing functional groups [16]. This results in enhancement of the wettability, adhesion, roughness, and reactivity properties. The plasma treatment methods vary from corona (atmospheric pressure) to vacuum-based RF plasma. All types were found to have similar effectiveness on the enhancement of the surface properties of polymers [17], even when the ionized gas species varied [18]. For nonthermal, low-temperature, and cold plasma, no heat is generated or required, and they only affect a layer of a few tens nanometers depth on the surface. Therefore, these methods are widely used in medical applications and medicine in general [16,19–21]. Plasma is also known to be safe for the environment and for human health, as it kills microorganisms and cleans medical equipment, with no significant impact on human cells [21–23].

In food packaging, microbial and bacterial fouling is a critical concern, and a large amount of research has been devoted to the use of antibacterial packaging materials to prevent the vulnerability and susceptibility of food to any type of microorganism and to increase the shelf life of the food [1,24,25]. Some antibacterial techniques involve mixing of the antibacterial agent within the polymeric material to generate biomaterials [26,27]. However, this process is not suitable for all packaging materials because it changes the main functional and mechanical properties of the materials and decreases their stability. Other studies have applied the antibacterial agents by surface grafting or surface tangling, with attachment on the surface achieved by chemical grafting, coating, or plasma treatment [28,29]. To enhance the antibacterial properties, different chemicals or nature-based compounds, such as chitosan [30–33] and alkyl pyridiniums [34], were grafted onto polymer surfaces and proved to reduce the number of, or entirely kill, bacteria (both gram-positive and gram-negative species). Other chemicals known as organic acids, which are mostly used as preservatives, were tested for their antibacterial activation when attached to polymers. Polyacrylic acid, for instance, was grafted with chitosan on LDPE and tested against the strong bacteria *Escherichia coli* (*E. coli*) [15]. These acids showed successful results in decreasing and eliminating the presence of bacteria on the LDPE surface.

Ascorbic acid (ASA), commonly known as vitamin C, is an organic acid known to be an antioxidant. It possesses two hydroxyl groups that can be deprotonated and is an effective radical scavenger [35,36]. It is an essential vitamin needed to maintain body health [37]. There are different studies on the antimicrobial effect of ascorbic acid that provide positive results. It was found to inhibit bacterial growth and prevent biological infections [38–40]. Further investigations found antiviral [40,41] as well as antifungal activities [40], alone or in combination with additional agents, to exert a synergetic effect. Tests with *E. coli*, *Staphylococcus aureus* (*S. aureus*), and some other types of bacteria [42–45] demonstrated the very effective ability of vitamin C to inhibit the growth of, and kill, bacterial colonies by penetrating bacterial walls, affecting the metabolism with no harmful effect on human cells [43]. ASA has the ability to enter a cell and modify its redox reaction through its hydroxyl groups, which eventually leads to the inhibition of microorganism growth; thus, ASA can be considered a good antimicrobial agent [46,47].

In this study, ASA was used for the preparation of a progressive LDPE surface through plasma-assisted grafting, which has excellent antimicrobial properties. The antibacterial effectiveness of the LDPE surface modified by ASA was tested against *E. coli* and *S. aureus*.

2. Materials and Methods

2.1. Materials

Commercial grade low-density polyethylene (LDPE) FE8000 was supplied in pellet form by Qatar Petrochemical Company (QAPCO, Doha, Qatar). Thin homogeneous films approximately 0.4 mm thick were prepared by compression molding using an industrial mounting press machine (Carver, Wabash, IN, USA). The pellets were melted at 160 °C and compressed for 2 min using a force of 2 tons, while maintaining the set temperature to obtain a film with the desired smooth surface. The samples were then cooled to room temperature by water. The LDPE films were cleaned by acetone to remove any additives, residuals, or any possible contaminations from the molding process that might affect the surface properties, and were then dried in an air atmosphere for 20 min at room temperature. Small strips (5 cm × 1 cm) were cut out and directly used for the surface treatment and subsequent analyses.

Ethylene glycol (>98% FLUKA, Morris Plains, NJ, USA), formamide (>98% FLUKA, Merelbeke, Belgium), ultra-pure water (prepared by Purification System Direct Q3, Millipore Corporation, Molsheim, France), and acetone (99.9% Scharlau, Barcelona, Spain) were used as testing liquids for wettability analyses.

L-ascorbic acid (>99.0% Research-Lab, Uran Islampur, India) molecular weight = 176.14 g/mol was used as an antimicrobial agent.

2.2. Plasma Treatment of LDPE

Plasma treatment of LDPE films was performed using a Venus 75-HF enclosed low-temperature plasma-generating system (Plasma Etch Inc., Carson, CA, USA). Plasma-excited species were generated using a radio-frequency (RF) generator operating at a frequency of 13.56 MHz. The chamber of the plasma system was evacuated to a pressure level of approximately 0.2 Torr using a rotary vacuum pump before plasma ignition. Optimization of the treatment process was carried out by varying the nominal power, treatment time, and working gas to obtain the maximum level of hydrophilicity on the LDPE surface. The applied nominal power varied from 50 W to 120 W, and the treatment time ranged from 10 s to 180 s at a constant optimal nominal power of 80 W. The gas flow rate was 10 cm^3/min. The film surfaces were treated from both sides in air.

2.3. Antibacterial Agent Grafting

Immediately after the plasma treatment, the LDPE samples were immersed in a 10 vol % aqueous solution of ASA. The immersion process was continuous for 24 hours at 24 °C to achieve radical grafting. ASA is converted to an ascorbate radical by electron donation to a radical [48], namely, the alkoxyl radical present in the plasma-treated LDPE surface created by the decomposition of hydroperoxide. The ascorbate radicals can then interact with the double bonds present in plasma-treated LDPE created by disproportionation reactions; therefore, ASA can be covalently grafted onto the LDPE surface (Figure 1). After the grafting process, the LDPE samples were thoroughly washed with water and ethanol to remove weakly bound or unreacted ASA from the LDPE surface.

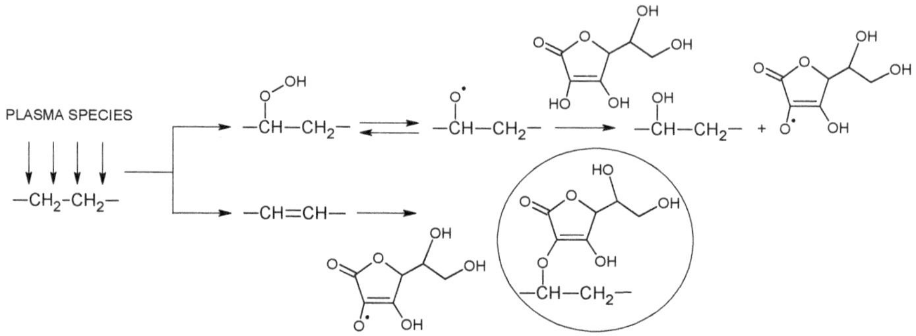

Figure 1. Scheme of ascorbic acid (ASA) grafting on low-density polyethylene (LDPE) via plasma treatment.

2.4. Hydroperoxide Determination

Iodometric titration was performed to determine the concentration of all hydroperoxide species accumulated on the surface of LDPE after plasma treatment. Plasma-treated LDPE samples were placed into a covered Erlenmeyer flask, which was filled with 50 mL of glacial acetic acid. An excess (1.0 g) of sodium iodide was added, and the flask was purged with argon gas for 15 min to eliminate interactions with air. After 15 min, the well-stirred mixture became yellow (oxidation of iodide to iodine by hydroperoxides incorporated on the LDPE surface) and was titrated with a 0.0005 M sodium thiosulfate pentahydrate aqueous solution. The reactions were carried out in an argon atmosphere and protected from light. The hydroperoxide concentration on the LDPE surface was calculated per area considering two treated sides of the LDPE samples. The titration was repeated 3 times to obtain average values and to ensure reliable results.

2.5. Surface Wettability Measurements

The changes in hydrophilicity induced by plasma treatment of LDPE films were evaluated by static contact angle measurements using the sessile drop method. An OCA35 surface free energy analysis system (DataPhysics, Filderstadt, Germany) equipped with a CCD camera was employed for this purpose. Water, formamide, and ethylene glycol were used as testing liquids to evaluate the total surface free energy and polar and dispersive components using the conventional Owens–Wendt–Rabel–Kaelble method. A droplet of approximately 3 μL of each testing liquid was placed on the air-facing samples. The contact angle was calculated after approximately 3 s to allow thermodynamic equilibrium between the liquid and the sample interface to be reached. The reported value for each testing liquid corresponds to the mean of at least five measurements taken on different parts of the substrate surface.

2.6. Graft Yield Analysis

Graft yield measurements were used to prove the grafting of ASA on the LDPE surface. The graft yield of modified LDPE was calculated by gravimetric measurements. The graft yield (GY) was calculated by Equation (1):

$$GY[\%] = ((W_2 - W_1)/W_1) \cdot 100\% \tag{1}$$

where W_1 and W_2 represent the weights of the LDPE samples before and after the modification.

2.7. Film Thickness Investigation

Thickness measurements were carried out by an F20-UVX film thickness analyzer (Filmetrics, San Diego, CA, USA) to analyze the thickness of plasma-affected and ASA-modified layers of the LDPE surface. The film thickness value was evaluated based on the differences in reflectance (%) between reference and measured samples in wavelength range of 190–1700 nm. LDPE substrate (4.5 mm thick, with refractive index of 1.5) and LDPE substrate (4.5 mm thick, with refractive index of 1.4, considering polar functional groups) were used as reference samples for the thickness measurements of plasma-treated and ASA-modified LDPE layers. Analysis of plasma-treated and ASA-modified LDPE samples was performed in air atmosphere. The spectrum was analyzed by varying the measured parameters to obtain the best fit between the theoretical and measured data using FILMeasure software, v7.19.0. Readings from five different areas were captured for each sample, and a mean value was evaluated.

2.8. Peel Test

A 90° peel test was performed to measure the adhesion characteristics of LDPE samples in terms of the peel resistance using a Lloyd 1K Lf plus-UTM standard testing machine (Lloyd Instruments, West Sussex, UK). Samples 19 mm in width and 6 cm in length were attached on a polypropylene tape containing poly(2-ethylhexyl acrylate) adhesive (Scotch tape). The test was undertaken with Scotch tape pressed on top of the treated LDPE surface. The unbonded end of the testing tape was peeled off at 90° at a crosshead speed of 10 mm/min. The test was stopped after 6 min when the tape was complexly detached from the LDPE surface, and 6 separate readings were carried out to obtain average values of the peeling force.

2.9. Surface Chemistry Characterization

Fourier transform infrared spectroscopy with attenuated total reflectance (FTIR-ATR) was used to qualitatively investigate the chemical composition changes of plasma-treated LDPE surfaces. An FTIR Spectrometer Frontier (PerkinElmer, Waltham, MA, USA) equipped with a ZnSe crystal was used for these analyses to capture data from a penetration depth of 1.66 μm. Spectra in the wavenumber range of 4000–550 cm^{-1} were obtained using an average of 8 scans, with a resolution of 4 cm^{-1}.

The chemical composition changes caused by corona treatment of the LDPE surface were quantified by X-ray photoelectron spectroscopy (XPS). An AXIS XPS system (Kratos Analytical, Manchester, UK) was used for this study. The XPS system contains a spherical mirror analyzer and a delay-line detector for fast screening of the chemical composition, ensuring high spectral resolution and sensitivity. This system allows the analysis of data at a sampling depth of 1–10 nm.

2.10. Surface Morphology Analysis

The surface morphology of LDPE samples was analyzed by scanning electron microscopy (SEM). This technique allowed us to obtain information about surface morphology changes after each modification step. For this purpose, a Nova Nano SEM 450 microscope (FEI, Hillsboro, OR, USA) was employed. A thin Au layer a few nanometers thick was sputter-coated on the LDPE samples to obtain high-resolution images with high magnification (20,000×) and to avoid the accumulation of electrons on the measured layer.

Detailed information about the three-dimensional changes in the surface topography of the LDPE samples was obtained using atomic force microscopy (AFM). An MFP-3D AFM device (Asylum Research, Abingdon, Oxford, UK) was employed in these experiments. Scanning was carried out under ambient conditions by a silicon probe (Al reflex-coated Veeco model, OLTESPA, Olympus, Tokyo, Japan) in tapping mode in air (AC mode), allowing images with a surface area of 1×1 μm^2 to be obtained. Moreover, the roughness parameter value (Ra) was calculated from AFM images obtained from Z-sensor.

2.11. Antibacterial Tests

A modified ISO 22196, an internationally recognized test method was used to evaluate the antibacterial activity of modified plastic materials (and other nonporous surfaces of products) to inhibit the growth of, or kill, test microorganisms [49]. The LDPE samples were first disinfected by UV radiation and then placed in sterile Petri dishes. This was followed by inoculation of the samples (25×25 mm^2) using 0.1 ml of standardized bacteria suspension of *S. aureus* (CCM 4516, 1.8×10^6 cfu/mL) and *E. coli* (CCM 4517, 1.4×10^7 cfu/mL). The samples were covered by disinfected polypropylene foil (20×20 mm^2) with 70% ethanol. Incubation of the inoculated samples was performed at 95% of relative humidity at 35 °C for 24 hours. The polypropylene foil was then removed, and LDPE samples were imprinted on plate count agar (3 times on different areas) and incubated at 35 °C for 24 hours. Then, the results were read, and the increase in the number of bacterial colonies was evaluated based on scaling from 0 to 5, where 0 represents the best antimicrobial effect, with no growth of bacteria colonies. An additional incubation at 35 °C for 24 hours was followed by final reading and evaluation of the results. All of these analyses were performed using 3 different LDPE samples to ensure reliable antimicrobial efficiency results.

3. Results

3.1. Hydroperoxide Concentration

Plasma treatment was used as an effective tool for generation of active species in the LDPE surface necessary for the subsequent grafting process by ASA. As the plasma treatment introduces polar functional groups onto the surface by radicalization, different kinds of functional groups can be found (mainly oxygen-containing groups). Through exposure to air during and after plasma, most of the free radicals are converted into peroxides [50]. However, it is difficult to distinguish the amount of peroxide functionalities in either the infrared (IR) spectroscopy or XPS O1s shift spectra; thus, a classic quantification method by iodometric titration according to Wagner and Thelen [51,52] was used to obtain valid concentration values. In Figure 2, LDPE samples were treated with air plasma at different exposure times. By applying iodometric titration, it was found that the hydroperoxide concentration increased as the treatment time increased from 10 s (7.6×10^{-8} mol/cm^2) to 60 s (9.0×10^{-8} mol/cm^2) of exposure to air plasma. An additional increase in treatment time did not lead to another increase in hydroperoxide concentration. This proves that exposing the polymer samples for a longer time does not increase the formation of hydroperoxides; thus, at the optimum time, the surface would be saturated with a certain amount of peroxides [50]. From this observation, an optimum time for achieving the maximum hydroperoxide concentration was evaluated, and LDPE samples were treated with plasma at 60 s prior to the ASA grafting process.

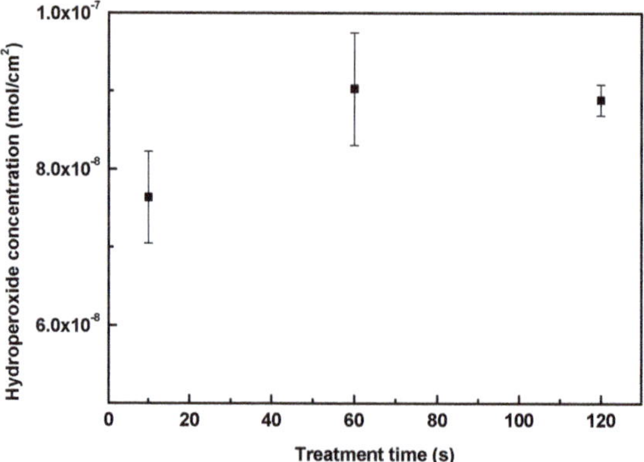

Figure 2. The peroxide concentration of plasma-treated LDPE samples.

3.2. Surface Wettability Analysis

The changes in the surface wettability of the modified samples were analyzed through contact angle measurements, which are shown in Figures 3 and 4 and summarized in Tables 1 and 2. Surface free energy and wettability are indicators of the ability of the liquid surface to be attached to the solid surface. This indicates that the lower the contact angle of a sample is, the higher its wettability is. To evaluate surface free energy and its components, water (surface free tension = 72.1 mN/m, polar component = 52.2 mN/m and dispersive component = 19.9 mN/m) [53], ethylene glycol (surface free tension = 48.0 mN/m, polar component = 19.0 mN/m, and dispersive component = 29.0 mN/m) [54] and formamide (surface free tension = 56.9 mN/m, polar component = 33.4 mN/m, and dispersive component = 23.5 mN/m) [53] were used. Untreated LDPE has hydrophobic properties and is chemically inert; thus, its wettability is low under basic conditions. Its water contact angel was 95.7°, with a low total surface free energy (29.3 mJ/m^2) and insignificant polar component (1.9 mJ/m^2). These results refer to the hydrocarbon skeleton –C–H, which has poor reactivity, and thus, no polarity was observed.

The initiation of reactions on the surface by RF plasma improved the polarity by inducing the surface through radicalization. The introduction of new oxygen-containing functional groups helped to increase the total surface free energy to 49.0 mJ/m^2, and the contact angle of water decreased to almost half, with a value of 50.0°. The wettability increased as a result of the new polar functional groups on the surface. Plasma treatment affected the LDPE surface only at very small depth (28.2 nm), which was confirmed by film thickness measurements. Grafting of the antibacterial agent enhanced the polarity even further because of its effective side and defined structure attachment. The ASA-grafted LDPE surface using plasma treatment exhibited the lowest value for the contact angle of water (32.3°); therefore, the highest value for the total surface free energy (67.0 m J/m^2) and its polar component (63 mJ/m^2) was achieved. The introduced functional groups on the LDPE surface were able to effectively react and form new bonds with ASA, and the graft yield was 0.4%, indicating a multilayer formation of ASA. This was confirmed also by film thickness measurements, where the film thickness was 10.1 nm. The graft yield and film thickness analyses confirmed a formation of ASA multilayered structures on the LDPE surface. The effect of plasma treatment on the grafting of ASA onto the LDPE surface was shown, in comparison with untreated LDPE subjected to modification by ASA with subsequent thorough washing. In this case, the contact angle and the total surface free energy were similar to those of untreated LDPE.

Table 1. The contact angles and graft yields of LDPE samples.

LDPE	Water (°)	Ethylene Glycol (°)	Formamide (°)	GY (%)	Film Thickness (nm)
Untreated (A)	95.7 (±3.0)	67.7 (±1.2)	76.5 (±1.8)	-	-
Plasma-treated (B)	50.0 (±1.6)	16.6 (±1.7)	11.1 (±1.8)	-	28.2 (±4.0)
A + ASA	98.6 (±1.4)	65.8 (±1.2)	76.4 (±1.2)	0.0	-
B + ASA	32.3 (±6.9)	25.5 (±2.0)	25.3 (±2.5)	0.4	10.1 (1.0)

Table 2. The surface free energy of LDPE samples.

LDPE	Dispersive (mJ/m^2)	Polar (mJ/m^2)	Total Surface Free Energy (mJ/m^2)
Untreated (A)	27.5	1.9	29.3
Plasma-treated (B)	19.6	29.4	49.0
A + ASA	21.2	3.2	24.3
B + ASA	3.7	63.3	67.0

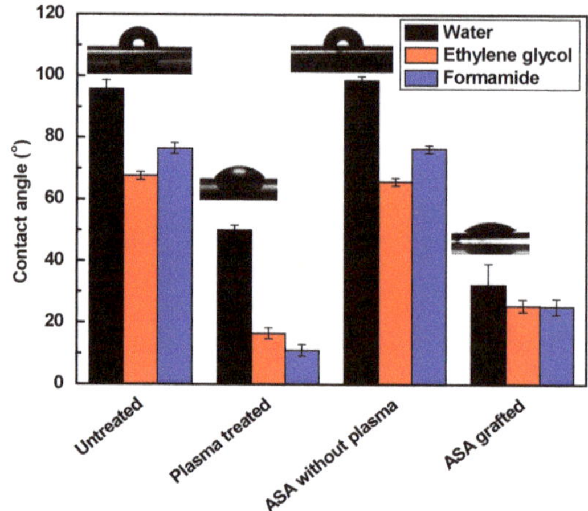

Figure 3. The contact angles of testing liquids on LDPE samples.

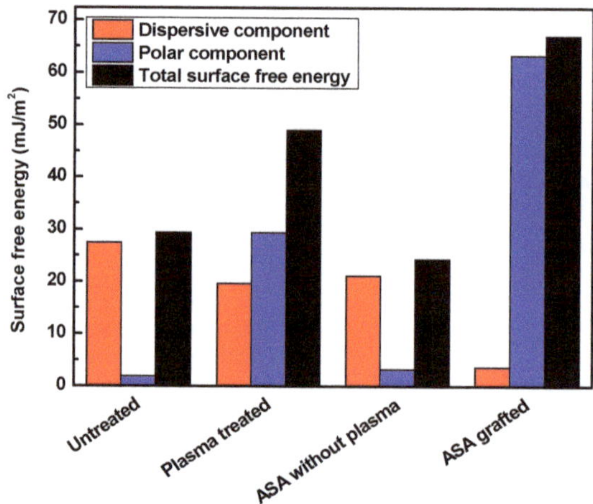

Figure 4. The surface free energy of LDPE samples.

3.3. Adhesion Analysis

Adhesion properties depend on the wettability and surface morphology (roughness) of a material surface. Roughness occurs as a result of physicochemical interactivity or chemical composition on the surface. Adhesion properties can be effectively analyzed by measurements of the peeling resistance. Figure 5 shows the changes in peeling resistance and the changes in adhesion of LDPE after plasma treatment and ASA grafting. Higher resistance induces higher adhesion, which is an outcome of a rougher surface and better wettability. The untreated LDPE surface exhibited relatively poor adhesion, and therefore, the peel resistance reached a value of 40.5 N/m because of the smooth surface and low wettability. The plasma-treated samples showed significant enhancement in the adhesion of the LDPE surface. The peel resistance increased significantly to 83.5 N/m after plasma treatment. This increase in peel resistance was affected mainly by the increase in the wettability and surface roughness caused by the incorporation of polar functional groups and etching reactions, respectively. ASA grafting onto the LDPE surface led to even higher peel resistance (97.3 N/m) because the highest wettability was achieved. ASA was also subjected to LDPE modification with and without application of plasma treatment to study the effect of plasma treatment on the covalent grafting of ASA on the LDPE surface. The untreated LDPE with ASA showed similar peel resistance (45.2 N/m) to the untreated LDPE sample, indicating the lack of ASA after thorough washing.

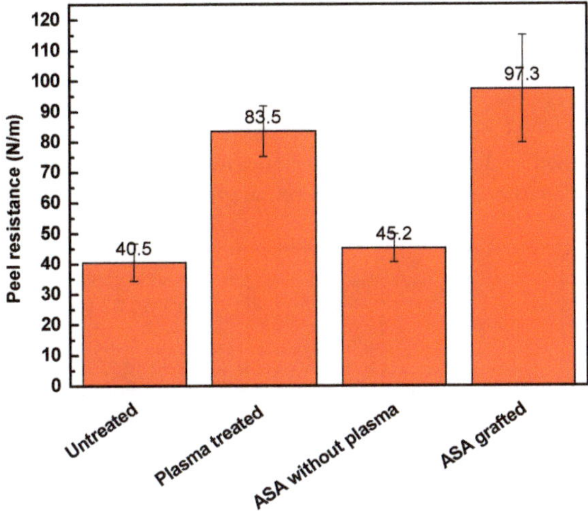

Figure 5. Peel resistance of LDPE samples.

3.4. Chemical Composition Investigation

The chemical composition of the LDPE samples after each modification step was analyzed using Fourier transform infrared (FTIR) spectroscopy. The FTIR spectra of untreated, plasma-treated, and modified LDPE samples are shown in Figure 6. The FTIR spectrum of untreated LDPE is characterized by specific absorbance bands attributed only to nonpolar hydrocarbons in the main chain and branches. These bands represent carbon–carbon and carbon–hydrogen vibrations. Thus, –C–H stretching vibrational bands are observed at 2915 cm^{-1} and 2847 cm^{-1}, and bending and scissoring vibrations are observed at 1473 cm^{-1} and 717 cm^{-1}, respectively. Plasma treatment of LDPE was responsible for the introduction of new functional groups by radicalizing some carbons on the polyethylene surface. This process allows the interaction between the plasma-activated surface and air elements. Oxygen-containing groups are the main functional groups following these interactions. However, these new functional groups were not clearly detected by FTIR spectroscopy because of the relatively high penetration depth of the IR beam when using a ZnSe crystal (1.66 μm) compared with the thickness of the plasma-affected layer, which was only a few tens of nm. The incorporation of new oxygen-containing groups was clearly confirmed by XPS analyses. ASA is an organic compound with ether, carbonyl, and 4-hydroxyl groups apparent in the oxygen-containing spectral region. New vibrational absorbance bands appeared in the FTIR spectrum of LDPE grafted by ASA, where –OH was represented by a broad absorbance band between 3650 cm^{-1} and 3150 cm^{-1}, C=O was observed at 1623 cm^{-1}, –COOH was observed at 1776 cm^{-1}, and C–O–C was observed at 1060 cm^{-1}. This could indicate the presence of ASA on the LDPE surface after the modification process.

For the quantification of the chemical composition of LDPE samples, the XPS technique was employed. The XPS spectra of LDPE samples after each modification step are shown in Figure 7. The XPS spectrum of untreated LDPE consists mainly of the C1s peak, with 98 at.% at a binding energy of ~285 eV. It also contains low-intensity O1s (2.6 at.%) and N1s (0.2 at.%) peaks associated with oxygen- and nitrogen-containing functional groups, originating from processing additives or intermediates coming from air interactions with the LDPE surface. Moreover, other peaks were observed in XPS spectra, which are associated with auger electrons, and therefore they were disregarded from total atomic %. Plasma treatment of the LDPE surface was responsible for the incorporation of new functional groups, as indicated by a significant increase in the intensity of the O1s peak at a binding energy of ~530 eV, achieving a value of 12.4 at.%. In addition, the intensity of the N1s peak increased in the XPS spectrum of LDPE after plasma treatment because of the incorporation of some nitrogen-containing groups (C–N or C–NH$_3^+$). This led to a reduction in the intensity of the C1s peak to 86.5 at.% due to the removal of some carbons during etching, radicalization, and replacement with oxygen-containing groups. ASA grafting on the LDPE surface led to an increase in the at.% of the O1s peak, which was attributed mainly to the oxygen-containing groups in ASA. The ASA-grafted LDPE samples showed an O1s peak with 15.8 at.%, whereas the intensity of the C1s peak decreased to 83.1 at.%. Furthermore, the N1s intensity remained unchanged in comparison with that of the plasma-treated LDPE sample (1.1%), as ASA does not contain any nitrogen functional groups.

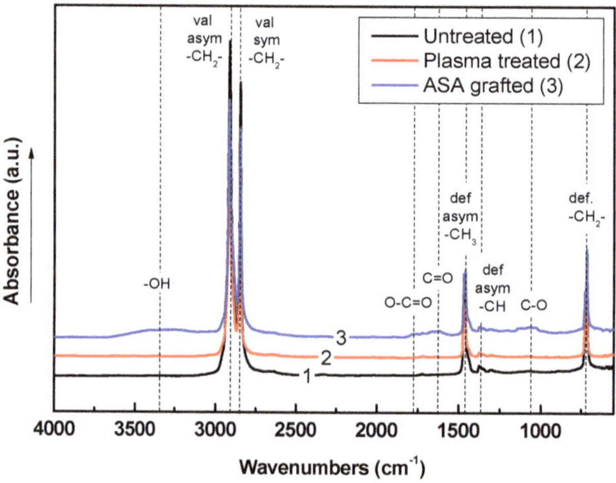

Figure 6. Fourier transform infrared spectroscopy (FTIR) spectra of LDPE samples.

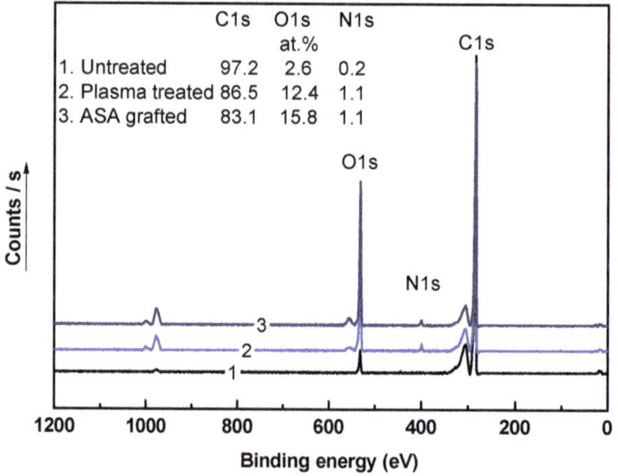

Figure 7. X-ray photoelectron spectroscopy (XPS) spectra of LDPE samples.

3.5. Surface Morphology Analysis

The changes in the surface morphology and topography of the LDPE samples after each modification step were studied through SEM and AFM, respectively. SEM and AFM images of LDPE samples are shown in Figure 8. The untreated LDPE surface exhibits characteristic texture and morphology originating from the production process. Plasma treatment did not cause any significant changes in the surface morphology obtained by SEM from larger surface areas. On the other hand, the LDPE samples grafted with ASA experienced clear changes in their surface morphology, showing bulges and valleys in the functionalized regions on the surface as grafting took place.

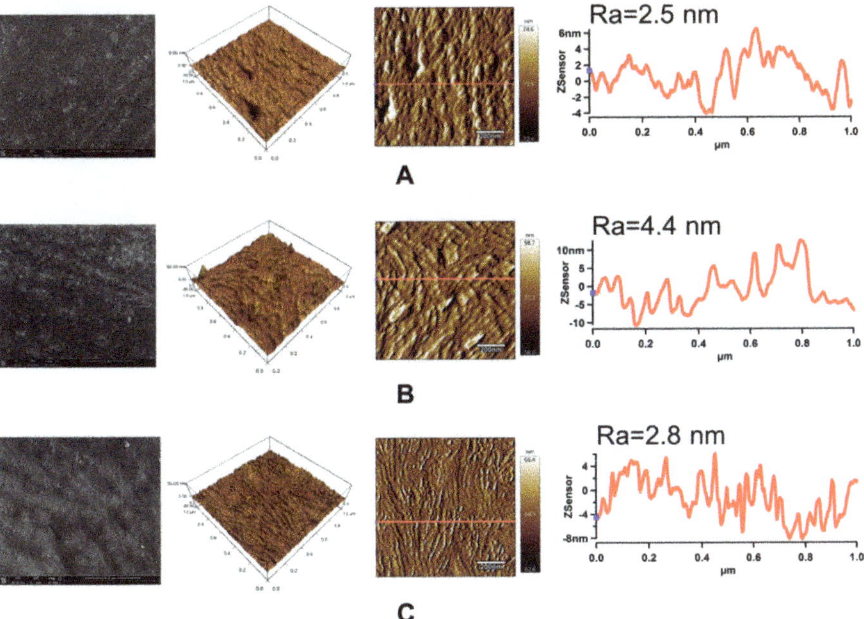

Figure 8. Scanning electron microscopy (SEM), 3D height. and amplitude atomic force microscopy (AFM) images with line profiles (Z-sensor) of LDPE: (**A**) untreated; (**B**) plasma-treated; (**C**) ASA-grafted. Note: Ra represents the roughness parameter.

AFM was used to determine the topographical and roughness changes (Ra) in the LDPE surface that occurred after plasma treatment and ASA grafting in the small surface area ($1 \times 1\ \mu m^2$). Moreover, this technique was used to obtain line profiles, clearly indicating specific nanopattern dimensions. As reported in Figure 8, the Ra value of the untreated LDPE surface was only 2.5 nm. An application of plasma treatment led to an increase in surface roughness, while Ra increased by almost 80% to 4.4 nm. The increase in roughness can be attributed to the etching process during plasma treatment, which led to nanosurface topography changes. Grafting of ASA onto the LDPE surface resulted in a less rough surface (Ra = 2.8 nm) in the small surface area, but with a specific texture belonging to the created ASA layer.

3.6. Antibacterial Analysis

The antimicrobial activities of the LDPE tested against gram-positive *S. aureus* and gram-negative *E. coli* using intensive microbial activity assays are summarized in Table 3 and are shown in Figure 9. The untreated LDPE showed no resistance or inhibition to bacterial growth. This was because of its poor inhibition properties resulting from the chemical composition of LDPE. Plasma treatment of

LDPE was responsible for the low resistance to bacterial growth on the surface. On the other hand, ASA grafted on the LDPE surface showed a high ability for inhibition of *S. aureus*. ASA proved to be highly active against *S. aureus*, similar to its use alone in free form [19,23,38]. ASA could have affected the protein on the bacterial wall, affecting bacterial growth because of its ability to lower the pH and cause instability of bacterial cell membranes [55]. After ASA was grafted onto the LDPE surface, the inhibition activity against *S. aureus* was successful, with over 80%–90% total inhibition. Interestingly, the inhibition of *E. coli* growth was not as intense, although a clear reduction in colony growth and reproduction was observed, as shown in Figure 9.

Table 3. Antimicrobial activity of LDPE samples.

LDPE	Increase in Bacterial Colonies [1]	
	S. aureus	*E. coli*
Untreated (A)	4, 4–5, 4–5	4, 4, 4–5
Plasma-treated (B)	5, 5, 5	5, 5, 5
B + ASA	0, 1, 1	4, 4, 4

[1] The scale for assessing the growth of bacterial colonies: 0—without growth; 1—detectable amount (single colony); 2—detectable amount (combined colony); 3—second imprint, distinguishable colonies, third imprint can be detected; 4—third imprint, distinguishable colonies; 5—overgrown, continuous growth.

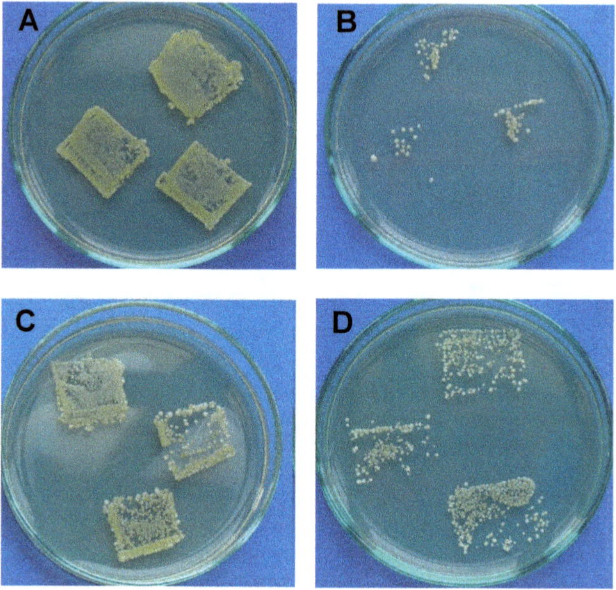

Figure 9. Example of total microbial counts of LDPE samples on plate count agar with inoculated bacteria: (**A**) untreated (*S. aureus*); (**B**) ASA-grafted (*S. aureus*); (**C**) untreated (*E. coli*); (**D**) ASA-grafted (*E. coli*).

4. Conclusions

In this study, ascorbic acid (ASA or vitamin C) was grafted onto an LDPE surface via plasma treatment in order to improve the antimicrobial effect. Plasma treatment was effectively used as a radical initiator with subsequent incorporation of ASA, which served as an antimicrobial agent, on the LDPE surface. This modification was confirmed by enhanced wettability and adhesion properties. The presence of ASA on the LDPE surface after the grafting process was confirmed by chemical composition analyses. Chemical composition and surface morphology or topography analyses were used to confirm the presence of ASA on the LDPE surface. The significant antimicrobial effect of such modified LDPE against gram-positive *S. aureus* was demonstrated, with an inhibition efficiency of over 80–90%.

Author Contributions: Conceptualization, A.P.; methodology, A.P.; validation, A.P.; formal analysis, A.P.; investigation, S.H., D.V., and A.P.; resources, M.L., I.K., and P.H.; data curation, A.P.; writing—original draft, S.H.; writing—review and editing, A.P., M.L., and I.K.; visualization, A.P. and S.H.; supervision, A.P.; project administration, A.P.; funding acquisition, A.P. and I.K.

Funding: This publication was made possible by Award JSREP07-022-3-010 and NPRP10-0205-170349 from the Qatar National Research Fund (a member of The Qatar Foundation). The statements made herein are solely the responsibility of the authors.

Acknowledgments: The authors gratefully acknowledge Gas Processing Center (GPC), Qatar University, for carrying out the XPS analysis.

Conflicts of Interest: The authors declare no conflict of interest.

References

1. Han, J.H.; Floros, J.D. Casting Antimicrobial Packaging Films and Measuring Their Physical Properties and Antimicrobial Activity. *J. Plast. Film Sheeting* **1997**, *13*, 287–298. [CrossRef]
2. De Azeredo, H.M.C. Nanocomposites for food packaging applications. *Food Res. Int.* **2009**, *42*, 1240–1253. [CrossRef]
3. Pankaj, S.; Bueno-Ferrer, C.; Misra, N.; Milosavljević, V.; O'Donnell, C.; Bourke, P.; Keener, K.; Cullen, P. Applications of cold plasma technology in food packaging. *Trends Food Sci. Technol.* **2014**, *35*, 5–17. [CrossRef]
4. Dowling, D.; Tynan, J.; Ward, P.; Hynes, A.; Cullen, J.; Byrne, G. Atmospheric pressure plasma treatment of amorphous polyethylene terephthalate for enhanced heatsealing properties. *Int. J. Adhes. Adhes.* **2012**, *35*, 1–8. [CrossRef]
5. Kavc, T.; Kern, W.; Ebel, M.F.; Svagera, R.; Pölt, P. Surface Modification of Polyethylene by Photochemical Introduction of Sulfonic Acid Groups. *Chem. Mater.* **2000**, *12*, 1053–1059. [CrossRef]
6. Pukánszky, B.; Fekete, E. Adhesion and Surface Modification. In *Mineral Fillers in Thermoplastics I: Raw Materials and Processing*; Springer: Berlin/Heidelberg, Germany, 2007; pp. 109–153.
7. Chanunpanich, N.; Ulman, A.; Strzhemechny, Y.M.; Schwarz, S.A.; Janke, A.; Braun, H.G.; Kraztmuller, T. Surface Modification of Polyethylene through Bromination. *Langmuir* **1999**, *15*, 2089–2094. [CrossRef]
8. Griesser, H.J.; Da, Y.; Hughes, A.E.; Gengenbach, T.R.; Mau, A.W.H. Shallow reorientation in the surface dynamics of plasma-treated fluorinated ethylene-propylene polymer. *Langmuir* **1991**, *7*, 2484–2491. [CrossRef]
9. Liston, E.; Martinu, L.; Wertheimer, M. Plasma surface modification of polymers for improved adhesion: A critical review. *J. Adhes. Sci. Technol.* **1993**, *7*, 1091–1127. [CrossRef]
10. Albertsson, A.C. (Ed.) *Long Term Properties of Polyolefins*; Advances in Polymer Science; Springer: Berlin/Heidelberg, Germany, 2004; Volume 169, ISBN 978-3-540-40769-0.
11. Desai, S.M.; Singh, R.P. Surface Modification of Polyethylene. In *Long Term Properties of Polyolefins*; Springer: Berlin/Heidelberg, Germany, 2004; pp. 231–294.
12. Allen, K.W. Polymer surface modification: Relevance to adhesion. *Polym. Int.* **2002**, *42*, 235.
13. Biederman, H. *Plasma Polymer Films*; Imperial College Press and Distributed by World Scientific Publishing Co.: Singapore, 2012; ISBN 978-1-86094-467-3.
14. Pascual, M.; Balart, R.; Sánchez, L.; Fenollar, O.; Calvo, O. Study of the aging process of corona discharge plasma effects on low density polyethylene film surface. *J. Mater. Sci.* **2008**, *43*, 4901–4909. [CrossRef]
15. Popelka, A.; Novák, I.; Lehocký, M.; Junkar, I.; Mozetič, M.; Kleinová, A.; Janigová, I.; Slouf, M.; Bílek, F.; Chodák, I. A new route for chitosan immobilization onto polyethylene surface. *Carbohydr. Polym.* **2012**, *90*, 1501–1508. [CrossRef]
16. Graves, D.B. Low temperature plasma biomedicine: A tutorial review. *Phys. Plasmas* **2014**, *21*, 80901. [CrossRef]
17. Popelka, A.; Novák, I.; Al-maadeed, M.A.S.A.; Ouederni, M.; Krupa, I. Effect of corona treatment on adhesion enhancement of LLDPE. *Surf. Coat. Technol.* **2018**, *335*, 118–125. [CrossRef]
18. Abusrafa, A.E.; Habib, S.; Krupa, I.; Ouederni, M.; Popelka, A. Modification of Polyethylene by RF Plasma in Different/Mixture Gases. *Coatings* **2019**, *9*, 145. [CrossRef]

19. Kong, M.G.; Kroesen, G.; Morfill, G.; Nosenko, T.; Shimizu, T.; Van Dijk, J.; Zimmermann, J.L. Plasma medicine: An introductory review. *New J. Phys.* **2009**, *11*, 115012. [CrossRef]

20. Bardos, L.; Barankova, H. Cold atmospheric plasma: Sources, processes, and applications. *Thin Solid Films* **2010**, *518*, 6705–6713. [CrossRef]

21. Helgadóttir, S.; Pandit, S.; Mokkapati, V.R.S.S.; Westerlund, F.; Apell, P.; Mijakovic, I. Vitamin C Pretreatment Enhances the Antibacterial Effect of Cold Atmospheric Plasma. *Front. Cell. Infect. Microbiol.* **2017**, *7*, 43. [CrossRef] [PubMed]

22. De Geyter, N.; Morent, R. Nonthermal Plasma Sterilization of Living and Nonliving Surfaces. *Annu. Rev. Biomed. Eng.* **2012**, *14*, 255–274. [CrossRef]

23. Joshi, S.G.; Paff, M.; Friedman, G.; Fridman, G.; Fridman, A.; Brooks, A.D. Control of methicillin-resistant Staphylococcus aureus in planktonic form and biofilms: A biocidal efficacy study of nonthermal dielectric-barrier discharge plasma. *Am. J. Infect. Control* **2010**, *38*, 293–301. [CrossRef] [PubMed]

24. Bazaka, K.; Jacob, M.V.; Chrzanowski, W.; Ostrikov, K. Anti-bacterial surfaces: Natural agents, mechanisms of action, and plasma surface modification. *RSC Adv.* **2015**, *5*, 48739–48759. [CrossRef]

25. Weng, Y.M.; Chen, M.J.; Chen, W. Antimicrobial Food Packaging Materials from Poly (ethylene-co-methacrylic acid). *LWT* **1999**, *32*, 191–195. [CrossRef]

26. Ahmed, I.; Ready, D.; Wilson, M.; Knowles, J.C. Antimicrobial effect of silver-doped phosphate-based glasses. *J. Biomed. Mater. Res. Part A* **2006**, *79*, 618–626. [CrossRef] [PubMed]

27. Valappil, S.P.; Pickup, D.M.; Carroll, D.L.; Hope, C.K.; Pratten, J.; Newport, R.J.; Smith, M.E.; Wilson, M.; Knowles, J.C. Effect of Silver Content on the Structure and Antibacterial Activity of Silver-Doped Phosphate-Based Glasses. *Antimicrob. Agents Chemother.* **2007**, *51*, 4453–4461. [CrossRef] [PubMed]

28. Wei, Q.; Haag, R. Universal polymer coatings and their representative biomedical applications. *Mater. Horiz.* **2015**, *2*, 567–577. [CrossRef]

29. Muñoz-Bonilla, A.; Fernández-García, M. Polymeric materials with antimicrobial activity. *Prog. Polym. Sci.* **2012**, *37*, 281–339. [CrossRef]

30. Aider, M. Chitosan application for active bio-based films production and potential in the food industry: Review. *LWT* **2010**, *43*, 837–842. [CrossRef]

31. Leceta, I.; Guerrero, P.; De La Caba, K. Functional properties of chitosan-based films. *Carbohydr. Polym.* **2013**, *93*, 339–346. [CrossRef]

32. Theapsak, S.; Watthanaphanit, A.; Rujiravanit, R. Preparation of Chitosan-Coated Polyethylene Packaging Films by DBD Plasma Treatment. *ACS Appl. Mater. Interfaces* **2012**, *4*, 2474–2482. [CrossRef]

33. Correia, V.G.; Ferraria, A.M.; Pinho, M.G.; Aguiar-Ricardo, A. Antimicrobial Contact-Active Oligo (2-oxazoline) s-Grafted Surfaces for Fast Water Disinfection at the Point-of-Use. *Biomacromolecules* **2015**, *16*, 3904–3915. [CrossRef]

34. Tiller, J.C.; Lee, S.B.; Lewis, K.; Klibanov, A.M. Polymer surfaces derivatized with poly (vinyl-N-hexylpyridinium) kill airborne and waterborne bacteria. *Biotechnol. Bioeng.* **2002**, *79*, 465–471. [CrossRef]

35. Ardjani, T.E.A.; Alvarez-Idaboy, J.R. Radical scavenging activity of ascorbic acid analogs: Kinetics and mechanisms. *Theor. Chem. Acc.* **2018**, *137*, 69. [CrossRef]

36. Niki, E. Action of ascorbic acid as a scavenger of active and stable oxygen radicals. *Am. J. Clin. Nutr.* **1991**, *54*, 1119–1124. [CrossRef] [PubMed]

37. Liu, K.; Yuan, C.; Chen, Y.; Li, H.; Liu, J. Combined effects of ascorbic acid and chitosan on the quality maintenance and shelf life of plums. *Sci. Hortic.* **2014**, *176*, 45–53. [CrossRef]

38. Verghese, R.J.; Ramya, S.; Kanungo, R. In vitro Antibacterial Activity of Vitamin C and in Combination with Ciprofloxacin against Uropathogenic Escherichia coli. *J. Clin. Diagn. Res.* **2017**, *11*, 1–5. [CrossRef]

39. Tajkarimi, M.; Ibrahim, S.A. Antimicrobial activity of ascorbic acid alone or in combination with lactic acid on Escherichia coli O157:H7 in laboratory medium and carrot juice. *Food Control* **2011**, *22*, 801–804. [CrossRef]

40. Hemilä, H. Vitamin C and infections. *Nutrients* **2017**, *9*, 339. [CrossRef]

41. Vrijsen, R.; Everaert, L.; Boeyé, A. Antiviral Activity of Flavones and Potentiation by Ascorbate. *J. Gen. Virol.* **1988**, *69*, 1749–1751. [CrossRef]

42. Verghese, R.; Mathew, S.; David, A. Antimicrobial activity of Vitamin C demonstrated on uropathogenic Escherichia coli and Klebsiella pneumoniae. *J. Curr. Res. Sci. Med.* **2018**, *3*, 88–93.

43. Kallio, J.; Jaakkola, M.; Mäki, M.; Kilpeläinen, P.; Virtanen, V. Vitamin C Inhibits Staphylococcus aureus Growth and Enhances the Inhibitory Effect of Quercetin on Growth of Escherichia coli In Vitro. *Planta Med.* **2012**, *78*, 1824–1830. [CrossRef]

44. Li, S.; Taylor, K.B.; Kelly, S.J.; Jedrzejas, M.J. Vitamin C Inhibits the Enzymatic Activity of *Streptococcus pneumoniae* Hyaluronate Lyase. *J. Biol. Chem.* **2001**, *276*, 15125–15130. [CrossRef]

45. Isela, S.R.; Sergio, N.; José, M.J. Ascorbic Acid On Oral Microbial Growth and Biofilm. *Pharma Innov. J.* **2013**, *2*, 103–109.

46. Eddy, B.P.; Ingram, M. Interactions between ascorbic acid and bacteria. *Bacteriol. Rev.* **1953**, *17*, 93–107. [PubMed]

47. Du, J.; Cullen, J.J.; Buettner, G.R. Ascorbic acid: Chemistry, biology and the treatment of cancer. *Biochim. Biophys. Acta BBA Bioenerg.* **2012**, *1826*, 443–457. [CrossRef] [PubMed]

48. Pal, D.; Nimse, S.B. Free radicals, natural antioxidants, and their reaction mechanisms. *RSC Adv.* **2015**, *5*, 27986–28006.

49. JIS Z 2801/ISO 22196—Microbe Investigations (MIS). Available online: https://www.microbe-investigations.com/testing-methods/jis-z-2801-iso-22196/ (accessed on 6 August 2019).

50. Suzuki, M.; Kishida, A.; Iwata, H.; Hata, Y.; Ikada, Y. Graft copolymerization of Acrylamide onto a polythylene surface pretreated with a glow discharge. *Macromolecules* **1986**, *19*, 1804–1808. [CrossRef]

51. Wagner, C.D.; Smith, R.H.; Peters, E.D. Determination of Organic Peroxides—Evaluation of a Modified Iodometric Method. *Anal. Chem.* **1947**, *19*, 976–979. [CrossRef]

52. Thelen, H.; Kaufmann, R.; Klee, D. Development and characterization of a wettable surface modified aromatic polyethersulphone using glow discharge induced HEMA-graft polymerisation. *Anal. Bioanal. Chem.* **1995**, *353*, 290–296. [CrossRef]

53. Busscher, H.; Van Pelt, A.; De Boer, P.; De Jong, H.; Arends, J. The effect of surface roughening of polymers on measured contact angles of liquids. *Colloids Surf.* **1984**, *9*, 319–331. [CrossRef]

54. Erbil, H.Y. Surface tension of polymers. In *CRC Handbook of Surface and Colloid Chemistry*; Birdi, K.S., Ed.; CRC Press: Boca Raton, FL, USA, 2015; pp. 265–312.

55. Jay, J.M. *Modern Food Microbiology*; Springer: Berlin/Heidelberg, Germany, 2012; ISBN 978-0-387-23180-8.

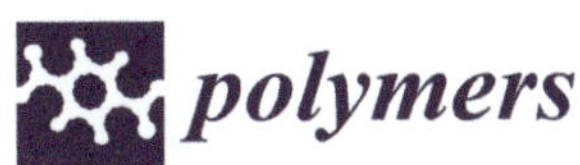 *polymers*

Article

Fast Surface Hydrophilization via Atmospheric Pressure Plasma Polymerization for Biological and Technical Applications

Hana Dvořáková, Jan Čech *, Monika Stupavská, Lubomír Prokeš, Jana Jurmanová, Vilma Buršíková, Jozef Ráheľ and Pavel Sťahel

Department of Physical Electronics, Faculty of Science, Masaryk University, Kotlarska 2, 611 37 Brno, Czech Republic; hana.dvorakoval@mail.muni.cz (H.D.); stupavska@mail.muni.cz (M.S.); prokes@chemi.muni.cz (L.P.); janar@physics.muni.cz (J.J.); vilmab@physics.muni.cz (V.B.); rahel@mail.muni.cz (J.R.); pstahel@physics.muni.cz (P.S.)
* Correspondence: cech@physics.muni.cz

Received: 6 August 2019; Accepted: 30 September 2019; Published: 4 October 2019

Abstract: Polymeric surfaces can benefit from functional modifications prior to using them for biological and/or technical applications. Surfaces considered for biocompatibility studies can be modified to gain beneficiary hydrophilic properties. For such modifications, the preparation of highly hydrophilic surfaces by means of plasma polymerization can be a good alternative to classical wet chemistry or plasma activation in simple atomic or molecular gasses. Atmospheric pressure plasma polymerization makes possible rapid, simple, and time-stable hydrophilic surface preparation, regardless of the type and properties of the material whose surface is to be modified. In this work, the surface of polypropylene was coated with a thin nanolayer of plasma-polymer which was prepared from a low-concentration mixture of propane-butane in nitrogen using atmospheric pressure plasma. A deposition time of only 1 second was necessary to achieve satisfactory hydrophilic properties. Highly hydrophilic, stable surfaces were obtained when the deposition time was 10 seconds. The thin layers of the prepared plasma-polymer exhibit highly stable wetting properties, they are smooth, homogeneous, flexible, and have good adhesion to the surface of polypropylene substrates. Moreover, they are constituted from essential elements only (C, H, N, O). This makes the presented modified plasma-polymer surfaces interesting for further studies in biological and/or technical applications.

Keywords: polymer surface; polymer modification; deposition; plasma polymer; hydrophilization; superhydrophilic layers; atmospheric pressure plasma; polypropylene; surface free energy

---

1. Introduction

Since their first adoption, polymers have found broad uses, ranging from technical applications to biological/medical applications [1–3]. Bulk polymer materials have convenient chemical and mechanical properties, and can be easily fabricated and deployed. But the effort towards the enlargement of the application basis of polymer materials poses demand on the modification of their surface properties. This ranges from improving surface wettability, the functionalization of surfaces, or deposition of barrier layers, to the fabrication of biocompatible surfaces [4–12]. In research focused on biological and/or medical applications of polymers, two principal approaches can be identified. The first one is focused on the synthesis of tailored, "bulk" polymers (see, e.g., [13]). The second is focused on the modification of the surface properties of "standard" polymers (see e.g., [12,14–16]).

We can find research on biological applications studying surfaces, polymers, or composites with (highly) hydrophobic as well as hydrophilic properties [17–20]. And, as stated in [21], discussion is ongoing about, e.g., the dependence of bacterial adhesion upon wettability. The positive influence of the hydrophilization of polymer surfaces on biocompatibility was reported, e.g., in studies on the functionalization of polyestersulfon (PES) membranes for dialysis [15], composites of polymethylmethacrylate and hydroxyapatite in dental implants, [17], and the improvement of cell adhesion on the hydrophilic, plasma-oxidized surface of poly(lactide-co-glycolide) (PLGA) [22]. The positive effect of highly hydrophilic surfaces on the antifouling properties, as well as protein resistance, was also reported [5,23–25]. Such antifouling surfaces can be adopted in, e.g., medical [23] or marine applications, such as coatings or sensors [26,27]. In this paper, we would like to introduce a method for the fast and permanent surface hydrophilization of a polymer surface. The method is based on the plasma deposition of plasma-polymeric nanolayers constituted from essential elements only (C, H, N, O).

Common materials like polyethylene (PE), polypropylene (PP), polyvinylchloride (PVC), or polymethylmethacrylate (PMMA) exhibit rather hydrophobic surface properties. The modification of a polymer surface to gain a hydrophilic nature can be done using wet chemistry methods, which could benefit greatly from plasma surface pre-treatment (e.g. immobilizing standard surfactants [28]). The alternative approach for improving wettability can benefit from the lack of wet-chemistry processing. This approach is based on the utilization of gas discharge (plasma) for surface modifications of solid-state surfaces (see, e.g. [29,30]). Direct plasma treatment in atomic gasses (i.e. oxygen, nitrogen, argon, air, etc.) was reported in works [31–33], where the introduction of hydrophilic functional groups to the surface was reported. The influence of plasma etching on the surface topography was also reported [29,34,35]. However, there is a significant drawback to the described techniques. The resulting wetting properties depend strongly on treated material properties and treatment (discharge) conditions [36]. A more significant disadvantage is the so-called aging effect, which is the gradual disappearance of improved wetting properties due to ongoing post-treatment surface reactions during sample storage in the ambient gas [37]. Highly-stable hydrophilic or super-hydrophilic surfaces can be prepared using plasma, e.g., by nano-texturizing the surface accompanied by the delivery of new hydrophilic functional groups. However, long processing times [38,39] in low-pressure plasmas are necessary for this method. Further methods of surface activation are plasma-initiated graft polymerization [40–46] and plasma polymerization [47–59].

Plasma polymerization is the process of creating a highly-branched polymer by plasma-initiated polymerization of the gas precursor. Typically, plasma polymer is created as a thin layer consisting of short chains with random organization and a high degree of crosslinking [48,60,61]. Low pressure plasma deposition systems were first developed for the preparation of homogeneous and uniform organosilicon, halocarbon, or hydrocarbon plasma polymer thin films. The list of investigated applications includes optical, anti-reflection, abrasion-resistant, and low-surface energy coatings, barrier layers, contact lubricants, dielectric layers, or intermediate adhesive and anticorrosive layers [56,62–67]. Low deposition and production rates at low pressure plasmas have motivated the current interest in developing more efficient methods for plasma polymerization using atmospheric pressure (AP) discharges. Depositions at atmospheric pressure allow easy, fast, and continuous processing due to the application of open systems without the need of an expensive vacuum system. The maximum deposition rates of AP systems can be as high as several tens of nm per second [68]. The main drawbacks of AP systems are inadequate uniformity of the deposited films and high precursor consumption.

Even an extremely thin plasma-deposited polymeric layer should be sufficient to provide a significant increase in surface wettability. Therefore, in combination with a suitably-selected AP system, unusually short deposition times can be achieved to significantly improve the aging-free wettability of the coated surface.

The presented fast and cost-effective method for preparing highly hydrophilic surfaces is based on short-time plasma polymerization at atmospheric pressure. We used a low concentration of propane-butane (P-B) diluted in nitrogen as a carrier gas. This gas mixture was known to increase the adhesion of polyester cords to rubber matrices [69]. In addition, this gas mixture is not toxic, nor is it very inexpensive. Deposited thin films are stable, flexible, and they are constituted from the essential elements only (C, H, N, O). Therefore, the presented method can be utilized in biological or technical applications. As potential applications, we can mention the cost-effective production of intermediate layers, biocompatible surfaces that could be utilized in medical applications (e.g. bandages, plasters, device coatings), and marine/water management applications (biofouling control coatings).

The wettability of deposited thin films was characterized by means of the water contact angle (WCA) measurement. The surface free energy (SFE) of thin films was derived from measurements of the WCA and diiodomethane contact angle (DCA). X-ray photoelectron spectroscopy (XPS) and Fourier transform infrared spectroscopy using attenuated total reflection (FTIR-ATR) were used to determine the chemical composition of the deposited layers. Scanning electron microscopy (SEM) imaging was used to characterize the surface morphology.

2. Materials and Methods

Commercial polypropylene (PP-H, TUPLEX, Brno, Czech Republic) foil with a thickness of 2 mm, manufactured by extrusion, was selected as a model substrate. Polypropylene (PP) samples with surface areas of 25×95 mm^2 were washed with isopropyl alcohol and dried for 72 hours at room temperature before processing. The water contact angle of the untreated sample was 92° and its surface free energy was 31.9 mJ.m^{-2}. A commercial mixture of propane and butane (mass percent composition: 84% propane, 15% butane, and 1% of C2 and C5 hydrocarbons) admixed into nitrogen with a purity of 99.999% was used in this study.

Figure 1 depicts the experimental configuration of the plasma deposition setup and sample position during the plasma treatment. The flow and composition of the processing gas were controlled by two mass flow meters. The volume concentration of P-B was set at 0.4, 0.8, and 1.2%, and a total gas flow of 3 l.min^{-1} was kept in all cases. The processing gas mixture was injected into the plasma reactor directly in a so-called T-configuration. The so-called Diffuse Coplanar Surface Barrier Discharge (DCSBD) [70] was used to generate plasma for plasma-polymerization. The plasma was generated as a thin layer above the DCSBD surface, using sine-wave high voltage with a frequency of 30 kHz and with a specific power density of 3.75 W.cm^{-2}. The PP foil sample was attached to the holder, a moving with a velocity of 15 cm s^{-1} along the DCSBD surface to simulate the conditions of continuous processing. The distance between the sample and the DCSBD electrode was set at 0.1 mm; the treatment times ranged from 0.5 s to 20 s.

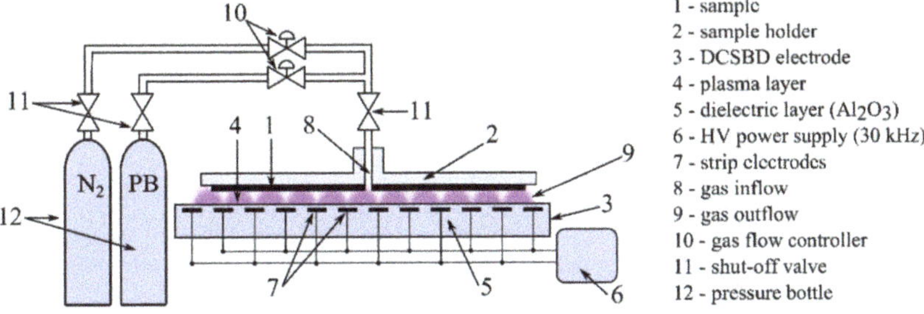

Figure 1. Schematic representation of the experimental set-up.

The wettability of sample surfaces was determined by measuring the static contact angle using the sessile drop method. The contact angles of two standard liquids (deionized water and diiodomethane) were measured and the surface free energy was evaluated by the Owens, Wendt, Rabel, and Kaelble (OWRK) model [71,72]. Sessile drop of a volume of 1 µl was dropped onto the measured surface and analyzed using See System E (Advex Instruments, Brno, Czech Republic). The sessile drop projection was acquired and the contact angles (CAs) were determined. Then, a statistical analysis of the contact angle data, including a surface energy evaluation, was performed using the selected OWRK model.

Sets of 10–15 drops of each test liquid/surface combination were used for statistical processing and, prior to CA analysis, the sessile drop was allowed to reach CA equilibrium (ca 30 s).

The chemical composition of the untreated and plasma-treated samples was evaluated via FTIR-ATR and XPS. These analyses were performed on samples treated for 15 seconds. Infrared spectra were obtained with FTIR spectrometer Tensor 27 equipped with a single reflection diamond ATR accessory Platinum ATR (Bruker Optics, Ettlingen, Germany). The XPS measurements were performed on the ESCALAB 250Xi (Thermo Fisher Scientific, Waltham, MA, USA). The X-Ray source was a micro-focused monochromatic Al Kα X-Ray source operating at 200 W (650 microns spot size). The measurements were done under the conditions of 50 eV pass energy and a resolution of 1 eV for a survey and of 20 eV pass energy and a resolution of 0.1 eV for high-resolution spectra. The analysis was carried under an ultrahigh vacuum of 10-9 mbar, at room temperature. To avoid surface charging, an electron neutralizer was used. All binding energies were referenced with respect to C-C/C-H at a binding energy of 284.8 eV.

The mechanical properties of the layers were studied by means of nanoindentation using a Hysitron TI950 (Bruker Corporation, Billerica, MA, USA) nanoindenter equipped with a Berkovich diamond indenter. Qualitative bending tests were carried out on coated PP foils to study the bending resistance of the samples. The procedure of the qualitative bending test is shown in Figure 2.

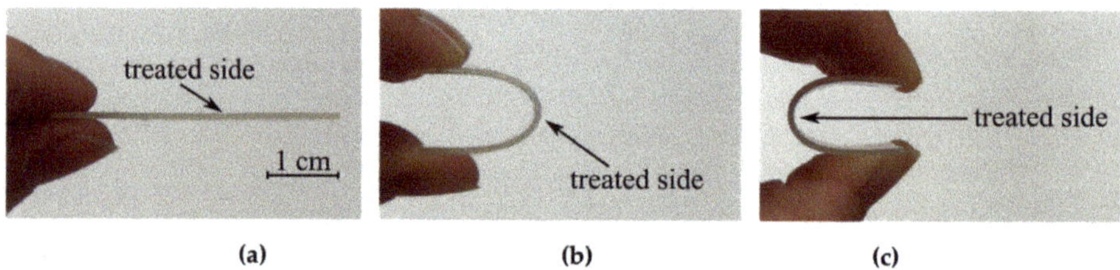

Figure 2. Illustration of the qualitative bending test: pictures (a) to (c) shows the bending procedure.

The surface morphology before and after the bending test was studied using a Tescan MIRA 3 scanning electron microscope. All samples were coated with a thin conductive carbon layer prior to the observation. Images were taken at magnifications of 50,000 and 100,000 times at a working voltage of 15 kV. Study of the sample surface after the bending test allowed us to study the bending resistance as well as the adhesion of the deposited layers to the PP substrate. To measure the film thickness, scalpel sections were made in the most deformed areas of the layers after bending, which enabled a partial removal of the layer from the substrate surface. The thickness was then measured using the SEM on the side portions of the fragments.

3. Results and Discussion

The presence of a highly-hydrophilic plasma polymer layer was evident already after short processing times. Figure 3 shows the water contact angle (WCA) and surface free energy (SFE) values as a function of the deposition time for three different processing gas compositions (precursor volume concentration 0.4, 0.8 and 1.2%). The value of WCA dropped sharply already during the first 7 seconds of plasma treatment, then remained constant. Surprisingly, WCA values for samples deposited from the lower P-B concentration in nitrogen converged to the final value more rapidly than in the case of higher precursor deposition mixtures. The values of WCA and SFE of samples deposited for 4 seconds or more were almost independent of P-B concentration. Longer deposition times caused only an increase in the layer thickness. For thicker layers, the scatter of the measured values of WCA and SFE decreased, suggesting that the homogeneity of the deposited layer was increased.

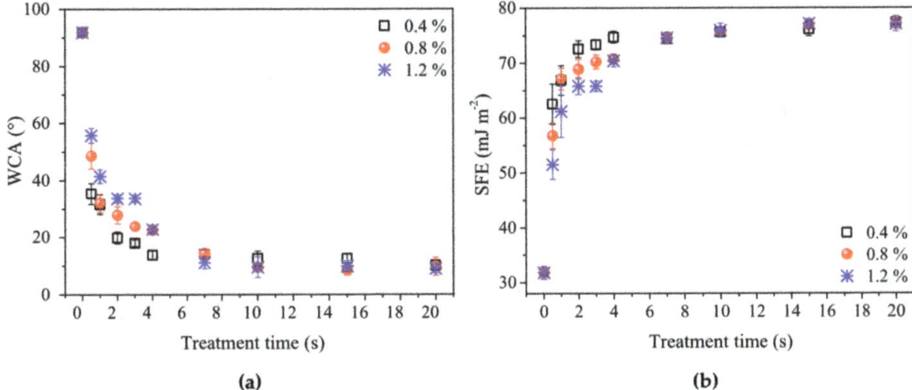

Figure 3. Dependences of surface parameters values on the plasma treatment time: (**a**) water contact angle (WCA); (**b**) surface free energy (SFE). Values for three different compositions of processing gas, i.e. 0.4, 0.8, and 1.2% of P-B in nitrogen, are shown.

The minimum WCA of 8.7° and the maximum SFE of 77 mJ.m^{-2} were obtained for 20 seconds of plasma treatment in a processing gas of 1.2% precursor concentration. Nevertheless, a WCA value of about 9° was achieved already after 10 seconds of plasma treatment for all tested precursor concentrations. According to [73–75], the criteria for superhydrophilic properties are as follows: either a complete wetting of the surface occurs, or the WCA is close to zero on such surfaces. The latter means, in practice, that the WCA is too low for optical recognition, i.e. according to [73–75], the WCA < 5–10°. According to these criteria, the prepared surfaces could be classified as superhydrophilic. The SFE components, as well as the diiodomethane contact angles, are given in the supplemental data, Figures S1–S4.

Figure 4 illustrates the visual appearance of the significant decrease of WCA from 92° to 9° after 15 seconds of plasma treatment in nitrogen with a 0.8% admixture of P-B. Without plasma polymerization, the typical values of WCA for plasma-activated polypropylene were within the range of 20° to 70° [76–86]. In general, plasma polymerization makes it possible to achieve much lower values of WCA. For example, the plasma polymerization of acrylic acid (AA), one of the most widely used precursors for hydrophilic modifications, produces surfaces with WCA values of 8° to 43° [45,52–55]. Using a plasma-initiated graft polymerization of AA on polyethylene terephthalate, a surface with a WCA of 5° with great time stability was prepared; however, the whole treatment procedure took more than five hours [44]. In this context, our results obtained by plasma deposition from the P-B mixture were very encouraging.

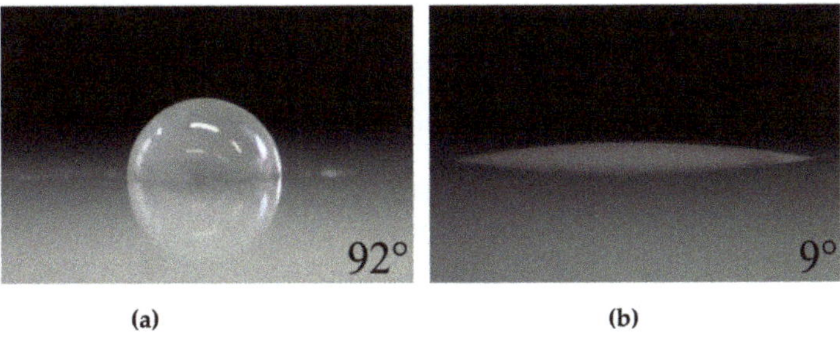

Figure 4. The images of water sessile drop on: (**a**) untreated sample; and (**b**) on the sample after 15 seconds of plasma treatment. Images shown for a processing gas with a volume concentration of P-B of 0.8%.

Figure 5 shows the aging characteristics of WCA (a) and SFE (b) for samples treated for technologically-feasible times of 4 seconds. The changes were monitored for 21 days of storage under ambient atmosphere conditions. The WCA increased by 2–3° and the SFE decreased by 4–6 mJ.m^{-2} during the first 7 days of aging; afterwards, their values remained constant. This shows great time stability, especially when compared to the pure plasma activation, in which the wettability improvement decays within a few hours or days [78,79,84,86]. The comparison with other plasma-polymerized or plasma-initiated grafted layers revealed at least equal or slightly better aging characteristics of our P-B layer [40,44,45,59]. The aging of plasma-modified surfaces is a well-known phenomenon described, for example, in [87]. The reversion of the plasma-modified surface properties towards unmodified values was found for both types of modification, i.e., hydrophilization and hydrophobization (see, e.g. [86,88,89]). The aging effect is explained in [87], in which Johansson stated in that thermodynamically driven reorientation, migration of low-molecular-weight additives from the bulk of the plastic to the surface, and airborn contamination of surface should be considered for the aging effect. For the presented plasma polymer thin films, the latter dominates the SFE decrease, explaining the observed decrease from the SFE maximum value being only 7%; see Figure 5. Another mechanism was also reported, i.e., a correlation of surface charges and wetting properties was found [90]. The aging of the surface modification could therefore be influenced also by the plasma-modified electrical properties of the surfaces [91–93].

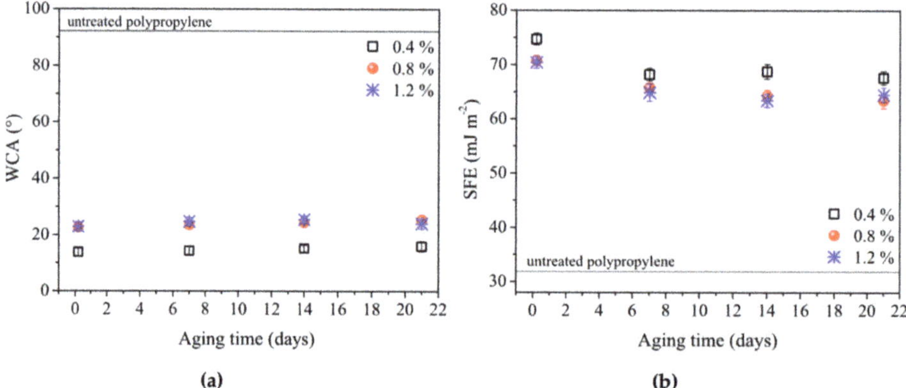

Figure 5. Changes of values (a) WCA and (b) SFE due to aging effect. Values for three different compositions of processing gas, i.e., 0.4, 0.8, and 1.2% of P-B in nitrogen, are shown. A treatment time of 4 seconds was kept in all cases.

ATR-FTIR analysis was made on plasma-treated and untreated samples. Spectra were normalized between the minimum and maximum intensity values. The normalized spectra of untreated and plasma-treated samples are compared in Figure 6. In the spectra were identified typical C-H absorption bands related to an isotactic polypropylene substrate [94,95]: 2950, 2917, 2837, 1453, 1376, 1359, 1329, 1304, 1167, 1153, 997, 972, 940, 899, 840, and 808 cm^{-1}. In the infrared spectra of the treated samples, two additional wide spectral bands were found. The first one, between 3500 and 3000 cm^{-1}, is related to hydroxyl groups of carboxylic acids and a N-H group of amines or amides [96,97]. The second band, between 1730 and 1650 cm^{-1}, reflects the presence of carbonyl group of aldehydes, ketones, carboxylic acids and amides, conjugated carbonyls, and imines. Carbonyl groups are often observed in the infrared spectra of polypropylene treated in air plasma [98–100] or in methane/oxygen plasma [101]. They can also originate from the post-treatment effect due to atmospheric humidity [57,95,100,102]. The same functional groups were also identified in the XPS spectra.

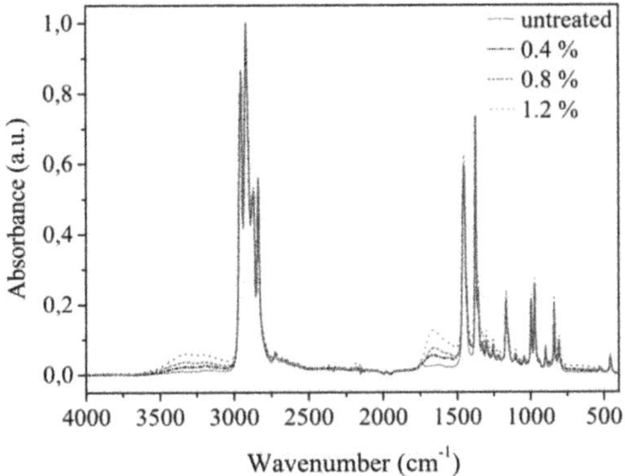

Figure 6. Infrared spectrum of untreated and plasma-treated samples for varied compositions of processing gas, i.e., 0.4, 0.8, and 1.2% of P-B in nitrogen. A treatment time of 15 seconds was kept in all cases.

Due to a high penetration depth (>1 µm) and low layer thickness (maximum around 70 nm), the increase of nitrogen and oxygen group absorption with the increase of precursor concentration is most likely related more to the increase in layer thickness than to real chemical changes in the plasma polymer (the layer thickness is discussed below, see Table 3).

A considerably better detection of surface chemistry is provided by XPS. This method revealed plasma-induced changes in the PP surface composition with a probing depth of 3–10 nm. The estimated minimum film thickness after 15 seconds of plasma treatment was 15 nm; therefore, we assume that the properties of the plasma polymer were measured without the effect of the underlying PP substrate. Atomic compositions for reference and plasma-modified PP surfaces are presented in Table 1. The XPS survey spectrum of untreated PP surface shows that the dominant signals are from C and O, see Figure S5. The initial PP polymer surface contains a low quantity of oxygen (7 at %).

Table 1. Atomic concentrations (at %) and element ratios of untreated and plasma-modified surfaces for three different compositions of processing gas, i.e., 0.4, 0.8, and 1.2% of P-B in nitrogen. XPS spectra were measured immediately after plasma treatment and after 14 days of aging in ambient air. A treatment time of 15 seconds was kept in all cases. The percentages shown are the mean values of three determinations.

Propane-Butane Concentration (%)	Aging Time (Days)	Atomic Composition (at %)			Ratio		
		C	O	N	O/C	N/C	(O+N)/C
untreated PP	-	93 ± 2	7 ± 2	0	0.08	0	0.08
0.4	-	68 ± 2	12 ± 1	20 ± 1	0.18	0.29	0.47
0.8	-	68 ± 1	9 ± 2	23 ± 2	0.13	0.34	0.47
1.8	-	75 ± 1	8 ± 1	17 ± 2	0.11	0.23	0.33
0.4	14	68 ± 1	13 ± 1	19 ± 1	0.19	0.28	0.46
0.8	14	69 ± 1	11 ± 1	20 ± 1	0.16	0.29	0.45
1.2	14	76 ± 1	10 ± 2	14 ± 1	0.13	0.18	0.32

The plasma modification resulted in the reduction of the C1s peak to approx. 70%. As the concentration of the precursor increased, the intensity of the C peak decreased, and the nitrogen content increased, indicating the growth of the nitrogen containing surface layer. The XPS data of all PP modified surfaces show nitrogen incorporation into the deposited carbon layer. The most pronounced difference in the surface composition of the untreated PP sample and the coated PP surface is the appearance of a significant N1s peak, see Figure S5.

The high resolution C1s peak was fitted with 6 principal components: C–C/C–H (binding energy at 284.8 eV), C–N (285.7 eV), C–O (286.3 eV), C=N (287.1 eV), C=O (287.8 eV), and O–C=O/C–CO–N (289.1 eV), which is consistent with the FTIR-ATR measurements and previous reports [103,104]. The results of the fits are presented in Table 2 and Figure 7. As shown, the most significant changes after the plasma treatment for PP are the significant increase of the C–N component and the decrease of the C–C/C–H component. The plasma modification resulted in a massive reduction of the hydrocarbon peak to 35% for 1.2% of P-B. Components which correspond to hydrophilic functional groups containing nitrogen (C–N, C=N and C=O–N) were introduced on the PP surface by means of aforementioned plasma modification.

Table 2. Percentage (%) of peak areas (functional groups) in C1s peaks of untreated and plasma-modified surfaces for three different compositions of processing gas, i.e., 0.4, 0.8, and 1.2% of P-B in nitrogen. XPS spectra were measured immediately after plasma treatment and after 14 days of aging in ambient air. A treatment time of 15 seconds was kept in all cases.

Propane-Butane Concentration (%)	Aging Time (Days)	Functional Groups (%)					
		C–C/C–H	C–O	C=O	COO/C–CO–N	C–N	C=N
untreated PP	-	87	9	2	2	0	0
0.4	-	45	9	10	3	21	12
0.8	-	36	12	8	1	29	14
1.2	-	35	12	5	2	36	10
0.4	14	48	8	10	5	17	12
0.8	14	36	12	9	2	28	13
1.2	14	36	12	6	3	32	10

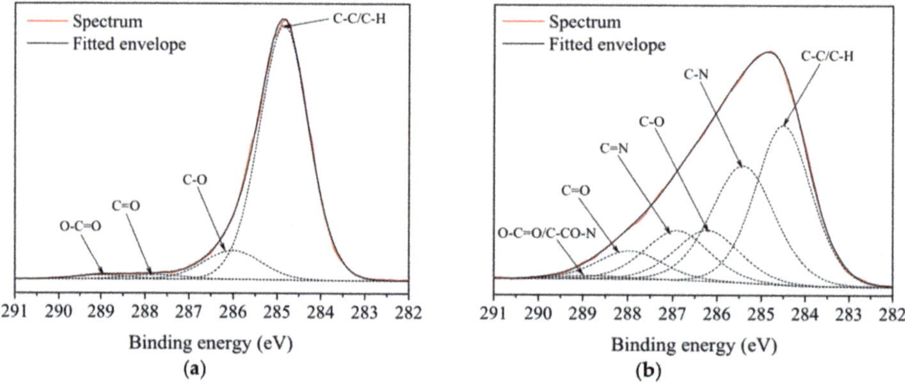

Figure 7. C1s high resolution scan of: (a) untreated sample; and (b) sample treated for 10 seconds in processing gas with volume concentration of precursor of 0.8%.

The most significant change, i.e., 36%, was seen in the increase in the C–N component (1.2% P-B). In the case of plasma-treated PP samples, the peak areas of C–O, C=O, and O–C=O all increased by a factor 1.5–2 compared to the untreated PP. The relatively low oxygen content is not surprising, since the processing gas did not contain oxygen. Oxygen can be incorporated into the structure of the plasma polymer due to post-treatment reactions with ambient atmosphere or air humidity [68,70], or it may be due to a contamination which occurred while using the open plasma reactor. Table 2 shows that the proportion of C–C/C–H bonds decreased with increasing concentrations of precursor in the processing gas, while the proportion of C–N bonds increased. These results correspond to the WCA (Figure 3a); however, it must be noted that the differences in the value of WCA were very small, i.e., within the measurement errors.

As can be seen from the data in Tables 1–3, the composition and surface chemistry did not change significantly in the two weeks after plasma treatment, which is in a good agreement with the time stability of the wetting properties (see Figure 4).

Table 3. Dependence of thickness of coatings of samples deposited for 20 s and calculated deposition rate on the volume concentration of P-B in processing gas.

Propane-Butane Concentration (%)	Thickness (nm)	Deposition Rate (nm.s^{-1})
0.4	23 ± 4	1.2 ± 0.2
0.8	44 ± 2	2.2 ± 0.1
1.2	94 ± 7	4.7 ± 0.4

Figure 8 shows SEM images representing the changes in the surface morphology after plasma treatment and after the bending test. Fine structures on the surface are an artefact from the carbon coating. The surface of plasma-treated samples was smooth, homogenous, and without pinholes or protrusions for all treatment times and for all concentrations of precursor. In fact, there were no differences between the surface topography of the untreated and plasma-treated samples (Figure 8a–e). On the other hand, significant differences were observed between samples subjected to the bending test (Figure 8f–j). The untreated sample showed only a local accumulation of material on the surface and the creation of wrinkles. In contrast, the layers deposited for 20 seconds were intensively wrinkled (and partially cracked) after the severe bending test (see Figure 2). The degree of wrinkling and cracking increased with the concentration of precursor, as shown in Figure 8h–j, mainly because of the increase in film thickness. No cracks were observed on the sample deposited for 7 seconds at 0.4% P-B concentration; however, significant wrinkles were created on the surface after the bending test. With an increase of P-B concentration, the degree of wrinkling and cracking increased, as the film thickness became comparable to 20 s deposition conditions; see Figure S6 and deposition rates in Table 3.

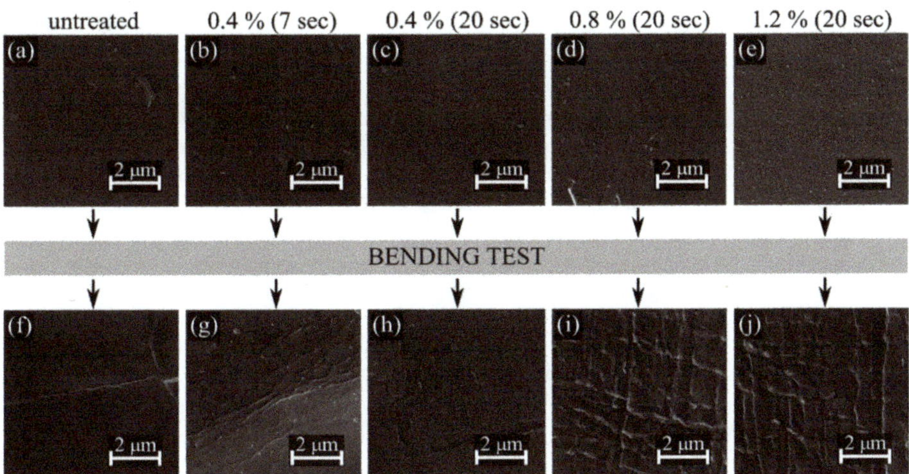

Figure 8. Surface topography of untreated and plasma-treated surface before the bending test; top row (**a**) to (**e**); and after the bending test: bottom row (**f**) to (**j**). The treatment time was 7, resp. 20 seconds and the composition of processing gas was 0.4, 0.8, resp. 1.2% of P-B in nitrogen.

The layers were not delaminated from the PP substrate except for the regions scratched with the scalpel (Figure 9a–c). The fragments separated by the scalpel (Figure 9d–f) were used for the measurement of layer thickness and for the linear estimation of the average deposition rate. The measured thickness of coatings, growth rate, and the concentration of precursor are summarized in Table 3.

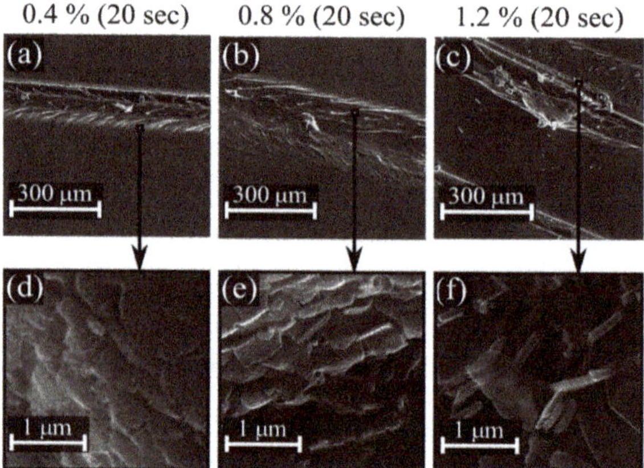

Figure 9. Surface topography of the scratched area for varied composition of processing gas, i.e., 0.4, 0.8, resp. 1.2% of P-B in nitrogen. A treatment time of 20 seconds was kept in all cases. The bottom row represents a magnified view of the surface from the area indicated on the corresponding top row images.

The prepared thin plasma polymer layer is smooth and homogeneous, as shown in Figure 8. The layer thickness increased with the processing time and precursor concentration. The peel test showed that the layer exhibited excellent adhesion to the substrate and, in the case of the lower thickness, is relatively flexible and resistant to delamination and cracking, even in the case of severe mechanical stress.

4. Conclusions

A new hydrophilization technique based on a plasma deposition of a thin film from mixtures of propane-butane with nitrogen at atmospheric pressure was proposed and successfully tested. Unlike simple plasma treatment, the observed high surface free energy values are due to the properties of the deposited plasma-polymer nanolayer. Therefore, the wettability improvement does not depend on the substrate material, and the aging of the surface modification is highly reduced. The deposited polymer-like layers have proven to be very homogeneous, showing uniform thickness. The measured surface free energy of the coatings was in the range of 60–77 mJ.m^{-2}, depending on the coating process conditions. FTIR spectra showed the organic film structure and the presence of C–N and C–C/C–H bonds, as well as C–O, C=O, and O–C=O bonds. The results of XPS analysis were in good accordance with the FTIR observations, confirming also the presence of C–O, C=O, and O–C=O bonds. The basic character of the film surface determined by the surface energy measurement agrees well with the nitrogen-containing hydrophilic groups detected in the surface structure of the films. The films exhibited homogeneous coverage of polymeric substrate and highly hydrophilic properties. With WCA < 10°, the films could be classified as superhydrophilic [73–75]. Moreover, the hydrophilized surfaces were composed of essential elements only (C, N, O, H), and as such, their properties could be of interest for utilization in biological, as well as technical, applications.

Supplementary Materials: The supplementary material is available online at http://www.mdpi.com/2073-4360/ 11/10/1613/s1, Figure S1: Dependences of diiodomethane contact angle (DCA) values on the plasma treatment time. Values given for three different compositions of processing gas, i.e. 0.4, 0.8 and 1.2% of P-B in nitrogen are shown, Figure S2: Dependences of surface free energy (SFE) values and corresponding LW, resp. AB components on the plasma treatment time for a gas mixture of 0.4% of P-B in nitrogen, Figure S3: Dependences of surface free energy (SFE) values and corresponding LW, resp. AB components on the plasma treatment time for a gas mixture of 0.8% of P-B in nitrogen, Figure S4: Dependences of surface free energy (SFE) values and corresponding LW, resp. AB components on the plasma treatment time for a gas mixture of 1.2% of P-B in nitrogen, Figure S5: XPS survey spectra of the surface of 'as-received' and after thin film deposition for 10 seconds in a gas mixture of 0.8% of P-B in nitrogen is given. The background of the spectra used for computational peak analysis is provided, Figure S6 Surface topography of plasma-treated sample after bending test is given for 1.2% P-B in nitrogen and 7 seconds deposition time.

Author Contributions: In alphabetical order: Conceptualization, V.B., J.R. and P.S.; Formal analysis, J.J., L.P. and M.S.; Funding acquisition, V.B. and P.S.; Investigation, V.B., J.Č., H.D., J.J., L.P., M.S. and P.S.; Methodology, H.D., L.P., J.R. and P.S.; Supervision, P.S.; Writing—original draft, V.B., J.Č. and H.D.; Writing—review & editing, V.B., J.Č. and P.S. re.

Funding: This research has been supported by the project LO1411 (NPU I) funded by Ministry of Education Youth and Sports of Czech Republic and by Czech Science Foundation under project GACR 19-15240S.

Conflicts of Interest: The authors declare no conflict of interest. The funders had no role in the design of the study; in the collection, analyses, or interpretation of data; in the writing of the manuscript, or in the decision to publish the results.

References

1. Young, R.J.; Lovell, P.A. *Introduction to Polymers*; CRC Press: Boca Raton, FL, USA, 2011; ISBN 9781439894156.

2. Olatujni, O. *Natural Polymers*; Olatunji, O., Ed.; Springer International Publishing: Cham, Switzerland, 2016; ISBN 978-3-319-26412-7.

3. Mascia, L. *Polymers in Industry from A to Z: A Concise Encyclopedia*; WILEY-VCH Verlag: Weinheim, Germany, 2012; ISBN 9783527644049.

4. Nemani, S.K.; Annavarapu, R.K.; Mohammadian, B.; Raiyan, A.; Heil, J.; Haque, M.A.; Abdelaal, A.; Sojoudi, H. Surface Modification of Polymers: Methods and Applications. *Adv. Mater. Interfaces* **2018**, *5*, 1801247. [CrossRef]

5. Fabbri, P.; Messori, M. Surface Modification of Polymers. In *Modification of Polymer Properties*; Jasso-Gastinel, C.F., Kenny, J.M., Eds.; Elsevier: Amsterdam, The Netherlands, 2017; pp. 109–130. ISBN 978-0-323-44353-1.

6. Iqbal, M.; Dinh, D.K.; Abbas, Q.; Imran, M.; Sattar, H.; Ul Ahmad, A. Controlled Surface Wettability by Plasma Polymer Surface Modification. *Surfaces* **2019**, *2*, 349–371. [CrossRef]

7. Gotoh, K.; Yasukawa, A.; Kobayashi, Y. Wettability characteristics of poly(ethylene terephthalate) films treated by atmospheric pressure plasma and ultraviolet excimer light. *Polym. J.* **2011**, *43*, 545–551. [CrossRef]

8. Šimor, M.; Ráhel', J.; Černák, M.; Imahori, Y.; Štefečka, M.; Kando, M. Atmospheric-pressure plasma treatment of polyester nonwoven fabrics for electroless plating. *Surf. Coatings Technol.* **2003**, *172*, 1–6. [CrossRef]

9. Messori, M.; Toselli, M.; Pilati, F.; Fabbri, E.; Fabbri, P.; Pasquali, L.; Nannarone, S. Prevention of plasticizer leaching from PVC medical devices by using organic–inorganic hybrid coatings. *Polymer (Guildf)* **2004**, *45*, 805–813. [CrossRef]

10. Oehr, C. Plasma surface modification of polymers for biomedical use. *Nucl. Instruments Methods Phys. Res. Sect. B Beam Interact. Mater. Atoms* **2003**, *208*, 40–47. [CrossRef]

11. Klee, D.; Höcker, H. Polymers for Biomedical Applications: Improvement of the Interface Compatibility. In *Biomedical Applications Polymer Blends. Advances in Polymer Science*; Eastmond, G.C., Höcker, H., Klee, D., Eds.; Springer Berlin Heidelberg: Berlin, Heidelberg, 1999; Volume 149, pp. 1–57.

12. Yoshida, S.; Hagiwara, K.; Hasebe, T.; Hotta, A. Surface modification of polymers by plasma treatments for the enhancement of biocompatibility and controlled drug release. *Surf. Coatings Technol.* **2013**, *233*, 99–107. [CrossRef]

13. Kulshrestha, A.S.; Mahapatro, A. *Polymers for Biomedical Applications*; American Chemical Society.: Washington, DC, USA, 2008; pp. 1–7.

14. Lin, P.; Lin, C.-W.; Mansour, R.; Gu, F. Improving biocompatibility by surface modification techniques on implantable bioelectronics. *Biosens. Bioelectron.* **2013**, *47*, 451–460. [CrossRef]

15. Irfan, M.; Idris, A. Overview of PES biocompatible/hemodialysis membranes: PES–blood interactions and modification techniques. *Mater. Sci. Eng. C* **2015**, *56*, 574–592. [CrossRef]

16. Macgregor, M.; Vasilev, K. Perspective on Plasma Polymers for Applied Biomaterials Nanoengineering and the Recent Rise of Oxazolines. *Materials (Basel)* **2019**, *12*, 191. [CrossRef]

17. Tihan, T.G.; Ionita, M.D.; Popescu, R.G.; Iordachescu, D. Effect of hydrophilic–hydrophobic balance on biocompatibility of poly(methyl methacrylate) (PMMA)–hydroxyapatite (HA) composites. *Mater. Chem. Phys.* **2009**, *118*, 265–269. [CrossRef]

18. Stallard, C.P.; McDonnell, K.A.; Onayemi, O.D.; O'Gara, J.P.; Dowling, D.P. Evaluation of Protein Adsorption on Atmospheric Plasma Deposited Coatings Exhibiting Superhydrophilic to Superhydrophobic Properties. *Biointerphases* **2012**, *7*, 31. [CrossRef] [PubMed]

19. Mredha, M.T.I.; Pathak, S.K.; Tran, V.T.; Cui, J.; Jeon, I. Hydrogels with superior mechanical properties from the synergistic effect in hydrophobic–hydrophilic copolymers. *Chem. Eng. J.* **2019**, *362*, 325–338. [CrossRef]

20. Kocijan, A.; Conradi, M.; Hočevar, M. The Influence of Surface Wettability and Topography on the Bioactivity of TiO2/Epoxy Coatings on AISI 316L Stainless Steel. *Materials (Basel)* **2019**, *12*, 1877. [CrossRef]

21. Schmalz, G.; Galler, K.M. Biocompatibility of biomaterials—Lessons learned and considerations for the design of novel materials. *Dent. Mater.* **2017**, *33*, 382–393. [CrossRef] [PubMed]

22. Wan, Y.; Qu, X.; Lu, J.; Zhu, C.; Wan, L.; Yang, J.; Bei, J.; Wang, S. Characterization of surface property of poly(lactide-co-glycolide) after oxygen plasma treatment. *Biomaterials* **2004**, *25*, 4777–4783. [CrossRef] [PubMed]

23. Francolini, I.; Vuotto, C.; Piozzi, A.; Donelli, G. Antifouling and antimicrobial biomaterials: An overview. *APMIS* **2017**, *125*, 392–417. [CrossRef]

24. Utrata-Wesolek, A. Antifouling surfaces in medical application. *Polimery* **2013**, *58*, 685–695. [CrossRef]

25. Jiménez-Pardo, I.; van der Ven, L.; van Benthem, R.; de With, G.; Esteves, A. Hydrophilic Self-Replenishing Coatings with Long-Term Water Stability for Anti-Fouling Applications. *Coatings* **2018**, *8*, 184. [CrossRef]

26. Buskens, P.; Wouters, M.; Rentrop, C.; Vroon, Z. A brief review of environmentally benign antifouling and foul-release coatings for marine applications. *J. Coatings Technol. Res.* **2013**, *10*, 29–36. [CrossRef]

27. Donnelly, B.; Bedwell, I.; Dimas, J.; Scardino, A.; Tang, Y.; Sammut, K. Effects of Various Antifouling Coatings and Fouling on Marine Sonar Performance. *Polymers (Basel)* **2019**, *11*, 663. [CrossRef] [PubMed]

28. Černák, M.; Kováčik, D.; Ráhel', J.; St'ahel, P.; Zahoranová, A.; Kubincová, J.; Tóth, A.; Černáková, L. Generation of a high-density highly non-equilibrium air plasma for high-speed large-area flat surface processing. *Plasma Phys. Control. Fusion* **2011**, *53*, 124031. [CrossRef]

29. Palumbo, F.; Di Mundo, R.; Cappelluti, D.; D'Agostino, R. SuperHydrophobic and SuperHydrophilic Polycarbonate by Tailoring Chemistry and Nano-texture with Plasma Processing. *Plasma Process. Polym.* **2011**, *8*, 118–126. [CrossRef]

30. Pappas, D. Status and potential of atmospheric plasma processing of materials. *J. Vac. Sci. Technol. A Vac. Surf. Film* **2011**, *29*, 020801. [CrossRef]

31. Noeske, M.; Degenhardt, J.; Strudthoff, S.; Lommatzsch, U. Plasma jet treatment of five polymers at atmospheric pressure: Surface modifications and the relevance for adhesion. *Int. J. Adhes. Adhes.* **2004**, *24*, 171–177. [CrossRef]

32. Lehocký, M.; Drnovská, H.; Lapčíková, B.; Barros-Timmons, A.M.; Trindade, T.; Zembala, M.; Lapčík, L. Plasma surface modification of polyethylene. *Colloid. Surf. A* **2003**, *222*, 125–131. [CrossRef]

33. Dorai, R.; Kushner, M.J. A model for plasma modification of polypropylene using atmospheric pressure discharge. *J. Phys. D Appl. Phys.* **2003**, *36*, 666. [CrossRef]

34. Fricke, K.; Steffen, H.; von Woedtke, T.; Schröder, K.; Weltmann, K.-D. High Rate Etching of Polymers by Means of an Atmospheric Pressure Plasma Jet. *Plasma Process. Polym.* **2011**, *8*, 51–58. [CrossRef]

35. Egitto, F.D. Plasma etching and modification of organic polymers. *Pure Appl. Chem.* **1990**, *62*, 1699–1708. [CrossRef]

36. Cho, J.S.; Kim, K.H.; Han, S.; Beag, Y.W.; Koh, S.K. Hydrophilic surface formation on polymers by ion-assisted reaction. *Prog. Org. Coat.* **2003**, *48*, 251–258. [CrossRef]

37. Gerenser, L.J. XPS studies of in situ plasma-modified polymer surfaces. *J. Adhes. Sci. Technol.* **1993**, *7*, 1019–1040. [CrossRef]

38. Tsougeni, K.; Petrou, P.S.; Tserepi, A.; Kakabakos, S.E.; Gogolides, E. Nano-texturing of poly (methyl methacrylate) polymer using plasma processes and applications in wetting control and protein adsorption. *Microelectron. Eng.* **2009**, *86*, 1424–1427. [CrossRef]

39. Skarmoutsou, A.; Charitidis, C.A.; Gnanappa, A.K.; Tserepi, A.; Gogolides, E. Nanomechanical and nanotribological properties of plasma nanotextured superhydrophilic and superhydrophobic polymeric surfaces. *Nanotechnology* **2012**, *23*, 505711. [CrossRef] [PubMed]

40. Gupta, B.; Plummer, C.; Bisson, I.; Frey, P.; Hilborn, J. Plasma-induced graft polymerization of acrylic acid onto poly (ethylene terephthalate) films: Characterization and human smooth muscle cell growth on grafted films. *Biomaterials* **2002**, *23*, 863–871. [CrossRef]

41. Qiu, Y.X.; Klee, D.; Plüster, W.; Severich, B.; Höcker, H. Surface modification of polyurethane by plasma-induced graft polymerization of poly (ethylene glycol) methacrylate. *J. Appl. Polym. Sci.* **1996**, *61*, 2373–2382. [CrossRef]

42. Wavhal, D.S.; Fisher, E.R. Hydrophilic modification of polyethersulfone membranes by low temperature plasma-induced graft polymerization. *J. Membrane Sci.* **2002**, *209*, 255–269. [CrossRef]

43. Ulbricht, M.; Belfort, G. Surface modification of ultrafiltration membranes by low temperature plasma II. Graft polymerization onto polyacrylonitrile and polysulfone. *J. Membrane Sci.* **1996**, *111*, 193–215. [CrossRef]

44. Sun, J.; Yao, L.; Sun, S.; Gao, Z.Q.; Qiu, Y.P. Effect of storage condition and aging on acrylic acid inverse emulsion surface-grafting polymerization of PET films initiated by atmospheric pressure plasmas. *Surf. Coat. Technol.* **2011**, *205*, 2799–2805. [CrossRef]

45. Johnsen, K.; Kirkhorn, S.; Olafsen, K.; Redford, K.; Stori, A. Modification of polyolefin surfaces by plasma-induced grafting. *J. Appl. Polym. Sci.* **1996**, *59*, 1651–1657. [CrossRef]

46. Bahners, T.; Prager, L.; Pender, A.; Gutmann, J.S. Super-wetting surfaces by plasma- and UV-based grafting of micro-rough acrylate coating. *Prog. Org. Coat.* **2013**, *76*, 1356–1362. [CrossRef]

47. Sciarratta, V.; Vohrer, U.; Hegemann, D.; Müller, M.; Oehr, C. Plasma functionalization of polypropylene with acrylic acid. *Surf. Coat. Technol.* **2003**, *174*, 805–810. [CrossRef]

48. Biederman, H. *Plasma Polymer Films*; Imperial College Press: London, UK, 2004; ISBN 1–86094–467–1.

49. Foest, R.; Kindel, E.; Ohl, A.; Stieber, M.; Weltmann, K.D. Non-thermal atmospheric pressure discharges for surface modification. *Plasma Phys. Contr. F.* **2005**, *47*, B525. [CrossRef]

50. Malmsten, M.; Johansson, J.-Å.; Burns, N.L.; Yasuda, H.K. Protein adsorption at n-butane plasma polymer surfaces. *Colloid. Surface. B* **1996**, *6*, 191–199. [CrossRef]

51. Merche, D.; Vandencasteele, N.; Reniers, F. Atmospheric plasmas for thin film deposition: A critical review. *Thin Solid Films* **2012**, *520*, 4219–4236. [CrossRef]

52. Morent, R.; De Geyter, N.; Trentesaux, M.; Gengembre, L.; Dubruel, P.; Leys, C.; Payen, E. Stability study of polyacrylic acid films plasma-polymerized on polypropylene substrates at medium pressure. *Appl. Surf. Sci.* **2010**, *257*, 372–380. [CrossRef]

53. Ward, L.J.; Schofield, W.C.E.; Badyal, J.P.S.; Goodwin, A.J.; Merlin, P.J. Atmospheric pressure plasma deposition of structurally well-defined polyacrylic acid films. *Chem. Mater.* **2003**, *15*, 1466–1469. [CrossRef]

54. Beck, A.J.; Short, R.D.; Matthews, A. Deposition of functional coatings from acrylic acid and octamethylcyclotetrasiloxane onto steel using an atmospheric pressure dielectric barrier discharge. *Surf. Coat. Technol.* **2008**, *203*, 822–825. [CrossRef]

55. Topala, I.; Dumitrascu, N.; Popa, G. Properties of the acrylic acid polymers obtained by atmospheric pressure plasma polymerization. *Nucl. Instrum. Meth. B* **2009**, *267*, 442–445. [CrossRef]

56. Morent, R.; De Geyter, N.; Van Vlierberghe, S.; Dubruel, P.; Leys, C.; Gengembre, L.; Schacht, E.; Payen, E. Deposition of HMDSO-based coatings on PET substrates using an atmospheric pressure dielectric barrier discharge. *Prog. Org. Coat.* **2009**, *64*, 304–310. [CrossRef]

57. Yasuda, H.; Bumgarner, M.O.; Marsh, H.C.; Morosoff, N. Plasma polymerization of some organic compounds and properties of the polymers. *J. Polym. Sci. Pol. Chem. Edition* **1976**, *14*, 195–224. [CrossRef]

58. Heyse, P.; Dams, R.; Paulussen, S.; Houthoofd, K.; Janssen, K.; Jacobs, P.A.; Sels, B.F. Dielectric barrier discharge at atmospheric pressure as a tool to deposit versatile organic coatings at moderate power input. *Plasma Process. Polym.* **2007**, *4*, 145–157. [CrossRef]

59. Hossain, M.M.; Hegemann, D.; Fortunato, G.; Herrmann, A.S.; Heuberger, M. Plasma Deposition of Permanent Superhydrophilic a-C: H: N Films on Textiles. *Plasma Process. Polym.* **2007**, *4*, 471–481. [CrossRef]

60. Shen, M.; Bell, A.T. A Review of Recent Advances in Plasma Polymerization. *ACS Sym. Ser.* **1979**, *108*, 1–33.

61. Friedrich, J. Mechanisms of plasma polymerization-reviewed from a chemical point of view. *Plasma Process. Polym.* **2011**, *8*, 783–802. [CrossRef]

62. Biederman, H. Polymer films prepared by plasma polymerization and their potential application. *Vacuum* **1987**, *37*, 367–373. [CrossRef]

63. Serra, R.; Zheludkevich, M.L.; Ferreira, M.G.S. Influence of the RF plasma polymerization process on the barrier properties of coil-coating. *Prog. Org. Coat.* **2005**, *53*, 225–234. [CrossRef]

64. Lee, H.R.; Kim, D.J.; Lee, K.H. Anti-reflective coating for the deep coloring of PET fabrics using an atmospheric pressure plasma technique. *Surf. Coat. Tech.* **2001**, *142*, 468–473. [CrossRef]

65. Múgica-Vidal, R.; Alba-Elías, F.; Sainz-García, E.; Ordieres-Meré, J. Atmospheric plasma-polymerization of hydrophobic and wear-resistant coatings on glass substrates. *Surf. Coat. Tech.* **2014**, *259*, 374–385. [CrossRef]

66. Vautrin-Ul, C.; Boisse-Laporte, C.; Benissad, N.; Chausse, A.; Leprince, P.; Messina, R. Plasma-polymerized coatings using HMDSO precursor for iron protection. *Prog. Org. Coat.* **2000**, *38*, 9–15. [CrossRef]

67. Ko, Y.M.; Choe, H.C.; Jung, S.C.; Kim, B.H. Plasma deposition of a silicone-like layer for the corrosion protection of magnesium. *Prog. Org. Coat.* **2013**, *76*, 1827–1832. [CrossRef]

68. Massines, F.; Sarra-Bournet, C.; Fanelli, F.; Naudé, N.; Gherardi, N. Atmospheric pressure low temperature direct plasma technology: Status and challenges for thin film deposition. *Plasma Process. Polym.* **2012**, *9*, 1041–1073. [CrossRef]

69. Amir, I.; Hudec, I.; Volovič, M. Surface modification of textile reinforming material by plasma treatment and plasma polymerization. *Chem. Listy* **2009**, *103*, s100–s101.

70. Šimor, M.; Ráheľ, J.; Vojtek, P.; Černák, M.; Brablec, A. Atmospheric-pressure diffuse coplanar surface discharge for surface treatments. *Appl.Phys. Lett.* **2002**, *81*, 2716–2718. [CrossRef]

71. Kaelble, D.H. Dispersion-Polar Surface Tension Properties of Organic Solids. *J. Adhesion* **1970**, *2*, 66–81. [CrossRef]

72. Owens, D.K.; Wendt, R.C. Estimation of the surface free energy of polymers. *J. Appl. Polym. Sci.* **1969**, *13*, 1741–1747. [CrossRef]

73. Drelich, J.; Chibowski, E. Superhydrophilic and superwetting surfaces: Definition and mechanisms of control. *Langmuir* **2010**, *26*, 18621–18623. [CrossRef] [PubMed]

74. Jaleh, B.; Parvin, P.; Wanichapichart, P.; Saffar, A.P.; Reyhani, A. Induced super hydrophilicity due to surface modification of polypropylene membrane treated by O_2 plasma. *Appl. Surf. Sci.* **2010**, *257*, 1655–1659. [CrossRef]

75. Okabe, Y.; Kurihara, S.; Yajima, T.; Seki, Y.; Nakamura, I.; Takano, I. Formation of super-hydrophilic surface of hydroxyapatite by ion implantation and plasma treatment. *Surf. Coat. Technol.* **2005**, *196*, 303–306. [CrossRef]

76. Carrino, L.; Moroni, G.; Polini, W. Cold plasma treatment of polypropylene surface: A study on wettability and adhesion. *J. Mater. Process. Tech.* **2002**, *121*, 373–382. [CrossRef]

77. Terpilowski, K.; Rymuszka, D.; Holysz, L.; Chibowski, E. Changes in wettability of polycarbonate and polypropylene pretreated with oxygen and argon plasma. In Proceedings of the 8th International Conference on Material Technologies and Modeling MMT-2014, Ariel, Izrael, 28 July 2014; Zinigrad, M., Ed.; 2014; pp. 155–165, ISBN 978-965-91944-2-1.

78. Slepička, P.; Vasina, A.; Kolská, Z.; Luxbacher, T.; Malinský, P.; Macková, A.; Švorčík, V. Argon plasma irradiation of polypropylene. *Nucl. Instrum. Meth. B* **2010**, *268*, 2111–2114. [CrossRef]

79. Pandiyaraj, K.N.; Selvarajan, V.; Deshmukh, R.R.; Gao, C. Modification of surface properties of polypropylene (PP) film using DC glow discharge air plasma. *Appl. Surf. Sci.* **2009**, *255*, 3965–3971. [CrossRef]

80. Kwon, O.J.; Myung, S.W.; Lee, C.S.; Choi, H.S. Comparison of the surface characteristics of polypropylene films treated by Ar and mixed gas (Ar/O_2) atmospheric pressure plasma. *J. Colloid Interf. Sci.* **2006**, *295*, 409–416. [CrossRef] [PubMed]

81. Harth, K.; Hibst, H. Surface modification of polypropylene in oxygen and nitrogen plasmas. *Surf. Coat. Technol.* **1993**, *59*, 350–355. [CrossRef]

82. Kwon, O.J.; Tang, S.; Myung, S.W.; Lu, N.; Choi, H.S. Surface characteristics of polypropylene film treated by an atmospheric pressure plasma. *Surf. Coat. Technol.* **2005**, *192*, 1–10. [CrossRef]

83. Massines, F.; Gouda, G.; Gherardi, N.; Duran, M.; Croquesel, E. The role of dielectric barrier discharge atmosphere and physics on polypropylene surface treatment. *Plasmas Polym.* **2001**, *6*, 35–49. [CrossRef]

84. Carrino, L.; Polini, W.; Sorrentino, L. Ageing time of wettability on polypropylene surfaces processed by cold plasma. *J. Mater. Process. Tech.* **2004**, *153*, 519–525. [CrossRef]

85. Massines, F.; Gouda, G. A comparison of polypropylene-surface treatment by filamentary, homogeneous and glow discharges in helium at atmospheric pressure. *J. Phys. D Appl. Phys.* **1998**, *31*, 3411. [CrossRef]

86. Morent, R.; De Geyter, N.; Leys, C.; Gengembre, L.; Payen, E. Study of the ageing behaviour of polymer films treated with a dielectric barrier discharge in air, helium and argon at medium pressure. *Surf. Coat. Technol.* **2007**, *201*, 7847–7854. [CrossRef]

87. Johansson, K.S. Surface Modification of Plastics. In *Applied Plastics Engineering Handbook*; Kutz, M., Ed.; Plastics Design Library; Elsevier: Amsterdam, The Netherlands, 2017; pp. 443–487. ISBN 978-0-323-39040-8.

88. Shao, T.; Zhang, C.; Long, K.; Zhang, D.; Wang, J.; Yan, P.; Zhou, Y. Surface modification of polyimide films using unipolar nanosecond-pulse DBD in atmospheric air. *Appl. Surf. Sci.* **2010**, *256*, 3888–3894. [CrossRef]

89. Zhang, C.; Zhou, Y.; Shao, T.; Xie, Q.; Xu, J.; Yang, W. Hydrophobic treatment on polymethylmethacrylate surface by nanosecond-pulse DBDs in CF4 at atmospheric pressure. *Appl. Surf. Sci.* **2014**, *311*, 468–477. [CrossRef]

90. Yablokov, M.; Gilman, A.; Kuznetsov, A. MODIFICATION OF WETTABILITY OF POLYMER SURFACES BY PLASMA. In Proceedings of the 21st Symposium on Application of Plasma Processes Book of Contributed Papers; Medvecká, V., Országh, J., Papp, P., Matějčík, Š., Eds.; Department of Experimental Physics, Faculty of Mathematics, Physics and Informatics, Comenius University in Bratislava; Society for Plasma Research and Applications in cooperation with Library and Publishing Centre CU: Bratislava, Slovakia, 2017; pp. 19–26.

91. Kong, F.; Chang, C.; Ma, Y.; Zhang, C.; Ren, C.; Shao, T. Surface modifications of polystyrene and their stability: A comparison of DBD plasma deposition and direct fluorination. *Appl. Surf. Sci.* **2018**, *459*, 300–308. [CrossRef]

92. Zhang, C.; Lin, H.; Zhang, S.; Xie, Q.; Ren, C.; Shao, T. Plasma surface treatment to improve surface charge accumulation and dissipation of epoxy resin exposed to DC and nanosecond-pulse voltages. *J. Phys. D. Appl. Phys.* **2017**, *50*, 405203. [CrossRef]

93. Zhang, C.; Ma, Y.; Kong, F.; Yan, P.; Chang, C.; Shao, T. Atmospheric pressure plasmas and direct fluorination treatment of Al2O3-filled epoxy resin: A comparison of surface charge dissipation. *Surf. Coatings Technol.* **2019**, *362*, 1–11. [CrossRef]

94. Luongo, J.P. Infrared study of polypropylene. *J. Appl. Polym. Sci.* **1960**, *3*, 302–309. [CrossRef]

95. Urbaniak-Domagala, W. The use of the spectrometric technique FTIR-ATR to examine the polymers surface. In *Advanced Aspects of Spectroscopy*; INTECH Open Access Publisher: Rijeka, Croatia, 2012.

96. Wexler, A.S. Integrated Intensities of Absorption Bands in Infrared Spectroscopy. *Appl. Spectrosc. Rev.* **1967**, *1*, 29–98. [CrossRef]

97. Klages, C.P.; Hinze, A.; Khosravi, Z. Nitrogen Plasma Modification and Chemical Derivatization of Polyethylene Surfaces—An In Situ Study Using FTIR-ATR Spectroscopy. *Plasma Process. Polym.* **2013**, *10*, 948–958. [CrossRef]

98. Sellin, N.; de, C.; Campos, J.S. Surface composition analysis of PP films treated by corona discharge. *Mater. Res.* **2003**, *6*, 163–166. [CrossRef]

99. Morent, R.; De Geyter, N.; Leys, C.; Gengembre, L.; Payen, E. Comparison between XPS- and FTIR-analysis of plasma-treated polypropylene film surfaces. *Surf. Interface Anal.* **2008**, *40*, 597–600. [CrossRef]

100. Urbaniak-Domagała, W.; Wrzosek, H.; Szymanowski, H.; Majchrzycka, K.; Brochocka, A. Plasma Modification of Filter Nonwovens Used for the Protection of Respiratory Tracts. *Fibres Tex. East. Eur.* **2010**, *18*, 94–99.

101. Tsai, C.Y.; Juang, R.S.; Huang, C. Surface Modification of Polypropylene Membrane by RF Methane/Oxygen Mixture Plasma Treatment. *Jpn. J. Appl. Phys.* **2011**, *50*, 08KA02. [CrossRef]

102. Guruvenket, S.; Rao, G.M.; Komath, M.; Raichur, A.M. Plasma surface modification of polystyrene and polyethylene. *Appl. Surf. Sci.* **2004**, *236*, 278–284. [CrossRef]

103. Mutel, B.; Grimblot, J.; Dessaux, O.; Goudmand, P. XPS investigations of nitrogen-plasma-treated polypropylene in a reactor coupled to the spectrometer. *Surf. Interface Anal.* **2000**, *30*, 401–406. [CrossRef]

104. Klemberg-Sapieha, J.E.; Küttel, O.M.; Martinu, L.; Wertheimer, M.R. Dual-frequency N2 and NH3 plasma modification of polyethylene and polyimide. *J. Vac. Sci. Technol. A* **1991**, *9*, 2975–2981. [CrossRef]

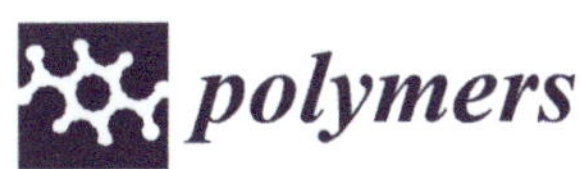 *polymers*

Article

Electrospinning of Hyaluronan Using Polymer Coelectrospinning and Intermediate Solvent

Lenka Vítková [1], Lenka Musilová [1,2], Eva Achbergerová [3], Antonín Minařík [1,2], Petr Smolka [1,2], Erik Wrzecionko [1,2] and Aleš Mráček [1,2,*]

[1]   Department of Physics and Materials Engineering, Faculty of Technology, Thomas Bata University in Zlín, Vavrečkova 275, 760 01 Zlín, Czech Republic; l_davidova@utb.cz (L.V.); lmusilova@utb.cz (L.M.); minarik@utb.cz (A.M.); smolka@utb.cz (P.S.); wrzecionko@utb.cz (E.W.)
[2]   Center of Polymer Systems, Thomas Bata University in Zlín, tř. Tomáše Bati 5678, 760 01 Zlín, Czech Republic
[3]   CEBIA-Tech, Faculty of Applied Informatics, Thomas Bata University in Zlín, Nad Stráněmi 4511, 760 05 Zlín, Czech Republic; achbergerova@utb.cz
*   Correspondence: mracek@utb.cz; Tel.: +420-733-690-668

Received: 9 August 2019; Accepted: 8 September 2019; Published: 18 September 2019

Abstract: In the current study, we present methods of sodium hyaluronate, also denoted as hyaluronan (HA), nanofiber fabrication using a direct-current (DC) electric field. HA was spun in combination with poly(vinyl alcohol) (PVA) and polyethylene oxide (PEO) and as a pure polymer. Nonaggressive solvents were used due to the possible use of the fibers in life sciences. The influences of polymer concentration, average molecular weight (M_w), viscosity, and solution surface tension were analyzed. HA and PVA were fluorescent-labeled in order to examine the electrospun structures using fluorescence confocal microscopy. In this study, two intermediate solvent mixtures that facilitate HA electrospinning were found. In the case of polymer co-electrospinning, the effect of the surfactant content on the HA/PVA electrospinning process, and the effect of HA M_w on HA/PEO nanofiber morphology, were examined, respectively.

Keywords: electrospinning; hyaluronan; poly(vinyl alcohol); polyethylene oxide; nanofibers; intermediate solvent; fluorescence confocal microscopy

---

1. Introduction

Electrospinning, nowadays a well-established fiber-fabrication method first described in 1902 by Cooley and Morton [1,2], is based on several electrohydrodynamic phenomena [3]. The method utilizes Taylor cones, the product of electric field-induced instabilities in liquid bodies stabilized by capillary forces. However, if electric forces overcome capillary forces, a liquid jet is ejected, which is then subjected to elongation at high rates, causing a decrease in diameter up to micron fractions. Due to the large specific surface of the polymer jet, rapid evaporation of the solvent occurs, leading to the solidification of the polymer jet in the form of a nanofiber. Several instabilities may occur and disrupt the electrospinning process, causing particle formation (so-called electrospraying), bead-on-string structure formation, or branching. The preliminary cause of these phenomena is Rayleigh instability, which is surface tension-driven and electrostatically hindered. Similarly to Taylor cone formation prior to electrospinning, cone-shaped undulations may be formed on the cylindrical jet, leading to jet collapse if the charge per unit area is small, providing electrospraying or bead-on-string structured fibers, or undulation stabilization and elongation, giving branched fibers [3,4]. This technique inherently gives nonwoven mats composed of infinite fibers. In practice, DC electric fields of up to tens of kilovolts magnitude are used in most cases.

Nanofibrous materials have uses in many industry fields, for example, in porous materials [5–7], the fuel cell industry [8,9], petroleum engineering [10,11], and particularly biomedical applications such as wound dressing [12,13], drug delivery, [14,15], or tissue engineering [16,17]. The most popular synthetic polymers for electrospinning include polyethylene oxide (PEO), poly(lactide), and polyethyleneimine [4,18,19]. Regarding natural polymers, proteins, such as silk fibroin [20], or polysaccharides, such as alginate, cellulose, or chitosan, can be used [14]. The presence of a nanostructure was proven to enhance cell proliferation [21], which is why combined 3D printing and electrospinning techniques are widely studied for potential uses in tissue engineering (see, e.g., Mori et al. (2018) [22]).

Hyaluronan (HA) is a polysaccharide abundant in the extracellular matrix of living organisms. Its primary structure is linear, and its secondary structure is typically a twisted ribbon. Due to a rather stiff backbone chain caused by the disaccharide structure, internal hydrogen bonds, and interactions with solvents, the ternary structure is an expanded random coil [23,24]. It was experimentally proven that the presence of ions can influence coil diameter [25]. The coil structure is capable of absorbing approximately 1000 times its weight of water [26]. This provides solutions of HA with extraordinarily high viscosity at low concentrations, as well as shear-thinning behavior. Its excellent biocompatibility and solubility in water makes it a popular choice in biomedical applications [27]. HA melt processing is impossible due to its instability at high temperatures [28].

The main complications in HA solution electrospinning are the high surface tension of HA aqueous solutions, extremely high viscosity at low concentrations, preventing the formation of highly concentrated solutions, and the low evaporation rate of water. Attempts to fabricate HA nanofibers using electrospinning have been made by many researchers. A common approach to electrospinning of polymers with low spinnability is the use of a highly spinnable polymer that then serves as a dragging polymer [20,29]. Under certain conditions, this approach provides core–shell nanofibers, as was demonstrated by Ma et al. (2017) [30] using chitosan and an HA solution. To overcome the problem of high surface tension, the use of surfactants [31] or a different solvent choice is possible [27,32,33]. According to Malkin et al. (2017), a change in solvent also has a positive effect on spinnability due to introducing a polymer–solvent demixing solidification mechanism [34]. Previously, it was assumed that electrospinning is only possible above critical concentration, i.e., polymer concentration corresponding to one entanglement per chain [3]. However, using solutions of PEO and polyethylene glycol (PEG), Yu et al. (2006) demonstrated that electrospinning is possible below critical concentration [35]. It was argued that the ability to form smooth fibers via electrospinning is governed by solution elasticity [36–38]. On the other hand, Shenoy et al. (2005) performed electrospinning experiments on several polymer solutions, and concluded that complete stabilization of the electrospinning process is provided by a minimum of 2.5 entanglements per chain [39]. Malkin et al. (2017) argued that stabilization of the electrospinning process can be achieved at concentrations below critical if an intermediate solvent is used [34]. Experiment evidence suggests great contribution of solution elasticity, interaction parameters, surface tension, and conductivity to the electrospinning process. Ambient parameters, such as temperature and humidity, need to be taken into account as well [40]. Although there have been attempts for analysis of electrospinning jet behavior [4,41], so far none are comprehensive enough to account for all influence.

In the current study, nanofibrous mats containing HA were obtained using electrospinning. Two approaches were employed: HA co-electrospinning in a blend with highly spinnable polymers PVA and PEO, respectively, and the use of an intermediate solvent. As intermediate solvents, mixtures containing water and isopropanol (IPA), and water, ethanol EtOH, and methanol (MeOH) were used. The solvent mixtures were found with the aid of a Teas graph, incorporating the method described in Luo et al. (2010) [42]. We attempted to offer insight on the influence of shear viscosity and polymer-chain conformation in the solution on the electrospinning process. Furthermore, HA and PVA were fluorescent-labeled, which allowed the products to be observed by fluorescence confocal microscopy.

2. Materials and Methods

2.1. Materials and Chemicals

HA of M_w 243 kDa, 370 kDa, 600 kDa, and 1180 kDa was purchased from Contipro a.s. Demineralized (DEMI) water was prepared using the Milipore Direct-Q 3UV system. PVA of M_w 89–98 kDa, 99+% hydrolyzed, PEO of M_w 300 and 600 kDa, respectively, EtOH absolute Spectranal, IPA puriss p.a., ACS reagent, disodium hydrogen phosphate dodecahydrate ≥99%, 4-acetamido-TEMPO free radical, 97%, dimethyl sulfoxide (DMSO) ACS reagent, ≥99,9%, Nile Blue A, dye content ≥75%, NaBH$_3$CN reagent grade, 95%, pyridine anhydrous 99.8%, dibutiltin dilaurate, 95%, fluorescein isothyocyanite isomer (FITC), ≥90% and benzethonium chloride (BEC), ≥97% were purchased from Sigma Aldrich. MeOH p.a. was purchased from Lach:Ner. Sodium bromide pure was purchased from Lachema a.s. Sodium hypochlorite solution pure was purchased from Penta. NaCl PharmaGrade was purchased from SAFC. NaHCO$_3$ ACS Grade was purchased from VWR.

2.2. HA Fluorescent Labeling

In order to prepare Nile Blue A labeled-HA (600 kDa), HA was initially oxidized according to a previously published method, Huerta-Angeles et al. (2012) [43], followed by fluorescent labeling described in Šmejkalová et al. (2017) [44].

Initially, HA (1 g) was dissolved in 10 mL of DEMI water. To the HA solution, sodium bromide (0.129 g) and disodium hydrogen phosphate (0.771 g) were added. The reaction mixture was cooled to 5 °C, followed by the addition of 4-acetamido-TEMPO (5 mg) and 450 μL of sodium hypochlorite. The reaction was carried out for 45 min under nitrogen atmosphere at 5 °C. The oxidized HA was dialyzed against DEMI water for 3 days and freeze-dried (yield: 96%). In the second step, an aqueous solution (2 wt.%) of oxidized HA (0.5 g) was stirred with Nile Blue A (92 mg) predissolved in DMSO (5 mL) for 5 h. Subsequently, NaBH$_3$CN (79 mg) was added to this reaction mixture which was then stirred over night at room temperature. The crude product was precipitated by NaCl solution and IPA, and the remaining Nile Blue A was washed out using IPA. The product was then dialyzed against 0.5 wt.% NaCl and 0.5 wt.% NaHCO$_3$ aqueous solutions for 2 days and against DEMI water for 3 days. The final product was obtained in a form of a blue lyophilisate (yield 78%).

2.3. PVA Fluorescent Labeling

PVA was labeled with FITC following the procedure published by Kaneo et al. (2005) [45]. Briefly, PVA (2.5 g) was dissolved in DMSO (66.6 mL) and pyridine (416.6 μL) under stirring at 80 °C for 24 h. FITC (83 mg) and dibutiltin dilaurate (31 μL) were added to the PVA solution and the reaction was carried out for 2 h at 95 °C in darkness. The crude product was precipitated and washed with IPA, followed by dialysis against DEMI water and lyophilization. The yield of the reaction was 88%.

2.4. Solutions Preparation

HA of respective M_w was dissolved in binary and ternary solvent mixtures at 50 °C under vigorous stirring for 48 h regardless the HA M_w and solvent mixture, to obtain completely homogenized solution. The solvent mixtures chosen for the experiments were H$_2$O:IPA in 10:7 weight ratio, and H$_2$O:EtOH:MeOH in 5:5:1 weight ratio.

HA/PVA blend solutions with BEC surfactant were prepared in the following way; 2 wt.% HA 600 kDa solution and 1 wt.% PVA 89–98 kDa solution were prepared separately by dissolving the respective polymers in DEMI water for 24 h at elevated temperature (50 °C for HA, and 80 °C for PVA). BEC aqueous solutions of the following concentrations: 1 wt.%, 2 wt.%, 5 wt.%, and 10 wt.% were prepared separately as well. The final solutions were prepared by mixing 2.5 g of HA solution with 2 g of PVA solution. After the components were properly mixed, 0.03 g of BEC solution of respective concentration was added and the solution was mixed properly. Slight turbidity appearance was observed upon the surfactant addition. For the purpose of confocal microscopy, 4% of the respective polymer content was replaced by fluorescent labeled analogue.

HA/PEO 2 wt.% blend solutions were prepared by mixing the polymers at 1:1 weight ratio and dissolving them in DEMI water by stirring vigorously at room temperature for 48 h. HA M_w used were 243, 370, and 600 kDa. PEO M_w used were 300 and 600 kDa. For the purpose of confocal microscopy, 4% of HA content was replaced by Nile Blue A labeled HA.

2.5. Electrospinning Equipment

A homemade electrospinner consisting of high DC voltage power supply Spellman SL150, a grounded metal collector, 40.3 mm in diameter, and a simple metal rod spinneret, 8 mm in diameter, were used in the study (see Figure 1). The tip-to-collector distance was kept at 76 mm. Experiments were conducted in air atmosphere at room temperature and humidity, and normal pressure. The fibers were collected using a recycled paper substrate to ensure good adhesion.

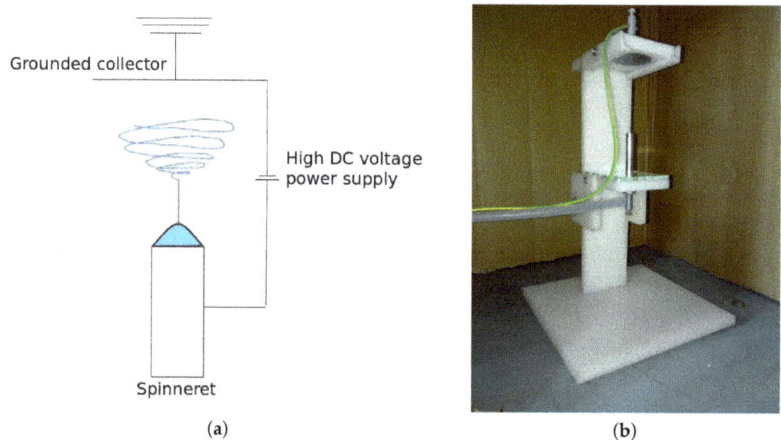

Figure 1. Electrospinning device: (**a**) device scheme and (**b**) device used in experiments.

2.6. Characterization

Dynamic viscosity was determined using a Malvern Kinexus Pro+ rotational rheometer with cup-and-bob geometry. The measurements were conducted at 25 °C at 11 different shear rates ranging from 0.1 to 10 s^{-1}.

Portable conductometer Mettler Toledo Seven2Go Pro was used to determine the conductivity of the solutions. Each solution was measured 3 times at room temperature.

Surface tension was determined by a pendant drop method using a Krüss Drop Shape Analyzer DSA 100. Three separate drops of each sample were measured. Each drop was measured 30 times with a 1 s delay between the measurements. Dixon's Q-test was used to exclude the outliers. The measurement was conducted at 25 °C in an air atmosphere.

The fiber morphology analysis was done using a Phenom Pro X Scanning Electron Microscope (SEM) in the backscattered electron mode. The samples were sputtered with a layer of gold prior to the analysis. Acceleration voltage was 10 kV. Optical analysis of the images was done using ImageJ software.

An Olympus FLUOVIEW FV3000 Laser Scanning Microscope was used for fluorescence confocal microscopy. Excitation wavelenghts available were 405, 488, 561, and 640 nm. Wavelength ranges 600–640 nm and 450–520 nm were, respectively, used as emission spectra for HA labeled by Nile Blue A, and PVA labeled by FTIC. The immersion objective ($Z = 60$) with numerical aperture $A = 1.35$ was used for nanofibers observation.

3. Results and Discussion

The electrospinning process is highly dependent on the intrinsic properties of the spinning solution. The most prevalent were polymer M_w, concentration and polydispersity, all of which were reflected in viscosity, and also surface tension and conductivity [40]. Higher conductivity was presumed to facilitate stability in the spinning process [4], while high surface tension prevented electrospinning onset [3]. Measurement of shear viscosity was done at low shear rates in a narrow range, since the formation of a Taylor cone in sufficiently conductive fluids typically occurs without inducing high shear rates [46], and solutions are considered Newtonian liquids in this part of the process.

3.1. HA/PVA Blend Aqueous Solutions

In the past, electrospinning HA/PVA aqueous solutions was not possible without the addition of a small amount of surfactant. BEC was chosen due to the coil-shrinking effect on HA conformation [24], which we assumed to be beneficial in terms of electrospinning.

The surface tension of HA/PVA blend aqueous solutions was lower than the HA aqueous solutions (around 70 mN·m^{-1}, see Jurošková (2017) [47]), which is likely the result of surfactant BEC addition. Surface tension decreases with increasing of BEC content. Solution conductivity was increased by the increase of BEC content as a result of BEC ionic nature (see Table 1).

Table 1. Characteristics of Hyaluronan (HA)/Poly(vinyl alcohol) (PVA) blend solution with benzethonium chloride (BEC).

BEC Content [wt.%]	Surface Tension [mN·m^{-1}]	Conductivity [μS·cm^{-1}]
0.006	43.7 ± 0.9	1473 ± 7
0.013	42.6 ± 0.2	1496 ± 2
0.033	41.4 ± 0.6	1513 ± 4
0.065	41.1 ± 0.2	1546 ± 4

HA/PVA blend aqueous solutions show the highest conductivity of the spinnable solutions used in the current study (see Tables 1 and 3–5). This is likely the synergic effect of dissociation of HA and PVA in water, and the addition of ionic surfactant to the solution.

Low polymer concentration causes the viscosity of the solutions to be low as well. As apparent from Figure 2, there was a quick drop of viscosity present upon addition of 0.065 wt.% of BEC.

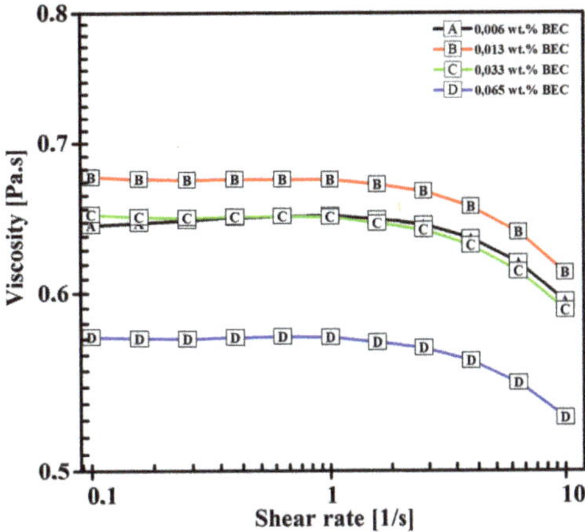

Figure 2. Viscosity of HA/PVA blend solutions with BEC as a function of shear rate.

0.065 wt.% of BEC was close to critical aggregation concentration [24] and HA coil shrinkage was expected, causing significant increase in turbidity of the solution, which was observed during the preparation, and it was in agreement with findings of Gřundělová et al. (2013) [24]. It is safe to assume that BEC effectively created an intermediate solvent to HA, and electrospinning HA is therefore encouraged not only by mixing with highly spinnable PVA, but also by the intermediate solvent effect. Due to HA precipitation, further increase of BEC content would be counterproductive.

The effect of surfactant content on electrospun-structure morphology and the electrospinning process was examined. As the BEC content increased, the minimum spinning voltage decreased (see Table 2) due to the decrease in surface tension.

Table 2. Morphology analysis of HA/PVA blend solutions with BEC electrospinning products.

BEC Content [wt.%]	Spinning Voltage [kV]	Product Form	Beads Diameter [μm]	Fibers Diameter [μm]
0.006	20.4	Elongated beads	0.1–0.5	/
0.013	18.8	Elongated beads	0.3–0.7	/
0.033	17.5	Bead-on-string	0.1–0.8	0.03–0.06
0.065	15.9	Bead-on-string	0.2–0.6	0.06–0.1

With the increase of BEC content, product morphology shifted from elongated beads to bead-on-string structured fibers (see Figure 3).

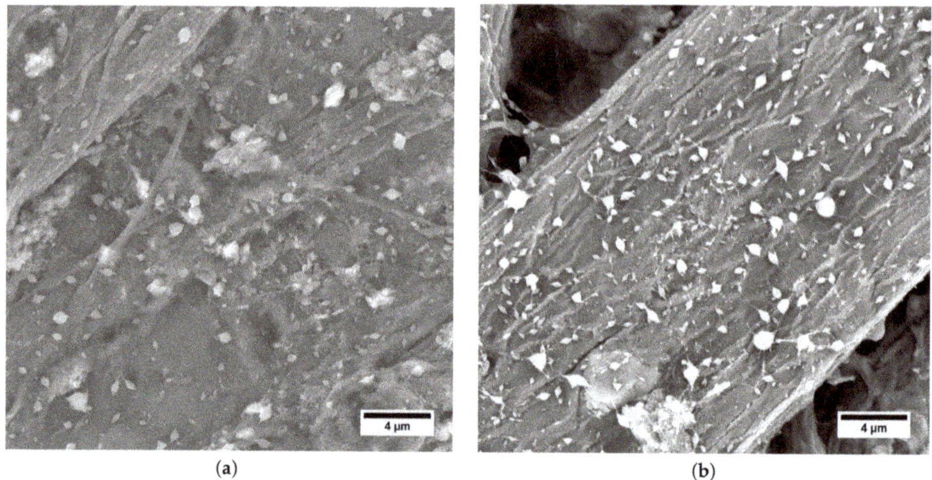

(a)

(b)

Figure 3. Scanning Electron Microscopy (SEM) micrographs of electrospun structures obtained from HA/PVA blend solutions with BEC. BEC content (**a**) 0.033 wt.%. and (**b**) 0.065 wt.%.

The increase in stability may be the result of increased conductivity of the solution, as argued by Reneker and Yarin (2008) [4], or the increase in polymer–polymer interactions, i.e., lowering solvent quality, which would be in agreement with the findings of Malkin et al. (2017) [34]. However, electrospinning smooth fibers was not achieved by this method. There were multiple reasons, such as insufficient solution elasticity, difference in HA and PVA viscoelastic behavior, or uneven BEC distribution, leading to formation of clumps of the respective polymers. Higher BEC content also led to the formation of multiple Taylor cones, therefore increasing the yield of the process. To proceed on this subject, it would be possible to choose a different surfactant with higher critical aggregation concentration, or alter the HA/PVA ratio in the solution in a way to increase the elongation elasticity. Both approaches would certainly lead to a better understanding of the co-electrospinning phenomenon, and might lead to smooth nanofiber production. The electrospun structures did not exceed 1 μm in diameter (see Table 2).

Nile Blue A labeled HA 600 kDa and FITC labeled PVA 89–98 kDa blend aqueous solution containing 0.065 wt.% of BEC was electrospun, and the products were observed by fluorescence confocal microscopy. It is clear from Figure 4 that both polymers are present jointly in fibers, as well as beads. However, due to insufficient magnification it was not possible to determine the respective position of the polymers within the structures. The absence of some structures when illuminated by a different wavelength suggested fluctuation in the contents of the respective polymers throughout spinning.

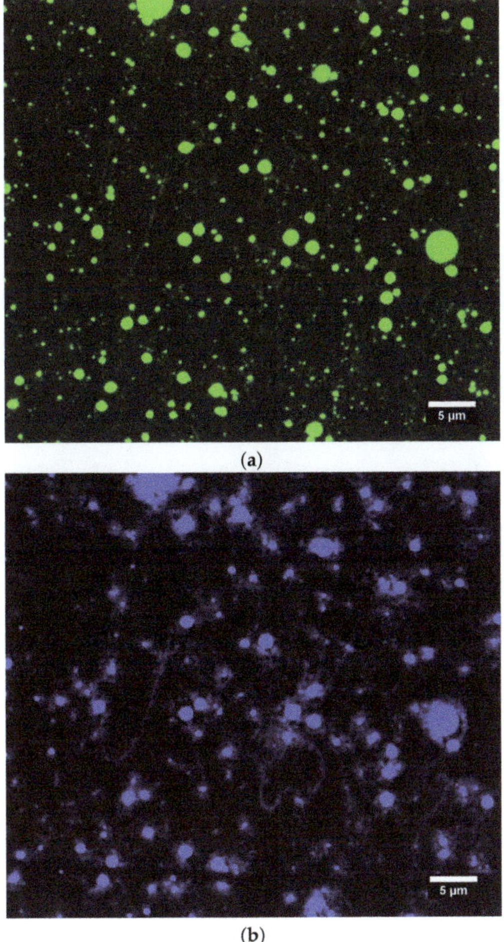

Figure 4. *Cont.*

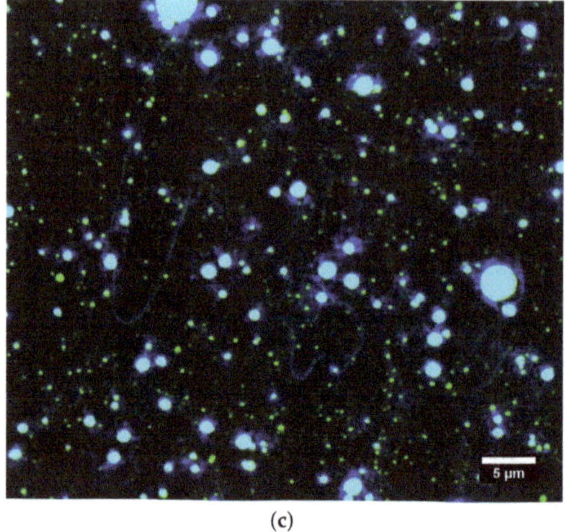

(c)

Figure 4. Fluorescence confocal microscope micrographs of electrospun structures obtained from Nile Blue A labeled HA 600 kDa and FITC labeled PVA 89–98 kDa aqueous solution with BEC content 0.065 wt.%. (**a**) Nile Blue A labeled HA visible. Emission spectrum 600–640 nm. (**b**) FITC labeled PVA visible. Emission spectrum 450–520 nm. (**c**) Both fluorescent labeled polymers visible—combined emission spectra.

3.2. HA/PEO Blend Aqueous Solutions

HA/PEO blend solutions were spun in order to produce HA containing nanofibers from aqueous solutions without use of any additional substances, such as surfactants or salts. PEO served as elasticity and shear viscosity mediator, as well as a highly spinnable polymer for co-electrospinning. In order to examine the influence of HA M_w on the electrospinning process, PEO was used in such M_w and concentration that would facilitate the electrospinning of all HA M_w chosen. The PEO M_w and concentration were found experimentally.

Surface tension of HA/PEO blend aqueous solutions was significantly higher than the one of HA in intermediate solvents solutions (see Tables 3–5) due to use of water as a solvent, but still considerably lower than HA aqueous solutions [47], caused by the surface tension-enhancing effect of HA being hindered by the presence of PEO, which induced a decrease in surface tension of the aqueous solutions instead [48].

Table 3. Characteristics of HA/Polyethylene oxide (PEO) aqueous blend solutions.

PEO M_w [kDa]	HA M_w [kDa]	Surface Tension [mN·m^{-1}]	Conductivity [μS·cm^{-1}]
	243	46 ± 2	1284 ± 8
300	370	49.3 ± 0.2	1255 ± 7
	600	49 ± 2	1241 ± 6
	243	51 ± 2	1312 ± 4
600	370	52 ± 2	1258 ± 3
	600	54 ± 1	1213 ± 3

The effect of respective polymers M_w on the surface tension was inconclusive, as a result of the low polymer concentration used (2 wt.%). Conductivity of the solutions was significantly higher than that of the pure HA solutions (see Tables 3–5), despite the lower concentration of HA. This can be attributed to the higher dissociation of HA in water than in solvents containing alcohols. As a result of the increase in molar fraction with the decrease in M_w while the same weight fraction of a polymer was kept, solution conductivity slightly decreased with the increase of HA M_w. As the M_w decreased, the effect of the end groups also gained significance and, in the case of HA and PEO, contributed to conductivity as well.

Table 4. Characteristics of HA solutions in H_2O:Isopropanol (IPA) in 10:7 weight ratio solvent mixtures.

M_w [kDa]	Concentration [wt.%]	Surface Tension [mN·m^{-1}]	Conductivity [μS·cm^{-1}]
	3.2	27 ± 1	854 ± 2
600	2.3	25 ± 0.5	568 ± 1
	1.3	26.2 ± 0.3	465 ± 1
	2.9	24 ± 3	682 ± 2
1180	1.2	28 ± 2	386 ± 2
	1.0	27.1 ± 0.5	343 ± 2

Table 5. Characteristics of HA solutions in H_2O:Ethanol (EtOH):Methanol (MeOH) in 5:5:1 weight ratio solvent mixtures.

M_w [kDa]	Concentration [wt.%]	Surface Tension [mN·m^{-1}]	Conductivity [μS·cm^{-1}]
	2.8	32.3 ± 0.9	740 ± 8
600	2.4	30.5 ± 0.5	748 ± 1
	0.7	30.2 ± 0.8	249 ± 1
	2.3	27.4 ± 0.8	704 ± 1
1180	2.2	28 ± 2	616 ± 1
	1.5	27.6 ± 0.9	459 ± 1

Solution viscosity dropped as the M_w of the respective polymers decreased (see Figure 5). Viscosity was significantly lower than that in the case of solutions of HA in intermediate solvents, which was one of the purposes of adding PEO into an HA aqueous solution.

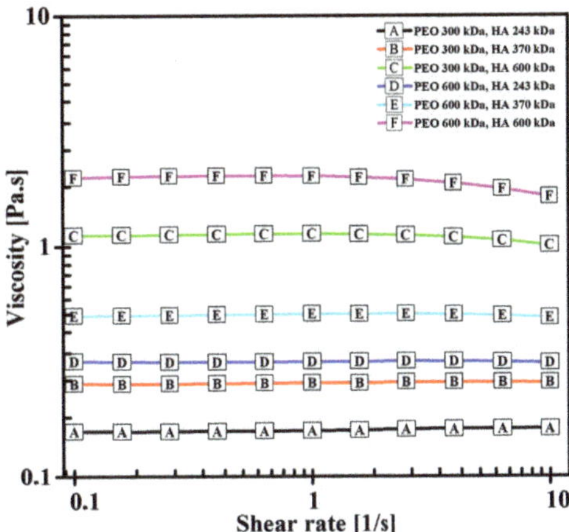

Figure 5. Viscosity of HA/PEO blend solutions as a function of shear rate.

In order to examine the effect of HA M_w on the morphology of structures electrospun from HA/PEO blend aqueous solutions, the processing parameters, i.e., tip-to-collector distance and spinning voltage, were kept constant for each series of samples. All of the solutions gave bead-on-string structured fibers, which was a result of Raileygh instability acting on a conductive liquid jet in a strong electric field [4]. The structures did not exceed 1 μm in size (see Table 6).

Table 6. Morphology analysis of HA/PEO blend-solution electrospinning products.

PEO M_w	HA M_w	Spinning Voltage [kV]	Product Form	Beads Diameter [µm]	Fibers Diameter [µm]
	243	18.2	Bead-on-string	0.25–0.5	0.02–0.05
300	370	18.2	Bead-on-string	0.3–0.5	0.03–0.07
	600	18.2	Bead-on-string	0.2–0.6	0.04–0.1
	243	24.5	Bead-on-string	0.2–0.7	0.05–0.1
600	370	24.5	Bead-on-string	0.3–0.6	0.05–0.1
	600	24.5	Bead-on-string	0.3–0.6	0.03–0.09

As can be seen in Figure 6, the shape of the beads was influenced by the M_w of PEO. If PEO 300 kDa (Figure 6a–c) was used, the beads were almost spherical, while for PEO 600 kDa (Figure 6d–f) strong elongation of the beads was apparent.

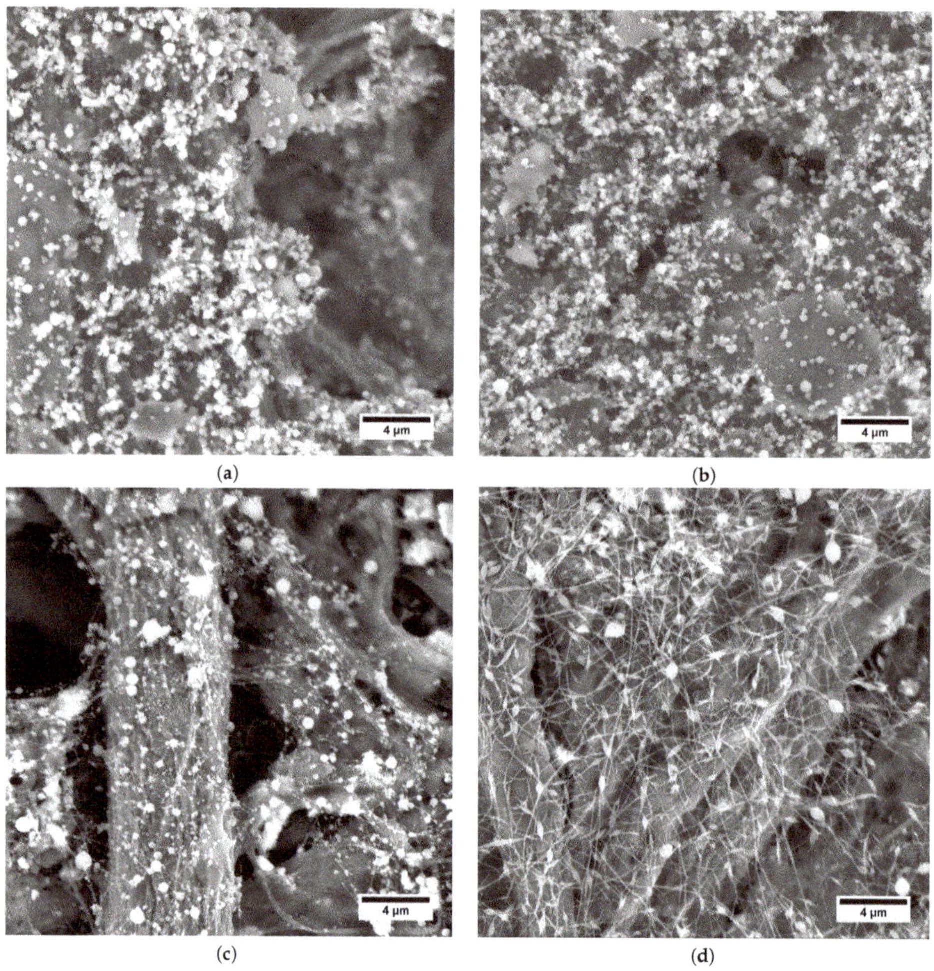

(a)

(b)

(c)

(d)

Figure 6. *Cont.*

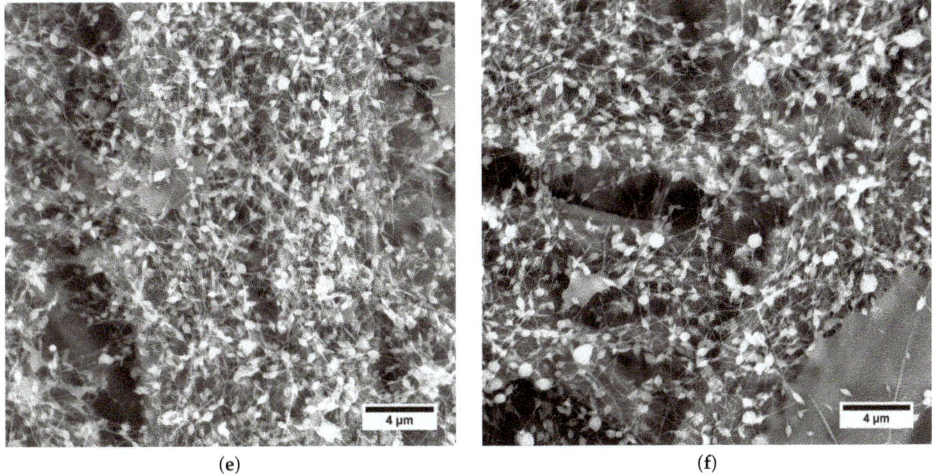

(e)   (f)

Figure 6. SEM micrographs of electrospun structures obtained from HA/PEO blend solutions. (**a**) HA 243 kDa, PEO 300 kDa. (**b**) HA 370 kDa, PEO 300 kDa. (**c**) HA 600 kDa, PEO 300 kDa. (**d**) HA 243 kDa, PEO 600 kDa. (**e**) HA 370 kDa, PEO 600 kDa. (**f**) HA 600 kDa, PEO 600 kDa.

The significance of morphology difference could even suggest different bead origin. Another explanation is a significant shift in solution elasticity induced by the different M_{w} of PEO. A great difference in viscoelasticity of the respective polymers also contributed to the uneven distribution of polymers throughout the spinning via a phenomenon known as polymer wrapping in coextrusion [49]. Further experimental examination of this phenomenon is needed in order to fully understand the causes. No effect of HA M_{w} on HA/PEO fiber morphology was found in the current study, as it was likely hindered by the significant PEO content in the used samples.

Electrospinning of HA/PEO blend solution containing Nile Blue A labeled HA 600 kDa and PEO 600 kDa allowed us to investigate the obtained structures by using fluorescence confocal microscopy. This technique proved the presence of HA in both fibers and beads (see Figure 7). On account of PEO's nonfluorescence, it is not visible in the figure.

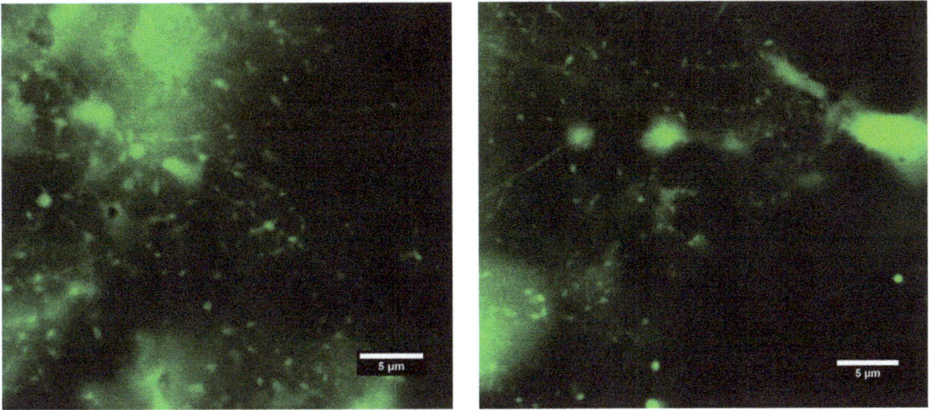

Figure 7. Fluorescence confocal microscope micrographs of electrospun structures obtained from the Nile Blue A Labeled HA 600 kDa and PEO 600 kDa aqueous solution. Emission spectrum: 600–640 nm.

3.3. HA Solutions in Intermediate Solvents

Two solvent mixture systems were chosen for the experiments: H$_2$O:IPA in 10:7 weight ratio and H$_2$O:EtOH:MeOH in 5:5:1 weight ratio. These were chosen with the aid of a Teas graph (see Figure 8) in such manner that they would lower the surface tension compared to water solutions, and encourage polymer–polymer interactions over polymer–solvent interactions, leading to smooth nanofiber production.

Electrospinning solutions of three different concentrations were considered for each solvent mixture and HA M_w, respectively, with the intention to find upper and lower limiting concentration for electrospinning.

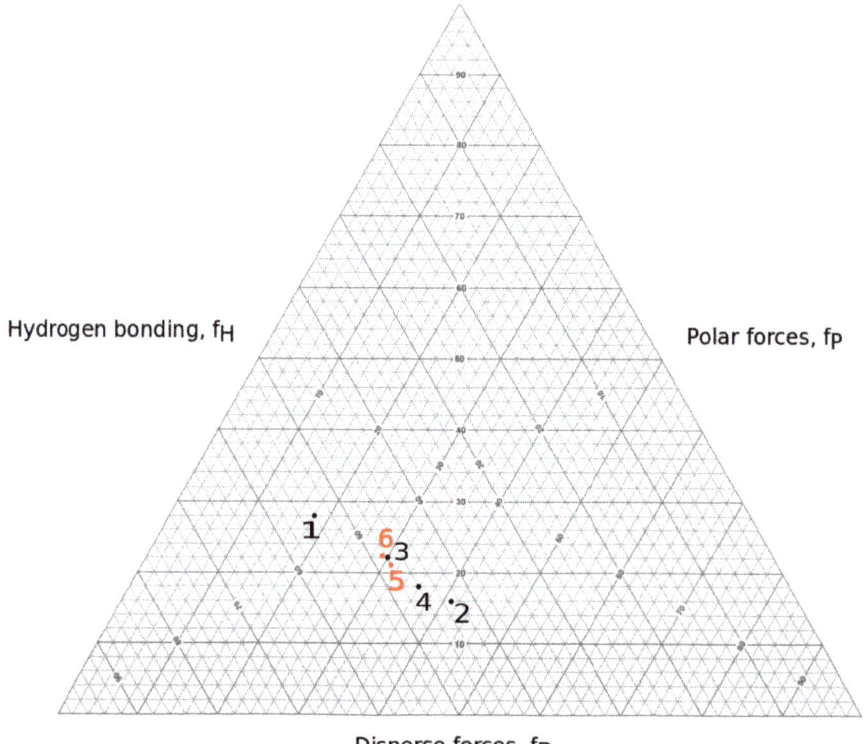

Figure 8. Solvent–mixture representation in Teas graph. 1: water; 2: IPA; 3: MeOH; 4: EtOH; 5: H2O:IPA 10:7; 6: H2O:EtOH:MeOH 5:5:1.

The use of H$_2$O:IPA an H$_2$O:EtOH:MeOH mixed solvents led to a significant reduction of surface tension compared to the aqueous HA solutions, see Tables 4 and 5 [47]. H$_2$O:EtOH:MeOH solutions showed slightly higher surface tension than that of the H$_2$O:IPA solutions.

No significant effect of HA concentration or M_w on surface tension was found in the current study, which was due to very low polymer concentrations and the narrow range of concentrations used. As a consequence of HA's ionic nature, solution conductivity was decreased as the concentration of the polymer was decreased. Electrospinning at very low concentration could therefore be hindered by two mechanisms—insufficient polymer chain entanglement, caused by low polymer concentration, and instability of the cylindrical jet, due to a decrease in conductivity.

The shear viscosity of the solutions decreased with the decrease in both concentration, and M_w of the polymer, with the difference being in the range of several orders of magnitude, as is evident from Figure 9.

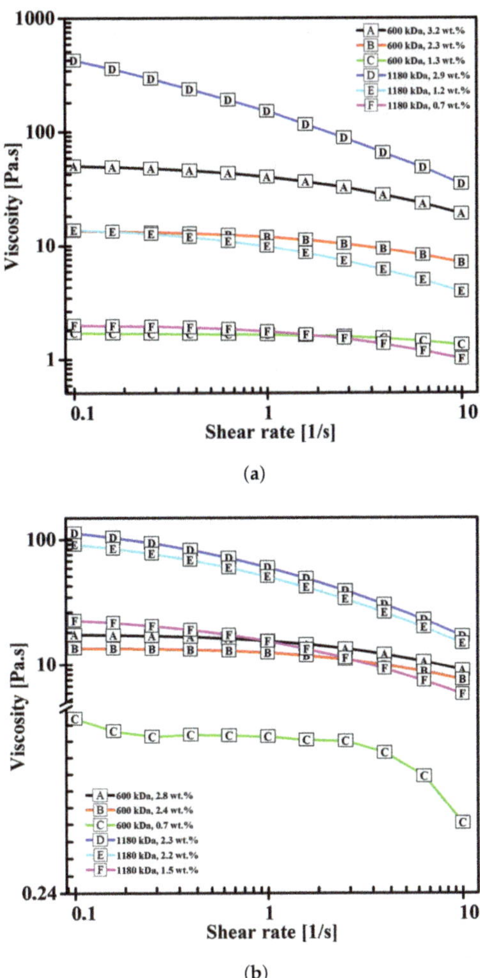

Figure 9. Viscosity of HA solutions in intermediate solvents as a function of shear rate. (**a**) H_2O:IPA in 10:7 weight ratio solvent mixture. (**b**) H_2O:EtOH:MeOH in 5:5:1 weight ratio solvent mixture.

Viscosity is sometimes considered the determining parameter of spinnability via electrostatic force [18]. The findings of this study contradict such assumption as overly simplified, which is in agreement with Yarin et al. (2001) [46], who claimed that shear viscosity was insignificant in terms of Taylor cone formation. The maximum viscosity of a spinnable solution can differ as much as ten times if different M_w of the same polymer are used. We assumed that the determining parameter was polymer chain entanglement, which is affected by the polymer chain conformation in given solvent, the ionic strength of the solution, polymer concentration, and other parameters. Further investigation on this subject is necessary.

Regardless the solvent mixture, spinning voltage decreased with the decrease of concentration (see Tables 7 and 8), which can possibly be explained by a shift in solution viscoelasticity, causing the critical instability wavelength leading to Taylor cone formation to increase [3], thus consuming less energy and lowering spinning voltage. Since surface tension does not change with concentration, its effect on spinning voltage can be neglected.

Table 7. Morphology analysis of H_2O:IPA in 10:7 weight ratio HA solutions electrospinning products.

HA M_w [kDa]	Concentration [wt.%]	Spinning Voltage [kV]	Product Form	Particles Diameter [μm]	Fibers Diameter [μm]
	3.2	20.5	Spherical Particles; Fibers	0.3–0.6	0.04–0.1
600	2.3	20.0	Spherical Particles; Fibers	0.3–1.0	0.06–0.1
	1.3	16.2	Spherical Particles; Fibers	0.4–1.2	0.05–0.1
	2.9	24.0	Fibers	/	0.05–0.09
1180	1.2	16.3	Spherical Particles; Fibers	0.3–0.8	0.06–0.1
	1.0	19.0	Spherical Particles	0.7–1.1	/

Table 8. Morphology analysis of H_2O:EtOH:MeOH in 5:5:1 weight ratio HA solutions electrospinning products.

HA M_w [kDa]	Concentration [wt.%]	Spinning Voltage [kV]	Product Form	Particles Diameter [μm]	Fibers Diameter [μm]
	2.8	19.1	Spherical Particles; Fibers	0.4–1.0	0.05–0.07
600	2.4	16.5	Spherical Particles; Fibers	0.2–0.8	0.05–0.07
	0.7	14.9	Spherical Particles	0.3–1.2	/
	2.3	29.2	/	/	/
1180	2.2	22.9	Fibers	/	0.05–0.08
	1.5	19.2	Fibers	/	0.05–0.1

No influence of the concentration on the electrospun structure morphology was found, which was the result of little difference in the concentrations of the respective samples. However, the concentration clearly governed the transition between electrospinning and electrospraying, which is explained by polymer chain entanglement according to Shenoy et al. (2005) [39] or by polymer solution elasticity according to Yu et al. (2006) [35]. The mechanism could not clearly be determined from the experiments conducted in the current study. The obtained fibers did not exceed 100 nm in diameter in any of the cases, making them promising in terms of biomedicine. The spherical particle diameter was mostly in the 1 to 0.1 μm range.

In the case of each solvent mixture, HA 600 kDa showed more tendency to undergo instabilities, which resulted in a combination of electrospinning and electrospraying (Figures 10a–c and 11a–b), whereas HA 1180 kDa was able to provide smooth fibers (Figures 10d and 11c,d). It could be assumed that a higher HA M_w was more favorable in terms of electrospinning as a result of the higher elasticity of the solution.

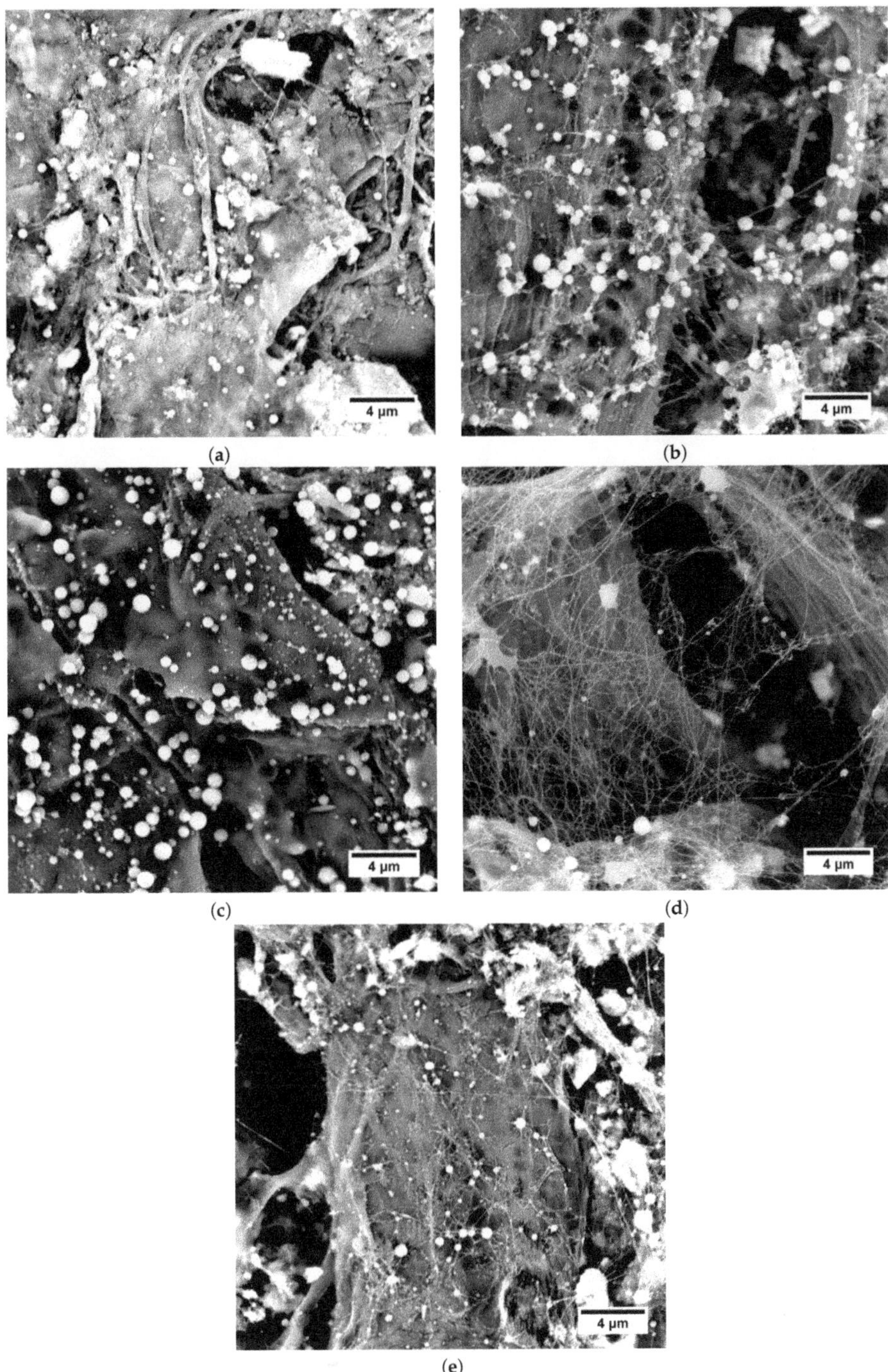

Figure 10. SEM micrographs of electrospun structures obtained from H_2O:IPA in 10:7 weight ratio solutions. (**a**) 3.2 wt.% HA M_w 600 kDa; (**b**) 2.3 wt.% HA M_w 600 kDa; (**c**) 1.3 wt.% HA M_w 600 kDa; (**d**) 2.9 wt.% HA M_w 1180 kDa; (**e**) 1.2 wt.% HA M_w 1180 kDa.

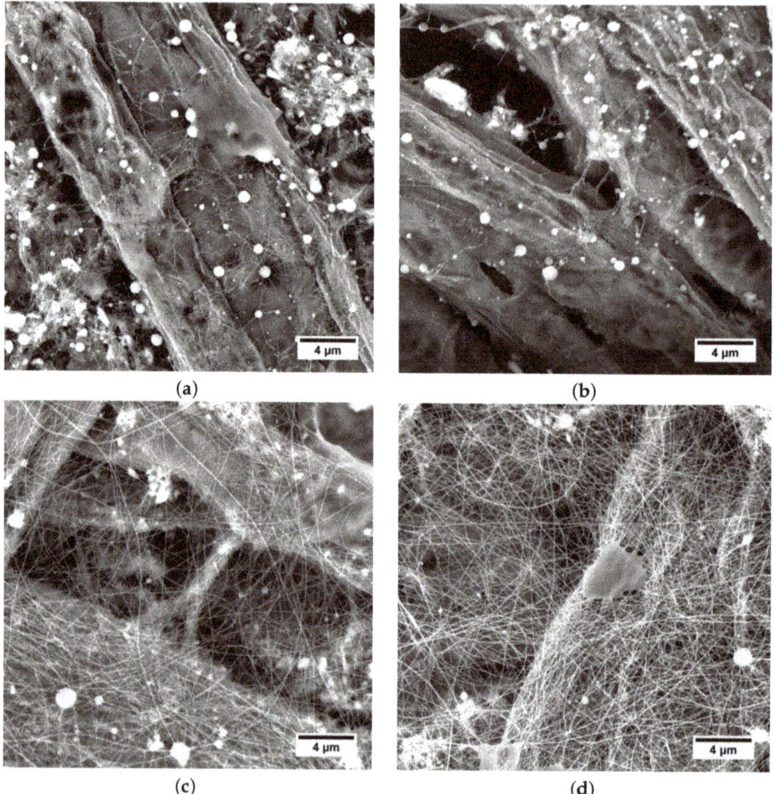

Figure 11. SEM micrographs of electrospun structures obtained from H_2O:EtOH:MeOH in 5:5:1 weight ratio solutions. (**a**) 2.8 wt.% HA M_w 600 kDa; (**b**) 2.4 wt.% HA M_w 600 kDa; (**c**) 2.2 wt.% HA M_w 1180 kDa; (**d**) 1.5 wt.% HA M_w 1180 kDa.

We can also see that instabilities in the form of electrospraying and branching were more frequent for solutions using H_2O:IPA in 10:7 weight ratio solvent mixture (Figure 10). Solvent mixture H_2O:EtOH:MeOH in 5:5:1 weight ratio provided higher stability to the electrospinning process. As the conductivity of the solutions using respective solvent mixtures was comparable (see Tables 4 and 5), the stabilization mechanism was presumably more complex than stabilization by charge density suggested by Reneker and Yarin (2008) [4]. Polymer chain conformation and interaction parameters can be expected to have a significant influence on electrospinning phenomena stabilization, but further investigation is necessary in order to fully understand the process.

Although the conductivity of HA solutions in intermediate solvents was significantly lower than those of aqueous blend solutions with PVA or PEO, smooth fibers were only obtained from a certain HA in intermediate solvent solutions. Our conclusion is that although conductivity does have a positive effect on electrospun jet stabilization, as stated by Reneker and Yarin (2008) [4], the influence of polymer viscoelasticity needs to be taken into account, as suggested by Stepanyan et al. (2014) and Palangetic et al. (2014) [36,37]. In the case of blend-solution electrospinning, the situation is complicated due to the difference in the viscoelastic properties of the polymers, which are simultaneously drawn at high elongation rates. A non-negligible effect of interaction parameters was present as well, because an intermediate solvent has a great influence on the solution behavior in a strong electric field, as was demonstrated in this study, as well as previously [34,39,42].

4. Conclusions

Electrospinning of biocompatible and biodegradable polymers is a desirable technique for use in biomedicine and life sciences. Production of HA nanofibers is a challenging task due to the extremely high viscosity and high surface tension of aqueous solutions.

In this study, nanofibers containing HA were obtained by solution electrospinning. Two approaches to the problem were chosen: co-electrospinning of aqueous blend solutions of HA/PVA and HA/PEO, respectively, and use of the intermediate solvent for pure HA solutions electrospinning. The choice of materials was done with regard to potential uses for cell cultivation. To facilitate fiber formation in HA/PVA blend solutions, the addition of BEC was necessary. Both HA/PEO and HA/PVA blend solutions provided bead-on-string structured fibers. As intermediate solvents, H_2O:IPA in a 10:7 weight ratio and H_2O:EtOH:MeOH in a 5:5:1 weight ratio were chosen. Both solvent mixtures facilitate the electrospinning of HA of M_w 600 and 1180 kDa. Lower M_w solutions had higher tendency to form spherical particles. There is clear correlation between the decrease in solution surface tension and the decrease in spinning voltage can be seen in the results, but no significant impact of these parameters on the fiber diameter was found. Variation in electrospun-structure dimensions and morphology was intensely associated with the change in M_w of the polymers. It was experimentally demonstrated that shear viscosity cannot be used as a sole determining parameter of solution spinnability, as there are differences as high as ten times the order of magnitude for spinnable solutions that differ only in polymer M_w.

The best results were achieved with the HA 1180 kDa solution in H_2O:EtOH:MeOH 5:5:1 at concentrations of 2.2 wt.% and 1.5 wt.%, as these provided smooth fibers. Fiber diameter did not exceed 100 nm for any sample that provided fibers,which makes them promising in terms of tissue engineering.

Author Contributions: Conceptualization, L.V. and A.M. (Aleš Mráček); methodology, L.V., L.M., E.A., P.S., E.W., and A.M. (Antonín Minařík); validation, L.V. and A.M. (Aleš Mráček); formal analysis, L.V.; investigation, L.V. and A.M. (Aleš Mráček); resources, L.V. and A.M. (Aleš Mráček); writing—original draft preparation, L.V.; writing—review and editing, L.V. and A.M. (Aleš Mráček); visualization, L.V. and A.M. (Aleš Mráček); supervision, A.M. (Aleš Mráček)

Funding: The research was funded by the Ministry of Education, Youth, and Sports of the Czech Republic, Program NPU I (LO1504); the European Regional Development Fund (No. CZ.1.05/2.1.00/19.0409); as well as by TBU (Nos. IGA/FT/2018/011 and IGA/FT/2019/012), funded from resources for specific university research. The work of author Eva Achbergerová was supported by the European Regional Development Fund under project CEBIA-Tech Instrumentation No. CZ.1.05/2.1.00/19.0376.

Conflicts of Interest: The authors declare no conflicts of interest.

Abbreviations

The following abbreviations are used in this manuscript:

HA	Hyaluronan
DC	Direct current
PVA	Poly(vinyl alcohol)
PEO	Polyethylene oxide
M_w	Average molecular weight
PEG	Polyethylene glycol
EtOH	Ethanol
IPA	Isopropanol
MeOH	Methanol
DEMI	Demineralized
DMSO	Dimethyl sulfoxide
FITC	Fluorescein isothyocyanite isomer
BEC	Benzethonium chloride
SEM	Scanning Electron Microscope/Scanning Electron Microscopy

References

1. Cooley, J. Apparatus for Electrically Dispersing Fluids. U.S. Patent 692631, 4 February 1902.
2. Morton, W. Method of Dispersing Fluids. U.S. Patent 705691, 29 July 1902.
3. Lukáš, D.; Sarkar, A.; Martinová, L.; Vodseďálková, K.; Lubasová, D.; Chaloupek, J.; Pokorný, P.; Mikeš, P.; Chvojka, J.; Komárek, M. Physical principles of electrospinning (electrospinning as a nano-scale technology of the twenty-first century). *Text. Prog.* **2009**, *41*, 59–140, doi:10.1080/00405160902904641. [CrossRef]
4. Reneker, D.; Yarin, A. Electrospinning jets and polymer nanofibers. *Polymer* **2008**, *49*, 2387–2425, doi:10.1016/j.polymer.2008.02.002. [CrossRef]
5. Xiao, B.; Wang, W.; Zhiang, X.; Long, G.; Fan, J.; Chen, H.; Deng, L. A novel fractal solution for permeability and Kozeny-Carman constant of fibrous porous media made up of solid particles and porous fibers. *Powder Technol.* **2019**, *349*, 92–98, doi:10.1016/j.powtec.2019.03.028. [CrossRef]
6. Liang, M.; Fu, C.; Xiao, B.; Luo, L.; Wang, Z. A fractal study for the effective electrolyte diffusion through charged porous media. *Int. J. Heat Mass Transf.* **2019**, *137*, 365–371, doi:10.1016/j.ijheatmasstransfer.2019.03.141. [CrossRef]
7. Xiao, B.; Zhang, X.; Jiang, G.; Long, G.; Wang, W.; Zhang, Y.; Liu, G. Kozeny-Carman constant for gas flow through fibrous porous media by fractal-Monte Carlo simulations. *Fractals* **2019**, *27*, 1950062, doi:10.1142/S0218348X19500622. [CrossRef]
8. Xiao, B.; Wang, W.; Zhang, X.; Long, G.; Chen, H.; Deng, L. A novel fractal model for relative permeability of gas diffusion layer in proton exchange membrane fuel cell with capillary pressure effect. *Fractals* **2019**, *27*, 1950012, doi:10.1142/S0218348X19500129. [CrossRef]
9. Liang, M.; Liu, Y.; Xiao, B.; Yang, S.; Wang, Z.; Han, H. An analytical model for the transverse permeability of gas diffusion layer with electrical double layer effects in proton exchange membrane fuel cells. *Int. J. Hydrog. Energy* **2018**, *43*, 17880–17888, doi:10.1016/j.ijhydene.2018.07.186. [CrossRef]
10. Long, G.; Xu, G. The Effects of Perforation Erosion on Practical Hydraulic-Fracturing Applications. *SPE Prod. Oper.* **2017**, *22*, 645–659, doi:10.2118/185173-PA. [CrossRef]
11. Long, G.; Liu, S.; Xu, G.; Wong, S.; Chen, H.; Xiao, B. A Perforation-Erosion Model for Hydraulic-Fracturing Applications. *SPE Prod. Oper.* **2018**, *33*, 770–783, doi:10.2118/174959-PA. [CrossRef]
12. Sofi, H.; Akram, T.; Tamboli, A.; Majeed, A.; Shabir, N.; Sheikh, F. Novel lavender oil and silver nanoparticles simultaneously loaded onto polyurethane nanofibers for wound-healing applications. *Int. J. Pharm.* **2019**, *569*, 118590–118590, doi:10.1016/j.ijpharm.2019.118590. [CrossRef]
13. Afshar, S.; Rashedi, S.; Nazockdast, H.; Ghazalian, M. Preparation and characterization of electrospun poly(lactic acid)-chitosan core–shell nanofibers with a new solvent system. *Int. J. Biol. Macromol.* **2019**, *138*, 1130–1137, doi:10.1016/j.ijbiomac.2019.07.053. [CrossRef] [PubMed]
14. Lee, K.Y.; Jeong, L.; Kang, Y.O.; Lee, S.J.; Park, W.H. Electrospinning of polysaccharides for regenerative medicine. *Adv. Drug Deliv. Rev.* **2009**, *61*, 1020–1032, doi:10.1016/j.addr.2009.07.006. [CrossRef] [PubMed]
15. Hampejsova, Z.; Batek, J.; Sirc, J.; Hobzova, R.; Bosakova, Z. Polylactide/polyethylene glycol fibrous mats for local paclitaxel delivery: comparison of drug release into liquid medium and to HEMA-based hydrogel model. *Monatsh. Chem.* **2019**, doi:10.1007/s00706-019-02469-5. [CrossRef]
16. Yin, Y.; Zhao, X.; Xiong, J. Modeling Analysis of Silk Fibroin/Poly(ε-caprolactone) Nanofibrous Membrane under Uniaxial Tension . *Nanomaterials* **2019**, *9*. [CrossRef] [PubMed]
17. Mombini, S.; Mohammadnejad, J.; BakhsHandeh, B.; Narmani, A.; Nourmohammadi, J.; Vahdat, S.; Zirak, S. Chitosan-PVA-CNT nanofibers as electrically conductive scaffolds for cardiovascular tissue engineering. *Int. J. Biol. Macromol.* **2019**, *140*, 278–287, doi:10.1016/j.ijbiomac.2019.08.046. [CrossRef]
18. Agarwal, S.; Wendorff, J.; Greiner, A. Use of electrospinning technique for biomedical applications. *Polymer* **2008**, *49*, 5603–5621, doi:10.1016/j.polymer.2008.09.014. [CrossRef]
19. Son, W.; Youk, J.; Lee, T.; Park, W. The effects of solution properties and polyelectrolyte on electrospinning of ultrafine poly(ethylene oxide) fibers. *Polymer* **2004**, *45*, 2959–2966, doi:10.1016/j.polymer.2004.03.006. [CrossRef]
20. Jin, H.; Fridrikh, S.; Rutledge, G.; Kaplan, D. Electrospinning Bombyx mori Silk with Poly (ethylene oxide). *Biomacromolecules* **2002**, *3*, 1233–1239, doi:10.1021/bm025581u. [CrossRef] [PubMed]

21. Chung, S.; Son, S.; Min, J. The nanostructure effect on the adhesion and growth rates of epithelial cells with well-defined nanoporous alumina substrates. *Nanotechnology* **2010**, *21*, 1–7, doi:10.1088/0957-4484/21/12/125104. [CrossRef]

22. Mori, A.; Fernandéz, M.; Blunn, G.; Tozzi, G.; Roldo, M. 3D Printing and Electrospinning of Composite Hydrogels for Cartilage and Bone Tissue Engineering. *Polymers* **2018**, *10*, 285, doi:10.3390/polym10030285. [CrossRef]

23. Ingr, M.; Kutálková, E.; Hrnčiřík, J. Hyaluronan random coils in electrolyte solutions—A molecular dynamics study. *Carbohydr. Polym.* **2017**, *170*, 289–295, doi:10.1016/j.carbpol.2017.04.054. [CrossRef] [PubMed]

24. Gřundělová, L.; Mráček, A.; Kašpárková, V.; Minařík, A.; Smolka, P. The influence of quarternary salt on hyaluronan conformation and particle size in solution. *Carbohyd. Polym.* **2013**, *98*, 1039–1044, doi:10.1016/j.carbpol.2013.06.057. [CrossRef] [PubMed]

25. Musilová, L.; Kašpárková, V.; Mráček, A.; Minařík, A.; Minařík, M. The behaviour of hyaluronan solutions in the presence of Hofmeister ions: A light scattering, viscometry and surface tension study. *Carbohydr. Polym.* **2019**, *212*, 395–402, doi:10.1016/j.carbpol.2019.02.032. [CrossRef] [PubMed]

26. Cowman, M.; Matsuoka, S. Experimental approaches to hyaluronan structure. *Carbohyd. Res.* **2005**, *340*, 791–809, doi:10.1016/j.carres.2005.01.022. [CrossRef] [PubMed]

27. Um, I.; Fang, D.; Hsiao, B.; Okamoto, A.; Chu, B. Electro-Spinning and Electro-Blowing of Hyaluronic Acid. *Biomacromolecules* **2004**, *5*, 1428–1436, doi:10.1021/bm034539b. [CrossRef] [PubMed]

28. Caspersen, M.; Roubroeks, J.; Qun, L.; Shan, H.; Fogh, J.; RuiDong, Z.; Tommeraas, K. Thermal degradation and stability of sodium hyaluronate in solid state. *Carbohyd. Polym.* **2014**, *107*, 25–30, doi:10.1016/j.carbpol.2014.02.005. [CrossRef]

29. Xie, J.; Hsieh, Y. Ultra-high surface fibrous membranes from electrospinning of natural proteins: casein and lipase enzyme. *J. Mater. Sci.* **2003**, *38*, 2125–2133, doi:10.1023/A:1023763727747. [CrossRef]

30. Ma, H.; Chen, G.; Zhang, J.; Liu, Y.; Nie, J.; Ma, G. Facile fabrication of core–shell polyelectrolyte complexes nanofibers based on electric field induced phase separation. *Polymer* **2017**, *110*, 80–86, doi:10.1016/j.polymer.2016.12.062. [CrossRef]

31. Uppal, R.; Ramaswamy, G.; Arnold, C.; Goodband, R.; Wang, Y. Hyaluronic acid nanofiber wound dressing-production, characterization, and in vivo behavior. *J. Biomed. Mater. Res.—Part B: Appl. Biomater.* **2011**, *97 B*, 20–29, doi:10.1002/jbm.b.31776. [CrossRef]

32. Liu, Y.; Ma, G.; Fang, D.; Xu, J.; Zhang, H.; Nie, J. Effects of solution properties and electric field on the electrospinning of hyaluronic acid. *Carbohydr. Polym.* **2011**, *83*, 1011–1015, doi:10.1016/j.carbpol.2010.08.061. [CrossRef]

33. Brenner, E.; Schiffman, J.; Thompson, E.; Toth, L.; Schauer, C. Electrospinning of hyaluronic acid nanofibers from aqueous ammonium solutions. *Carbohydr. Polym.* **2012**, *87*, 926–929, doi:10.1016/j.carbpol.2011.07.033. [CrossRef]

34. Malkin, A.; Semakov, A.; Skvortsov, I.; Zatonskikh, P.; Kulichikhin, V.; Subbotin, A.; Semenov, A. Spinnability of Dilute Polymer Solutions. *Macromolecules* **2017**, *50*, 8231–8244, doi:10.1021/acs.macromol.7b00687. [CrossRef]

35. Yu, J.H.; Fridrikh, S.V.; Rutledge, G.C. The role of elasticity in the formation of electrospun fibers. *Polymer* **2006**, *47*, 4789–4797, doi:10.1016/j.polymer.2006.04.050. [CrossRef]

36. Stepanyan, R.; Subbotin, A.; Cuperus, L.; Boonen, P.; Dorschu, M.; Oosterlinck, F.; Butlers, M. Fiber diameter control in electrospinning. *Appl. Phys. Lett.* **2014**, *105*, doi:10.1063/1.4900778. [CrossRef]

37. Palangetic, L.; Reddy, N.; Srinivasan, S.; Cohen, R.E.; McKinley, G.H.; Clasen, C. Dispersity and spinnability: Why highly polydisperse polymer solutions are desirable for electrospinning. *Polymer* **2014**, *55*, 4920–4931, doi:10.1016/j.polymer.2014.07.047. [CrossRef]

38. Stepanyan, R.; Subbotin, A.V.; Cuperus, L.; Boonen, P.; Dorschu, M.; Oosterlinck, F.; Butlers, M.J.H. Nanofiber diameter in electrospinning of polymer solutions: Model and experiment. *Polymer* **2016**, *97*, 428–439, doi:10.1016/j.polymer.2016.05.045. [CrossRef]

39. Shenoy, S.; Bates, W.; Frish, H.; Wnek, G. Role of chain entanglements on fiber formation during electrospinning of polymer solutions: good solvent, no-specific polymer-polymer interaction limit. *Polymer* **2005**, *46*, 3372–3384, doi:10.1016/j.polymer.2005.03.011. [CrossRef]

40. Rogina, A. Electrospinning process: Versatile preparation method for biodegradable and natural polymers and biocomposite systems applied in tissue engineering and drug delivery. *Appl. Surf. Sci.* **2014**, *296*, 221–230, doi:10.1016/j.apsusc.2014.01.098. [CrossRef]

41. Miloh, T.; Spivak, B.; Yarin, A. Needleless electrospinning: Electrically driven instability and multiple jetting from the free surface of a spherical liquid layer. *J. Appl. Phys.* **2009**, *106*, 114910, doi:10.1063/1.3264884. [CrossRef]

42. Luo, C.; Nangrejo, M.; Edirisinghe, M. A novel method of selecting solvents for polymer electrospinning. *Polymer* **2010**, *51*, 1654–1662, doi:10.1016/j.polymer.2010.01.031. [CrossRef]

43. Huerta-Angeles, G.; Němcová, M.; Příkopová, E.; Šmejkalová, D.; Pravda, M.; Velebný, V. Reductive alkylation of hyaluronic acid for the synthesis of biocompatible hydrogels by click chemistry. *Carbohydr. Polym.* **2012**, *90*, 1704–1711, doi:10.1016/j.carbpol.2012.07.054. [CrossRef] [PubMed]

44. Šmejkalová, D.; Muthný, T.; Nešporová, K.; Hermannová, M.; Achbergerová, E.; Huerta-Angeles, G.; Svoboda, M.; Čepa, M.; Machalová, V.; Luptáková, D.; et al. Hyaluronan polymeric micelles for topical drug delivery. *Carbohydr. Polym.* **2017**, *156*, 86–96, doi:10.1016/j.carbpol.2016.09.013. [CrossRef] [PubMed]

45. Kaneo, Y.; Hashihama, S.; Kakinoki, A.; Tanaka, T.; Nakano, T.; Ikeda, Y. Pharmacokinetics and Biodisposition of Poly(vinyl alcohol) in Rats and Mice. *Drug Metab. Pharmacokin.* **2005**, *20*, 435–442, doi:10.2133/dmpk.20.435. [CrossRef]

46. Yarin, A.; Koombhongse, S.; Reneker, D. Taylor cone and jetting from liquid droplets in electrospinning of nanofibers. *J. Appl. Phys.* **2001**, *90*, 4836–4846, doi:10.1063/1.1408260. [CrossRef]

47. Jurošková, D. *Charakterizace Hyaluronanu Sodného ve VodnýCh Roztocích a Na fázových Rozhraních*; Department of Physics and Materials Engineering, Thomas Bata University in Zlín: Zlín, Czech Republic, 2017.

48. Kim, M.; Cao, B. Additional Reduction of Surface Tension of Aqueous Polyethylene Oxide (PEO) Solution at High Polymer Concentration. *Europhys. Lett.* **1993**, *24*, 229–234, doi:10.1209/0295-5075/24/3/012. [CrossRef]

49. Han, C. A Study of Coextrusion in a Circular Die. *J. Appl. Polym. Sci.* **1975**, *19*, 1875–1883. [CrossRef]

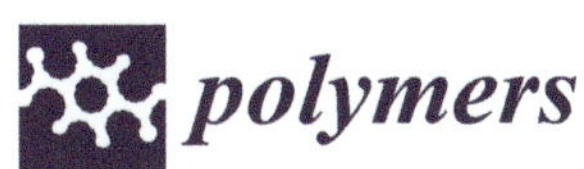 

Article

(PVA/Chitosan/Fucoidan)-Ampicillin: A Bioartificial Polymeric Material with Combined Properties in Cell Regeneration and Potential Antibacterial Features

Andres Bernal-Ballen [1,*], Jorge-Andres Lopez-Garcia [2] and Kadir Ozaltin [2]

[1] Grupo de Investigación en Ingeniería Biomédica, Vicerrectoría de Investigaciones, Universidad Manuela Beltrán, Avenida Circunvalar No. 60-00, Bogotá 110231, Colombia

[2] Centre of Polymer Systems, Tomas Bata University in Zlín, Tr. Tomase Bati 5678, 76001 Zlín, Czech Republic

* Correspondence: andres_bernal9@hotmail.com; Tel.: +51-1-5460600

Received: 17 July 2019; Accepted: 6 August 2019; Published: 9 August 2019

Abstract: Chitosan, fucoidan, and polyvinyl alcohol are categorized as polymers with biomedical applications. Ampicillin, on the other hand, is considered as an important antibiotic that has shown effectivity in both gram-positive and gram-negative micro-organisms. The aforementioned polymers possess unique properties that are considered desirable for cell regeneration although they exhibit drawbacks that can affect their final application. Therefore, films of these biomaterials were prepared and they were characterized using FTIR, SEM, XRD, degree of swelling and solubility, and MTT assay. The statistical significance of the experiments was determined using a two-way analysis of variance (ANOVA) with $p < 0.05$. The characterization techniques demonstrated that the obtained material exhibits properties suitable for cell regeneration, and that a higher concentration of natural polymers promotes cells proliferation to a greater extent. The presence of PVA, on the other hand, is responsible for matrix stability and dictates the degree of swelling and solubility. The SEM images demonstrated that neither aggregations nor clusters were formed, which is favorable for the biological properties without detrimental to the morphological and physical features. Cell viability was comparatively similar in samples with and without antibiotic, and the physical and biological properties were not negatively affected. Indeed, the inherent bactericidal effect of chitosan was reinforced by the presence of ampicillin. The new material is an outstanding candidate for cell regeneration as a consequence of the synergic effect that each component provides to the blend.

Keywords: bioartificial polymeric material; cell proliferation; chitosan; fucoidan; polyvinyl alcohol; ampicillin

---

1. Introduction

Biomedical engineering applies and develops therapies and technologies in order to support, repair or replace damaged cells, tissues, and organs, and it combines techniques from diverse disciplines such as physics, chemistry, biology, engineering, and medicine [1]. Thus, in the past decades a new generation of synthetic biodegradable polymers and analogous natural polymers have been developed for biomedical applications [2]. However, natural polymers do not have the appropriate mechanical properties whereas synthetics are deficient in terms of biocompatibility [3]. Thus, in order to overcome the poor performance of natural polymers, bioartificial polymeric materials (BAPMs) have been introduced [4–8]. These combine the biocompatibility of the biological component with the physical and mechanical properties of the synthetic ones [9]. BAPMs may be produced as hydrogels, films, scaffolds, and a great variety of potential applications have been reported including dialysis membranes, artificial skin, cardiovascular devices, implants, bandages, or even controlled drug-release systems [10,11].

Natural polymers possess several inherent advantages, such as bioactivity, the ability to present receptor-binding ligands to cells, susceptibility to cell-triggered proteolytic degradation, and natural remodeling. Synthetic biomaterials, on the other hand, are generally biologically inert, they have more predictable properties and batch-to-batch uniformity, and they have the unique advantage of having tailored property profiles for specific applications, hence, they are devoid of many of the disadvantages of natural polymers [2].

One of the main biopolymer used in biomedical applications is chitosan (CHI) [12–14]. This polysaccharide is primarily composed of β-(1–4)-linked D-glucosamine and N-acetyl glucosamine subunits. It is biodegradable, biocompatible, non-toxic, and it has a functional hydrophilic surface that promotes cell growth. Moreover, it exhibits antimicrobial activity, a characteristic that is useful for tissue engineering applications [15]. Reports that describe CHI used in cell adhesion, proliferation, and differentiation are plentiful in the literature [16,17]. Another important feature of CHI consists of its significant osteoconductivity, and minimal osteoinductivity. This induces the proliferation of osteoblasts, and neovascularization in vivo. For orthopedic applications, CHI can be prepared in various geometries such as sponges, fibers, films and other complex structures. In addition, this material has shown antibacterial, antifungal, hemostatic, analgesic, and mucoadhesive properties, all of which are relevant in drug delivery and tissue healing [18]. CHI satisfies most of the properties that a material should have as a candidate for tissue engineering applications [19].

Another relevant material in the field of biomedical application is fucoidan (FUC). This natural polymer is a marine-sourced sulphated polysaccharide extracted from brown algae, and it has numerous biological activities including anticancer, antitumor, antivirus, and anti-inflammatory properties [20,21]. Indeed, it is highly water soluble, biocompatible, biodegradable, and non-toxic [22] and it has demonstrated its antiviral activities both in vivo and in vitro. Since it presents low cytotoxicity compared with other antiviral drugs, FUC is currently used in clinical medicine [23]. FUC has been targeted in tissue engineering applications; also, it triggers the biological activities of alkaline phosphatase and osteocalcin, which are phenotypic markers for the early stages of osteoblast differentiation [16].

In spite of the merits that polysaccharides exhibit as biomaterials, they suffer from several drawbacks including variations in material properties based on their source, microbial contamination, uncontrolled water uptake, poor mechanical strength, and unpredictable degradation patterns [24]. Nonetheless, these polymers are readily recognized and accepted by the body due to their biochemical similarity with components of the human extracellular matrix [18].

Synthetic polymers are an appropriate counterpart for reducing the drawbacks of natural polymers. For that reason, these materials have become an obvious alternative for biomedical applications. There are three explanations for this: although most biologically derived biodegradable polymers possess good biocompatibility, some may trigger an immune response in the human body that could possibly be avoided by the use of an appropriate synthetic biopolymer; chemical modifications to biologically derived biodegradable polymers are difficult; and chemical modifications likely cause alteration in the bulk properties of biologically derived biodegradable polymers [25].

Polyvinyl alcohol (PVA), a representative water-soluble polymer, is considered as one of the most attractive biomedical polymers due to its synergistic properties such as biocompatibility, excellent mechanical strength, flexibility, thermal stability, and absence of toxicity [26]. A complete description of this material can be found in numerous scientific publications [26–32]. Indeed, blending PVA with a great variety of natural polymer is also reported [33–38]. These publications describe the physical, chemical, and biological interactions of PVA, and most of the cited publications mention the use of PVA in blends that have potential uses in the biomedical field.

The combination of natural and synthetic polymers has emerged as a plausible solution for overcoming the inherent weaknesses exhibited by natural materials [39]. Ideally, the blend is not detrimental to the mechanical properties of the synthetic polymer, as evidenced in a profusion of published literature [12,39–43]. In fact, it has been shown that the variation in the concentration of the blend is a crucial factor for obtaining a material with appropriate mechanical properties [12], without negatively affecting the biological properties [8]. Furthermore, if BAPMs had antibacterial properties this would be a desirable feature with positive repercussions in the bioengineering field. In this matter, a previous study has shown that a low concentration of ampicillin (AMP) in a matrix made of CHI and PVA showed moderate antibacterial activity against bacterial strains, even without adding an extremely high concentration of the antibiotic [3].

The combination of PVA and CHI has been fully described in the literature. There are also reports in which these polymers are combined with antibiotics [44]. Nonetheless, the system of PVA, CHI, FUC, and AMP has not been described to date, and we hypothesize that this BAPM is a new approach to cell regeneration because the BAPM combines the recognized biological properties of CHI and FUC, the outstanding mechanical features of PVA, and the antibacterial activity of AMP. Therefore, this work involves the preparation of a new BAPM consisting of PVA/CHI/FUC/AMP, the characterization of the obtained material, and an in vitro evaluation of its performance for cell culture.

2. Materials and Methods

2.1. General Information

Poly(vinyl alcohol) (PVA) (M_w = 47,000 g mol^{-1}), with a polymerization degree of 1000 and 98% of hydrolysis was provided by Fluka Analytica (Prague, The Czech Republic); chitosan (CHI) of medium molecular weight with a deacetylation degree of 75–85%, and glutaraldehyde (GLU) grade II at 25% in H_2O were purchased from Sigma Aldrich (Prague, The Czech Republic). Anticoagulant fucoidan (FUC) was obtained from *Fucus vesiculosus* and provided by Sigma Aldrich, St. Louis, MO, USA. Lactic acid (analytical grade) was produced by Lachema, Brno, The Czech Republic; and hydrochloric acid (analytical grade) was supplied by Penta, Prague, The Czech Republic. Sodium ampicillin (AMP) was produced by Farmalógica, S.A. (Bogota-Colombia) and donated to this research by the Hospital Cardiovascular del Niño de Cundinamarca (Soacha, Colombia). None of the reagents were subject to further processing.

2.2. Sample Preparation

A 1% w/v acid solution of CHI was prepared by the addition of the polymer to acetic acid at 0.5 M under mild manual stirring. Once the polymer was partially dissolved, the solution was stirred in a shaker (Multifunctional Orbital Biosan, PSU 20i, Latvia) for 24 h. An aqueous solution of PVA at 1% w/v was prepared in deionized water at 70 °C under vigorous magnetic stirring using a Heidolph MR Hei-Standard magnetic stirrer with heating (Heidolph Instruments GmbH, Schwabach, Germany). Natural and synthetic polymers solutions were crosslinked and plasticized using glutaraldehyde-hydrochloric acid (0.25 wt% and 1.2 wt%) and lactic acid (0.5 wt.%) as compared to the total amount of the polymer. The prepared solutions underwent degasification using an ultrasonic cleaning unit (S80 Elmasonic, Elma GmbH, with degasification function, Singen, Germany). An aqueous solution of fucoidan (FUC) was prepared by dissolving the polymer in distilled water to obtaining a 0.1% w/w solution. Ampicillin was dissolved in distilled water using a mild magnetic stirrer (Heidolph Instruments GmbH, Schwabach, Germany).

CHI and FUC were blended in a ratio of 1:1 v/v and the new solution was stirred for 30 min at 600 rpm. Once the solutions of natural polymers were obtained, the artificial polymer was added and stirred for 30 more minutes at 600 rpm followed by the addition of AMP at 1.0 wt% of the total amount of the polymers. The solutions were stirred for 10 min and films were obtained using the casting method and pouring 0.5 mL per cm^2 on plastic Petri dishes. Films were allowed to dry at 37 °C for a week in a no-air circulating oven. Table 1 shows the prepared films.

2.3. Fourier Transform Infrared Spectroscopy (ATR-FTIR)

Spectra for pure polymers and all blends were obtained using a Nicolet iS5 spectrometer (Thermo Fisher, Waltham, MA, USA) equipped with an attenuated total reflectance (ATR) accessory utilizing the Zn-Se crystal. Each spectrum represents 64 co-added scans referenced against an empty ATR cell spectrum. The spectra range was from 4000 to 650 cm^{-1} with a resolution of 1.92 cm^{-1}.

Table 1. List of the prepared BAPMs and the ratio of all the components.

Description
PVA
CHI
PVA/CHI/AMP 1:1
PVA/CHI/AMP 2:1
PVA/CHI/AMP 1:2
PVA/CHI 1:1
PVA/CHI 2:1
PVA/CHI 1:2
PVA/CHI/FUCOIDAN/AMP 1:1
PVA/CHI/FUCOIDAN/AMP 2:1
PVA/CHI/FUCOIDAN/AMP 1:2
PVA/CHI/FUCOIDAN 1:1
PVA/CHI/FUCOIDAN 2:1
PVA/CHI/FUCOIDAN 1:2

Note: The ratio indicates the relationship between synthetic (PVA) and natural (both CHI/FUC) polymers.

2.4. X-ray Photoelectron Spectroscopy (XPS)

Determination of the chemical composition of the blends was carried out by X-ray photoelectron spectroscopy (XPS). The Thermo Scientific K-Alpha XPS system (Thermo Scientific, UK) was used, which was equipped with monochromatic Al K-alpha X-ray source (1486.6 eV) with an X-ray beam of 400 mm in size at 6 mA 12 kV. The spectra were collected in the constant analyser energy mode with pass energy of 200 eV and Thermo Scientific Avantage 5.952 software (Thermo Fischer Scientific) was used for digital acquisition.

2.5. Scanning Electron Microscopy (SEM)

Micrographs of the prepared samples were taken by the scanning electron microscope Nova NanoSEM 450 (FEI, Hillsboro, OR, USA) equipped with a high vacuum detector and operated at 5 kV with a spot size of 2.5 A coating with a thin layer of gold/palladium was performed by a sputter coater SC 7620 (Quorum Technologies, Newhaven, East Sussex, UK).

2.6. Degree of Swelling and Solubility

The gravimetric method was used to obtain the degree of swelling and solubility. Disk specimens were dried in a desiccator for 24 h to obtain their dry weight (W_1). Then, the samples were immersed in deionized water at 20 °C for several different time periods (2, 4, 6, 8, 10, 15, 20, 30, 60, 90 min and 24 h). At the end of each period, the excess water was removed from the surface with filter paper and the specimens were weighed again. The last measure (24 h) was considered as the swelling limit (W_2) because in that period the equilibrium was reached. Finally, samples were dried until they achieved constant weight (W_3). The degree of swelling (DS) and degree of solubility (SD) were determined using Equations (1) and (2), respectively. The test was performed with five specimens for each sample (n = 5) to obtain statistically significant data:

$$DS = [(W_2 - W_1)/W_1] \times 100 \tag{1}$$

$$SD = [(W_1 - W_3)/W_1] \times 100 \tag{2}$$

2.7. Cell Proliferation Test (MTT)

Prior to the in vitro cytocompatibility test, samples were placed in 10×10 mm foil and subjected to UV-radiation source (wavelength of 258 nm) for 30 min. Primary mouse embryonic fibroblast cells (NIH/3T3, ATCC® CRL-1658™, USA) were used as a cell line. The ATCC-formulated Dulbecco's Modified Eagle's Medium (BioSera, France) containing 10% of calf serum (BioSera, France) and 100 U mL^{-1} Penicillin/Streptomycin (BioSera, France) was used as a culture medium. The cells were seeded onto the samples in a concentration of 2×10^4 cells per mm^2 and placed in an incubator for 72 h at 37 °C. All the tests were performed in triplicate.

After the incubation period of 72 h, the cell viability was tested using the MTT assay (Duchefa Biochemie, Netherlands). Firstly, the cells were washed with PBS (BioSera, France) and fresh medium with MTT with a concentration of 0.5 mg/mL was added. After 4 h, formed formazan crystals were dissolved in DMSO and the absorbance was measured at 570 nm and the reference wavelength was adjusted to 690 nm. The results are presented as reduction/augmentation of cell viability as a percentage when compared to cell cultivated on pure PLA.

2.8. Statistical Analysis

All experiments with quantitative measurements were performed on five specimens, except for the cell culture where three specimens were tested. The experimental values are reported in the form of average ± standard deviation. Results were statistically compared by applying a two-way analysis of variance (ANOVA) with $p < 0.05$ with SPSS software.

3. Results and Discussion

3.1. Attenuated Total Reflectance Fourier Transform Infrared Spectroscopy (ATR-FTIR)

Figure 1 shows the obtained spectra for the BAPMs as well as for the individual components. For PVA, an intense band between 3550 and 3200 cm^{-1} appeared and it corresponds to OH from the intermolecular and intramolecular hydrogen bonds. The detected transitions around 2900 cm^{-1} are associated to C–H from alkyl groups [45]. At 1423 cm^{-1} the –CH$_2$– bending appears, and at 1416 and 1327 cm^{-1} the band ascribed to CH and OH, respectively, is evident [45–47]. The peak at 848 cm^{-1} corresponds to skeletal C–H rocking of pure PVA [48].

The spectrum for FUC has been elucidated in the literature and these reports are in a good agreement with our results. The peaks at 3470 and 2936 cm^{-1} correspond to the stretching vibration of hydrogen bond for OH and C–H groups, respectively. The bands located between 1200 cm^{-1} and 970 cm^{-1} are caused by C–C and C–O stretching vibrations in the pyranoid ring and C–O–C stretching of the glycosidic bonds, whereas the absorption band located at 1240 cm^{-1} belongs to S=O stretching and evidences the presence of sulfate. An absorption peak at 1648 cm^{-1} is attributed to the C=O asymmetric stretching vibrations of the carbonyl groups and the peak centered in the region of 840 cm^{-1} and the shoulder at 820 cm^{-1} (C–S–O) suggest a complex pattern of substitution, primarily at the C–4 position with other substitutions at C–2 and/ or C–3 (equatorial positions) in lower amounts. The features at 622 cm^{-1} and 583 cm^{-1} can be ascribed to the asymmetric and symmetric O=S=O deformation of sulfates [49–53].

AMP shows an absorption peak in the region of 1730–1720 cm^{-1}, which is caused by C=O β-lactam stretching. The peaks at 1664 and 1560 cm^{-1} belong to C=O amide stretching and –NH amide groups, respectively [54].

PVA and CHI interactions are evidenced in the frequency observed in the region of 3388 cm^{-1} to 3427 cm^{-1} corresponding to intermolecular hydrogen bonds [36]. Blends of PVA/CHI show an absorption band located between 1640 and 1560 cm^{-1}, which is connected to symmetric stretching and bending of acetamide groups, respectively. The change in the characteristic shape of the CHI spectra, as well as the shifting of peak to a lower frequency range on account of hydrogen bonding between –OH of PVA and –OH or –NH$_2$ of chitosan were observed in the blended films. These results suggested the formation of hydrogen bonds between the CHI and PVA molecules [3].

The combination of CHI and FUC is mediated by an electrostatic interaction of the positively charged amino group in CHI and the negatively charged sulfate group in FUC [55]. For that reason, the characteristic peak of sulfate in FUC located at 1240 cm^{-1} is not evidenced in the samples containing FUC. The peak at 1725 cm^{-1} confirmed the formation of CHI/FUC complex. Moreover, the peak corresponding to the cross-linking primary amine of CHI is centered at 1530 cm^{-1} [56]

Peaks for CHI are distinguishable at particular wavenumbers. Thus, the NH group-stretching vibration appears at 3375 cm^{-1}, and the signal for OH is located at 3450 cm^{-1} [57]. The symmetric, and asymmetric –CH$_2$– stretching occurring in the pyranose ring are located at 2920, 2880, 1430, 1320, 1275 and 1245 cm^{-1} [58]. The amide I, II, and II vibrations are located at 1650, 1586 cm^{-1}, and 1322 cm^{-1}, respectively. The saccharine-related signals are at 1155 and 900 cm^{-1} [59,60]. The peaks at 1030 and 1080 cm^{-1} indicate the C–O stretching vibration [5].

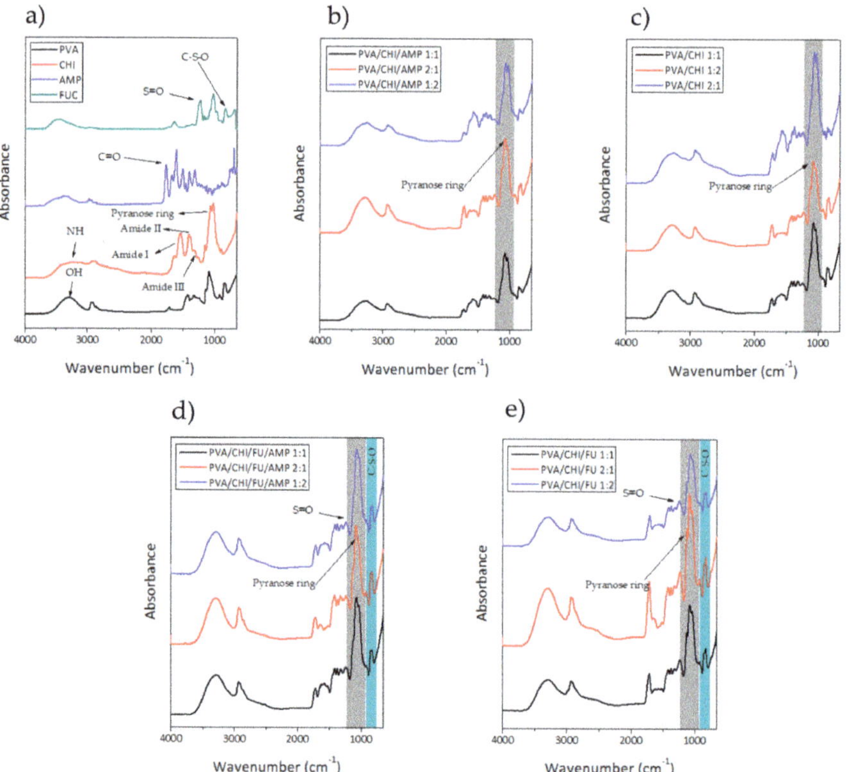

Figure 1. FTIR for the obtained BAPM. (**a**) Row materials; (**b**) PVA/CHI/AMP samples; (**c**) PVA/CHI samples; (**d**) PVA/CHI/FUC/AMP samples; and (**e**) PVA/CHI/FUC samples.

3.2. X-ray Photoelectron Spectroscopy (XPS)

The surface chemical compositions of the individual components are given in Table 2. The main components of PVA, CHI, FUC and AMP are carbon and oxygen, therefore the C1s and O1s have the highest atomic levels. Furthermore, samples with a higher ratio of CHI present a higher amount of carbon and a relatively lower amount of oxygen due to their molecular formula, except for the PVA/CHI/FUC 1:2 counterpart. The level of nitrogen shown for each sample is due to the presence of CHI and AMP. The samples that have a higher ratio of CHI show the highest N1s level, compared to other samples, except for the PVA/CHI/AMP 2:1 counterpart only. The level of sulfur stems from AMP and FUC due to their molecular formula. Thus, the S2p level observed for the samples containing AMP and FUC was expected. However, the level is low since both are additives and not the main components of the prepared films. The reason for the unexpected behavior of the PVA/CHI/FUC 1:2 sample with regard to the C1s level and the PVA/CHI/AMP 2:1 sample for N1s is unclear and needs to be studied further [61–65].

Table 2. List of the XPS spectra given in atomic percent of all the components.

Samples	C1s%	O1s%	N1s%	S2p%
PVA/CHI/AMP 1:1	69.3	26.0	4.6	0.1
PVA/CHI/AMP 2:1	77.1	21.3	1.6	0.1
PVA/CHI/AMP 1:2	70.3	24.7	4.8	0.2
PVA/CHI 1:1	67.1	28.1	4.7	-
PVA/CHI 2:1	70.0	26.2	3.7	-
PVA/CHI 1:2	67.6	27.5	4.9	-
PVA/CHI/FUC/AMP 1:1	69.2	26.3	4.4	0.2
PVA/CHI/FUC/AMP 2:1	72.5	25.6	2.0	-
PVA/CHI/FUC/AMP 1:2	69.8	25.9	4.4	-
PVA/CHI/FUC 1:1	70.2	26.2	3.6	0.1
PVA/CHI/FUC 2:1	68.6	27.0	4.1	0.2
PVA/CHI/FUC 1:2	75.7	20.8	3.3	0.1

Figure 2 shows the representative XPS spectra analysis for the components with and without AMP containing counterparts with a 1:2 ratio. As can be seen, all representative samples are similar for C1s, N1s and O1s spectra because the selected components have the same PVA/CHI ratio. The C1s spectra show three bonding states at 283.08 eV attributed to C–C and C–H bonds; at 285.08 eV attributed to C–NH and C–NH$_2$ bonds; and at 286.08 attributed to C–O–C=O and C–OH bonds, due to the presence of PVA and CHI. The O1s spectra exhibit a major bonding state at 536.08 eV attributed to C–O bond, due to the presence of PVA and CHI. The rather lower level of N1s spectra was detected at 399 eV attributed to CHI and AMP. The S2p spectra were not sufficient to appear in Figure 2, which was related to AMP and FUC, however, their amounts are presented in Table 2.

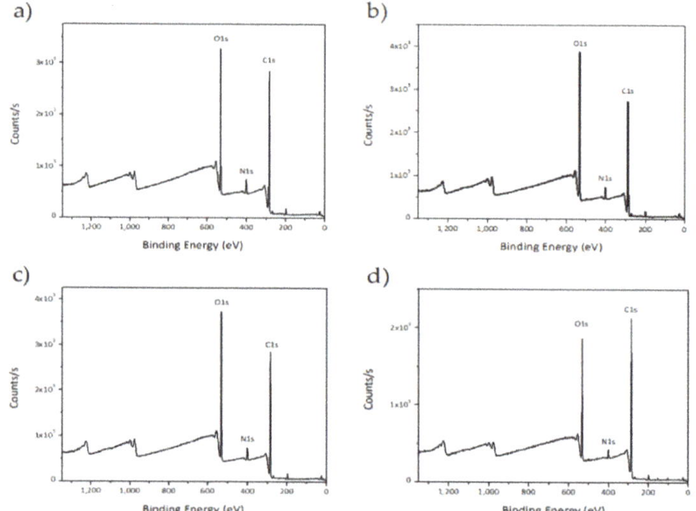

Figure 2. XPS spectra for (**a**) PVA/CHI/AMP 1:2; (**b**) PVA/CHI/ 1:2; (**c**) PVA/CHI/FUC/AMP 1:2; (**d**) PVA/CHI/FUC 1:2.

3.3. Scanning Electron Microscopy (SEM)

The surface morphology did not present any relevant feature and the obtained images exhibit a flat surface and neither particles nor clusters were observable (images are not shown in this manuscript). This is because polymers are compatible and one phase is formed during the preparation process (the FTIR spectra demonstrated all the characteristic peaks of FUC, CHI, and PVA during the blending process and revealed that FUC was successfully coupled to CHI/PVA material [36]). On the other hand, cross-sectional images (Figure 3) showed that FUC creates a porous structure. This is caused by the strong ionic interaction of CHI positively charged amino groups with FUC negatively charged sulfate groups, which caused precipitation of the polymer solutions and formation of porous networks. The structure is considered as appropriate for adhesion and cell proliferation [66].

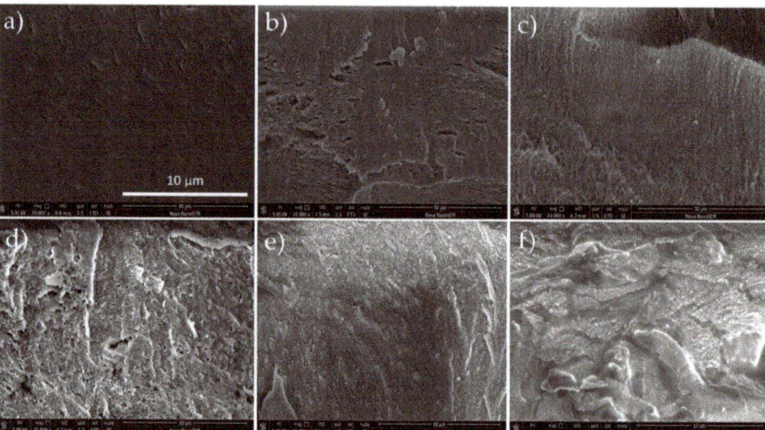

Figure 3. SEM images for BAPM in the presence of AMP. (**a**) PVA/CHI/AMP 1:1; (**b**) PVA/CHI/AMP 2:1; (**c**) PVA/CHI/AMP 1:2; (**d**) PVA/CHI/FUC/AMP 1:1; (**e**) PVA/CHI/FUC/AMP 2:1; and (**f**) PVA/CHI/FUC/AMP 1:2.

3.4. Degree of Swelling and Solubility

Materials in contact with living organisms require the ability to conduce and store water as an essential process to achieve proper cell signaling and nutrition [67]. For that reason, it is crucial to determine the degree of swelling and solubility of the obtained BAPM. The results are shown in Table 3. All the films swelled rapidly and reached equilibrium within the first 30 min of the experiment although relevant differences are caused by the presence of distinct components. PVA exhibits the highest degree of swelling because of the presence of hydrophilic groups, which are available for forming hydrogen bonding with water. Interactions between hydroxyl groups of PVA and amino or hydroxyl groups of CHI reduce the number of interactions between –OH groups and water [68], reducing the overall hydrophilicity of the system. Moreover, as the material was crosslinked and plasticized, there was a reduction of these groups and there were variations in the swelling. Although a reduction of swelling in systems made of PVA and CHI has been reported, and this is associated with the increase in the CHI ratio, in this study, PVA and CHI reacted with glutaraldehyde, and therefore the cross-linked blend becomes less capable of hydrogen bonding with water molecules due to acetalization and the formation of the Schiff base, which resulted in a decreased degree of swelling at equilibrium [69].

Table 3. Degree of swelling and solubility degree for the prepared BAPM.

MATERIAL	Degree of Swelling after 6 h [%] (SD)	Average Weight Loss [%] (SD)
PVA	1652 ± 337	36 ± 10
CHI	354 ± 45	32 ± 3
PVA/CHI/AMP 1:1	1364 ± 258	49 ± 7
PVA/CHI/AMP 2:1	1448 ± 148	52 ± 8
PVA/CHI/AMP 1:2	1301 ± 225	35 ± 10
PVA/CHI 1:1	1160 ± 244	52 ± 5
PVA/CHI 2:1	1263 ± 344	61 ± 3
PVA/CHI 1:2	1032 ± 108	36 ± 8
PVA/CHI/FUCOIDAN/AMP 1:1	1324 ± 208	36 ± 5
PVA/CHI/FUCOIDAN/AMP 2:1	1342 ± 76	54 ± 13
PVA/CHI/FUCOIDAN/AMP 1:2	1169 ± 303	31 ± 10
PVA/CHI/FUCOIDAN 1:1	1159 ± 173	36 ± 3
PVA/CHI/FUCOIDAN 2:1	1414 ± 404	38 ± 12
PVA/CHI/FUCOIDAN 1:2	1227 ± 44	38 ± 6

The presence of FUC increased the swelling of the material. The addition of negatively charged FUC increases the availability of free functional groups in the PVA/CHI/FUC blend. Hence, it exhibited a higher degree of swelling in comparison to the PVA/CHI material [36]. Although the swelling behavior of samples containing FUC is higher compared to the samples with its absence, the presence of plasticizer and crosslinker diminished the swelling as a result of the decline in the hydroxyl and amino groups. The surface generally increases upon swelling of the film, which makes it suitable for more cell adhesion and infiltration. Indeed, the presence of FUC in the BAPM resulted in increased surface area (rugosity on the surface), as was evidenced in SEM images. The water retention ability of the samples containing FUC was comparatively less than samples without FUC. This may be due to the fact that unbound water molecules are easily removed from the surface of FUC samples [70].

The water absorption capacity is increased with an increase in the ratio of CHI/FUC [71], however the presence of PVA is dominant as it has a high preference for water molecules due to the hydrophilic groups in the polymer. The addition of negatively charged FUC increases the availability of free functional groups in the blends, therefore, the swelling behavior of a material containing FUC is higher compared to a material without FUC. The presence of FUC in the films increases the surface area, which increases cell adhesion and infiltration [36].

3.5. Cell Proliferation (MTT Assay)

Stability is a crucial factor for cell adhesion and proliferation. These cells behave in culture in a similar way as they do in vivo, migrating towards the air interface to form the epithelial surface. The cells' behavior on the prepared films was evaluated by MTT assay and the results are given in Figure 4. After three days in culture, the results showed patterns that should be taken into account; for instance, systems 1:1 and 1:2 have higher proliferation than the samples where PVA is the main material of the formulation. In fact, the highest proliferation rates are on the 1:2 matrices.

The antibiotic seems to have no adverse effect on cell growth, since the system PVA/CHI/AMP evinced higher proliferation than the sample without ampicillin. The histograms show that the most thriving sample in terms of cell viability was the PVA/CHI/AMP 1:2. On the other hand, FUC seems to have a detrimental effect, owing to the low proliferation rates when this polysaccharide was added. Indeed, the lowest proliferation rates are observed on the films 2:1 with FUC. It should be noted that cell proliferation is highly reliant on surface topography, surface crystallinity, hydrophilic/hydrophobic character, roughness and chemical compositions. As mentioned above, the systems with FUC have porous structures and solid particles across the films, which are counterproductive for cell growth. Fibroblast tends to grow efficiently on hydrophilic surfaces, and their attachment is mainly related to carbonyl and carboxyl groups along with hydrogen bonding and van der Waals forces, which reinforce the linking between cell and films. This information is in agreement with the spectroscopic results, since polar entities are the mainly organic functionalities of the studied systems [72,73].

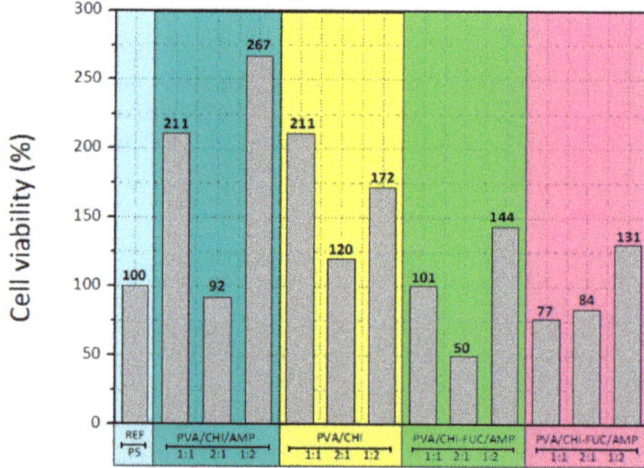

Figure 4. Cell viability of the BAPM (MTT assay).

4. Conclusions

Bioartificial polymeric materials are considered as a new class of constituents that combine the appropriate mechanical properties of artificial polymers with the satisfactory biological features of natural polymers. Thus, this research examined the development of a BAPM made of PVA, CHI, and FUC, and the incorporation of AMP as an antibacterial agent. The prepared films were tested and it was elucidated that the BAPM has potential for cell regeneration in vitro. The characterization techniques used here indicated that PVA brings water resistibility to the system, whereas CHI/FUC are responsible for creating a porous microstructure, which allows cells to adhere and grow within the matrix. The obtained information indicated that PVA, CHI, and FUC are compatible, as evidenced in FTIR spectra, as well as in SEM images.

Although PVA/CHI films behaved better than PVA/CHI/FUC in terms of cell viability, FUC does not inhibit cell growth. On the other hand, a higher concentration of CHI/FUC displayed better properties for cell regeneration, which is an indication that the inherent attributes of these natural polymers favor interactions between living organism, and therefore, it is plausible to assume that the obtained film is a credible alternative for cell cultures.

Although the presence of AMP reduces cell growth, the obtained results are promising because even though the obtained values are smaller, the potential to effectively control bacteria colonization brings alternative benefits to the film. Thus, these biomaterials are integral to the future of disciplines such as tissue engineering, and the ability to tailor cell regeneration is a desirable property in systems with potential use in living organisms.

Author Contributions: A.B.-B. and J.-A.L.-G. designed the study. A.B.-B., J.-A.L.-G. and K.O. were responsible for the characterization techniques. A.B.-B. wrote and edited the manuscript. J.-A.L.-G. and K.O. collaborated by reviewing the manuscript. All the authors approved the final version.

Funding: The authors would like to express their gratitude to the Ministry of Education, Youth and Sports of the Czech Republic, Program NPU I, LO1504.

Acknowledgments: Our gratitude goes to the program *Reconocimiento Docente* from *Universidad Manuela Beltran*; Also, we are grateful to Jose Leonardo Cely Andrade from the *Hospital Cardiovascular del Niño de Cundinamarca* (Soacha, Colombia) for providing the antibiotic for this study; and to the *Universidad Distrital Francisco José de Caldas* for the initial characterization techniques

Conflicts of Interest: The authors declare no conflict of interest.

References

1. Repanas, A.; Andriopoulou, S.; Glasmacher, B. The significance of electrospinning as a method to create fibrous scaffolds for biomedical engineering and drug delivery applications. *J. Drug Deliv. Sci. Technol.* **2016**, *31*, 137–146. [CrossRef]
2. Nair, L.S.; Laurencin, C.T. Biodegradable polymers as biomaterials. *Prog. Polym. Sci.* **2007**, *32*, 762–798. [CrossRef]
3. Bernal-Ballen, A.; Lopez-Garcia, J.; Merchan-Merchan, M.-A.; Lehocky, M. Synthesis and Characterization of a Bioartificial Polymeric System with Potential Antibacterial Activity: Chitosan-Polyvinyl Alcohol-Ampicillin. *Molecules* **2018**, *23*, 3109. [CrossRef] [PubMed]
4. Cascone, M.G.; Barbani, N.; P. Giusti, C.C.; Ciardelli, G.; Lazzeri, L. Bioartificial polymeric materials based on polysaccharides. *J. Biomater. Sci. Polym. Ed.* **2001**, *12*, 267–281. [CrossRef] [PubMed]
5. Pineda-Castillo, S.; Bernal-Ballén, A.; Bernal-López, C.; Segura-Puello, H.; Nieto-Mosquera, D.; Villamil-Ballesteros, A.; Muñoz-Forero, D.; Munster, L. Synthesis and Characterization of Poly (Vinyl Alcohol)-Chitosan-Hydroxyapatite Scaffolds: A Promising Alternative for Bone Tissue Regeneration. *Molecules* **2018**, *23*, 2414. [CrossRef]
6. Bernal, A.; Balkova, R.; Kuritka, I.; Saha, P. Preparation and characterisation of a new double-sided bio-artificial material prepared by casting of poly(vinyl alcohol) on collagen. *Polym. Bull.* **2013**, *70*, 431–453. [CrossRef]
7. Bernal, A.; Kuritka, I.; Saha, P. Preparation and characterization of poly(vinyl alcohol)-poly(vinyl pyrrolidone) blend: A biomaterial with latent medical applications. *J. Appl. Polym. Sci.* **2013**, *127*, 3560–3568. [CrossRef]
8. Bernal-Ballén, A.; Kuritka, I.; Saha, P. Preparation and characterization of a bioartificial polymeric material: Bilayer of cellulose acetate-PVA. *Int. J. Polym. Sci.* **2016**, *2016*. [CrossRef]
9. Giusti, P.; Lazzeri, L.; De Petris, S.; Palla, M.; Cascone, M.G. Collagen-based new bioartificial polymeric materials. *Biomaterials* **1994**, *15*, 1229–1233. [CrossRef]
10. Scotchford, C.A.; Cascone, M.G.; Downes, S.; Giusti, P. Osteoblast responses to collagen-PVA bioartificial polymers in vitro: The effects of cross-linking method and collagen content. *Biomaterials* **1998**, *19*, 1–11. [CrossRef]
11. Ahmad, S.I.; Hasan, N.; Zainul Abid, C.K.V.; Mazumdar, N. Preparation and characterization of films based on crosslinked blends of gum acacia, polyvinylalcohol, and polyvinylpyrrolidone-iodine complex. *J. Appl. Polym. Sci.* **2008**, *109*, 775–781. [CrossRef]
12. Sionkowska, A. Current research on the blends of natural and synthetic polymers as new biomaterials: Review. *Prog. Polym. Sci.* **2011**, *36*, 1254–1276. [CrossRef]
13. Kumar, M.N.V.R. A review of chitin and chitosan applications. *React. Funct. Polym.* **2000**, *46*, 1–27. [CrossRef]

14. Jayakumar, R.; Prabaharan, M.; Kumar, P.T.S.; Nair, S.V.; Tamura, H. Biomaterials based on chitin and chitosan in wound dressing applications. *Biotechnol. Adv.* **2011**, *29*, 322–337. [CrossRef] [PubMed]

15. Miguel, S.P.; Moreira, A.F.; Correia, I.J. Chitosan based-asymmetric membranes for wound healing: A review. *Int. J. Biol. Macromol.* **2019**. [CrossRef] [PubMed]

16. Lowe, B.; Venkatesan, J.; Anil, S.; Shim, M.S.; Kim, S.-K. Preparation and characterization of chitosan-natural nano hydroxyapatite-fucoidan nanocomposites for bone tissue engineering. *Int. J. Biol. Macromol.* **2016**, *93*, 1479–1487. [CrossRef]

17. Saravanan, S.; Vimalraj, S.; Thanikaivelan, P.; Banudevi, S.; Manivasagam, G. A review on injectable chitosan/beta glycerophosphate hydrogels for bone tissue regeneration. *Int. J. Biol. Macromol.* **2019**, *121*, 38–54. [CrossRef]

18. Shelke, N.B.; James, R.; Laurencin, C.T.; Kumbar, S.G. Polysaccharide biomaterials for drug delivery and regenerative engineering. *Polym. Adv. Technol.* **2014**, *25*, 448–460. [CrossRef]

19. LogithKumar, R.; KeshavNarayan, A.; Dhivya, S.; Chawla, A.; Saravanan, S.; Selvamurugan, N. A review of chitosan and its derivatives in bone tissue engineering. *Carbohydr. Polym.* **2016**, *151*, 172–188. [CrossRef]

20. Ozaltin, K.; Lehocky, M.; Humpolicek, P.; Pelkova, J.; Di Martino, A.; Karakurt, I.; Saha, P. Anticoagulant Polyethylene Terephthalate Surface by Plasma-Mediated Fucoidan Immobilization. *Polymers (Basel)* **2019**, *11*, 750. [CrossRef]

21. Venkatesan, J.; Singh, S.; Anil, S.; Kim, S.-K.; Shim, M. Preparation, characterization and biological applications of biosynthesized silver nanoparticles with chitosan-fucoidan coating. *Molecules* **2018**, *23*, 1429. [CrossRef] [PubMed]

22. Manivasagan, P.; Hoang, G.; Moorthy, M.S.; Mondal, S.; Doan, V.H.M.; Kim, H.; Phan, T.T.V.; Nguyen, T.P.; Oh, J. Chitosan/fucoidan multilayer coating of gold nanorods as highly efficient near-infrared photothermal agents for cancer therapy. *Carbohydr. Polym.* **2019**, *211*, 360–369. [CrossRef] [PubMed]

23. Li, B.; Lu, F.; Wei, X.; Zhao, R. Fucoidan: Structure and bioactivity. *Molecules* **2008**, *13*, 1671–1695. [CrossRef] [PubMed]

24. Bhatia, S. Natural polymers vs. synthetic polymer. In *Natural Polymer Drug Delivery Systems*; Springer: Berlin, Germany, 2016; pp. 95–118.

25. Tian, H.; Tang, Z.; Zhuang, X.; Chen, X.; Jing, X. Biodegradable synthetic polymers: Preparation, functionalization and biomedical application. *Prog. Polym. Sci.* **2012**, *37*, 237–280. [CrossRef]

26. Karimi, A.; Navidbakhsh, M. Mechanical properties of PVA material for tissue engineering applications. *Mater. Technol.* **2014**, *29*, 90–100. [CrossRef]

27. Paradossi, G.; Cavalieri, F.; Chiessi, E.; Spagnoli, C.; Cowman, M.K. Poly (vinyl alcohol) as versatile biomaterial for potential biomedical applications. *J. Mater. Sci. Mater. Med.* **2003**, *14*, 687–691. [CrossRef] [PubMed]

28. Kumar, A.; Han, S.S. PVA-based hydrogels for tissue engineering: A review. *Int. J. Polym. Mater. Polym. Biomater.* **2017**, *66*, 159–182. [CrossRef]

29. Goodship, V.; Jacobs, D.K. *Polyvinyl Alcohol: Materials, Processing and Applications*; Smithers Rapra Technology: Shrewsbury, Shropshire, UK, 2009; Volume 16.

30. Stammen, J.A.; Williams, S.; Ku, D.N.; Guldberg, R.E. Mechanical properties of a novel PVA hydrogel in shear and unconfined compression. *Biomaterials* **2001**, *22*, 799–806. [CrossRef]

31. Yang, J.M.; Su, W.Y.; Leu, T.L.; Yang, M.C. Evaluation of chitosan/PVA blended hydrogel membranes. *J. Membr. Sci.* **2004**, *236*, 39–51. [CrossRef]

32. Sudhamani, S.R.; Prasad, M.S.; Sankar, K.U. DSC and FTIR studies on gellan and polyvinyl alcohol (PVA) blend films. *Food Hydrocoll.* **2003**, *17*, 245–250. [CrossRef]

33. Santos, C.; Silva, C.J.; Buttel, Z.; Guimaraes, R.; Pereira, S.B.; Tamagnini, P.; Zille, A. Preparation and characterization of polysaccharides/PVA blend nanofibrous membranes by electrospinning method. *Carbohydr. Polym.* **2014**, *99*, 584–592. [CrossRef] [PubMed]

34. Degirmenbasi, N.; Kalyon, D.M.; Birinci, E. Biocomposites of nanohydroxyapatite with collagen and poly (vinyl alcohol). *Colloids Surf. B Biointerfaces* **2006**, *48*, 42–49. [CrossRef] [PubMed]

35. Jayasekara, R.; Harding, I.; Bowater, I.; Christie, G.B.Y.; Lonergan, G.T. Preparation, surface modification and characterisation of solution cast starch PVA blended films. *Polym. Test.* **2004**, *23*, 17–27. [CrossRef]

36. Zhang, W.; Zhao, L.; Ma, J.; Wang, X.; Wang, Y.; Ran, F.; Wang, Y.; Ma, H.; Yu, S. Electrospinning of fucoidan/chitosan/poly (vinyl alcohol) scaffolds for vascular tissue engineering. *Fibers Polym.* **2017**, *18*, 922–932. [CrossRef]

37. Priya, B.; Gupta, V.K.; Pathania, D.; Singha, A.S. Synthesis, characterization and antibacterial activity of biodegradable starch/PVA composite films reinforced with cellulosic fibre. *Carbohydr. Polym.* **2014**, *109*, 171–179. [CrossRef] [PubMed]

38. Kamoun, E.A.; Chen, X.; Eldin, M.S.M.; Kenawy, E.-R.S. Crosslinked poly (vinyl alcohol) hydrogels for wound dressing applications: A review of remarkably blended polymers. *Arab. J. Chem.* **2015**, *8*, 1–14. [CrossRef]

39. Abbasian, M.; Massoumi, B.; Mohammad-Rezaei, R.; Samadian, H.; Jaymand, M. Scaffolding polymeric biomaterials: Are naturally occurring biological macromolecules more appropriate for tissue engineering? *Int. J. Biol. Macromol.* **2019**. [CrossRef]

40. Williams, D.F. On the nature of biomaterials. *Biomaterials* **2009**, *30*, 5897–5909. [CrossRef]

41. Ratner, B.D.; Bryant, S.J. Biomaterials: Where we have been and where we are going. *Annu. Rev. Biomed. Eng.* **2004**, *6*, 41–75. [CrossRef]

42. Islam, A.; Yasin, T.; Gull, N.; Khan, S.M.; Munawar, M.A.; Shafiq, M.; Sabir, A.; Jamil, T. Evaluation of selected properties of biocompatible chitosan/poly (vinyl alcohol) blends. *Int. J. Biol. Macromol.* **2016**, *82*, 551–556. [CrossRef]

43. Baldwin, A.D.; Kiick, K.L. Polysaccharide-modified synthetic polymeric biomaterials. *Pept. Sci. Orig. Res. Biomol.* **2010**, *94*, 128–140. [CrossRef] [PubMed]

44. Ozaltin, K.; Lehocky, M.; Humpolicek, P.; Vesela, D.; Mozetic, M.; Novak, I.; Saha, P. Preparation of active antibacterial biomaterials based on sparfloxacin, enrofloxacin, and lomefloxacin deposited on polyethylene. *J. Appl. Polym. Sci.* **2018**, *135*, 46174. [CrossRef]

45. Mansur, H.S.; Sadahira, C.M.; Souza, A.N.; Mansur, A.A.P. FTIR spectroscopy characterization of poly (vinyl alcohol) hydrogel with different hydrolysis degree and chemically crosslinked with glutaraldehyde. *Mater. Sci. Eng. C* **2008**, *28*, 539–548. [CrossRef]

46. Kumar, G.N.H.; Rao, J.L.; Gopal, N.O.; Narasimhulu, K.V.; Chakradhar, R.P.S.; Rajulu, A.V. Spectroscopic investigations of Mn 2+ ions doped polyvinylalcohol films. *Polym. (Guildf.)* **2004**, *45*, 5407–5415. [CrossRef]

47. Bernal, A.; Kuritka, I.; Kasparkova, V.; Saha, P. The effect of microwave irradiation on poly(vinyl alcohol) dissolved in ethylene glycol. *J. Appl. Polym. Sci.* **2013**, *128*, 175–180. [CrossRef]

48. Hema, M.; Selvasekarapandian, S.; Arunkumar, D.; Sakunthala, A.; Nithya, H. FTIR, XRD and ac impedance spectroscopic study on PVA based polymer electrolyte doped with NH4X (X = Cl, Br, I). *J. Non. Cryst. Solids* **2009**, *355*, 84–90. [CrossRef]

49. Pielesz, A.; Biniaś, W. Cellulose acetate membrane electrophoresis and FTIR spectroscopy as methods of identifying a fucoidan in Fucusvesiculosus Linnaeus. *Carbohydr. Res.* **2010**, *345*, 2676–2682. [CrossRef]

50. Perumal, R.K.; Perumal, S.; Thangam, R.; Gopinath, A.; Ramadass, S.K.; Madhan, B.; Sivasubramanian, S. Collagen-fucoidan blend film with the potential to induce fibroblast proliferation for regenerative applications. *Int. J. Biol. Macromol.* **2018**, *106*, 1032–1040. [CrossRef] [PubMed]

51. Ho, T.T.M.; Bremmell, K.E.; Krasowska, M.; MacWilliams, S.V.; Richard, C.J.E.; Stringer, D.N.; Beattie, D.A. In situ ATR FTIR spectroscopic study of the formation and hydration of a fucoidan/chitosan polyelectrolyte multilayer. *Langmuir* **2015**, *31*, 11249–11259. [CrossRef]

52. Rodriguez-Jasso, R.M.; Mussatto, S.I.; Pastrana, L.; Aguilar, C.N.; Teixeira, J.A. Microwave-assisted extraction of sulfated polysaccharides (fucoidan) from brown seaweed. *Carbohydr. Polym.* **2011**, *86*, 1137–1144. [CrossRef]

53. Lim, S.J.; Aida, W.M.W.; Maskat, M.Y.; Mamot, S.; Ropien, J.; Mohd, D.M. Isolation and antioxidant capacity of fucoidan from selected Malaysian seaweeds. *Food Hydrocoll.* **2014**, *42*, 280–288. [CrossRef]

54. Hussein-Al-Ali, S.H.; El Zowalaty, M.E.; Hussein, M.Z.; Geilich, B.M.; Webster, T.J. Synthesis, characterization, and antimicrobial activity of an ampicillin-conjugated magnetic nanoantibiotic for medical applications. *Int. J. Nanomed.* **2014**, *9*, 3801. [CrossRef] [PubMed]

55. Huang, Y.-C.; Chen, J.-K.; Lam, U.-I.; Chen, S.-Y. Preparing, characterizing, and evaluating chitosan/fucoidan nanoparticles as oral delivery carriers. *J. Polym. Res.* **2014**, *21*, 415. [CrossRef]

56. Yu, S.-H.; Wu, S.-J.; Wu, J.-Y.; Wen, D.-Y.; Mi, F.-L. Preparation of fucoidan-shelled and genipin-crosslinked chitosan beads for antibacterial application. *Carbohydr. Polym.* **2015**, *126*, 97–107. [CrossRef] [PubMed]

57. Pawlak, A.; Mucha, M. Thermogravimetric and FTIR studies of chitosan blends. *Thermochimica Acta* **2003**, *396*, 153–166. [CrossRef]

58. Marsano, E.; Vicini, S.; Skopińska, J.; Wisniewski, M.; Sionkowska, A. Chitosan and poly(vinyl pyrrolidone): Compatibility and miscibility of blends. *Macromol. Symp.* **2004**, *218*, 251–260. [CrossRef]

59. Oliveira, J.M.; Rodrigues, M.T.; Silva, S.S.; Malafaya, P.B.; Gomes, M.E.; Viegas, C.A.; Dias, I.R.; Azevedo, J.T.; Mano, J.F.; Reis, R.L. Novel hydroxyapatite/chitosan bilayered scaffold for osteochondral tissue-engineering applications: Scaffold design and its performance when seeded with goat bone marrow stromal cells. *Biomaterials* **2006**, *27*, 6123–6137. [CrossRef]

60. Li, M.; Cheng, S.; Yan, H. Preparation of crosslinked chitosan/poly(vinyl alcohol) blend beads with high mechanical strength. *Green Chem.* **2007**, *9*, 894. [CrossRef]

61. Garcia-Cruz, L.; Casado-Coterillo, C.; Iniesta, J.; Montiel, V.; Irabien, A. Chitosan: Poly (vinyl) alcohol composite alkaline membrane incorporating organic ionomers and layered silicate materials into a PEM electrochemical reactor. *J. Membr. Sci.* **2016**, *498*, 395–407. [CrossRef]

62. Awada, H.; Daneault, C. Chemical Modification of Poly (vinyl alcohol) in Water. *Appl. Sci.* **2015**, *5*, 840–850. [CrossRef]

63. Jin, L.; Bai, R. Mechanisms of lead adsorption on chitosan/PVA hydrogel beads. *Langmuir* **2002**, *18*, 9765–9770. [CrossRef]

64. Maachou, H.; Genet, M.J.; Aliouche, D.; Dupont-Gillain, C.C.; Rouxhet, P.G. XPS analysis of chitosan–hydroxyapatite biomaterials: From elements to compounds. *Surf. Interface Anal.* **2013**, *45*, 1088–1097. [CrossRef]

65. Padavan, D.T.; Hamilton, A.M.; Millon, L.E.; Boughner, D.R.; Wan, W. Synthesis, characterization and in vitro cell compatibility study of a poly (amic acid) graft/cross-linked poly (vinyl alcohol) hydrogel. *Acta Biomater.* **2011**, *7*, 258–267. [CrossRef] [PubMed]

66. Puvaneswary, S.; Talebian, S.; Raghavendran, H.B.; Murali, M.R.; Mehrali, M.; Afifi, A.M.; Kasim, N.H.B.A.; Kamarul, T. Fabrication and in vitro biological activity of β-TCP-Chitosan-Fucoidan composite for bone tissue engineering. *Carbohydr. Polym.* **2015**, *134*, 799–807. [CrossRef] [PubMed]

67. Mi Zo, S.; Singh, D.; Kumar, A.; Cho, Y.W.; Oh, T.H.; Han, S.S. Chitosan-hydroxyapatite macroporous matrix for bone tissue engineering. *Curr. Sci.* **2012**, 1438–1446.

68. Berger, J.; Reist, M.; Mayer, J.M.; Felt, O.; Gurny, R. Structure and interactions in chitosan hydrogels formed by complexation or aggregation for biomedical applications. *Eur. J. Pharm. Biopharm.* **2004**, *57*, 35–52. [CrossRef]

69. Kim, J.H.; Kim, J.Y.; Lee, Y.M.; Kim, K.Y. Properties and swelling characteristics of cross-linked poly (vinyl alcohol)/chitosan blend membrane. *J. Appl. Polym. Sci.* **1992**, *45*, 1711–1717. [CrossRef]

70. Venkatesan, J.; Bhatnagar, I.; Kim, S.-K. Chitosan-alginate biocomposite containing fucoidan for bone tissue engineering. *Mar. Drugs* **2014**, *12*, 300–316. [CrossRef]

71. Sezer, A.D.; Hatipoglu, F.; Cevher, E.; Oğurtan, Z.; Bas, A.L.; Akbuğa, J. Chitosan film containing fucoidan as a wound dressing for dermal burn healing: Preparation and in vitro/in vivo evaluation. *Aaps Pharmscitech* **2007**, *8*, E94–E101. [CrossRef]

72. Chung, I.-C.; Li, C.-W.; Wang, G.-J. The influence of different nanostructured scaffolds on fibroblast growth. *Sci. Technol. Adv. Mater.* **2013**, *14*, 44401. [CrossRef]

73. López-Garcia, J.; Lehocky, M.; Humpolíček, P.; Sáha, P. HaCaT keratinocytes response on antimicrobial atelocollagen substrates: Extent of cytotoxicity, cell viability and proliferation. *J. Funct. Biomater.* **2014**, *5*, 43–57. [CrossRef] [PubMed]

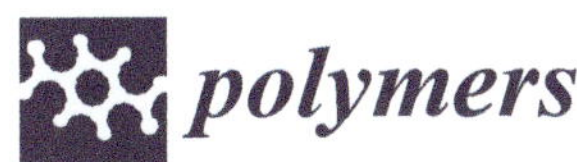 

Article

UV Light Assisted Coating Method of Polyphenol Caffeic Acid and Mediated Immobilization of Metallic Silver Particles for Antibacterial Implant Surface Modification

Ji Yeon Lee [1], Ludwig Erik Aguilar [2,*], Chan Hee Park [2,3,*] and Cheol Sang Kim [2,3,*]

[1] Department of Mechanical Design Engineering, Graduate School, Chonbuk National University, Jeonju 54896, Korea

[2] Department of Bionanosystem Engineering, Chonbuk National University, Jeonju 54896, Korea

[3] Division of Mechanical Design Engineering, Chonbuk National University, Jeonju 54896, Korea

* Correspondence: leaguilar@jbnu.ac.kr (L.E.A.); biochan@jbnu.ac.kr (C.H.P.); chskim@jbnu.ac.kr (C.S.K.)

Received: 28 May 2019; Accepted: 12 July 2019; Published: 18 July 2019

Abstract: Titanium implants are extensively used in biomedical applications due to their excellent biocompatibility, corrosion resistance, and superb mechanical stability. In this work, we present the use of polycaffeic acid (PCA) to immobilize metallic silver on the surface of titanium materials to prevent implant bacterial infection. Caffeic acid is a plant-derived phenolic compound, rich in catechol moieties and it can form functional coatings using alkaline buffers and with UV irradiation. This combination can trigger oxidative polymerization and deposition on the surface of metallic substrates. Using PCA can also give advantages in bone implants in decreasing inflammation by decelerating macrophage and osteoclast activity. Here, chemical and physical properties were investigated using FE-SEM, EDS, XPS, AFM, and contact angle. The in vitro biocompatibility and antibacterial studies show that PCA with metallic silver can inhibit bacterial growth, and proliferation of MC-3T3 cells was observed. Therefore, our results suggest that the introduced approach can be considered as a potential method for functional implant coating application in the orthopedic field.

Keywords: titanium implants; polycaffeic acid; metallic silver particles; anti-bacterial properties

1. Introduction

In the orthopedic field, peri-implant infection is one of the most serious complications leading to 14% of all implant failures. It can lead to prolonged hospitalization, financial burden, and even death [1,2]. Infection at the periphery of implants is usually due to two reasons: The attachment and formation of biofilm by microorganisms through colonization and the compromised immune ability at the interface of the implant and tissue [3–5]. Microorganisms are able to attach to the implant, especially biocompatible implants due to protein membrane formation [6–9]. Usually, the human body is able to defend itself from infections, depending on the immune response system [10,11]. However, in order to prevent infection during and after the placement of implants, stringent aseptic surgical conditions are used, and antibiotics are administered. However, these approaches have obvious shortcomings, and research has thus been carried out on finding a way to modify implanted materials at the surface level. Therefore, in order to avoid the formation of biofilm, which ultimately prevents infection, the antibacterial properties of the materials used in implants need to be increased.

Currently, titanium (Ti) alloys are one of the most common materials used in many biomedical applications such as dental and orthopedic implants because of its biocompatibility, resistance to corrosion, high strength-to-weight ratio, good tolerance, and presence of reactive titanium oxide surface layer [12–16]. Since titanium was first used for implant fabrication in 1930, it has been shown to be superior to stainless steel or cobalt alloys. Titanium (Ti) and its various alloys, which are mixed with vanadium or aluminum, were first used in surgery in 1950. Because of its mechanical properties (modulus and hardness), Ti alloys play an important role in orthopedic surgery and have been used continuously since these properties were first revealed [17–19]. Moreover, the stiffness of Ti actively adjusts bone cell phenotypic specification [20–22].

Despite the excellent physical properties of titanium, infection is still a major issue. To address this, several studies have attempted to improve the antibacterial capability of titanium via surface modification and coating using functional materials such as polymeric coating [23–25]. Many studies have explored several surface coatings, examples of which are the use of catechol-containing moieties such as multiple 3,4-dihydroxyphenylalanine (DOPA) derived from mussels, which strongly binds to proteins and is capable of surface coating functionalization, and the development of a special wetting surface [26–28]. In our study, we used poly-caffeic acid (PCA), which is a naturally derived phenolic compound that can form a multifunctional coating on various substrates by polymerization under a mild alkali condition. When used as a biomedical applied coating material, PCA is inherently better than polydopamine due to its various properties such as antioxidant, anticancer, anti-inflammatory, antihypertensive, and antibacterial agent. It is, therefore, a promising candidate material for surface coating on titanium implants [29–32].

Phenolic molecules like polycaffeic acid are also proven to have in situ reduction properties and are widely used for the green synthesis of metallic nanomaterials especially in the formation of metallic silver [33–35]. Metallic silver is known as an antibacterial agent, and several hypotheses have been presented about its mechanisms in inhibiting bacterial growth [36–39]. In this current study, we utilized the inherent property of PCA as a reduction agent to reduce Ag ions to metallic Ag particles, in order to create a multifunctional surface to mitigate the occurrence of bacterial infection in metallic implant procedures. Several theories have been presented on the antibacterial effects of Ag [40,41]: (1) Ag adheres to the bacterial cell wall and cause a structural change in the bacterial cell, which induces cell death. Alternatively, Ag comes into contact with the bacteria so that the action of the free radicals on the Ag causes holes in the cell, which then dies. (2) Ag emits silver ions that bind to the thiol groups and cause damage to the cell. (3) Silver is weakly acidic and naturally reacts with cells that are slightly alkaline due to the characteristic of acid to react with the alkali, eventually leading to the death of the bacterial cells. (4) Ag that enters into the bacteria affects the signal transduction due to the phosphorylation of the protein substrate, which inhibits signal transduction and induces growth arrest. Here, we focus on the prevention of peri-implant infection by coating on a titanium surface using a combination of two different antibacterial materials for a synergistic enhancing effect of anti-inflammation using a simple method. This work will explore the potential application of this approach in the orthopedic field.

2. Materials and Methods

2.1. Materials

Ti foil (0.127 mm thick, annealed, 99%) was obtained from Alfa Aesar (Daejon, Korea). The foil was cut into small pieces with dimensions of 2.5 cm × 5 cm. Silver nitrate ($AgNO_3$, MW 169.87 g/mol) was purchased from SHOWA chemicals (Gyoda, Japan). Chemical products such as Caffeic Acid (CA) (3,4-Dihydroxybenzeneacrylic acid, MW 180.16 g/mol), and sodium hydroxide (NaOH, MW 40.00 g/mol, pellets (anhydrous)) were purchased from Sigma Aldrich Chemical Co. Ltd. (Seoul, Korea) and used without further purification. All aqueous solutions were prepared by ultrapure water purified using a Milli-Q UV-Plus water purification system (Millipore, Bedford, MA, USA). The resistivity of the water was >1018 MΩcm^{-1}.

2.2. Preparation of the Polycaffeic Acid Coated-Metallic Silver Functionalized Titanium Samples

Each Ti piece cut to the same size was chemically etched to remove the passive oxide layer and impurities on the surface. The Ti pieces were then submerged into a solution of hydrofluoric acid (7 wt.%) and nitric acid (12%) in distilled water (DW, 81 wt.%) for approximately 20 s followed by instant sonication first in ethanol and then in DW for ten minutes. These Ti pieces were then dried under a nitrogen atmosphere. As shown in Figure 1A, after soaking, one of the prepared Ti pieces was submerged into the CA solution (0.09 g of CA powder with 30 mL of 0.1 mM pH10 carbonate-bicarbonate buffer solution) under UV irradiation (200–300 nm wavelength) at 95% power setting using Omnicure® Series 1500 lamp for 4 h (distance between top of the solution and tip where light is emitted is 5 cm) with slow stirring at room temperature. For the metallic silver in-situ immobilization using a polyphenol reduction method, the Ti pieces after coating with PCA were fully submerged in a silver nitrate solution (20 mL of 4 mM $AgNO_3$ solution with DW (which has a pH of 8.5–9.0) using 1 M NaOH solution) for 24 h at room temperature with continuous stirring. The samples were then washed again in DW and air-dried at room temperature.

2.3. Characterization of Surface Properties

The morphology of the coating was observed using field emission scanning electron microscopy (FE-SEM, SUPRA 40VP, Carl Zeiss, Germany) at accelerating voltages of 15.0 and 5.0 kV. The samples were sputter-coated under argon to render them electrically conductive. The surface topography of the prepared samples was visualized using atomic force microscopy (AFM, Multimode-8, Bruker, Germany) operating in tapping mode with a scan rate of 0.7535 Hz. Energy dispersive X-ray spectroscopy (EDS) coupled with FE-SEM was employed to analyze the elemental composition of the specimens' surface. Elemental mapping was then performed to confirm the distribution of elements on the surface of each sample. X-ray photoelectron spectroscopy (XPS, AXIS-NOVA, Kratos Inc., San Diego, CA, USA) data were obtained with a monochromatized Al K-alpha X-ray irradiation source. To determine the surface hydrophobic-philicity of the samples, a water droplet of 5 µL was carefully dropped onto the sample's surface, and the water contact angle was measured at 1 and 5 s intervals using a contact angle meter (GBX, Digidrop, France) at ambient temperature.

2.4. Determination of Antibacterial Activity

Three types of bacterial colonies were prepared to test the antibacterial property of the coating surface. Gram-negative bacteria *Escherichia coli* (*E. coli*), gram-positive bacteria *Staphylococcus Aureus* (*S. aureus*) and *Pseudomonas aeruginosa* (*P. aeruginosa*) were separately thawed in lysogeny broth (LB) and were incubated separately for 24 h in a shaking incubator maintained at 37 °C temperature. Using the spread plate method, after mixing the prepared LB solution with the thawed bacteria 1 mL of bacterial suspension containing around 10^5 colony forming units for each bacteria, the same amount of bacterial solution was suspended on the nutritive agar plate, and the samples were carefully placed on the inoculated plates. After spreading the bacteria on the plate, three different types of titanium samples (etched Ti, Ti–PCA, and Ti–PCA–Ag) were placed on these plates in a triangular orientation and incubated at 37 °C for 24 h. The titanium samples were sterilized under ultra-violet (UV) radiation for twelve hours before placing on agar Petri plates. All specimens were tested for antibacterial activities using the growth inhibition study method, which involves measuring the % inhibition area (mm) of no bacterial growth around each sample. All antibacterial activity tests were performed in triplicate ($n = 3$ per group per type of bacteria).

The % inhibition was calculated using the following equation.

$$\% \text{ inhibition} = \frac{\text{area inhibited (mm)} - \text{size of the sample (mm)}}{\text{Size of the sample (mm)}} \times 100 \tag{1}$$

2.5. Elution of Ag from the Ti–PCA–Ag Sample

To determine the amount of Ag being eluted from Ti–PCA–Ag sample, we have submerged a round sample with a circumference of 10 mm in distilled water (1 mL). The eluted Ag was then detected using inductive coupled plasma optical emission spectroscopy (ICP-OES) (iCAP-7000 qTegra, Thermo Fisher Scientific, Waltham, MA, USA). Three different time points were used on day 1, 2, and 3 ($n = 3$).

2.6. In Vitro Biocompatibility Study

Biocompatibility testing was conducted using a pre-osteoblast (MC-3T3) cell line obtained from normal mouse clone 4, ATCCR CRL-5293. The medium for the growth of the cell line was alpha-Minimal Essential Medium (αMEM, Sigma Chemical Co., St. Louis, MO, USA) supplemented with 10% fetal bovine serum (FBS, Sigma Chemical Co., St. Louis, MO, USA) and 1% penicillin/streptomycin (100X). The cells were cultivated in the cell culture plate (SPL Life Science, Gyeonggi-do, Korea) and placed in an incubator at 37 °C with 5% CO_2 and 95% humidity. Smaller circular samples of 6.5 mm in diameter were cut from each sample and fixed in the wells of a 96-well cell culture plate (SPL life science, Gyeonggi-do, Korea). After sterilization under UV radiation for 3 h and washing three times with phosphate buffered saline (PBS), the cells with a density of 2×10^3 cells/well were seeded on the samples by pipetting onto the center and placing into an incubator for five days. During the five days, the cell culture medium was replaced every 24 h. The cell biocompatibility of the samples was measured using a cell counting kit-8 (CCK-8) after one, three, and five days of cell culture. At the appointed time, the solution was added to each well and was incubated for 2 h in a 37 °C incubator with 5% CO_2 and 95% humidity. The measurement of the absorbance was checked at 450 nm using a microplate reader (Tecan, Austria).

The cells were then subsequently imaged using confocal laser microscopy (Zeiss, SR CLSM, LSM 880, Carl Zeiss, Germany). The cell morphology and distribution of MC-3T3 cells on each sample (etched Ti, Ti–PCA and Ti–PCA–Ag) were evaluated after five days using super-resolution confocal laser scanning microscope (SR CLSM, LSM 880, Carl Zeiss, Germany). Beforehand, the MC-3T3 cells on samples were fixed in 4% paraformaldehyde solution overnight at 4 °C and then washed three times with 1% PBS (phosphate buffer saline). After that, the MC-3T3 cells were immersed in 0.5% Triton X-100 for 2 min at room temperature and then washed two times with 1% PBS. After blocking with 1% human serum albumin HSA/PBS dilution for 30 min, the cells in each of the samples were stained with Actin-green and DAPI for actin filament staining and cell nuclei visualization, respectively.

2.7. Image and Statistical Analysis

All images were analyzed using ImageJ software. The results are presented as the mean ± standard deviation (SD) for n = 3. Statistical significance was assessed by determining the level of significance with a one-way ANOVA followed by a Tukey test for means comparison using the OriginPro 8.5 software ($p < 0.05$) (OriginLab Corporation, Northampton, MA, USA).

3. Results and Discussion

3.1. Confirmation of Coating and Surface Morphology Analysis

As seen in Figure 1A, the coating process followed a simple UV assisted polymerization process of caffeic acid on the surface of the Titanium foil, the immobilization of metallic silver on the surface was also followed right after the coating of PCA on top of the Ti substrate, Ag ions can be immobilized and reduced to metallic silver particles in situ by using the catechol moiety of the PCA [4]. It has been recently discovered that natural polyphenols can be deposited via electromagnetic radiation (UV-light) to various substrates like glass and stainless steel [26]. After the UV-light irradiation of caffeic acid over the titanium substrate, the apparent difference in color among the four Ti pieces was observed, as shown in Figure 1B. Visual inspection initially indicates that the coating was successfully deposited on the surface of the titanium. The leftmost pure Ti specimen has a matte light gray color, while the Ti that was chemically etched has a darker color due to the removal of the oxide layer on its surface. In the case of the Ti specimen coated with PCA, the material developed a visible yellowish-gray coating over the Ti specimen. Unlike the previous test piece, the surface of the specimen with metallic silver on the PCA coating is shiny rather than matte.

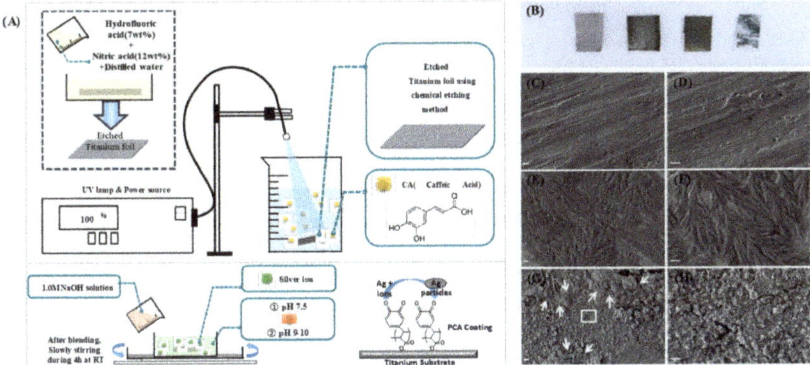

Figure 1. (**A**) Schematics of the process of coating polycaffeic acid (PCA) on etched titanium (upper image) and synthesis of metallic silver on the Ti–PCA substrate (lower image) and (**B**) photo and (**C–H**) FE-SEM image in two magnitudes (scale bar X 5.0 k = 1 μm, 20.0 k = 5 μm) of titanium foil according to each process.

In order to confirm the coating on the surface of the Ti samples, we employed field emission scanning electron microscopy and the FE-SEM images in Figure 1C–H shows the surface morphology of each Ti sample in different conditions. Etched Ti samples are shown to be featureless (Figure 1C–D), while in the case of the Ti–PCA specimen, the polymerized caffeic acid (PCA) formed leaf-like structures and covered almost the entire specimen (Figure 1E–F). The PCA coating was formed by gathering several thousand strings of strands of about 1 μm in length. In Figure 1G–H, the specimen is covered with metallic Ag particles with traces of the PCA remaining after the process (indicated by arrows). The Ag reduction properties of PCA are further shown in the FESEM images as the particles had been deposited in situ. The potential of catechol moieties as a redox agent has been explored and the overall mechanism for reducing Ag ions to metallic Ag is generally through the formation of quinone and semiquinone in the polycaffeic acid chemical structure at pH levels higher than 9.45. The highly electronegative oxygen in the quinones can attract and stabilize the Ag ions and form metallic silver on the surface of the polycaffeic acid coating.

Another novel feature of this method is the use of plant-derived polyphenol. Caffeic acid has been shown to have antibacterial, antiviral, anti-cancer, and anti-inflammatory effects [42–44]. In comparison to polydopamine, coatings based on plant phenolic compounds have been found to have faster adhesion rate, affordable cost, excellent availability, and good structural diversity [45,46]. The use of the UV-assisted polymerization of CA also enhances this coating technique. It has thus been demonstrated that UV irradiation can control the polymerization of plant-derived phenolics via induction of reactive oxygen species singlet oxygen (O_2), superoxide radicals (O_2^-) or highly reactive hydroxyl radicals (OH) in the presence of traces of O_2. Furthermore, the method is simple enough to be used on a larger scale, and the materials used are widely available and simple to set-up.

3.2. Surface Topology Analysis

We can further determine whether the coating was successfully applied to the sample surface by comparing the roughness of the samples' surfaces through AFM imaging (2D and 3D). AFM imaging can also indicate the relative surface height of the samples. Examining the 2D and 3D and FESEM images in Figure 2A,D, it can be seen that the surface treated by etching has a thin hatch shape running through a single direction. In Figure 2B,E, the 2D and 3D images show that the coating of PCA on the surface shown an increased surface roughness and a shape resembling the edge of a leaf. As shown in Figure 2C,F, the 2D and 3D AFM images confirm that metallic silver particles are present and had adhered well due to the ridge-like shape formed of the surface of the Ti substrate. The summary of the surface roughness for each sample (etched Ti, Ti–PCA, and Ti–PCA–Ag) are tabulated in Table 1. Showing that the roughness value for each substrate differ with Ti–PCA having the highest roughness value followed by Ti–PCA–Ag. These results further corroborate the initial findings that the coating process was successfully applied by comparing the surface topology difference of each Ti samples.

Table 1. Atomic force microscopy (AFM) analysis survey of the surface roughness of different modified titanium samples.

Samples	Rms (nm)	Ra (nm)	Rmax (nm)
Etched Ti	2.469 ± 1.2	3.280 ± 1.1	10.49
Ti–PCA	25.3845 ± 2.7	39.6855 ± 2.8	163.3
Ti–PCA–Ag	8.6545 ± 0.9	18.2215 ± 1.7	56.67

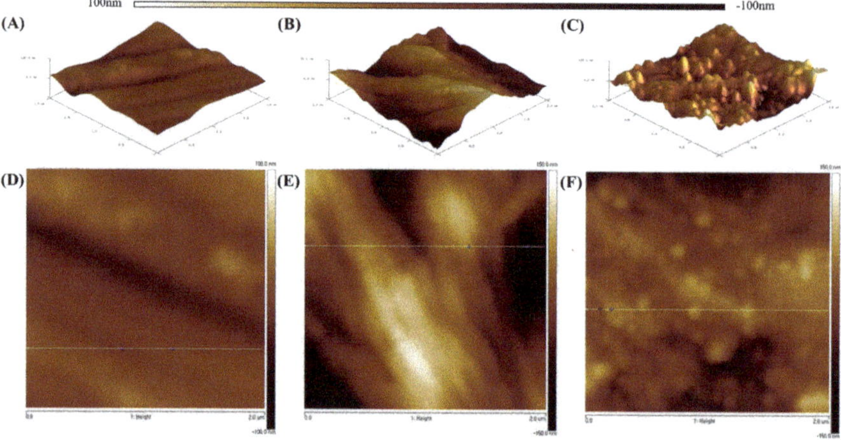

Figure 2. 3D and 2D surface AFM images (2 μm × 2 μm) and lateral profiles of each surface: (**A,D**) Etched Ti, (**B,E**) Ti–PCA, and (**C,F**) Ti–PCA–Ag.

3.3. Surface Chemical Analysis

EDS, elemental mapping, and XPS were performed in order to verify the FESEM results and to confirm the surface deposition of the PCA and metallic silver on the Ti substrate. As shown in Figure 3A–C, the surface chemical composition changed as the carbon content increased (6.55% and 18%) and oxygen content increased (29.81% and 25.51%) with the Ti–PCA and Ti–PCA–Ag samples, respectively (Figure 3B,C). In contrast to Figure 2A, the chemically etched titanium sample (etched Ti) comprises only of Ti and O elements. The elemental mapping, as seen in Figure 3E,F, verifies the distribution of these elements in the given samples. Ag element is also uniformly distributed on the surface of the samples, as evidenced by the purple color distribution. The XPS spectra are shown in Figure 3D verifying the presence of the metallic silver with a tall Ag3d peak in the survey scan. In addition, the 6 eV peak difference verifies that the Ag element is in a particle zero valence state. As shown in the spectrum, two bands at ca. 368 and 374 eV were observed, which were ascribed to the $Ag3d_{5/2}$ and $Ag3d_{3/2}$ binding energies, respectively.

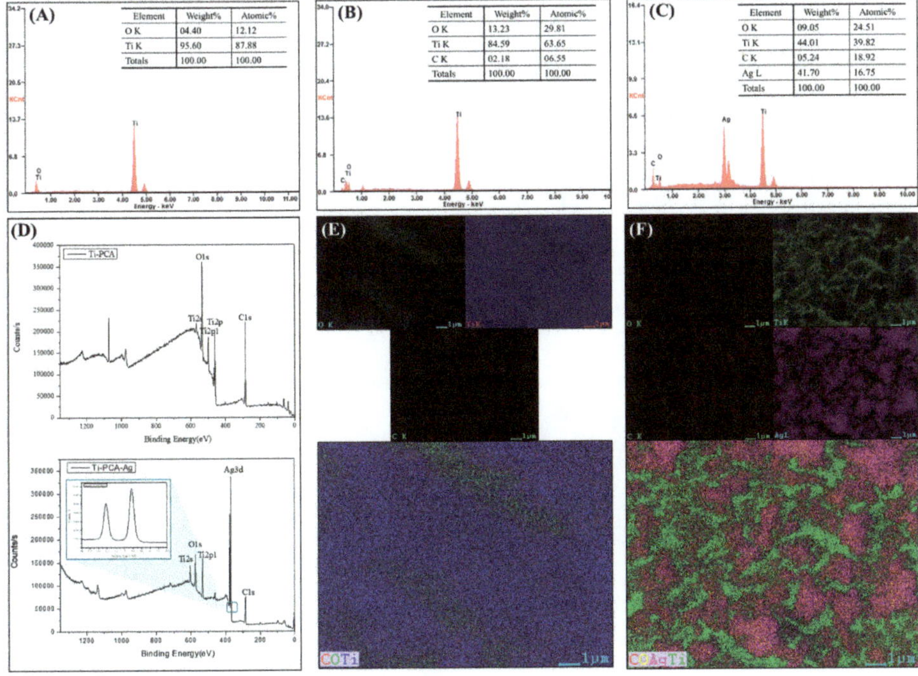

Figure 3. Energy dispersive X-ray spectroscopy (EDS) spectra showing the main components present on the surface of (**A**) Etched Ti, (**B**) Ti–PCA, and (**C**) Ti–PCA–Ag. (**D**) XPS spectra of the Ti–CA (**upper**) and Ti–PCA–Ag (**below**) (inset: XPS spectra Ag3d scan). Mapping of the elemental species on the surface of (**E**) Ti–PCA and (**F**) Ti–PCA–Ag.

3.4. Physical Surface Characteristic (Wettability)

In order to determine the hydrophilic-phobicity of the substrates, contact angle measurements were executed. In principle, droplets with a contact angle of more than 90° are classified as hydrophobic, and those with a contact angle below 90 °C are classified as hydrophilic. A water contact angle with a value close to 0° is superhydrophilic and that with a value above 150° is termed superhydrophobic. Among the many surface modifications of Ti implants, some focused on the increasing surface hydrophilicity because it can influence the bonding strength, the number of proteins bound to a surface, the interactions of the cell with the surface of the implant, the rate of osseointegration, and bacterial adhesion with subsequent biofilm formation [47]. When examining Figure 4, the contact angles of DW after one second were 58.37 ± 0.9°, 21.8 ± 0.2°, 48.2 ± 0.2°, and after 5 s the values are reduced to 56.77 ± 0.61°, 14.5 ± 0.14°, 45.2 ± 0.3° for etched Ti, Ti–PCA, Ti–PCA–Ag, respectively. Ti–PCA turns out to be the most hydrophilic amongst the groups due to the presence of the residual OH⁻ groups on its surface. Ti–PCA–Ag on the other had less OH⁻ groups on its surface due to the reduction of Ag ions to Ag particles and this exhausted some of the PCA coatings, thus making it less hydrophilic than Ti–PCA.

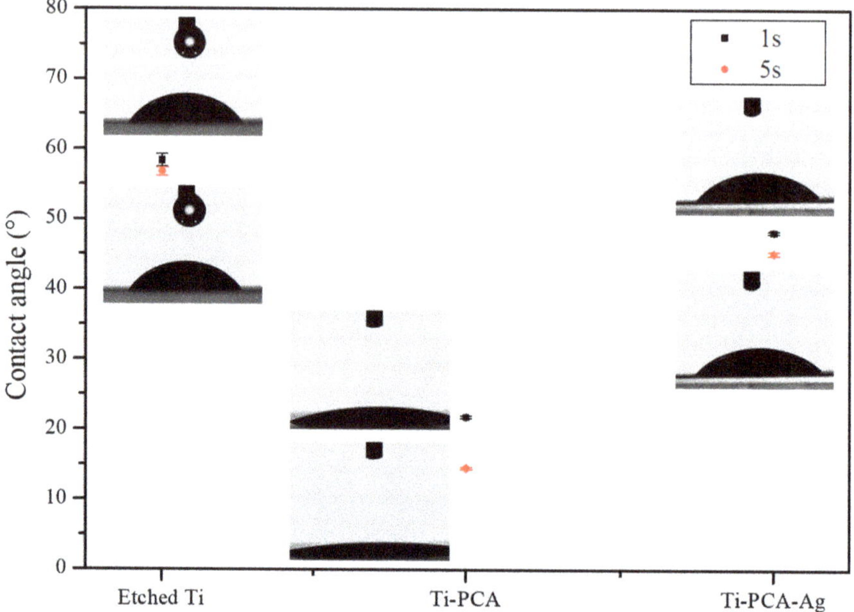

Figure 4. Contact angle values of the Etched Ti (control), Ti–PCA, and Ti–PCA–Ag measured in 1 and 5 s (*n* = 3).

3.5. Bacteria Inhibition Test

Until now, increasing the antibacterial properties of Ti are being explored considering that it is one of the most common materials for biomedical implants. In this study, by using three types of bacteria (*E. coli*, *S. aureus*, and *P. aeruginosa*), we demonstrated the antibacterial effect of the coated samples. The result of the test demonstrated that the Ti–PCA–Ag sample has sufficient anti-bacterial efficacy. Especially, among the three strains, *S. aureus* showed the greatest effect due to the Ag particle's ability to disrupt the cell wall synthesis of gram-positive bacteria. Moreover, it was found that even with only PCA coating, a weak antibacterial effect can be obtained, as reported with other contemporary works, polyphenols can also disrupt bacterial membrane and can lead to increased permeability and susceptibility to damage [41,47]. Moreover, in the case of Ti–PCA–Ag, the effect of silver ions released from metallic Ag particles destroys the biofilm formed by the microorganisms. In addition, the interaction between silver and the protein in the bacteria is considered to be an important mechanism for the antibacterial properties of metallic silver particles. The results in Figure 5A,B shows that the antibacterial effect for all three strains of bacteria is more pronounced on samples containing metallic silver on its surface. Thus, we can attribute the main antibacterial property of coated Ti substrate from the metallic silver and not the PCA coating. The main purpose of the PCA coating is to reduce and immobilize, in-situ the present Ag ions to form metallic silver particles. The mechanism behind the reducing properties of PCA is due to the affinity of metallic ions to the quinone and semiquinone functional moieties of PCA [33]. The ions can agglomerate and form particles on top of the PCA coating creating an enhanced antibacterial surface compared to just using PCA alone. We expect the main antibacterial mechanism of Ti–PCA–Ag is due to the elution of Ag ions from the coating, and this was observed via ICP-OES (Figure 6C). Ag 328 was eluted at the highest concentration during the first day of submersion in distilled water at 0.007 ± 0.0033 ppm. Additionally, as time goes by, the concentration of the Ag 328 in the media has reduced to 0.003 ± 0.0009 ppm (day 2) and 0.0001 ± 0.00004 ppm (day

3). This means that the Ag ions are mostly released after three days and could, therefore, disrupt initial bacterial colonization on the implant substrate.

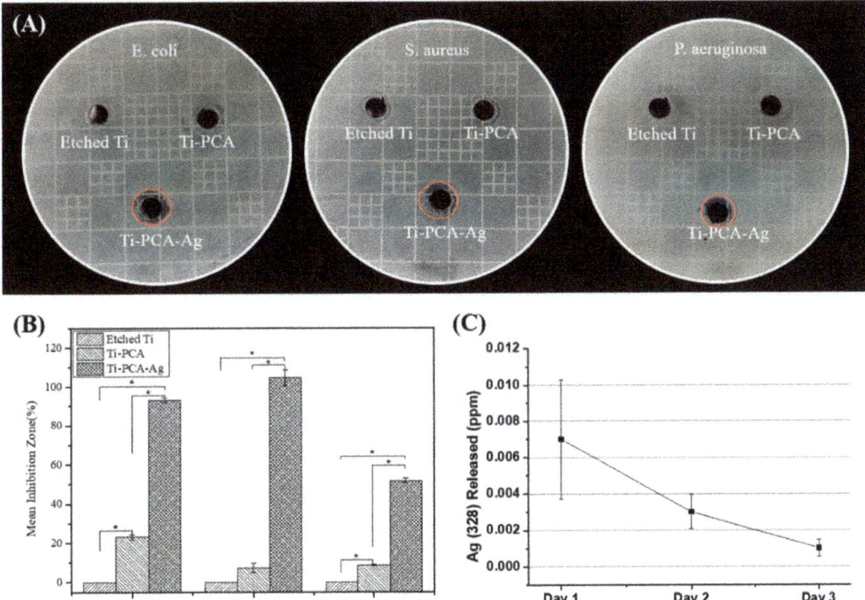

Figure 5. Representative images of antibacterial activities using (**A**) *E. coli* (**left**), *S. aureus* (**middle**), and *P. aeruginosa* (**right**) after 24 h after adhered specimens. (**B**) Graph of inhibition zone measured by Image J software ($n = 3$ and * indicates statistical significance ($p < 0.05$) measured by one way ANOVA Tukey test). (**C**) Ag 328 elution from the Ti–PCA–Ag detected via inductively coupled plasma atomic emission spectroscopy (ICP-OES) ($n = 3$).

3.6. Biocompatibility on MC-3T3 Pre-Osteoblast Cells

To determine the biocompatibility of the specimens, in vitro cytocompatibility tests were conducted using MC-3T3 pre-osteoblast cell line and the amount of actively metabolic cells were measured using CCK-8 assay (Figure 6). Initially, from the statistical analysis of the results for each specimen, we obtained lesser osteocyte activity for the Ti–PCA sample as a whole for days 1 and 3 compared to Etched Ti and the Ti–PCA–Ag samples. This could be due to the pro-oxidant effect of PCA towards the MC-3T3 cells. However, an upward trend was observed and cell proliferation was increased from days 1, 3, and 5. Moreover, on the fifth day it has almost the same absorbance value as that of the other groups, showing that adhesion of PCA has a positive influence on the cell and cannot be seen as cytotoxic. This is further corroborated by the confocal imaging of MC-3T3 cells after five days of incubation. Cells have adhered and exhibited robust filamentous growth on each of the samples. Ti–PCA–Ag samples also had an upward trend on all time points and comparable to the values of etched Ti. Combining this result with the antibacterial inhibition test, we, therefore, believe that the Ti–PCA–Ag sample is not only suitable for use as an implant material but is also effective in preventing infection and other diseases in peri-implant applications that are still considered to be problematic to date. Furthermore, we anticipate that it will be possible to improve the quality of implants by applying Ti–PCA–Ag modifications to a variety of metallic osteo prosthesis used in situations requiring antimicrobial action.

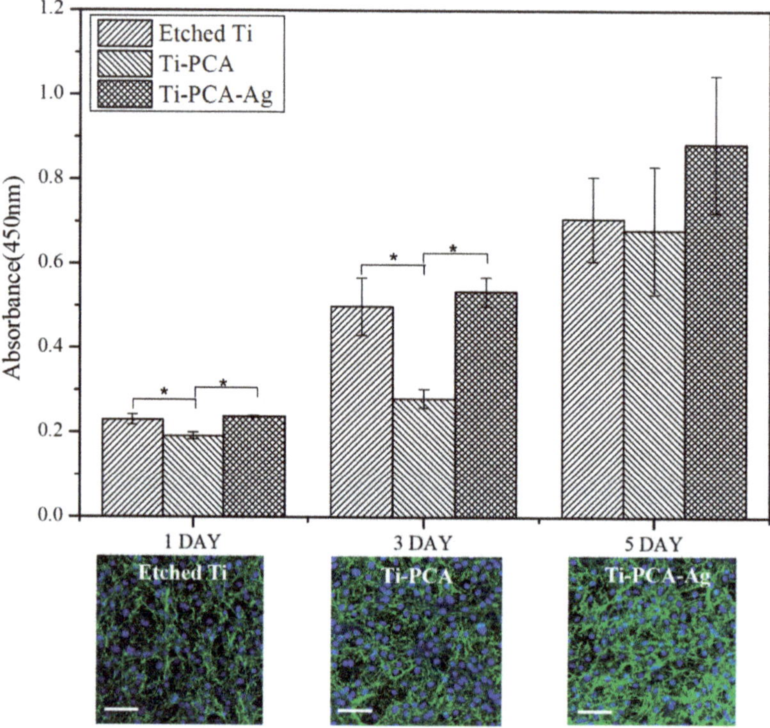

Figure 6. In vitro biocompatibility test using CCK-8 assay of MC-3T3 osteoblast cells. The data are reported as the mean ± standard deviation ($n = 3$ and * indicates statistical significance ($p < 0.05$) measured by one way ANOVA Tukey test) (inset: confocal images of MCT-3T3 cells after five days of incubation) (scale bar = 100 μm).

4. Conclusions

In summary, we successfully coated Ti substrates with PCA and metallic silver using a facile UV light assisted method for implant applications. We confirm that the coating process was successful with SEM and AFM analysis. At the same time, we verified the deposition of PCA and metallic silver by confirming the composition of the elements through XPS, EDS, and mapping methods and the physical property (hydrophilicity) of the samples were verified using water contact angle measurements. In addition, Ti–PCA–Ag exhibited antibacterial and biocompatibility properties thus we believe that this method can be used in modifying titanium for osteoprosthesis and other biomedical applications.

Author Contributions: J.Y.L. executed all experiments, interpreted the data and wrote the manuscript. L.E.A. conceptualized the research topic and designed the experimental protocols and interpretation of data. C.H.P and C.S.K. provided guidance in interpretation of data and financial support.

Funding: This work was supported by a grant from the National Research Foundation (NRF) of Korea, funded by the Korean Government (Project No. 2018R1D1A1B07048782) (Project No. S2718885), Ministry of Education, Science and Technology (Project no. 2019R1A2C1003988) and (NRF-2017-Fostering Core Leaders of the Future Basic Science Program/Global Ph.D. Fellowship Program). It was also partially supported by a grant from the Basic Science Research Program through the National NRF of Korea by the Ministry of Education, Science and Technology (Project no. 2016R1A2A2A07005160).

Conflicts of Interest: The authors declare no conflict of interest.

References

1. Persson, G.R.; Samuelsson, E.; Lindahl, C.; Renvert, S. Mechanical non-surgical treatment of peri-implantitis: A single-blinded randomized longitudinal clinical study. II. Microbiological results. *J. Clin. Periodontol.* **2010**, *37*, 563–573. [CrossRef] [PubMed]
2. Roy, M.; Bandyopadhyay, A.; Bose, S. Induction plasma sprayed nano hydroxyapatite coatings on titanium for orthopaedic and dental implants. *Surf. Coat. Technol.* **2011**, *205*, 2785–2792. [CrossRef] [PubMed]
3. Wang, J.; Zhou, H.; Guo, G.; Tan, J.; Wang, Q.; Tang, J.; Liu, W.; Shen, H.; Li, J.; Zhang, X. Enhanced anti-infective efficacy of zno nanoreservoirs through a combination of intrinsic anti-biofilm activity and reinforced innate defense. *ACS Appl. Mater. Interfaces* **2017**, *9*, 33609–33623. [CrossRef] [PubMed]
4. Vögeling, H.; Duse, L.; Seitz, B.S.; Plenagl, N.; Wojcik, M.; Pinnapireddy, S.R.; Bakowsky, U. Multilayer bacteriostatic coating for surface modified titanium implants. *Phys. Status Solidi (A)* **2018**, *215*, 1700844. [CrossRef]
5. Hoyos-Nogués, M.; Buxadera-Palomero, J.; Ginebra, M.-P.; Manero, J.M.; Gil, F.; Mas-Moruno, C. All-in-one trifunctional strategy: A cell adhesive, bacteriostatic and bactericidal coating for titanium implants. *Colloids Surf. B Biointerfaces* **2018**, *169*, 30–40. [CrossRef] [PubMed]
6. Jahed, Z.; Shahsavan, H.; Verma, M.S.; Rogowski, J.L.; Seo, B.B.; Zhao, B.X.; Tsui, T.Y.; Gu, F.X.; Mofrad, M.R.K. Bacterial networks on hydrophobic micropillars. *ACS Nano* **2017**, *11*, 675–683. [CrossRef] [PubMed]
7. Kalan, L.; Grice, E.A. Fungi in the wound microbiome. *Adv. Wound Care* **2018**, *7*, 247–255. [CrossRef]
8. Del Pozo, J.L. Biofilm-related disease. *Expert Rev. Anti-Infect.* **2018**, *16*, 51–65. [CrossRef]
9. Wandiyanto, J.V.; Truong, V.K.; Al Kobaisi, M.; Juodkazis, S.; Thissen, H.; Bazaka, O.; Bazaka, K.; Crawford, R.J.; Ivanova, E.P. The fate of osteoblast-like mg-63 cells on pre-infected bactericidal nanostructured titanium surfaces. *Materials* **2019**, *12*, 1575.
10. Yadav, S.K.; Khan, G.; Bansal, M.; Thokala, S.; Bonde, G.V.; Upadhyay, M.; Mishra, B. Multiparticulate based thermosensitive intra-pocket forming implants for better treatment of bacterial infections in periodontitis. *Int. J. Biol. Macromol.* **2018**, *116*, 394–408. [CrossRef]
11. David, N.; Nallaiyan, R. Biologically anchored chitosan/gelatin-SrHAP scaffold fabricated on Titanium against chronic osteomyelitis infection. *Int. J. Biol. Macromol.* **2018**, *110*, 206–214. [CrossRef] [PubMed]
12. Palka, K.; Pokrowiecki, R. Porous Titanium implants: A review. *Adv. Eng. Mater.* **2018**, *20*, 1700648. [CrossRef]
13. Wally, Z.; van Grunsven, W.; Claeyssens, F.; Goodall, R.; Reilly, G. Porous Titanium for dental implant applications. *Metals* **2015**, *5*, 1902–1920. [CrossRef]
14. Heo, D.N.; Ko, W.K.; Lee, H.R.; Lee, S.J.; Lee, D.; Um, S.H.; Lee, J.H.; Woo, Y.H.; Zhang, L.G.; Lee, D.K.; et al. Titanium dental implants surface-immobilized with gold nanoparticles as osteoinductive agents for rapid osseointegration. *J. Colloid Interface Sci.* **2016**, *469*, 129–137. [CrossRef] [PubMed]
15. Bhattarai, D.; Aguilar, L.; Park, C.; Kim, C. A Review on properties of natural and synthetic based electrospun fibrous materials for bone tissue engineering. *Membranes* **2018**, *8*, 62. [CrossRef] [PubMed]
16. Chen, L.; Paulitsch, J.; Du, Y.; Mayrhofer, P.H. Thermal stability and oxidation resistance of Ti-Al-N coatings. *Surf. Coat Technol.* **2012**, *206–318*, 2954–2960. [CrossRef] [PubMed]
17. Farghali, R.A.; Fekry, A.M.; Ahmed, R.A.; Elhakim, H.K.A. Corrosion resistance of Ti modified by chitosan-gold nanoparticles for orthopedic implantation. *Int. J. Biol. Macromol.* **2015**, *79*, 787–799. [CrossRef]
18. Bartosik, M.; Daniel, R.; Zhang, Z.; Deluca, M.; Ecker, W.; Stefenelli, M.; Klaus, M.; Genzel, C.; Mitterer, C.; Keckes, J. Lateral gradients of phases, residual stress and hardness in a laser heated Ti0.52Al0.48N coating on hard metal. *Surf. Coat Technol.* **2012**, *206*, 4502–4510.
19. Duan, Y.; Liu, Y.; Li, J.; Wang, H.; Wen, S. Investigation on the nanomechanics of liposome adsorption on titanium alloys: temperature and loading effects. *Polymers* **2018**, *10*, 383. [CrossRef]
20. Shi, X.; Li, L.; Ostrovidov, S.; Shu, Y.; Khademhosseini, A.; Wu, H. Stretchable and micropatterned membrane for osteogenic differentiation of stem cells. *ACS Appl. Mater. Interfaces* **2014**, *6*, 11915–11923. [CrossRef]
21. Reilly, G.C.; Engler, A.J. Intrinsic extracellular matrix properties regulate stem cell differentiation. *J. Biomech.* **2010**, *43*, 55–62. [CrossRef]
22. Zheng, D.; Neoh, K.G.; Shi, Z.; Kang, E.T. Assessment of stability of surface anchors for antibacterial coatings and immobilized growth factors on titanium. *J. Colloid Interface Sci.* **2013**, *406*, 238–246. [CrossRef] [PubMed]

23. Sun, H.; Zhang, Y.; Dou, L.; Song, X.; Sun, C. Surface modification of pure titanium to improve its anti-inflammatory function. *Eur. J. Inflamm.* **2015**, *13*, 204–208. [CrossRef]
24. Kulkarni Aranya, A.; Pushalkar, S.; Zhao, M.; LeGeros, R.Z.; Zhang, Y.; Saxena, D. Antibacterial and bioactive coatings on titanium implant surfaces. *J. Biomed. Mater. Res. A* **2017**, *105*, 2218–2227. [CrossRef] [PubMed]
25. Godoy-Gallardo, M.; Wang, Z.; Shen, Y.; Manero, J.M.; Gil, F.J.; Rodriguez, D.; Haapasalo, M. Antibacterial Coatings on Titanium surfaces: A comparison study between in vitro single-species and multispecies biofilm. *ACS Appl. Mater. Interfaces* **2015**, *7*, 5992–6001. [CrossRef] [PubMed]
26. Aguilar, L.E.; Thomas, R.G.; Moon, M.J.; Jeong, Y.Y.; Park, C.H.; Kim, C.S. Implantable chemothermal brachytherapy seeds: A synergistic approach to brachytherapy using polymeric dual drug delivery and hyperthermia for malignant solid tumor ablation. *Eur. J. Pharm. Biopharm.* **2018**, *129*, 191–203. [CrossRef] [PubMed]
27. Aguilar, L.E.; Tumurbaatar, B.; Ghavaminejad, A.; Park, C.H.; Kim, C.S. functionalized non-vascular nitinol stent via electropolymerized polydopamine thin film coating loaded with bortezomib adjunct to hyperthermia therapy. *Sci. Rep.* **2017**, *7*, 9432. [CrossRef] [PubMed]
28. Ren, Y.; Zhao, X.; Liang, X.; Ma, P.X.; Guo, B. Injectable hydrogel based on quaternized chitosan, gelatin and dopamine as localized drug delivery system to treat Parkinson's disease. *Int. J. Biol. Macromol.* **2017**, *105*, 1079–1087. [CrossRef]
29. Behboodi-Sadabad, F.; Zhang, H.; Trouillet, V.; Welle, A.; Plumeré, N.; Levkin, P.A. UV-triggered polymerization, deposition, and patterning of plant phenolic compounds. *Adv. Funct. Mater.* **2017**, *27*, 1700127. [CrossRef]
30. Iacomino, M.; Paez, J.I.; Avolio, R.; Carpentieri, A.; Panzella, L.; Falco, G.; Pizzo, E.; Errico, M.E.; Napolitano, A.; del Campo, A.; et al. Multifunctional thin films and coatings from caffeic acid and a cross-linking diamine. *Langmuir* **2017**, *33*, 2096–2102. [CrossRef]
31. Kosova, F.; Kurt, F.O.; Olmez, E.; Tuğlu, I.; Arı, Z. Effects of caffeic acid phenethyl ester on matrix molecules and angiogenetic and anti-angiogenetic factors in gastric cancer cells cultured on different substrates. *Biotech. Histochem.* **2016**, *91*, 38–47. [CrossRef]
32. Aguilar, L.E.; Lee, J.Y.; Park, C.H.; Kim, C.S. Biomedical grade stainless steel coating of polycaffeic acid via combined oxidative and ultraviolet light-assisted polymerization process for bioactive implant application. *Polymers* **2019**, *11*, 584. [CrossRef] [PubMed]
33. Ferraris, S.; Miola, M.; Cochis, A.; Azzimonti, B.; Rimondini, L.; Prenesti, E.; Verné, E. In situ reduction of antibacterial silver ions to metallic silver nanoparticles on bioactive glasses functionalized with polyphenols. *Appl. Surf. Sci.* **2017**, *396*, 461–470. [CrossRef]
34. Qing, W.; Chen, K.; Wang, Y.; Liu, X.; Lu, M. Green synthesis of silver nanoparticles by waste tea extract and degradation of organic dye in the absence and presence of H_2O_2. *Appl. Surf. Sci.* **2017**, *423*, 1019–1024. [CrossRef]
35. Kim, M.; Jee, S.-C.; Shinde, S.K.; Mistry, B.M.; Saratale, R.G.; Saratale, G.D.; Ghodake, G.S.; Kim, D.-Y.; Sung, J.-S.; Kadam, A.A. Green-Synthesis of anisotropic peptone-silver nanoparticles and its potential application as anti-bacterial agent. *Polymers* **2019**, *11*, 271. [CrossRef] [PubMed]
36. Fordham, W.R.; Redmond, S.; Westerland, A.; Cortes, E.; Walker, C.; Gallagher, C.; Medina, C.; Waecther, F.; Lunk, C.; Ostrum, R.F.; et al. Silver as a bactericidal coating for biomedical implants. *Surf. Coat. Technol.* **2014**, *253*, 52–57. [CrossRef]
37. Yan, X.; He, B.; Liu, L.; Qu, G.; Shi, J.; Hu, L.; Jiang, G. Antibacterial mechanism of silver nanoparticles in Pseudomonas aeruginosa: Proteomics approach. *Metallomics* **2018**, *10*, 557–564. [CrossRef]
38. Romanò, C.L.; Scarponi, S.; Gallazzi, E.; Romanò, D.; Drago, L. Antibacterial coating of implants in orthopaedics and trauma: A classification proposal in an evolving panorama. *J. Orthop. Surg. Res.* **2015**, *10*, 1157. [CrossRef]
39. Besinis, A.; Hadi, S.D.; Le, H.R.; Tredwin, C.; Handy, R.D. Antibacterial activity and biofilm inhibition by surface modified titanium alloy medical implants following application of silver, titanium dioxide and hydroxyapatite nanocoatings. *Nanotoxicology* **2017**, *11*, 327–338. [CrossRef]
40. Rezk, A.I.; Unnithan, A.R.; Park, C.H.; Kim, C.S. Rational design of bone extracellular matrix mimicking tri-layered composite nanofibers for bone tissue regeneration. *Chem. Eng. J.* **2018**, *350*, 812–823. [CrossRef]

41. Kępa, M.; Miklasińska-Majdanik, M.; Wojtyczka, R.D.; Idzik, D.; Korzeniowski, K.; Smoleń-Dzirba, J.; Wąsik, T.J. Antimicrobial Potential of Caffeic Acid against Staphylococcus aureus Clinical Strains. *Biomed Res. Int.* **2018**, *2018*, 1–9. [CrossRef]
42. Arasoglu, T.; Derman, S.; Mansuroglu, B. Comparative evaluation of antibacterial activity of caffeic acid phenethyl ester and PLGA nanoparticle formulation by different methods. *Nanotechnology* **2016**, *27*, 025103. [CrossRef] [PubMed]
43. Pinho, E.; Soares, G.; Henriques, M. Evaluation of antibacterial activity of caffeic acid encapsulated by beta-cyclodextrins. *J. Microencapsul.* **2015**, *32*, 804–810. [CrossRef] [PubMed]
44. Akdeniz, B.; Sumnu, G.; Sahin, S. Microencapsulation of phenolic compounds extracted from onion (Allium cepa) skin. *J. Food Process. Preserv.* **2018**, *42*, e13648. [CrossRef]
45. Ballesteros, L.F.; Ramirez, M.J.; Orrego, C.E.; Teixeira, J.A.; Mussatto, S.I. Encapsulation of antioxidant phenolic compounds extracted from spent coffee grounds by freeze-drying and spray-drying using different coating materials. *Food Chem.* **2017**, *237*, 623–631. [CrossRef] [PubMed]
46. Gittens, R.A.; Scheideler, L.; Rupp, F.; Hyzy, S.L.; Geis-Gerstorfer, J.; Schwartz, Z.; Boyan, B.D. A review on the wettability of dental implant surfaces II: Biological and clinical aspects. *Acta Biomater.* **2014**, *10*, 2907–2918. [CrossRef] [PubMed]
47. Lima, V.N.; Oliveira-Tintino, C.D.; Santos, E.S.; Morais, L.P.; Tintino, S.R.; Freitas, T.S.; Geraldo, Y.S.; Pereira, R.L.S.; Cruz, R.P.; Menezes, I.R.A.; et al. Antimicrobial and enhancement of the antibiotic activity by phenolic compounds: Gallic acid, caffeic acid and pyrogallol. *Microb. Pathog.* **2016**, *99*, 56–61. [CrossRef] [PubMed]

Article

Antibacterial Activity and Cytotoxicity of Immobilized Glucosamine/Chondroitin Sulfate on Polylactic Acid Films

Ilkay Karakurt [1], Kadir Ozaltin [1], Daniela Vesela [1], Marian Lehocky [1,2,*], Petr Humpolíček [1,2] and Miran Mozetič [3]

[1] Centre of Polymer Systems, University Institute, Tomas Bata University in Zlín, Nam. T.G.M. 5555, 76001 Zlín, Czech Republic
[2] Faculty of Technology, Tomas Bata University in Zlín, Vavreckova 275, 76001 Zlín, Czech Republic
[3] Department of Surface Engineering, Jozef Stefan Institute, Jamova cesta 39, 1000 Ljubljana, Slovenia
* Correspondence: lehocky@post.cz; Tel.: +420-608-616-048

Received: 20 June 2019; Accepted: 12 July 2019; Published: 15 July 2019

Abstract: Polylactic acid (PLA) is one of the most produced polymeric materials, due to its exceptional chemical and mechanical properties. Some of them, such as biodegradability and biocompatibility, make them attractive for biomedical applications. Conversely, the major drawback of PLA in the biomedical field is their vulnerability to bacterial contamination. This study focuses on the immobilization of saccharides onto the PLA surface by a multistep approach, with the aim of providing antibacterial features and evaluting the synergistic effect of these saccharides. In this approach, after poly (acrylic acid) (PAA) brushes attached non-covalently to the PLA surface via plasma post-irradiation grafting technique, immobilization of glucosamine (GlcN) and chondroitin sulfate (ChS) to the PAA brushes was carried out. To understand the changes in surface properties, such as chemical composition, surface topography and hydrophilicity, the untreated and treated PLA films were analyzed using various characterization techniques (contact angle, scanning electron microscopy, X-ray photoelectron spectroscopy). In vitro cytotoxicity assays were investigated by the methyl tetrazolium test. The antibacterial activity of the PLA samples was tested against Escherichia coli and Staphylococcus aureus bacteria strains. Plasma-treated films immobilized with ChS and GlcN, separately and in combination, demonstrated bactericidal effect against the both bacteria strains and also the results revealed that the combination has no synergistic effect on antibacterial action.

Keywords: plasma treatment; surface modification; saccharide immobilization; antibacterial activity; count-plate method; cytotoxicity; chondroitin sulfate

---

1. Introduction

Among all sustainable polymers, polylactic acid (PLA) is one of the most promising and most produced bioplastics due to its superior chemical and mechanical properties, such as biodegradability, good biocompatibility, and low immunogenicity [1–3]. These remarkable properties make this aliphatic polyester attractive for medical and biological applications. The usage of PLA, especially in the biomedical field, is increasing constantly and these days PLA is used to produce blood contacting devices, such as implants for bone fixation, screws, pins, sutures, catheters, drug delivery systems, and so on [2–5]. However, there are a few drawbacks which limit its use in biomedical applications. For instance, due to the presence of pendant methyl groups in structure, PLA is relatively hydrophobic and has low surface energy [1]. The relatively hydrophobic nature of PLA may result in inefficient cell attachment and proliferation, and in some cases, this low cell affinity can lead to inflammation.

Another drawback is the lack of reactive functional groups, which reflects in impeding the specific cell interactions and limits PLA's usage in medical applications [2–6].

The future of PLA utilization in medical applications not only depends on overcoming intrinsic drawbacks, but also on appropriate modification of the surface to achieve new targeted properties or enhance the present features. In this regard, the most important challenges are the activation of the chemically inert PLA surface to reduce hydrophobicity and introduce a new molecules to possess significant physical and mechanical properties, in order to function properly when in contact with blood [7]. Therefore, the material needs appropriate surface treatments without affecting the bulk properties, due to the interactions, which mostly occur at the surface top layer. Some of the commonly used surface treatment methods are gamma irradiation, chemical vapour deposition, ozone treatment, wet chemical methods, UV irradiation, and plasma treatment [8–12]. As compared to other surface treatment methods, plasma technology has obvious advantages, such as non-toxic chemicals, heat-free processing, and modification without changing the bulk properties [13–16].

In the medical field, especially for implants, microbial colonization and biofilm formation are critical clinical problems, which may lead to long-term treatments, replacement of infected implants, amputations, and even patient death [17–19]. In a case when the antibacterial feature is needed, biomedical materials are coated or loaded with various bactericidal compounds. A considerable number of studies have been carried out to provide antibacterial features to polymeric substrates (PLA, PE, PS, PP) with grafting of various antibacterial moieties, such as enzymes [20], saccharides [21], metal ions and nanoparticles [22–24], Vitamin E [18], peptides [25,26], quaternary ammonium salts [27].

Amongst these antibacterial moieties, natural-origin saccharides, such as chondroitin sulfate (ChS) and glucosamine (GlcN) are attractive options, since the carbohydrate moieties can imprint both hydrophilic and biocompatibility properties to the material surface [28].

ChS is the most abundant mucopolysaccharide in all of the vertebrates [29]. This naturally occurring glycosaminoglycan (GAG) is an anionic polyelectrolyte. It has garnered much attention in recent years because of its promising properties, such as non-toxicity, biocompatibility, anti-inflammatory activity [30,31], and its potential as an antibacterial and antiviral agent [32]. In addition, the broad spectrum of the biological activities of ChS suggests a potential benefit to the biomedical field, for example, in osteoarthritis, interstitial cystitis and urinary tract infection treatments, cartilage tissue engineering [31], as dietary supplement, in the drug delivery systems [30,33], and in the development of orthopedic materials.

GlcN is a naturally-produced amino sugar, which is essential for glycoproteins, glycolipids, and glycosaminoglycans (GAG). This positively charged saccharide has anti-inflammatory, antioxidant, anti-ageing, anti-cancer, and antibacterial activities [34–37]. In recent studies, GlcN is observed to be a strong inhibitor to microbial growth, due to its sugar moiety [38,39]. Although the exact mechanism of antibacterial activity of GlcN is still unclear, it is probably caused by the disruption of bacterial cell wall via the free amino group.

GlcN is also used in biomedical applications in combination with ChS to achieve a synergistic effect [40,41]. For example, the efficacy of GlcN and ChS in treatment of symptomatic osteoarthritis of the knee [42], stimulation of vasculogenesis and angiogenesis [43], and treatment of Kashin-Beck disease [44] have been evaluated in some research. However, fewer researches have been conducted on the combination of GlcN and ChS than on either saccharide alone. In addition, the possible different effects of these saccharides and their combination on the antibacterial activity and cytotoxicity have not been fully clarified. To compare the efficacy of the two saccharides in cell viability and inhibition of the microbial growth more researches need to be done.

Thus, the present study aims to develop a PLA-based antibacterial platform using two common saccharides, which also have cell proliferation and adhesion features, and also to compare the antibacterial and cytotoxic activity of GlcN and CS, individually and in combination. While clinical trials and most in vitro studies have focused on the synergistic effects of glucosamine and CS in treatment of osteoarthritis, the present study focuses on the prevention of bacterial growth on polymer biomaterials by using saccharides. For this purpose, low-pressure radio frequency plasma (RF) method was utilized to activate the PLA surfaces and the molecules of GlcN and ChS were immobilized on the surface of the PLA films to improve in antibacterial activity. The bacterial adhesion of surface modified PLA films was systematically investigated by antibacterial activity against Escherichia coli (E. coli) and Staphylococcus aureus (S. aureus).

2. Materials and Methods

Poly(lactic acid) (PLA) 4032 D in pellets form was purchased from Nature Works (Blair, NE, USA). Acrylic acid (99%), sodium metabisulfite (99%), D-glucosamine hydrochloride (99%), chondroitin sulfate A sodium salt (60% balance is chondroitin sulfate C), and sodium hydroxide (98%) were analytical grade reagents and were supplied by Sigma Aldrich (St. Louis, MO, USA).

2.1. Preparation of PLA Films

PLA pellets were first dried in a desiccator at 60 °C overnight to eliminate humidity. Then, to obtain PLA sheets, the compression molding technique was used. PLA pellets were hot-pressed at 180 °C for 20min and then the molding plates were placed in another press for cooling. Sheets of PLA about 430 μm thick and 125 × 125 mm square shaped were obtained and then cut into samples of 25 × 25 mm for further surface treatments. The specimens were rinsed with a detergent solution and subsequently dried at room temperature prior to plasma exposure.

2.2. Plasma Treatment

The PLA films were treated using radio frequency (13.56 MHz), low-pressure plasma equipment (model: PICO Diener, Ebhausen, Germany). Briefly, the sample was first inserted into an evacuated to 60 Pa vacuum chamber and 20 sccm of air flowed into the chamber. Each side of the sample was treated by air plasma generated at power 50 W for 60 s.

2.3. PAA Grafting and Immobilization of Saccharides

Acrylic acid (AAc) solution was prepared by dissolving 10% of AAc in distilled water with sodium metabisulfite. PLA films were immersed immediately into this homopolymer solution after the irradiation step. Following the grafting reaction, the homopolymer was removed, then the samples were placed into the aqueous solution of 1 w% sodium hydroxide for neutralization.

For immobilization of antibacterial agents, AAc grafted samples were immersed into either 1 w% GlcN or 1 w% ChS solutions for 24 h. Thereafter, each sample was washed with deionised water. Subsequently, GlcN immobilized samples were lastly placed into 1 w% ChS solution. The films were then removed, washed in distilled water and left for drying overnight for further characterization. All the process is graphically presented in Figure 1.

2.4. Characterizations of the Samples

The elemental analysis of untreated and treated PLA films was conducted with X-ray photoelectron spectroscopy (XPS) using the Thermo Scientific K-Alpha XPS system (Thermo Fisher Scientific, Loughborough, UK). A monochromic Al Kα X-ray source (1486.6 eV) line was applied as the photoemission excitation with a 400μm spot size at a power of 72 W.

The change in hydrophobicity induced by surface modifications was analysed through static contact angle measurements by the Surface Energy Evaluation System (SEE System; Advex Instruments, Brno, Czech Republic). As a testing liquid, deionized water was used and digital images of a 2 μL water droplet on the surface were captured by the charged-coupled device (CCD) camera system. Three different spots on the droplet's image were determined and the contact angle value was obtained for each reading. Each representative contact angle was calculated by averaging at least 10 separate readings for each sample.

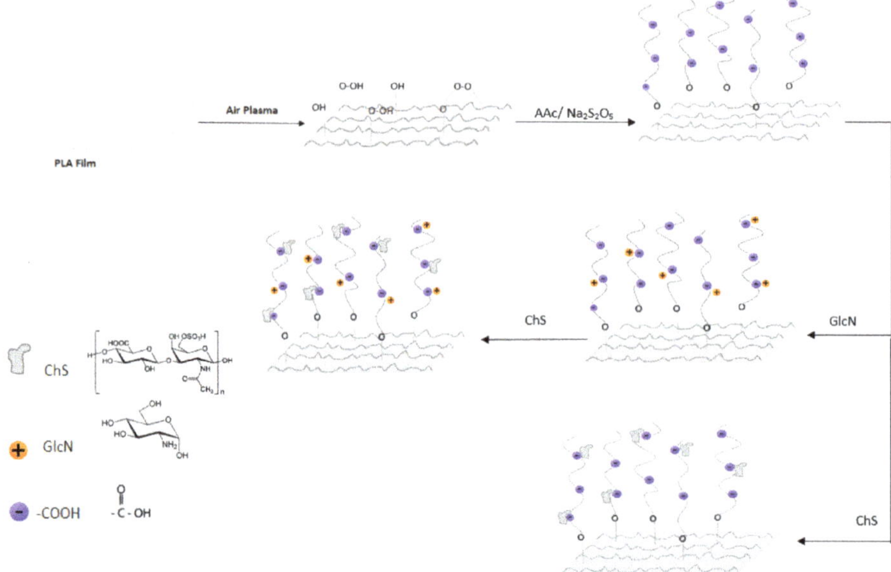

Figure 1. Plasma post-irradiation grafting of Acrylic acid (AAc) onto polylactic acid (PLA) surface followed by immobilization of glucosamine (GlcN) and chondroitin sulfate (ChS) molecules and only ChS immobilization.

The surface morphology of the specimens was observed using a NANOSEM 450 (FEI, Hillsboro, OR, USA) scanning electron microscope at an accelerating voltage of 5.0 kV after sputter coating with gold/palladium.

2.5. Cytotoxicity Testing

The proliferation of mouse embryonic fibroblast cells (ATCC CRL-1658[TM]; NIH/3T3) with the disinfected (30 min of exposure to a UV-radiation source) PLA samples were compared by MTT assay. The ATCC-formulated Dulbecco's Modified Eagle's Medium (Biosera, Nuaille, France) containing 10% of calf serum and $100\,U\,mL^{-1}$ penicillin/streptomycin (PAA Laboratories GmbH, Pasching, Austria) was used as the culture medium. The cells were seeded onto square samples (10 × 10 mm) in concentration 1×10^5 cells/mL in volume 200 µL for 1h incubation. The amount of 800 µL of culture medium was added after 1 h incubation and the cells were incubated at 37 ± 1 °C for 72 h. In this study, the cells cultured on polystyrene were used as a reference.

Cytotoxicity testing was conducted according to the international standard EN ISO 10993-5 with modification. After 72 h incubation, all culture medium was removed and incubated for 4 h with 900µL fresh medium which included 100 µL tetrazolium dye MTT solution $(5\,mg\,mL^{-1})$. Following the removal of all the medium, dimethyl sulfoxide (Merck, Darmstadt, Germany) was added for dissolution of the formed formazan crystals on the sample surfaces. The absorbance was measured at a wavelength of 570 nm (test) and 690 nm (reference) by using an Infinite M200 Pro NanoQuant microplate reader (Tecan, Zürich, Switzerland). Cell viability was shown as the percentage of viable cells after exposure to the tested solutions relative to the reference (100% viability).

2.6. Antibacterial Testing

The antibacterial properties of the PLA samples were quantitatively assessed according to an adapted method from the ISO 22196:2007 protocol [45]. As bacteria strains, Staphylococcus aureus (CCM 4516) and Escherichia coli (CCM 4517) were used.

Briefly, PLA films were sterilized by immersion in 70% ethanol for 24 h. Test bacteria were transferred to nutrient broth (NB) and for using as test inoculations, the number of bacteria was adjusted to certain concentrations, which is between 2.5×10^5 cells/mL and 1×10^6 cells/mL, with dilution. Both treated and untreated samples (25 mm × 25 mm) were placed, in groups of tree, into Petri dishes and 0.4 mL of the test inoculum was pipetted onto each sample, which was then covered with a polypropylene (PP) film (20 × 20 mm). The Petri dishes were incubated for 24 h at 35 °C. After the incubation, both the samples and PP covers were washed with 10 mL of neutralizing broth. The recovered bacterial suspensions were diluted to 10-fold concentrations. Each diluted and undiluted bacterial solution was placed into Petri dishes with 15 mL of count plate agar. The number of bacteria colonies were counted after 24 h of incubation.

3. Results and Discussion

3.1. Surface Chemistry and Morphology

Through static contact angle measurements, surface modification, which induced a change in hydrophilicity, was analyzed (Figure 2). As can be seen, the static water contact angle of untreated PLA was 87°. This makes it highly hydrophobic and difficult for further surface treatment because of lacking polar and functional groups on the PLA surface. The value of the contact angle decreased significantly to 47° after plasma treatment as a result of the presence of plasma-induced hydrophilic oxidative functional groups and surface roughening. The GlcN immobilization to PLA resulted in an increased hydrophobicity but still remained lower than untreated PLA. In contrast, PLA was modified by the ChS and led to a lower contact angle value, which can be connected to the more hydrophilic character derived from hydroxyl, carbonyl and amine groups of ChS. The immobilization of ChS after GlcN decreased the hydrophobicity of only GlcN coated sample. The results preliminarily suggest successful modification of the polymer surfaces, which is supported by the difference between the morphology of untreated and treated PLA surfaces.

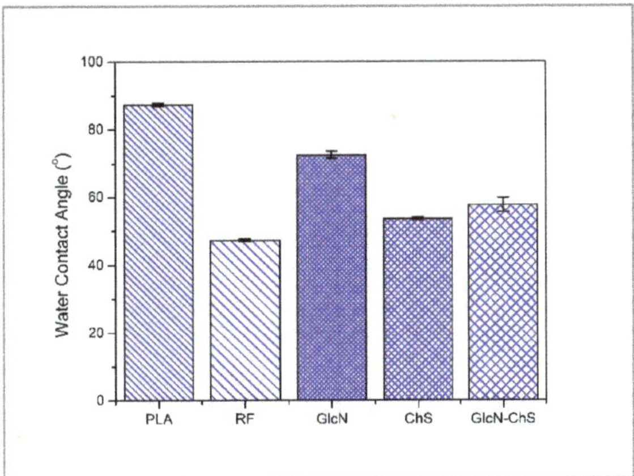

Figure 2. Water contact angles of untreated and treated PLA. Results represent mean value from three independent experiments with standard deviations.

Typical SEM micrographs of the pristine and modified PLA films are illustrated in Figure 3. The scale bar represents 20 µm. It can be seen that the surface of the untreated PLA sample is homogenous, and relatively smooth at a micrometric scale. Due to the compression molding technique, a minor fiber-like features are observed. After plasma treatment, the PLA surface possesses a rougher appearance, as a consequence of the surface functionalization and etching processes. This increased roughness is the desired surface condition for further immobilization steps. When the results are evaluated together with the contact angle values given in Figure 2, it can be said that plasma treatment leads to an improvement in hydrophilicity. Figure 3c–e present the images of the surfaces on which saccharide aggregates are clearly seen. ChS and GlcN were deposited as heterogeneous bidimensional layers on PLA surfaces.

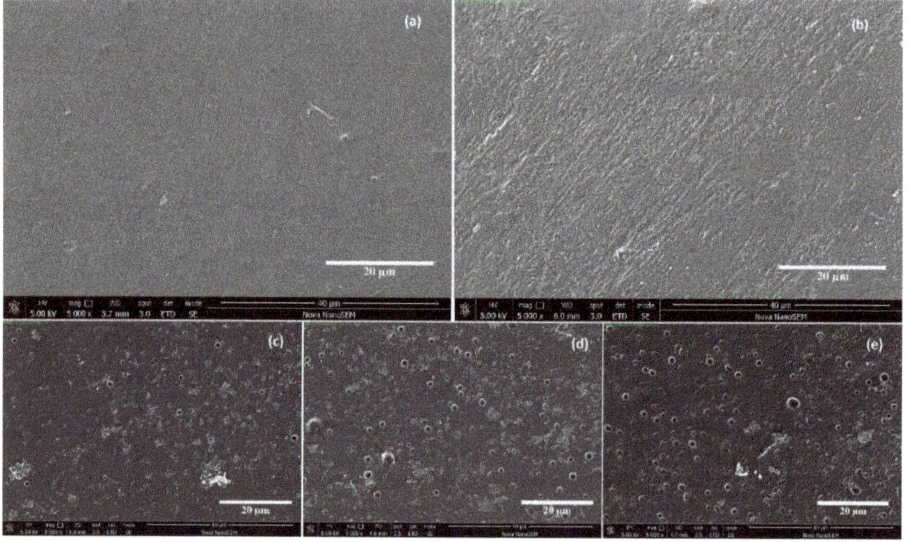

Figure 3. SEM images of the (**a**) untreated PLA; (**b**) plasma treated PLA; (**c**) ChS; (**d**) GlcN; (**e**) GlcN-ChS grafted films.

For the analysis of surface chemical compositions of the PLA surfaces, XPS measurements were carried out before and after the treatments. Figure 4 shows the XPS full spectra of the samples with their corresponding surface elemental compositions and N1 core level spectra. The XPS spectra (Figure 4A) for all samples show two main contributions corresponding to C(1s) at 285 eV and O(1s) located at 533 eV, due to the chemical structure of PLA. After air-plasma treatment, an increase in the peak intensity corresponding to the O(1s) transition can be clearly seen, due to the presence of oxide functional groups. In addition, a small peak that corresponds to the contribution of nitrogen, N(1s), with a binding energy of 400 eV is observed in the plasma treated sample. This nitrogen content stems from the air plasma application, which is mainly composed of oxygen (O) and nitrogen (N) radicals. After GlcN immobilization, it is expected that the nitrogen element appears on the PLA film surface. However, from the XPS spectra of GlcN immobilized PLA films (Figure 4A-d), no obvious N1s peak can be detected. This might result from the low amount of GlcN immobilized on the PLA surface. Figure 4B-c,B-e show the increase in the intensity of the nitrogen peak, which indicates ChS presence on the PLA surface.

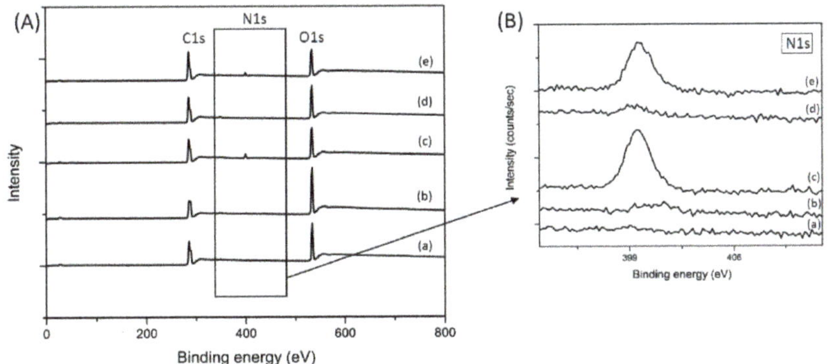

Figure 4. High resolution XPS spectra of (**A**) PLA films (a) untreated PLA; (b) plasma treated PLA; (c) ChS; (d) GlcN; (e) GlcN-ChS grafted films. (**B**) N1 core level spectra of PLA.

The elemental content of carbon, oxygen, nitrogen, and sulfur in each sample is summarized in Table 1. According to the chemical structure of PLA, the XPS analysis of the untreated films indicates that the surface is dominated by carbon (67.8%) and oxygen (31.9%) species. The increased O/C ratio in the RF samples indicates the presence of oxide functional groups just after the air plasma treatment. As shown in the table, untreated PLA films had no nitrogen and sulfur elements, whereas the nitrogen contents were observed for all the other samples, and sulfur contents were found for ChS and GlcN-ChS immobilized samples with the levels of 0.1% and 0.3%, respectively. These nitrogen and sulfur contents are the proof for successful activation of the PLA surface with plasma treatment and grafting of the saccharides.

Table 1. The atomic weight percentage of unmodified and modified polylactic acid (PLA) samples.

Sample Type	Composition (%)				Ratio
	C	**O**	**N**	**S**	**O/C**
PLA	67.8	31.9	-	-	0.47
RF	60.6	38.2	0.9	-	0.63
GlcN	69.4	29.6	0.5	-	0.44
ChS	66.3	28.9	4.4	0.1	0.43
GlcN-ChS	70.7	25.1	3.6	0.3	0.36

3.2. Cytotoxicity of PLA Films

Cell viability was quantified by an MTT assay as an indicator of mitochondrial succinate dehydrogenase enzyme activity. The viability of the reference corresponds to 100% survival of cells in the absence of the tested substances in the cultivation medium. According to the EN ISO 10993-5, values above 80% compared to reference are assigned to 'no cytotoxicity', values from 60% to 80% 'mild cytotoxicity', and values below 40% 'severe cytotoxicity' [46].

Cytotoxicity data are presented in Figure 5. All of the results obtained from the cytotoxicity assay resulted in significant readings as indicated by $p < 0.05$. It can be seen that all of the relative cytotoxicity values (%)-except for untreated PLA-are higher than 80%, independent of polymer modification. One possible reason for this phenomenon is the increased hydrophilicity of untreated PLA with various modifications (plasma treatment, grafting). The GlcN grafted samples display 85% cell viability, while the ChS immobilized samples have more than 120% viability of the cells. For the combination of these two saccharides, the cell viability increases to 148%, which shows better biocompatibility than either saccharide alone.

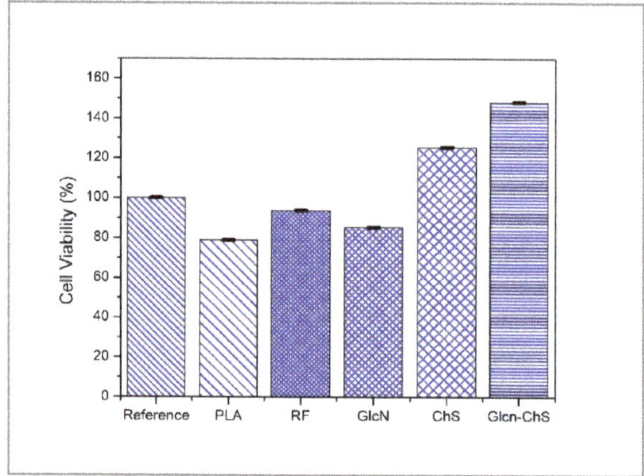

Figure 5. Cytotoxicity of untreated-treated PLA and polystyrene as reference. Error bars represent the standard deviation of the three independent experiments.

Other studies reported the evaluation of toxicity of chondroitin sulfate and glucosamine. Most of the studies claim little or no cytotoxicity of either GlcN or ChS for concentrations below 5.0 mg/mL [47,48]. One study reported the combination of GlcN and ChS resulted in an increase in cellular metabolic activity in chondrocytes monolayer cultures (cell viability 139%) upon 7 days of incubation, which is consistent with our findings [49]. In another study, an enhancement in cell proliferation was found with these two saccharides [50].

3.3. Antibacterial Performance of PLA Films

The antibacterial activity of the PLA films was evaluated against *Staphylococcus aureus* and *Escherichia coli* strains, and analyzed by comparing the number of viable cells in the agar plates after 24h of incubation time. As shown in Table 2, untreated PLA has antibacterial activity against neither bacteria strains. After plasma treatment (RF) a similar number of viable bacteria of untreated PLA is found, which indicates the absence of any bactericidal effect before surface coating with suitable agents. While the best antibacterial activity is observed with only ChS immobilized samples against E. coli bacteria strains (<1 means no colonies recovered), the highest antibacterial activity against the S. aureus strains is exerted by only GlcN attached PLA films with 1.9 cfu/cm^2. The difference in counts between GlcN combined with ChS immobilized samples and separately attached saccharides are small and this combination results in a destruction of more than 99.99% of the inoculation, which generally is accepted as the definition of bactericidal agents [51,52].

Table 2. The number of viable bacteria on the PLA films.

Sample Bacteria	Initial CFU	PLA	RF	GlcN	ChS	GlcN-ChS
S.aureus CCM 4516 N (cfu/cm^2)	2.0×10^6	1.8×10^5	2.1×10^5	1.9	8.8	1.1×10^1
E.coli CCM 4517 N (cfu/cm^2)	2.2×10^7	2.3×10^6	2.2×10^6	7.8	<1	7.7×10^1

4. Conclusions

The antibacterial surface modification of PLA films was achieved through the immobilization of GlcN and ChS on film surfaces via plasma treatment technique, followed by AAc grafting. The contact angle and XPS results verify successful immobilization of the saccharides. In the survey scan XPS spectra increases in characteristic elements (N and S) of ChS and GlcN were observed. SEM images showed that the saccharide aggregates partially covered the PLA film surfaces. The antibacterial testing results demonstrated that PLA films coated with ChS exhibited the highest antimicrobial activity against E. coli. Besides, only-GlcN immobilized PLA films showed the best bactericidal effect against S.aureus. When combined with ChS the degree of both bacteria growth inhibition was still up to 99.99%.

This study definitely proved that the developed GlcN/ChS coated PLA films are excellent bactericide agents against representative gram-positive and gram-negative bacteria. Furthermore, the combination of these two saccharides should be highlighted in the current study, due to increased cell viability, which could make it easier to bring the developed medical devices to the market.

Author Contributions: Conceptualization, I.K., K.O., P.H., M.M. and M.L.; Methodology, I.K. and K.O.; Surface analysis and chemical analysis, I.K. and K.O.; Antibacterial activity test performed, D.V. and I.K. Preparation of the manuscript, review and editing, I.K., K.O., P.H., M.M. and M.L. The manuscript was approved by all authors.

Funding: The authors thank the Internal Grant Agency of Tomas Bata University in Zlín (IGA/CPS/2019/004) for their financial support, the Czech Science Foundation (17-05095S), and the Slovenian Research Agency (L2-8179).

Acknowledgments: I.K. thanks the Internal Grant Agency of Tomas Bata University in Zlín (IGA/CPS/2019/004) for financial support. Authors are also grateful to Ing. Anežka Lengálová, Ph.D. for her kind guidance and helpful suggestions in the preparation of this paper. The authors thank the Czech Science Foundation (17-05095S) and the Slovenian Research Agency (L2-8179).

Conflicts of Interest: The authors declare no conflict of interest.

References

1. Vert, M. After soft tissues, bone, drug delivery and packaging, PLA aims at blood. *Eur. Polym. J.* **2015**, *68*, 516–525. [CrossRef]

2. He, C.; Chen, Q.; Yarmolenko, M.; Rogachev, A.; Piliptsou, D.; Jiang, X.; Rogachev, A. Structure and antibacterial activity of PLA-based biodegradable nanocomposite coatings by electron beam deposition from active gas phase. *Prog. Org. Coat.* **2018**, *123*, 282–291. [CrossRef]

3. Xiao, L.; Wang, B.; Yang, G.; Gauthier, M. Poly(Lactic Acid)-Based Biomaterials: Synthesis, Modification and Applications. In *Biomedical Science Engineering Technology*; Intech Open: Berlin, Germany, 2012.

4. Farah, S.; Anderson, D.G.; Langer, R. Physical and mechanical properties of PLA, and their functions in widespread applications—A comprehensive review. *Adv. Drug Deliv. Rev.* **2016**, *107*, 367–392. [CrossRef] [PubMed]

5. Swilem, A.E.; Lehocký, M.; Humpolíček, P.; Kucekova, Z.; Novák, I.; Mičušík, M.; Abd El-Rehim, H.A.; Hegazy, E.A.; Hamed, A.A.; Kousal, J. Description of D-glucosamine immobilization kinetics onto poly(lactic acid) surface via a multistep physicochemical approach for preparation of novel active biomaterials. *J. Biomed. Mater. Res. Part A* **2017**, *105*, 3176–3188. [CrossRef] [PubMed]

6. Stankevich, K.S.; Danilenko, N.V.; Gadirov, R.M.; Goreninskii, S.I.; Tverdokhlebov, S.I.; Filimonov, V.D. A new approach for the immobilization of poly(acrylic) acid as a chemically reactive cross-linker on the surface of poly(lactic) acid-based biomaterials. *Mater. Sci. Eng. C* **2017**, *71*, 862–869. [CrossRef] [PubMed]

7. Pandiyaraj, K.N.; Ferraria, A.M.; Rego, A.M.; Deshmukh, R.R.; Su, P.; Halleluyah, J.M.; Halim, A.S. Low-pressure plasma enhanced immobilization of chitosan on low-density polyethylene for bio-medical applications. *Appl. Surf. Sci.* **2015**, *328*, 1–12. [CrossRef]

8. Ozaltin, K.; Lehocky, M.; Humpolicek, P.; Pelkova, J.; Martino, A.D.; Karakurt, I.; Saha, P. Anticoagulant Polyethylene Terephthalate Surface by Plasma-Mediated Fucoidan Immobilization. *Polymers* **2019**, *11*, 750. [CrossRef] [PubMed]

9. Popelka, A.; Novák, I.; Lehocký, M.; Junkar, I.; Mozetič, M.; Kleinová, A.; Janigová, I.; Slouf, M.; Bílek, F.; Chodák, I. A new route for chitosan immobilization onto polyethylene surface. *Carbohydr. Polym.* **2012**, *90*, 1501–1508. [CrossRef]

10. Prat, R.; Shi, M.; Clouet, F. Interactions of Cold Plasmas with Polymers and Their Model Molecules: Degradation vs. Functionalzation. *J. Macromol. Sci. Part A* **1997**, *34*, 471–488. [CrossRef]

11. Bolbasov, E.; Rybachuk, M.; Golovkin, A.; Antonova, L.; Shesterikov, E.; Malchikhina, A.; Novikov, V.A.; Anissimov, Y.G.; Tverdokhlebov, S.I. Surface modification of poly(l-lactide) and polycaprolactone bioresorbable polymers using RF plasma discharge with sputter deposition of a hydroxyapatite target. *Mater. Lett.* **2014**, *132*, 281–284. [CrossRef]

12. Yu, D.; Lin, W.; Lin, C.; Yang, M. Cytocompatibility and Antibacterial Activity of a PHBV Membrane with Surface-Immobilized Water-Soluble Chitosan and Chondroitin-6-sulfate. *Macromol. Biosci.* **2006**, *6*, 348–357. [CrossRef]

13. Stoleru, E.; Dumitriu, R.P.; Munteanu, B.S.; Zaharescu, T.; Tănase, E.E.; Mitelut, A.; Ailiesei, G.L.; Vasile, C. Novel procedure to enhance PLA surface properties by chitosan irreversible immobilization. *Appl. Surf. Sci.* **2016**, *367*, 407–417. [CrossRef]

14. Saxena, S.; Ray, A.R.; Gupta, B. Chitosan immobilization on polyacrylic acid grafted polypropylene monofilament. *Carbohydr. Polym.* **2010**, *82*, 1315–1322. [CrossRef]

15. Chang, S.; Chian, C. Plasma surface modification effects on biodegradability and protein adsorption properties of chitosan films. *Appl. Surf. Sci.* **2013**, *282*, 735–740. [CrossRef]

16. Maslakci, N.N.; Ulusoy, S.; Oksuz, A.U. Investigation of the effects of plasma-treated chitosan electrospun fibers onto biofilm formation. *Sens. Actuators B Chem.* **2017**, *246*, 887–895. [CrossRef]

17. Ando, Y.; Miyamoto, H.; Noda, I.; Sakurai, N.; Akiyama, T.; Yonekura, Y.; Shimazaki, T.; Miyazaki, M.; Mawatari, M.; Hotokebuchi, T. Calcium phosphate coating containing silver shows high antibacterial activity and low cytotoxicity and inhibits bacterial adhesion. *Mater. Sci. Eng. C* **2010**, *30*, 175–180. [CrossRef]

18. Campoccia, D.; Visai, L.; Renò, F.; Cangini, I.; Rizzi, M.; Poggi, A.; Montanaro, L.; Rimondini, L.A.; Arciola, C.R. Bacterial adhesion to poly-(d,l)lactic acid blended with vitamin E: Toward gentle anti-infective biomaterials. *J. Biomed. Mater. Res. Part A* **2014**, *103*, 1447–1458. [CrossRef] [PubMed]

19. Barton, A.J.; Sagers, R.D.; Pitt, W.G. Bacterial adhesion to orthopedic implant polymers. *J. Biomed. Mater. Res.* **1996**, *30*, 403–410. [CrossRef]

20. Hawser, S.; Lociuro, S.; Islam, K. Dihydrofolate reductase inhibitors as antibacterial agents. *Biochem. Pharmacol.* **2006**, *71*, 941–948. [CrossRef]

21. Munteanu, N.S.; Pâslaru, E.; Zemljič, L.F.; Anamaria, S.; Pricope, G.M.; Vasile, C. Chitosan coatings applied to polyethylene surface to obtain food-packaging materials. *Cell. Chem. Technol.* **2014**, *48*, 565–575.

22. Machovsky, M.; Kuritka, I.; Bazant, P.; Vesela, D.; Saha, P. Antibacterial performance of ZnO-based fillers with mesoscale structured morphology in model medical PVC composites. *Mater. Sci. Eng. C* **2014**, *41*, 70–77. [CrossRef] [PubMed]

23. Badaraev, A.; Nemoykina, A.; Bolbasov, E.; Tverdokhlebov, S. PLLA scaffold modification using magnetron sputtering of the copper target to provide antibacterial properties. *Resour.-Eff. Technol.* **2017**, *3*, 204–211. [CrossRef]

24. Silva, F.D.; Cinca, N.; Dosta, S.; Cano, I.; Guilemany, J.; Caires, C.; Lima, A.R.; Silva, C.M.; Oliveira, S.L.; Caires, A.R.; et al. Corrosion resistance and antibacterial properties of copper coating deposited by cold gas spray. *Surf. Coat. Technol.* **2019**, *361*, 292–301. [CrossRef]

25. Li, X.; Li, P.; Saravanan, R.; Basu, A.; Mishra, B.; Lim, S.H.; Su, X.; Tambyah, P.A.; Leong, S.S. Antimicrobial functionalization of silicone surfaces with engineered short peptides having broad spectrum antimicrobial and salt-resistant properties. *Acta Biomater.* **2014**, *10*, 258–266. [CrossRef] [PubMed]

26. Santiago-Ortiz, L.; Hitchner, M.; Palmer, T.; Caputo, G.A. Characterization of a Histidine Containing Antimicrobial Peptide with pH Dependent Activity. *Biophys. J.* **2019**, *116*, 83a. [CrossRef]

27. Li, Z.; Yang, X.; Liu, H.; Yang, X.; Shan, Y.; Xu, X.; Shang, S.; Song, Z. Dual-functional antimicrobial coating based on a quaternary ammonium salt from rosin acid with in vitro and in vivo antimicrobial and antifouling properties. *Chem. Eng. J.* **2019**, *374*, 564–575. [CrossRef]

28. Russo, L.; Gloria, A.; Russo, T.; Damora, U.; Taraballi, F.; Santis, R.D.; Ambrosio, L.; Nicotra, F.; Cipolla, L. Glucosamine grafting on poly(ε-caprolactone): A novel glycated polyester as a substrate for tissue engineering. *RSC Adv.* **2013**, *3*, 6286. [CrossRef]

29. Dawlee, S.; Sugandhi, A.; Balakrishnan, B.; Labarre, D.; Jayakrishnan, A. Oxidized Chondroitin Sulfate-Cross-Linked Gelatin Matrixes: A New Class of Hydrogels. *Biomacromolecules* **2005**, *6*, 2040–2048. [CrossRef] [PubMed]

30. Yeh, M.; Cheng, K.; Hu, C.; Huang, Y.; Young, J. Novel protein-loaded chondroitin sulfate–chitosan nanoparticles: Preparation and characterization. *Acta Biomater.* **2011**, *7*, 3804–3812. [CrossRef] [PubMed]

31. Yuan, H.; Xue, J.; Qian, B.; Chen, H.; Zhu, Y.; Lan, M. Preparation and antifouling property of polyurethane film modified by chondroitin sulfate. *Appl. Surf. Sci.* **2017**, *394*, 403–413. [CrossRef]

32. Burge, K.Y.; Hannah, L.; Eckert, J.V.; Gunasekaran, A.; Chaaban, H. The Protective Influence of Chondroitin Sulfate, a Component of Human Milk, on Intestinal Bacterial Invasion and Translocation. *J. Hum. Lact.* **2019**, *35*, 538–549. [CrossRef]

33. Tóth, I.Y.; Illés, E.; Szekeres, M.; Tombácz, E. Preparation and characterization of chondroitin-sulfate-A-coated magnetite nanoparticles for biomedical applications. *J. Magn. Magn. Mater.* **2015**, *380*, 168–174. [CrossRef]

34. Dalirfardouei, R.; Karimi, G.; Jamialahmadi, K. Molecular mechanisms and biomedical applications of glucosamine as a potential multifunctional therapeutic agent. *Life Sci.* **2016**, *152*, 21–29. [CrossRef]

35. Jeong, K.; Ahn, K.; Lee, B.I.; Lee, C.; Kim, S. The mechanism of transglutaminase 2 inhibition with glucosamine: Implications of a possible anti-inflammatory effect through transglutaminase inhibition. *J. Cancer Res. Clin. Oncol.* **2009**, *136*, 143–150. [CrossRef]

36. Xing, R.; Liu, S.; Guo, Z.; Yu, H.; Li, C.; Ji, X.; Feng, J.; Li, P. The antioxidant activity of glucosamine hydrochloride in vitro. *Bioorg. Med. Chem.* **2006**, *14*, 1706–1709. [CrossRef]

37. Rozin, A.P. Glucosamine sulfate—Environmental antibacterial activity. *Clin. Rheumatol.* **2009**, *28*, 1221–1223. [CrossRef]

38. Appelt, H.R.; Oliveira, J.S.; Santos, R.C.; Rodrigues, O.E.; Santos, M.Z.; Heck, E.F.; Rosa, L.C. Synthesis and Antimicrobial Activity of Carbohydrate Based Schiff Bases: Importance of Sugar Moiety. *Int. J. Carbohydr. Chem.* **2013**, *2013*, 320892. [CrossRef]

39. Malik, S.; Singh, M.; Mathur, A. Antimicrobial Activity of Food Grade Glucosamine. *Int. J. Biotechnol. Bioeng. Res.* **2013**, *4*, 307–312.

40. Calamia, V.; Mateos, J.; Fernández-Puente, P.; Lourido, L.; Rocha, B.; Fernández-Costa, C.; Montell, E.; Verges, J.; Ruiz-Romero, J.; Blanco, F.J. A pharmacoproteomic study confirms the synergistic effect of chondroitin sulfate and glucosamine. *Sci. Rep.* **2014**, *4*, 5069. [CrossRef]

41. Lippiello, L.; Woodward, J.; Karpman, R.; Hammad, T.A. In Vivo Chondroprotection and Metabolic Synergy of Glucosamine and Chondroitin Sulfate. *Clin. Orthop. Relat. Res.* **2000**, *381*, 229–240. [CrossRef]

42. Glucosamine, Chondroitin Sulfate, and the Two in Combination for Painful Knee Osteoarthritis. *Obstet. Gynecol.* **2006**, *107*, 1415. [CrossRef]

43. De Souza Lins Borba, F.K.; Felix, G.L.; Costa, E.V.; Silva, L.; Dias, P.F.; Nogueira, R.D. Fractal analysis of extra-embryonic vessels of chick embryos under the effect of glucosamine and chondroitin sulfates. *Microvasc. Res.* **2016**, *105*, 114–118. [CrossRef]

44. Yue, J.; Yang, M.; Yi, S.; Dong, B.; Li, W.; Yang, Z.; Lu, Z.; Zhang, R.; Yong, J. Chondroitin sulfate and/or glucosamine hydrochloride for Kashin-Beck disease: A cluster-randomized, placebo-controlled study. *Osteoarthr. Cartil.* **2012**, *20*, 622–629. [CrossRef]

45. *ISO 22196:2007-Plastics-Measurement of Antibacterial Activity on Plastics Surfaces*; ISO: Geneva, Switzerland, 2007.

46. *ISO 10993-5:2009-Biological Evaluation of Medical Devices*; ISO: Geneva, Switzerland, 2009.

47. Graeser, A.; Giller, K.; Wiegand, H.; Barella, L.; Saadatmandi, C.B.; Rimbach, G. Synergistic Chondroprotective Effect of α-Tocopherol, Ascorbic Acid, and Selenium as well as Glucosamine and Chondroitin on Oxidant Induced Cell Death and Inhibition of Matrix Metalloproteinase-3—Studies in Cultured Chondrocytes. *Molecules* **2009**, *15*, 27–39. [CrossRef]

48. Lv, C.; Wang, L.; Zhu, X.; Lin, W.; Chen, X.; Huang, Z.; Huang, L.; Yang, S. Glucosamine promotes osteoblast proliferation by modulating autophagy via the mammalian target of rapamycin pathway. *Biomed. Pharmacother.* **2018**, *99*, 271–277. [CrossRef]

49. Bascoul-Colombo, C.; Garaiova, I.; Plummer, S.F.; Harwood, J.L.; Caterson, B.; Hughes, C.E. Glucosamine Hydrochloride but Not Chondroitin Sulfate Prevents Cartilage Degradation and Inflammation Induced by Interleukin-1α in Bovine Cartilage Explants. *Cartilage* **2015**, *7*, 70–81. [CrossRef]

50. Montell, E.; Contreras-Muñoz, P.; Torrent, A.; Varga, M.D.; Rodas, G.; Marotta, M. Mechanisms of action of chondroitin sulfate and glucosamine in muscle tissue: In vitro and in vivo results. a new potential treatment for muscle injuries? *Ann. Rheum. Dis.* **2018**, *77*, 1228–1229.

51. Pankey, G.A.; Sabath, L.D. Clinical Relevance of Bacteriostatic versus Bactericidal Mechanisms of Action in the Treatment of Gram-Positive Bacterial Infections. *Clin. Infect. Dis.* **2004**, *38*, 864–870. [CrossRef]

52. Augusta, S.; Gruber, H.F.; Streichsbier, F. Synthesis and antibacterial activity of immobilized quaternary ammonium salts. *J. Appl. Polym. Sci.* **1994**, *53*, 1149–1163. [CrossRef]

 *polymers*

Article

Growth on Metallo-Supramolecular Coordination Polyelectrolyte (MEPE) Stimulates Osteogenic Differentiation of Human Osteosarcoma Cells (MG63) and Human Bone Marrow Derived Mesenchymal Stem Cells

Janina Belka [1], Joachim Nickel [2,3],* and Dirk G. Kurth [1],*

[1] Lehrstuhl für Chemische Technologie der Materialsynthese, Universität Würzburg, Röntgenring 11, D-97070 Würzburg, Germany

[2] Lehrstuhl für Tissue Engineering und Regenerative Medizin, Universitätsklinikum Würzburg, Röntgenring 11, D-97070 Würzburg, Germany

[3] Fraunhofer ISC, Translationszentrum für Regenerative Therapien (TLZ-RT), Röntgenring 11, D-97070 Würzburg, Germany

* Correspondence: joachim.nickel@uni-wuerzburg.de (J.N.); dirk.kurth@matsyn.uni-wuerzburg.de (D.G.K.); Tel.: +49-(0)931-3184122 (J.N.); +49-(0)931-3182631 (D.G.K.)

Received: 24 May 2019; Accepted: 26 June 2019; Published: 27 June 2019

Abstract: Background: Culturing of cells is typically performed on standard tissue culture plates generating growth conditions, which in general do not reflect the native three-dimensional cellular environment. Recent investigations provide insights in parameters, which strongly affect the general cellular behavior triggering essential processes such as cell differentiation. The physical properties of the used material, such as stiffness, roughness, or topology, as well as the chemical composition of the cell-surface interface are shown to play a key role in the initiation of particular cellular responses. Methods: We extended our previous research, which identified thin films of metallo-supramolecular coordination polyelectrolytes (MEPEs) as substrate to trigger the differentiation of muscular precursor cells. Results: Here, we show that the same MEPEs similarly stimulate the osteogenic differentiation of pre-osteoblasts. Remarkably, MEPE modified surfaces also trigger the differentiation of primary bone derived mesenchymal stem cells (BMSCs) towards the osteogenic lineage. Conclusion: This result leads to the conclusion that these surfaces individually support the specification of cell differentiation toward lineages that correspond to the natural commitment of the particular cell types. We, therefore, propose that Fe-MEPEs may be used as scaffold for the treatment of defects at least in muscular or bone tissue.

Keywords: cell differentiation; metallo-supramolecular polymer; interface; iron metabolism

1. Introduction

In a previous study, we investigated the influence of Fe-MEPE modified surfaces on the differentiation of pre-myoblastic C2C12 cells in standard cell culture medium [1]. C2C12 cells differentiate towards myotubes after being seeded onto Fe-MEPE modified surfaces without the need to add agents that promote differentiation. In case of C2C12 cells, if grown on standard cell culture plates, myogenic differentiation is typically initiated by serum starvation [2]. To verify the hypothesis that culturing cells on Fe-MEPE modified surfaces might in general stimulate cell differentiation towards pre-determined differentiation lineages, we extended our investigations on differentiation of the human osteosarcoma cell line MG63 and that of primary human bone marrow derived mesenchymal stem cells (BMSCs).

Using standard cell culture plates differentiation of MG63 cells generally requires addition of special agents such as ascorbic acid and dexamethasone [3,4]. Standard protocols for the osteogenic differentiation of primary pluripotent BMSCs also utilize dexamethasone and ascorbic acid, but may also require addition of bone morphogentic protein (BMP)2 and β-glycerophosphate [5]. Dexamethasone primarily promotes cell proliferation whereas ascorbic acid induces expression of alkaline phosphatase (ALP) and osteocalcin [6,7]. Despite of their osseous origin, BMSCs can be differentiated towards different lineages in vitro such as osteoblasts, chondrocytes, astrocytes, neurons, skeletal, and cardiac muscles [8–11]. Even though these cells are pluripotent, they are committed to differentiate towards bone cells.

As a general sign of differentiation, cellular parameters such as cell growth and mitochondrial activities are assessed since high mitochondrial activities at low cell count indicates cell differentiation. The osteogenic differentiation process is characterized by several steps. At first, the cells adhere and proliferate. Subsequently, the extracellular matrix is formed, which is finally followed by mineralization. Collin et al. considered clustering of cells before matrix formation as a separate step [12].

During the formation of extracellular matrix the osteoblasts produce collagen type I, osteocalcin, osteopontin, and alkaline phosphatase (ALP), which drastically increases its activity at the beginning of bone matrix mineralization [13]. The high activity of ALP leads to an increased release of phosphate, which forms the mineral part of the bone with free calcium ions [14]. At onset of osteogenic differentiation, ALP is up-regulated, whereas osteocalcin is primarily expressed at a later stage upon mineralization. According to Collin et al., ALP is expressed at the beginning and at the end of these differentiation processes [12]. Owen et al. assume that ALP increases linearly during differentiation and drastically increases before mineralization [15]. With the beginning of mineralization, osteocalcin reaches its maximum of expression [16]. According to Bronckers et al. increased osteocalcin expression occurs in osteoblasts and osteocytes [17].

However, it is known that primary cells do not differentiate synchronously in vitro [18]. Thus, in a pool of cells, both ALP and osteocalcin expression may be detected at the same time. Additionally, cell differentiation is dependent on the site of cell collection, the number of residing cells, the gender of the donor, and on cell extraction and purification methods [19]. However, a detailed understanding of the orchestrated interaction and regulation of operating signaling pathways and regulating factors involved in these differentiation processes is not available yet.

Several studies on the effects of metal ions on cellular systems have been published in the past [20]. Concerning osseointegration metal ions were shown to play an important role in the processes of angiogenesis, osteogenesis, and mineralization of bone tissue. Above all, Cu(II)- and Co(II)-ions stimulate the secretion of certain growth factors, e.g., vascular endothelial growth factor (VEGF) via hypoxia inducible factor (HIF)-1, subsequently stimulating proliferation of endothelial cells and, thus, the formation of blood vessels [21]. The cations Zn(II) and Sr(II) instead stimulate osteoblast proliferation while inhibiting osteoclast activity [22,23]. These ions thus have a stimulating or anabolic effect on bone homeostasis. The cations Ca(II) and Mg(II) stimulate osteoblast proliferation. The latter also stimulates cell adhesion via binding of integrins to extracellular matrix proteins [24].

In this study, we investigated the effect of Fe-MEPE modified surfaces on the proliferation and differentiation of MG63 cells and BMSCs. As in our previous publication, the metallo-supramolecular coordination polyelectrolyte (MEPE) based on the ditopic ligand 1,4-bis (2,2′:6′,2′-terpyridin-4′-yl) benzene and Fe(II)-ions was used [1]. As several methods can be employed in order to deposit the positively charged Fe-MEPE on surfaces, we used the layer-by-layer deposition method and dip-coating from solution [25–30]. Due to the dynamic nature of the interaction of terpyridines and metal ions, such as Fe(II), Ni(II), or Co(II), MEPEs may represent ideal metal ion release systems. The released ions may stimulate specific cellular responses in target cells, which adhere to the modified surface.

2. Materials and Methods

2.1. Surface Modification

2.1.1. Layer by Layer

Layer-by-layer coating was carried out using the method described earlier [27]. Circular white borosilicate glass slides with a diameter of 18 mm were washed with ethanol and dried with compressed air before. The polyelectrolytes were dissolved in aqueous 0.1 M sodium acetate solution to improve the layer thickness by weakening the intra-molecular electrical repulsion of the polymer chains [31]. First, a layer of polyethyleneimine (PEI, Fluka) was deposited on the glass slide using a coating solution with a concentration of 10^{-2} M. For the formation of the second layer, a 1×10^{-3} M poly-(styrene sulfonate) (PSS) solution was used. The final MEPE-layer was adsorbed using a 2.1×10^{-3} M Fe-MEPE solution. An incubation time of 4 minutes was chosen. Between the coating steps, the substrates were rinsed with ultrapure water. After the application of the last layer, the samples were dried with compressed air [27,29]. The resulting LbL sample consisted of a PEI, a PSS, and a final Fe-MEPE Layer.

2.1.2. Dip Coating

Glass slides were washed with deionized water and finally with absolute ethanol. Fe-MEPE was dissolved in absolute ethanol with a concentration of 8.8×10^{-3} M [28]. Dip-coating was performed at ambient temperature by pulling the immersed glass slides at constant speed of 10 mm/min (Dip10) or 50 mm/min (Dip50) out of the coating solution. The slides were air-dried at ambient temperature.

2.2. Cell Culture and Biological Activity Testing

The MG63 (ATCC Number CRL-1427) and hMSCs (Donor 46: male, 73 years, Donor 54: female, 61 years, Donor 56: female, 75 years) cells were cultured in Dulbecco's Modified Eagle's Medium containing 10% FCS and 1% Penicilin/Streptomycin. For all experiments, 10,000 cells/cm^2 were seeded onto the different substrates. Analysis of cell viability and -proliferation of the MG63 was performed 3 and 5 days after seeding. The hMSCs were analyzed after 2, 5, 7, and 9 days. To differentiate the two parameters, proliferation and metabolic activity, the cells were incubated with Dulbecco's Modified Eagle's Medium containing WST-1 (Roche). The metabolic activity was measured with an ELISA microplate reader (TECAN infinite M200) at 450 nm. After measurement, the cells were trypsinized and the cell number determined using a Neubauer counting chamber.

2.3. RNA Isolation and cDNA Synthesis

For RNA isolation, MG63 cells and hMSCs were trypsinized, subsequently washed with PBS, and centrifuged. RNA from the pelleted cells was isolated using the RNeasy Micro Kit (Qiagen, D-40724, Germany) according to the manufacturer's instructions. RNA concentration and purity was determined by spectralphotometry at wavelengths of 260 nm and 280 nm, respectively. 500 ng of each RNA sample was reversely transcribed using the iScriptTM cDNA Syntheses Kit (BioRad, Hercules, CA, USA), according to the manufacturer's recommendations.

2.4. Quantitative RT-PCR

Differentially expressed genes were quantified by qRT-PCR. The used primers are: hGAPDHsense, 5′-TGACGCTGGGGCTGGCATTG-3′ and hGAPDHantisense, 5′-GCTCTTGCTGGGGCTGGTGG-3′, ALPsense, 5′-CTTGACCTCCTCGGAAGACACTC-3′ and ALPantisense 5′-GCCTGGTAGTTGTTGTG AGCATAG-3′, Osteocalcinsense 5′-TTGGACACAAAGGCTGCAC-3′, and Osteocalcinantisense, 5′-CTCACACTCCTCGCCCTATT-3′.

PCR-reactions were carried out using a C1000TM Thermal Cycler including a CFX96 real-time monitoring system according to Belka et al. [1].

2.5. Colorimetric Determination of Iron(II)

The iron content of the cells was determined by a colorimetric assay based on ferrozine complexation. With this method it is possible to detect iron contents with a lower limit of 0.2 nmol. In brief, Fe(III)- and Fe(II)-ions have at first to be removed from proteins and the cellular network. For that the cells were washed thoroughly and lysed before the iron could be removed quantitatively from iron-loaded proteins like ferritin or heme-proteins. Subsequently, all iron species were reduced since ferrozine only forms complexes with iron (II). Cell lysis and complex formation by ferrozine is described in further studies [1].

3. Results

3.1. Preparation and Characterization of Modified Surfaces for Cell Growth

It is well understood that the physico-chemical properties of an interface greatly affect interactions with cells [32]. Therefore, meticulous attention is paid to the preparation and characterization of modified surfaces. The surface modification and characterization follows the experimental protocol established in a previous study [1]. Two methods of surface modifications are employed for the current study. First, the layer by layer deposition is used to deposit the metallo-polymer on the surface [27,29,33]. The final films are composed of a polyethyleneimine primer, poly-(styrene sulfonate) (PSS), and finally Fe-MEPE. Second, Fe-MEPE is directly applied from solution onto the substrate by dip-coating [28]. The final layer thickness amounts to 25 ± 1 nm (FeDIp10) and 43 ± 2 nm (FeDip50). The roughness of the modified surfaces is in the range of 3–7 nm [1]. The mean values of the contact angles are 69 ± 2° for the Dip10 and 67 ± 2° for the Dip50 substrates, indicating that both substrates are hydrophilic (data not shown) [34]. The contact angle of the LbL surface is 57 ± 2° [1].

3.2. Cell-Substrate Interactions

First, we investigated the cell number and the metabolic activity of the cell line MG63 to assess the cytotoxicity of Fe-MEPE in relation to the TCPs-reference (tissue culture polystyrene) surface. In Figure 1, cell number (A) and activities (B) of MG63 cells grown on the functionalized surface are shown for day 3 and 5.

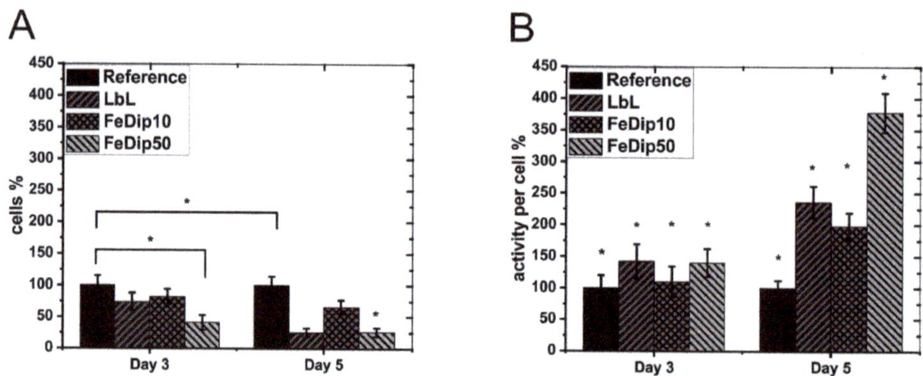

Figure 1. (**A**) Cell number and (**B**) metabolic activities of the used MG63 cells was determined at day 3 and 5 for the particular surface modifications. The cell number and the metabolic activity were set in relation to cells grown on tissue culture polystyrene (TCPS) substrates (reference). The experiments were performed at least two times in triplicate. * $p < 0.5$ (analyzed by one-way ANOVA with Tukey test).

As can be seen, application of Fe-MEPE modified surfaces had a significant influence on both, cell number and cell activity of the used osteosarcoma cells. At both days, the evaluated cell number for the functionalized substrates was below those of the reference substrate. The activity of the FeDip10 modified surface was 82 ± 12%, which is within the non-cytotoxicity limit defined for the range of 81–100%. In the case of the FeDip50 layer, a significantly reduced proliferation at day 3 can be observed, thus classifying this surface as cytotoxic with a degree of 3 [35]. Relative to the reference substrate, the cell number for the FeDip50 modified surface dropped from 41 ± 12% at day 3 to 26 ± 7% at day 5. The relative cell number on the LbL-modified surface decreased similarly from day 3 to day 5. Here, the values decreased from originally 74 ± 14% to 25 ± 7%. The FeDip10 modified surfaces showed moderately reduced cell proliferation from day 3 to 5, with values ranging from 82 ± 12% to 66 ± 11% compared to the reference substrate.

Comparing the cell activity on the basis of the mitochondrial activity per cell Figure 1B shows that the cells stopped proliferation in favor of metabolic activity, reaching 378 ± 31% at day 5 for FeDip50 modified surfaces compared to cells grown on the TCP reference surface. Cells grown on LbL modified surfaces showed an increase in metabolic activity of 139 ± 2 5% and FeDip10 modified surfaces of 98 ± 21%, respectively. These high metabolic cell activities suggest that cells grown on Fe-MEPE modified surfaces might have been stimulated to differentiate most likely to the osteogenic lineage. To confirm this hypothesis, alizarin red staining was carried out in order to detect a potentially increased Ca(II)-ion storage in the cells. Figure 2 shows microscopy images of the alizarin red stained cells at day 3.

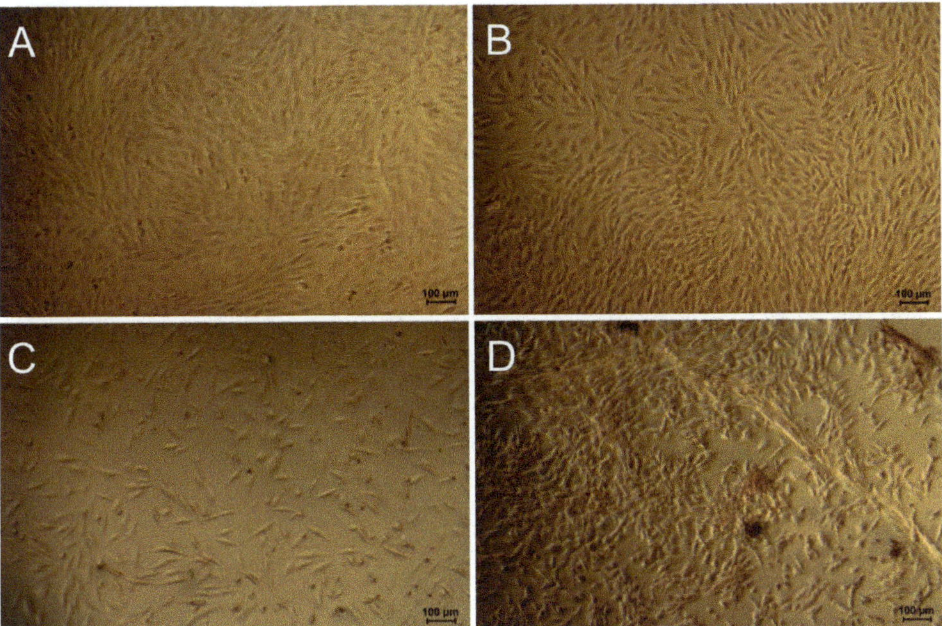

Figure 2. Light microscopy images of Alizarin red stained MG63 cells. (**A**) Cells grown on Reference substrate, (**B**) LbL, (**C**) FeDip10, and (**D**) FeDip50 at day 3.

The overall morphology and the cell number of cell grown on the reference substrates differed greatly as seen in the images taken from the individual surfaces. On the FeDip10 modified surfaces cells did not reach confluency, while cells grown on the reference substrate appeared confluent.

The low cell numbers are better visualized in Figure 3, showing Alizarin red stained cells at two different magnifications at day 5.

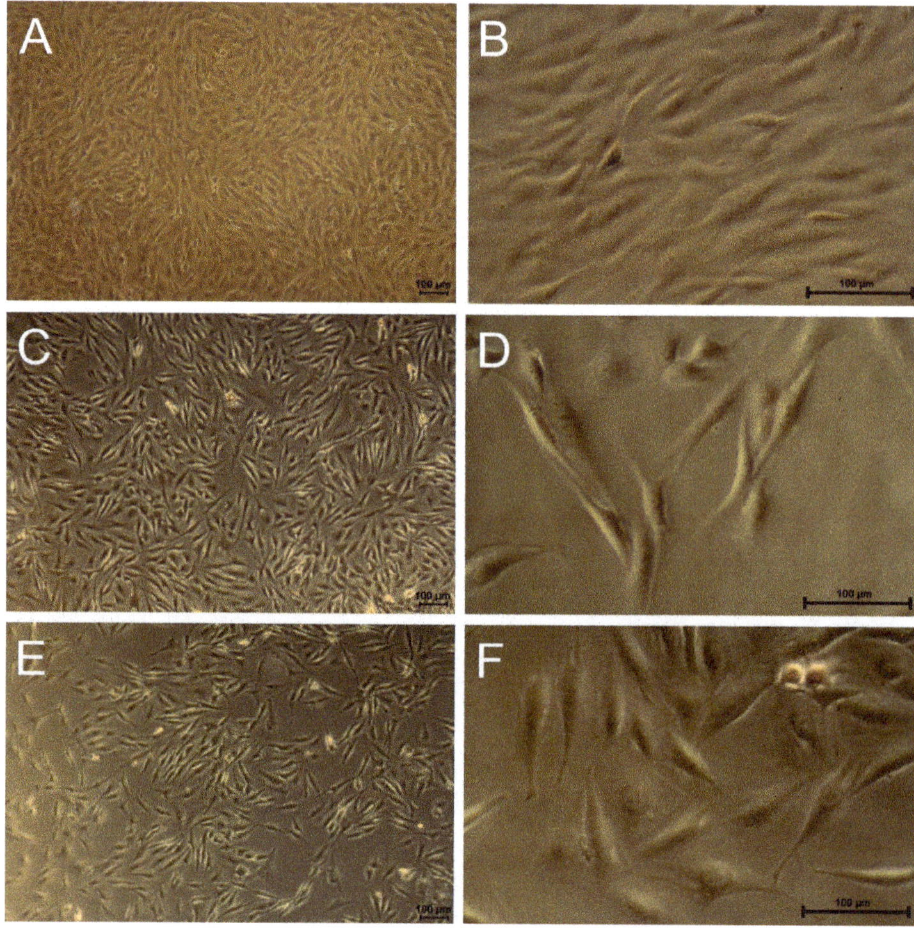

Figure 3. Light microscopy images at different magnification of Alizarin red stained MG63 cells grown on (**A,B**) reference substrates, (**C,D**) FeDip10, and (**E,F**) on FeDip50 substrates at day 5.

Based on cell morphology, cluster formation correlating with Alizarin red staining indicates osteogenic differentiation on the FeDip50 modified surfaces already on day 3, which represents the second differentiation stage according to Collin et al. [12].

By detachment and reseeding after growth for three days on Fe-MEPE modified surfaces, the already-formed clusters were disrupted, and thus, the cells had to re-initiate themselves. Due to the very low relative proliferation rates of the MG63 cells on Fe-MEPE modified substrates (see Figure 1) only small cell deposits on dip-coated Fe-MEPE modified surfaces can be seen at day 5. However, cells significantly differing in the overall morphology can be observed (Figure 3F), which were also characterized by increased Ca(II)-ion deposition (Figure 2D). On all Fe-MEPE modified surfaces, the cells adhere strongly with extended cell protrusions. As expected and in agreement with previous experiments, MG63 cells differentiate according to their initial origin (osteogenic) on Fe-MEPE modified surfaces [1]. In order to determine whether the used Fe-MEPE modified surfaces support in general differentiation process of already committed cells, we expanded our experiments by including human primary cells, so-called bone marrow derived mesenchymal stem cells (BMSCs) obtained from the spongiosa of femur bones.

As in the previously reported experiments, BMSCs are seeded and grown for up to 9 days on LbL, FeDip10 and FeDip50 modified surfaces. As in Figures 1 and 2, cell count and cell activity of BMSCs derived from three donors which are grown on LbL, FeDip10 and FeDip50 modified surfaces as shown in Figures 4 and 5.

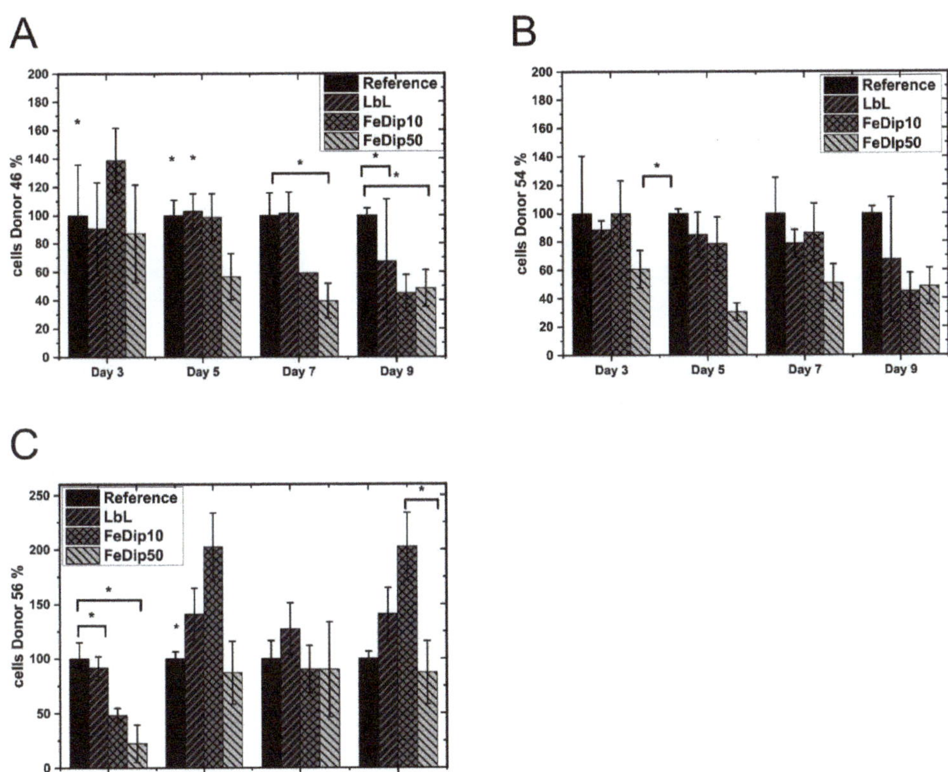

Figure 4. Cell growth of the used bone derived mesenchymal stem cells (BMSCs) from (**A**) Donor 46, (**B**) Donor 54, and (**C**) Donor 56 relative to that on the reference surface for the particular surface modifications at four different time points. All experiments were performed in triplicate. * $p < 0.5$ (analyzed by one-way ANOVA with Tukey test).

In general, similar results were obtained regarding proliferation of cells obtained from donor 46 and 54 on the differently modified surfaces compared to the reference surface. As shown for MG63 cells above, we note here the most prominent reduction in cell growth of 56 ± 16% for cells (donor 46) grown on FeDip50 modified substrates compared to the reference substrate. For donor 54, a reduction to 60 ± 13% compared to the reference substrate is observed. A reduction of proliferation for cells grown on LBL-modified surfaces as shown for MG63 cells could not be observed. In contrast, cells of donor 56 appear to proliferate much more on LBL- and Dip10-modified surfaces, whereas cell growth on FeDip50-modified surfaces is comparable to that of the reference substrate. Cells of this donor thus seem to differentiate less than cells of donor 46 and 54.

In order to proof the above mentioned hypothesis, the cell activity is tested by WST-1 assays. On the one hand, the observed low cell numbers might be caused by cell death or, on the other hand, are the result of a proliferative stop, which typically accompanies differentiation processes. If the cells are simply not viable due to a lack of cell adhesion or to a release of cytotoxic substances from the particular MEPE-modified surfaces, no increase in metabolic activity should be observed in these assays. As shown in Figure 5, the determined metabolic activity depicted as mitochondrial activity per cell, argue against cell death since values obtained from cells grown on the MEPE modified surfaces are throughout the duration of the experiment at least equal to those determined for cells grown on the reference substrates. For samples employing cells from donor 46 and 54 a clear trend towards an increased cell activity is apparent from day 5. At day 5, the relative activity of cells from donor 46, which are grown on FeDip50 modified substrates, are found to be 180 ± 67% compared to the reference surfaces. For cells derived from donor 54 this value rises to over 300% of that of the reference. In contrast, for cells grown on LbL-modified substrates the metabolic activities are not increased compared to the reference.

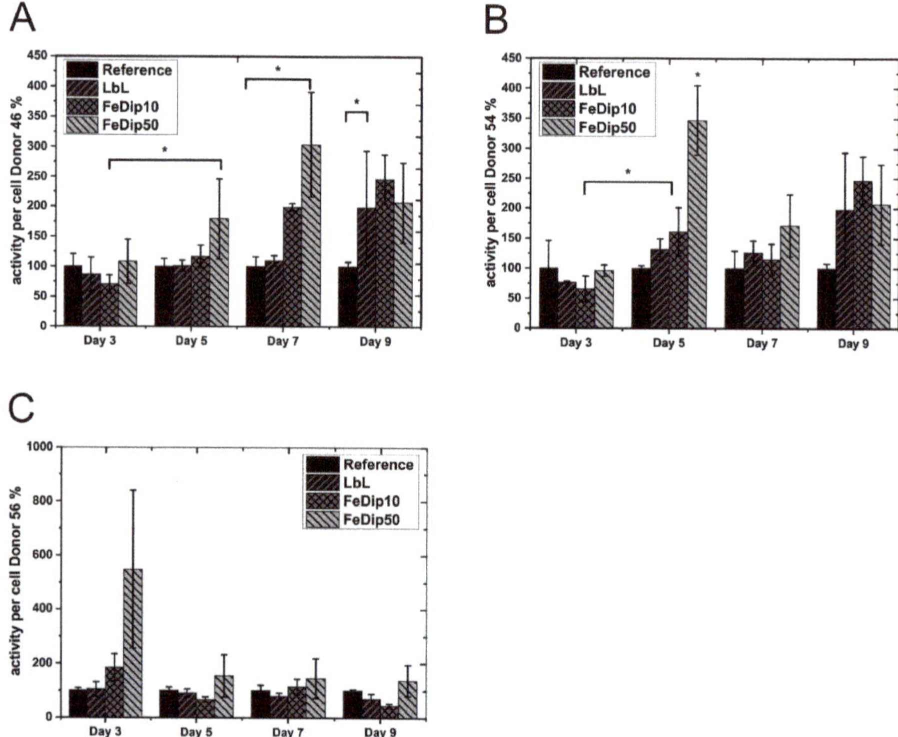

Figure 5. Relative Metabolic activities of the used hMSC cells from (**A**) Donor 46, (**B**) Donor 54, and (**C**) Donor 56 was determined at four different time points for the particular surface modifications in relation to cells grown on reference substrates. All experiments were performed in triplicate. * $p < 0.5$ (analyzed by one-way ANOVA with Tukey test).

Thus, based on these observations, we assume that cells, which are grown on the dip-coated surfaces, tend to differentiate. Due to the osteogeneous origin of the used BMSCs, a differentiation towards osteogenic lineages is expected, which can be assessed by specific assays such as Alizarin red staining (as shown for MG63 cells, see Figure 3) or by determination of alkaline phosphatase (ALP) activity.

Figure 6 shows microscopy images of Alizarin red stained cells of donor 46 and 54 on the reference and FeDip50 surface at day 5.

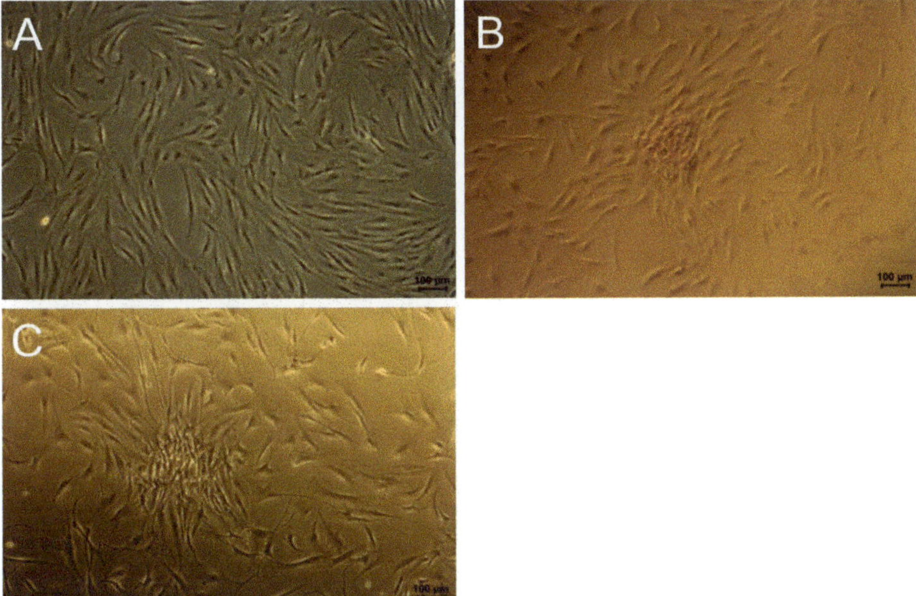

Figure 6. Examples of light microscopy images of Alizarin red stained BMSCs grown on (**A**) reference (Donor 409 46), (**B**) FeDip50 (Donor 46), and (**C**) FeDip50 (Donor 54) at day 5.

It becomes obvious that the cell morphology on the reference has not changed. Importantly, no staining by Alizarin red is detectable indicating that these cells reside in an undifferentiated state, which supports the data obtained for the cell activity (see Figure 5). In clear contrast, cells grown FeDip50 modified surfaces exhibit a different spindle-like phenotype. The cells are assembled in form of clusters, which are positively stained by Alizarin red. The cells on the reference surface appear denser than those grown on dip coated surface. This observation is in agreement with the determined cell numbers shown in Figure 4. Based on these stainings we conclude that the cells grown on FeDip50 modified surfaces have reached after stage 2 5 days differentiation in accordance with Collin et al. [12]. Thus, altered cellular activity and cell morphology indicate differentiation of the used cells towards the osteogenic lineage as supported by the positive alizarin red staining.

3.3. Iron Content

In order to show whether potential differences in the intracellular iron content may depend on the particular surfaces, the cells are removed from the substrates, washed, lysed, and the iron ions released from the carriers such as ferritin and hem-proteins are detected by photometric analysis. The relative content of iron ions of cells grown on the particular Fe-MEPE surfaces compared to the reference surface are shown in Figure 7.

In agreement to the data concerning cell number and activity, cells from donor 46 and 54 grown on Fe-MEPE modified surfaces show, as expected, an increased intracellular iron ion content over a prolonged time of 9 days compared to the reference. On the other hand, cells from donor 56 show an initial increase in iron ion content on day 3; however, starting from day 5, no significant differences in iron ion content compared to the control could be detected.

The influence of certain metal ions on BMSCs has already been shown. For instance, Yoshizawa et al. detected a rapid proliferation of BMSCs and an increase in extracellular matrix (ECM) mineralization in the presence of Mg(II)-ions in vitro [36]. On the other side, transition metal ions such as Mn(II), Fe(III), Co(II), Ni(II), and Cu(II), showed cytotoxic effects on osteoblastic cell lines such as MG63 cells already at a concentration of 0.1 mM [37]. Typically, osteogenic differentiation of BMSCs is initiated by addition of various agents such as dexamethasone, L-ascorbic acid-2-phosphates or ascorbic acid and β-glycerophosphates. As the aforementioned agents are absent in the experiments shown here, we assume that solely the Fe-MEPE modified surfaces trigger differentiation of the used BMSCs. However, the determination of the concentration of the active species acting on the cell in the interfacial region remains to be elucidated. As positive staining by Alizarin red already indicates differentiation of the investigated BMSCs towards osteogenic lineages this process is analyzed in more detail by investigating the expression profiles of characteristic marker genes such as ALP and Osteocalcin (OC) by qRT-PCR.

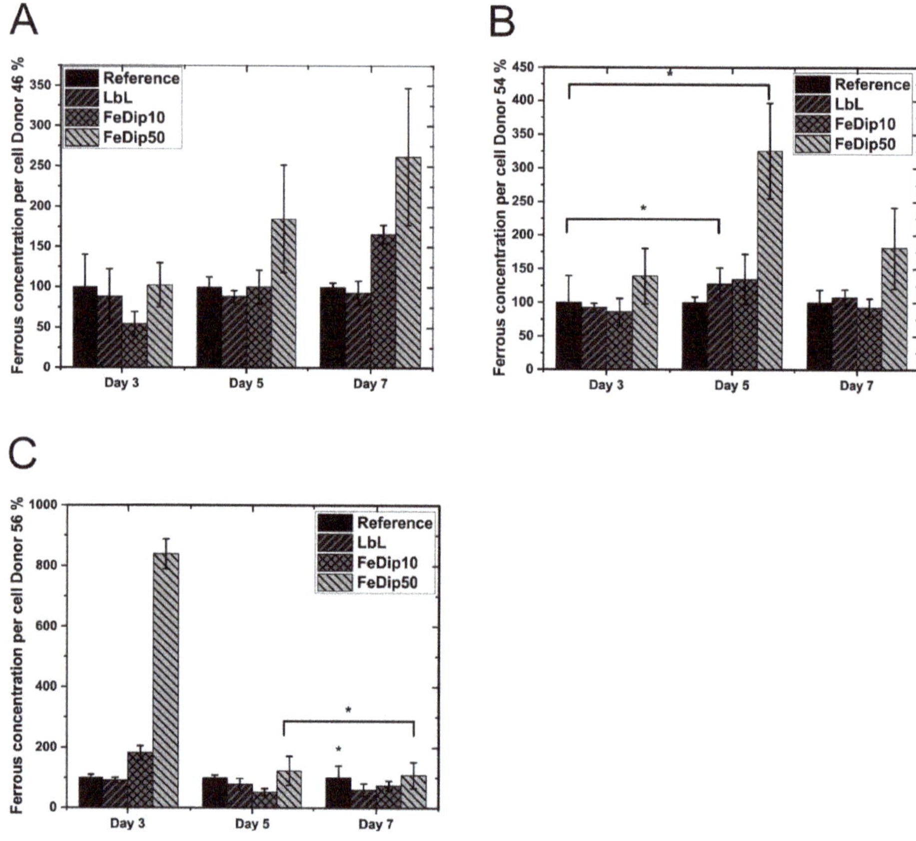

Figure 7. The bar diagrams represent the relative ferrous concentrations per cell for (**A**) donor 46, (**B**) donor 54, and (**C**) donor 56 grown on the indicated substrates. The values obtained for cells grown on TCPS substrates are set to 100% (the values for the references substrates also vary donor dependently. The absolute values for the reference substrates at day 3 are: donor 46 = 68.2 fMol/cell; donor 54 = 31 fMol/cell; donor 56 = 21.6 fMol/cell). All experiments were performed in triplicate. * $p <$ 0.5 (analyzed by one-way ANOVA with Tukey test).

3.4. Quantification of ALP and OC Gene Expression

BMSCs, when grown for 16 days in the aforementioned differentiation media, form a coherent network of ALP-positive cells, but when Co(II)-ions at a concentration of 40 μM to 100 μM are added, a decrease in the overall ALP activity is observed [38,39].

Figure 8 shows ALP and OC gene expression at day 3 and day 5 of the three donors.

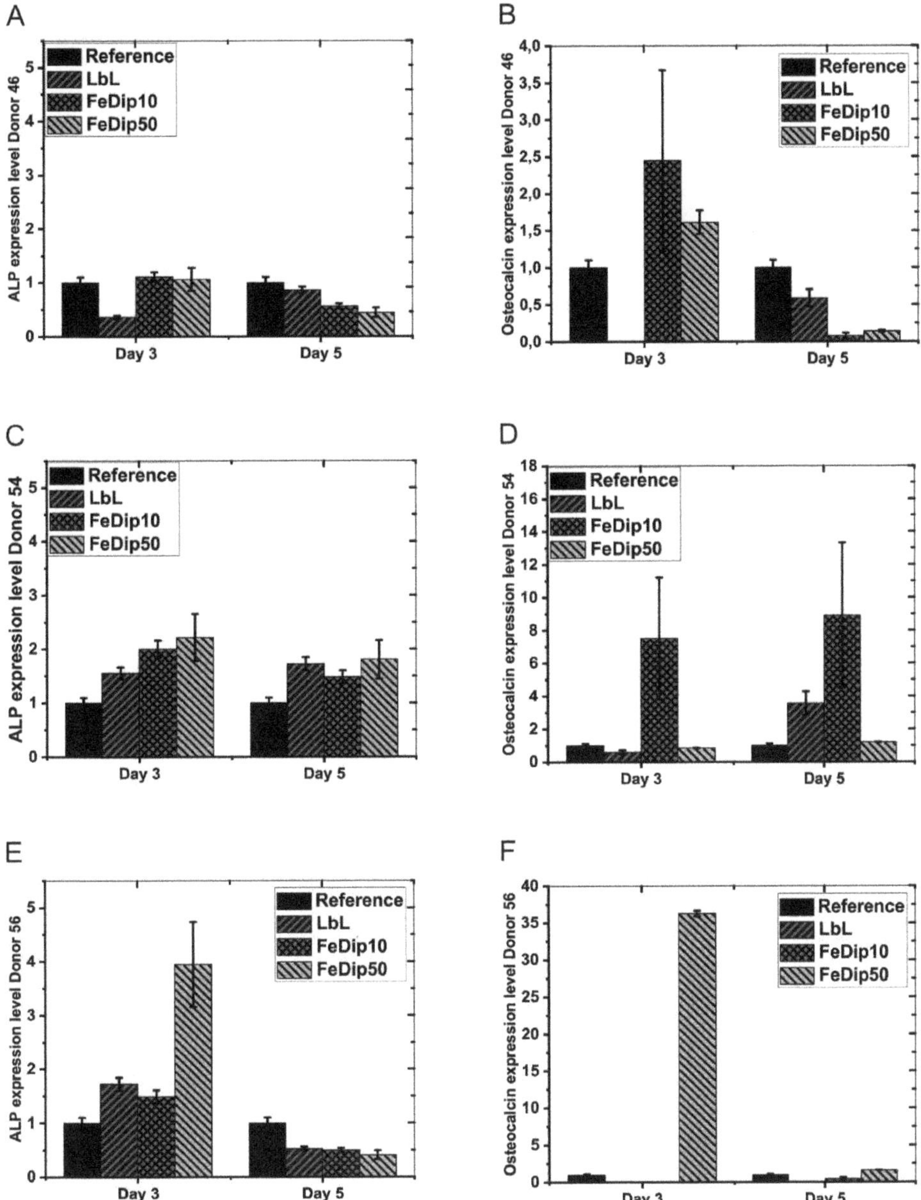

Figure 8. The bar diagrams represent values for relative ALP (**A,C,E**) and OC (**B,D,F**) expression levels of cells grown on the indicated substrates of Donor 46 (**A,B**), Donor 54 (**C,D**), and Donor 56 (**E,F**). All values are normalized to GAPDH expression levels. Subsequently the levels of ALP and OC expression determined for cells grown on reference substrates are set to 1. All experiments were performed in triplicate.

In contrast to the clear results obtained from experiments regarding cell number and cell activity those addressing ALP and OC expression need to be described and discussed in more detail. Unexpectedly, ALP expression in cells of donor 46 grown on FeDip modified surfaces is in general comparable to that of cells grown on the reference substrates. At day 5, the values for ALP activity decrease by half relative to the reference. For the LbL modified surface lower ALP expression is observed at day 3, but it rises almost to the reference value by day 5.

Since these findings are not in agreement with our data regarding cell number and activity, we also analyzed the expression of OC. Here, a significant increase in OC expression is observed for cells of donor 46 grown on FeDip10 and FeDip50 modified surfaces at day 3 if compared to the reference substrate. However, by day 5 OC expression in these cells declines strongly to values of approximately 10% of those of the reference. Instead, cells of donor 54 grown on the Fe-MEPE modified surfaces in general show higher ALP expression levels if compared to the controls. Astonishingly, cells of donor 56 show high expression of either ALP and OC but only if grown on FeDip50 modified surfaces at day 3, which is unexpected since data concerning cell number and activity identify cells of this donor to be the least differentiating ones.

Taken together, the results of these experiments suggest that Fe-MEPE-triggered differentiation strongly depends on the individual surface modification and, as expected, also on the individual donor thus preventing an exact determination of particular differentiation stages. Alkaline phosphatase, as already mentioned in the introduction, is primarily expressed upon osteogenic differentiation in two separated time frames. A first peak is observed at the beginning of osteogenic differentiation and a second at the beginning of mineralization. As shown by individual microscopy images, only sub-populations of cells can be stained by alizarin red implicating that cell differentiation, even in one specified sample, is non-homogenous. Thus, the data obtained from qRT-PCR experiments, representing the average ALP or OC expression levels of these non-homogenously differentiated cells, may not indicate a general differentiation stage.

However, it can be assumed that differentiation of at least some of the cells of donor 46 reached the second differentiation step already at day 3, which is confirmed by alizarin red staining (see Figure 8) but a comprehensive progress in differentiation till day 5 cannot be detected

Collin et al. observed an increase in osteocalcin expression by 30-fold during mineralization [12]. This considerable increase in osteocalcin gene expression might be achieved for BMSCs of donor 56 simply grown for 3 days on FeDip50-modified surfaces. Birgani et al. reported no increased ALP or osteocalcin gene expression until day 14 after metal ion addition. Moreover, Birgani et al. suggests that the ALP activities of the different donors can vary [40]. These different activities as a function of the BMSC donor have already been first described by Barbara et al already in 2004 but meanwhile also by many others [41–45]. The variety of gene expression between the different donors has also been observed in our study. But the duration of ALP and OC gene expression was reduced from 14 to only 3 days.

However, the results presented here indicate that osteogenic differentiation is initiated by Fe-MEPE modified surfaces and can be detected at day 3. In this study we detect ALP expressing cells already at day 5. Thus, the use of Fe-MEPE modified surfaces seems to trigger the osteogenic differentiation processes more rapidly.

4. Conclusions

The current investigation employing MG63 cells grown on Fe-MEPE modified substrates suggest initiation of osteogenic differentiation by both, high cell activity and altered morphology of the cells and/or cluster formation. Remarkably, LbL and FeDip50 modified surfaces show strongest effects on cell count and cell activity, which becomes visible already at day 3. Considering morphology, cells grown on LbL modified surfaces appear morphological similar and do not form cell clusters like those grown on the reference substrate. Based on these observations, MG63 cells seem to best differentiate to osteoplastic lineages if grown on these substrates, which is further supported by the occurrence of Alizarin red stained clusters (Figure 2).

Concerning BMSCs, our findings suggest that Fe-MEPE modified surfaces also stimulate osteogenic differentiation in these cells. Similar to MG63 cells, the Fe-MEPE modified surfaces suppress proliferation and promote differentiation of the used BMSCs to a variable extent which is dependent on the individual donor. Cells derived from two donors (46/54) differentiate better than cells of the third donor (56), which is supported by staining with Alizarin red. Clearly, BMSCs are stimulated for osteogenic differentiation, which appears donor and substrate specific. However, cell differentiation occurs non-coherently thus reflecting various differentiation stages already in one particular sample.

The effect of the modified surfaces on osteogenic differentiation can also be detected on mRNA level addressing well-known osteogenic marker genes, such as ALP and OC, although the data do not show the expected coherency. Nevertheless, all Fe-MEPE modified surfaces, which are investigated here, influence the osteogenic differentiation capacity of BMSCs without addition of agents inducing osteogenic differentiation. This remarkable result leads to the question of the underlying mechanism. It is well established, that metal ions affect cell differentiation and, therefore, most likely also a release of iron ions into the interfacial contact region between substrate and cells initiate cellular processes towards differentiation. The osteogenic properties of iron ions in context of BMSC differentiation has also been reported for iron oxide nanoparticles by Wang et al. [46,47]. Thus, our results indicate that Fe-MEPE functionalized surfaces may serve as innovative scaffolds for the treatment of bone defects.

Author Contributions: Conceptualization, J.N. and D.G.K.; methodology, J.N. and J.B.; validation, J.B.; investigation, J.B.; writing—original draft preparation, J.B.; writing—review and editing, J.N. and D.G.K.; supervision, J.N.; project J.N. and D.G.K.

Funding: This publication was funded by the German Research Foundation (DFG) and the University of Wuerzburg in the funding programme Open Access Publishing.

Conflicts of Interest: The authors declare no conflict of interest

References

1. Belka, J.; Weigel, T.; Berninger, A.-K.; Kurth, D.G.; Nickel, J. Growth and differentiation of myoblastic precursor cells on thin films of metallo-supramolecular coordination polyelectrolyte (mepe). *Adv. Mater. Interfaces* **2017**, *4*, 1600272. [CrossRef]

2. Bains, W.; Ponte, P.; Blau, H.; Kedes, L. Cardiac actin is the major actin gene-product in skeletal-muscle cell-differentiation in vitro. *Mol. Cell Biol.* **1984**, *4*, 1449–1453. [CrossRef] [PubMed]

3. Atmani, H.; Audrain, C.; Mercier, L.; Chappard, D.; Baslé, M.F. Phenotypic effects of continuous or discontinuous treatment with dexamethasone and/or calcitriol on osteoblasts differentiated from rat bone marrow stromal cells. *J. Cell. Biochem.* **2002**, *85*, 640–650. [CrossRef] [PubMed]

4. Carinci, F.; Pezzetti, F.; Spina, A.M.; Palmieri, A.; Laino, G.; De Rosa, A.; Farina, E.; Illiano, F.; Stabellini, G.; Perrotti, V.; et al. Effect of Vitamin C on pre-osteoblast gene expression. *Arch. Oral Boil.* **2005**, *50*, 481–496. [CrossRef] [PubMed]

5. Giannoudis, P.V.; Einhorn, T.A.; Schmidmaier, G.; Marsh, D. The diamond concept—Open questions. *Injury* **2008**, *39*, S5–S8. [CrossRef]

6. Bellows, C.; Heersche, J.; Aubin, J.E. Determination of the capacity for proliferation and differentiation of osteoprogenitor cells in the presence and absence of dexamethasone. *Dev. Boil.* **1990**, *140*, 132–138. [CrossRef]

7. Liu, P.; Oyajobi, B.O.; Russell, R.G.; Scutt, A. Regulation of osteogenic differentiation of human bone marrow stromal cells: Interaction between transforming growth factor-beta and 1,25(OH)(2) vitamin D(3) In vitro. *Calcif. Tissue Int.* **1999**, *65*, 173–180. [CrossRef]

8. Caplan, A.I. Mesenchymal stem-cells. *J. Orthop. Res.* **1991**, *9*, 641–650. [CrossRef]

9. Chen, J.; Li, Y.; Wang, L.; Lü, M.; Zhang, X.; Chopp, M. Therapeutic benefit of intracerebral transplantation of bone marrow stromal cells after cerebral ischemia in rats. *J. Neurol. Sci.* **2001**, *189*, 49–57. [CrossRef]

10. Ferrari, G.; Cusella, G.; Angelis, D.; Coletta, M.; Paolucci, E.; Stornaiuolo, A.; Cossu, G.; Mavilio, F. Muscle Regeneration by Bone Marrow-Derived Myogenic Progenitors. *Science* **1998**, *279*, 1528–1530. [CrossRef]

11. Toma, C.; Pittenger, M.F.; Cahill, K.S.; Byrne, B.J.; Kessler, P.D. Human Mesenchymal Stem Cells Differentiate to a Cardiomyocyte Phenotype in the Adult Murine Heart. *Circulation* **2002**, *105*, 93–98. [CrossRef] [PubMed]

12. Collin, P.; Wetterwald, A.; Boy-Lefevre, M.-L.; Fleisch, H.; Forest, N. Expression of collagen, osteocalcin, and bone alkaline phosphatase in a mineralizing rat osteoblastic cell culture. *Calcif. Tissue Int.* **1992**, *50*, 175–183. [CrossRef] [PubMed]

13. Zernik, J.; Twarog, K.; Upholt, W.B. Regulation of alkaline phosphatase and alpha2(I) procollagen synthesis during early intramembranous bone formation in the rat mandible. *Differentiation* **1990**, *44*, 207–215. [CrossRef] [PubMed]

14. Stein, G.S.; Lian, J.B. Molecular Mechanisms Mediating Proliferation/Differentiation Interrelationships During Progressive Development of the Osteoblast Phenotype. *Endocr. Rev.* **1993**, *14*, 424–442. [CrossRef] [PubMed]

15. Owen, T.A.; Aronow, M.; Shalhoub, V.; Barone, L.M.; Wilming, L.; Tassinari, M.S.; Kennedy, M.B.; Pockwinse, S.; Lian, J.B.; Stein, G.S. Progressive development of the rat osteoblast phenotype invitro—Reciprocal relationships in expression of genes associated with osteoblast proliferation and differentiation during formation of the bone extracellular-matrix. *J. Cell Physiol.* **1990**, *143*, 420–430. [CrossRef] [PubMed]

16. Ducy, P. CBFA1: A molecular switch in osteoblast biology. *Dev. Dyn.* **2000**, *219*, 461–471. [CrossRef]

17. Bronckers, A.L.; Gay, S.; Finkelman, R.D.; Butler, W.T. Developmental appearance of Gla proteins (osteocalcin) and alkaline phosphatase in tooth germs and bones of the rat. *Bone Miner.* **1987**, *2*, 361–373.

18. Dalton, S.; Coverdell, P.D. Linking the cell cycle to cell fate decisions. *Trends Cell Boil.* **2015**, *25*, 592–600. [CrossRef]

19. Muschler, G.F.; Boehm, C.; Easley, K. Aspiration to obtain osteoblast progenitor cells from human bone marrow: The influence of aspiration volume. *J. Bone Jt. Surgery-Am.* **1997**, *79*, 1699–1709. [CrossRef]

20. Glenske, K.; Donkiewicz, P.; Köwitsch, A.; Milosevic-Oljaca, N.; Rider, P.; Rofall, S.; Franke, J.; Jung, O.; Smeets, R.; Schnettler, R.; et al. Applications of Metals for Bone Regeneration. *Int. J. Mol. Sci.* **2018**, *19*, 826. [CrossRef]

21. Qiu, L.; Ding, X.; Zhang, Z.; Kang, Y.J. Copper Is Required for Cobalt-Induced Transcriptional Activity of Hypoxia-Inducible Factor-1. *J. Pharmacol. Exp. Ther.* **2012**, *342*, 561–567. [CrossRef] [PubMed]

22. Roy, M.; Fielding, G.A.; Bandyopadhyay, A.; Bose, S. Effects of Zinc and Strontium Substitution in Tricalcium Phosphate on Osteoclast Differentiation and Resorption. *Biomater. Sci.* **2013**, *1*, 74–82. [CrossRef] [PubMed]

23. Weng, L.; Boda, S.K.; Teusink, M.J.; Shuler, F.D.; Li, X.; Xie, J. Binary Doping of Strontium and Copper Enhancing Osteogenesis and Angiogenesis of Bioactive Glass Nanofibers while Suppressing Osteoclast Activity. *ACS Appl. Mater. Interfaces* **2017**, *9*, 24484–24496. [CrossRef] [PubMed]

24. Mourino, V.; Cattalini, J.P.; Boccaccini, A.R. Metallic ions as therapeutic agents in tissue engineering scaffolds: An overview of their biological applications and strategies for new developments. *J. R. Soc. Interface* **2012**, *9*, 401–419. [CrossRef] [PubMed]

25. Ai, H.; Jones, S.A.; Lvov, Y.M. Biomedical Applications of Electrostatic Layer-by-Layer Nano-Assembly of Polymers, Enzymes, and Nanoparticles. *Cell Biophys.* **2003**, *39*, 23–44. [CrossRef]

26. Decher, G. Fuzzy Nanoassemblies: Toward Layered Polymeric Multicomposites. *Science* **1997**, *277*, 1232–1237. [CrossRef]

27. Kurth, D.G.; Schütte, M. Layer-by-layer self-assembly of a metallo-supramolecular coordination polyelectrolyte studied by infrared spectroscopy, microgravimetry, and X-ray reflectance. *Macromol. Symp.* **2001**, *164*, 167–180. [CrossRef]

28. Schott, M.; Szczerba, W.; Kurth, D.G. Detailed Study of Layer-by-Layer Self-Assembled and Dip-Coated Electrochromic Thin Films Based on Metallo-Supramolecular Polymers. *Langmuir* **2014**, *30*, 10721–10727. [CrossRef]

29. Schütte, M.; Kurth, D.G.; Linford, M.R.; Cölfen, H.; Möhwald, H. Metallosupramolecular Thin Polyelectrolyte Films. *Angew. Chem. Int. Ed.* **1998**, *37*, 2891–2893. [CrossRef]

30. Yan, Y.; Huang, J. Hierarchical assemblies of coordination supramolecules. *Co-ord. Chem. Rev.* **2010**, *254*, 1072–1080. [CrossRef]

31. De Villiers, M.M.; Otto, D.P.; Strydom, S.J.; Lvov, Y.M. Introduction to nanocoatings produced by layer-by-layer (LbL) self-assembly. *Adv. Drug Deliv. Rev.* **2011**, *63*, 701–715. [CrossRef] [PubMed]

32. Gribova, V.; Auzely-Velty, R.; Picart, C. Polyelectrolyte multilayer assemblies on materials surfaces: From cell adhesion to tissue engineering. *Chem. Mater.* **2012**, *24*, 854–869. [CrossRef] [PubMed]

33. Schott, M.; Beck, M.; Winkler, F.; Lorrmann, H.; Kurth, D.G. Fabricating electrochromic thin films based on metallo-polymers using layer-by-layer self-assembly: An attractive laboratory experiment. *J. Chem. Educ.* **2015**, *92*, 364–367. [CrossRef]

34. Förch, R.; Schönherr, H.; Tobias, A.; Jenkins, A. *Surface Design: Applications in bioscience and nanotechnology*; Wiley-VCH: Weinheim, Germany, 2009.

35. DIN, E.I. Biological evaluation of medical devices—Part 5: Tests for in vitro cytotoxicity. *Part* **2009**, *1*, 10993.

36. Yoshizawa, S.; Brown, A.; Barchowsky, A.; Sfeir, C. Magnesium ion stimulation of bone marrow stromal cells enhances osteogenic activity, simulating the effect of magnesium alloy degradation. *Acta Biomater.* **2014**, *10*, 2834–2842. [CrossRef] [PubMed]

37. Hallab, N.J.; Vermes, C.; Messina, C.; Roebuck, K.A.; Glant, T.T.; Jacobs, J.J. Concentration- and composition-dependent effects of metal ions on human MG-63 osteoblasts. *J. Biomed. Mater. Res.* **2002**, *60*, 420–433. [CrossRef] [PubMed]

38. Jaiswal, N.; Haynesworth, S.E.; Caplan, A.I.; Bruder, S.P. Osteogenic differentiation of purified, culture-expanded human mesenchymal stem cells in vitro. *J. Cell. Biochem.* **1997**, *64*, 295–312. [CrossRef]

39. Schrock, K.; Lutz, J.; Mandl, S.; Hacker, M.C.; Kamprad, M.; Schulz-Siegmund, M. Co(ii)-mediated effects of plain and plasma immersion ion implanted cobalt-chromium alloys on the osteogenic differentiation of human mesenchymal stem cells. *J. Orthopaed Res.* **2015**, *33*, 325–333. [CrossRef]

40. Birgani, Z.T.; Malhotra, A.; Van Blitterswijk, C.A.; Habibovic, P. Human mesenchymal stromal cells response to biomimetic octacalcium phosphate containing strontium. *J. Biomed. Mater. Res. Part A* **2016**, *104*, 1946–1960. [CrossRef]

41. Barbara, A.; Delannoy, P.; Denis, B.; Marie, P. Normal matrix mineralization induced by strontium ranelate in MC3T3-E1 osteogenic cells. *Metabolism* **2004**, *53*, 532–537. [CrossRef]

42. Boanini, E.; Torricelli, P.; Fini, M.; Sima, F.; Serban, N.; Mihailescu, I.N.; Bigi, A. Magnesium and strontium doped octacalcium phosphate thin films by matrix assisted pulsed laser evaporation. *J. Inorg. Biochem.* **2012**, *107*, 65–72. [CrossRef] [PubMed]

43. Brennan, T.; Rybchyn, M.; Green, W.; Atwa, S.; Conigrave, A.; Mason, R.; Rybchyn, M. Osteoblasts play key roles in the mechanisms of action of strontium ranelate. *Br. J. Pharmacol.* **2009**, *157*, 1291–1300. [CrossRef] [PubMed]

44. Capuccini, C.; Torricelli, P.; Sima, F.; Boanini, E.; Ristoscu, C.; Bracci, B.; Socol, G.; Fini, M.; Mihailescu, I.; Bigi, A. Strontium-substituted hydroxyapatite coatings synthesized by pulsed-laser deposition: In vitro osteoblast and osteoclast response. *Acta Biomater.* **2008**, *4*, 1885–1893. [CrossRef] [PubMed]

45. Zhang, W.; Shen, Y.; Pan, H.; Lin, K.; Liu, X.; Darvell, B.W.; Lu, W.W.; Chang, J.; Deng, L.; Wang, D.; et al. Effects of strontium in modified biomaterials. *Acta Biomater.* **2011**, *7*, 800–808. [CrossRef] [PubMed]

46. Boda, S.K.; Thrivikraman, G.; Panigrahy, B.; Sarma, D.D.; Basu, B. Competing roles of substrate composition, microstructure, and sustained strontium release in directing osteogenic differentiation of hmscs. *ACS Appl. Mater. Interfaces* **2017**, *9*, 19389–19408. [CrossRef]

47. Wang, Q.; Chen, B.; Cao, M.; Sun, J.; Wu, H.; Zhao, P.; Xing, J.; Yang, Y.; Zhang, X.; Ji, M.; et al. Response of MAPK pathway to iron oxide nanoparticles in vitro treatment promotes osteogenic differentiation of hBMSCs. *Biomaterials* **2016**, *86*, 11–20. [CrossRef]

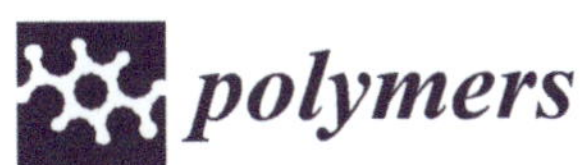 *polymers*

Article

Biotechnological Preparation of Gelatines from Chicken Feet

Pavel Mokrejš [1,*], Petr Mrázek [1], Robert Gál [2] and Jana Pavlačková [3]

1   Department of Polymer Engineering, Faculty of Technology, Tomas Bata University in Zlín, Vavrečkova 275, 760 01 Zlín, Czech Republic; p_mrazek@utb.cz
2   Department of Food Technology, Faculty of Technology, Tomas Bata University in Zlín, Vavrečkova 275, 760 01 Zlín, Czech Republic; gal@utb.cz
3   Department of Lipids, Detergents and Cosmetics Technology, Faculty of Technology, Tomas Bata University in Zlín, Vavrečkova 275, 760 01 Zlín, Czech Republic; pavlackova@utb.cz
*   Correspondence: mokrejs@utb.cz; Tel.: +420-576-031-230

Received: 15 May 2019; Accepted: 14 June 2019; Published: 18 June 2019

Abstract: In the European Union (EU), about five tons of poultry by-product tissues are produced every year. Due to their high collagen content, they represent a significant raw material source for gelatine production. The aim of the paper was the biotechnological preparation of gelatine from chicken feet. The influence of selected process factors on the gelatine yield, gel strength, viscosity, and ash of gelatine was observed; a two-level factor design of experiments with three variable process factors (enzyme addition, enzyme treatment time, and gelatine extraction time) was applied. After grinding and separating soluble proteins and fat, the purified raw material was treated in water at pH 7.5 with the addition of endoprotease at 23 °C and after thorough washing with water at 80 °C, gelatine was extracted. By the suitable choice of process conditions, gelatine with high gel strength (220–320 bloom), low ash content (<2.0%) and viscosity of 3.5–7.3 mPa·s can be prepared. The extraction efficiency was 18–38%. The presented technology is innovative mainly by the enzymatic processing of the source raw material, which is economically, technologically, and environmentally beneficial for manufacturers. Chicken gelatines are a suitable alternative to gelatines made from mammals or fish, and can be used in many food, pharmaceutical, and biomedical applications.

Keywords: biotechnology; by-products; chicken feet; extraction; food applications; gelatine; pharmaceutical applications; polymer biomaterials

---

1. Introduction

Gelatine is a significant, water-soluble protein that is obtained from collagenous raw materials by partial hydrolysis. The primary raw materials for gelatine production are pork and bovine skins/hides and bones [1]. In recent years, alternative sources of collagen, especially fish and by-products of the meat processing and poultry industry, have become more important for gelatine producers [2]. The reason for this is the growing global demand for gelatine, which is estimated to be around 451,000 tons for 2018: about a ¼ increase over six years [3]. Another impetus for the search for alternative sources of collagen is the growing demand for non-mammalian gelatines, especially from consumers from Islamic, Jewish, and Hindu countries. Also, the economic and ecological reasons for the consideration of by-products are forcing producers of such waste to seek ways of further re-utilization. The new application possibilities of gelatine made from alternative sources of collagen are also opening up.

Worldwide annual food waste production is estimated at up to 100 million tons. In the European Union (EU), the meat processing industry produces 16.5 million tons of waste, and 5.2 million tons of waste is generated in the fish production industry [4]. The United States (US) poultry industry produced 28.4 million tons of live weight of poultry in 2016; the largest proportion (24.9 million tons) was chicken [5]. Poultry meat accounts for 5.5% of total agricultural production and 12.7% of EU meat production. In 2014, 13 million tons of poultry meat was produced in the EU, 78% of which was chicken [6]. According to the FAO (Food and Agricultural Organization), the total consumption of poultry meat is growing by 3.6% per year. Even though in some countries animal by-products (e.g., kidneys, hearts, livers, lungs, stomachs, and tongues) are used for culinary purposes, approximately 20–30% of the live weight of poultry represents unused by-products. In some countries, for example, blood is used as an ingredient in food production and for the production of feeding meal. Feathers and other by-products are used for the production of feeding hydrolysates; part of the protein waste can also be composted or used for fuel production, and waste fats are used for soaps, biofuels, and lubricants [7–10].

Raw materials for the production of commercial gelatines are processed in low and high pH environments [1]. In the case of type A gelatine, the raw material is treated for 18–30 h in an acidic environment at pH 1.5–3.0; this procedure is suitable for pigskins. For type B gelatine, the raw material is treated in alkaline medium at pH 12.0 for several weeks to months; this procedure is mainly used for cowhides.

By-products from freshwater and saltwater fish rich in collagen (skins and bones) have long been known as the suitable raw materials for gelatines and collagen hydrolysates production. In the preparation of fish gelatines, the procedure is analogous to that of the production of gelatines from mammals. Most often, the raw material is treated for 12–48 h in an acidic environment at 5–10 °C, optionally at room temperature; weak solutions of acids (0.05–0.5 mol·L^{-1}) are used, especially acetic acid, citric acid, phosphoric acid, formic acid, hydrochloric acid, propionic acid, and optionally lactic acid. Depending on the type of the raw material (different amounts of intermolecular and intramolecular crosslinking of the collagen matrix) and the extraction conditions (water, usually at 45–55 °C for 3–12 h), gelatine (type A) is prepared with a yield of 5.5–66.0% [11–13]. Less common is the processing of the raw material in an alkaline environment, most often using Ca(OH)$_2$ (type B gelatine), as well as the use of protease [14–16]. Somewhat different is the industrial production of gelatines, soluble collagen, and hydrolysates from poultry tissues. On the global market, chicken products represent only a fraction of gelatines made from beef, pork, and fish. After 24 h of processing of fowl feet with weak acid solutions (0.5 M acetic acid, citric acid, hydrochloric acid, or lactic acid) and enzyme (pepsin) at 4 °C, soluble collagen is prepared with relatively low yields, 5.6–8.4% [17]. Using stronger solutions of the same acids (5.0%, v/v), acid-soluble collagen can be prepared under similar processing conditions (12–36 h incubation at 4–7 °C) with a yield of 7.9–31.2%, even without the use of pepsin [18]. Also, the preparation of gelatines, especially from chicken feet, skins, and tendons, is described. After the removal of soluble non-collagenous proteins and pigments, the feet are processed in an acidic environment using slightly stronger solutions (1.5–4.5%, v/v) of acetic acid, citric acid, or lactic acid at room temperature for 16–18 h. At moderate extraction temperatures (50–55 °C) and extraction times ranging from tens of minutes to hours, gelatine with a gel strength of 120–300 Bloom is prepared, which is comparable to commercial pork and fish gelatines. The disadvantage is the relatively low yields of the prepared gelatines, 6.0–14.5%, based on the dry weight of the source material [19,20]. After combined sulphuric acid and acetic acid treatment and 12-h extraction at 45 °C, 16% of high gel-strength gelatine (355 Bloom) can be obtained from chicken skins, which is more than in conventional commercial bovine gelatine [21]. The literature also describes the processing of other poultry by-products into gelatines. These are mainly chicken or turkey heads or duck feet. The processing is carried out according to an analogous procedure to that of chicken feet or skins. The alkaline method of poultry tissue processing is not widespread, but has been tried, for example, on chicken and turkey heads [22].

The processing of some by-products from poultry slaughter into gelatines have been satisfactorily described in the literature. However, all the techniques are based on the acidic or alkaline processing of the raw material; the enzymatic processing is only marginally mentioned. The use of enzymes brings with it many advantages, such as mild reaction conditions given by temperature and pH as well as low doses of enzymes used. The main goal of our paper is to propose a biotechnological method of preparing gelatines from chicken feet, through first treating the source material with a suitable proteolytic enzyme and then extracting the gelatine with hot water. This is followed by observing the influence of selected process factors on gelatine yield (percentage conversion of the source material to gelatines) and the quality of prepared gelatines (gel strength, viscosity, and ash content in gelatine). The specific hypothesis being tested is that chicken feet, after treatment with endoprotease and extraction with hot water, can be processed into gelatines whose properties are comparable with gelatines produced from pork and beef tissues.

2. Materials and Methods

A flow chart of biotechnological processing of chicken feet into gelatines is shown in Scheme 1.

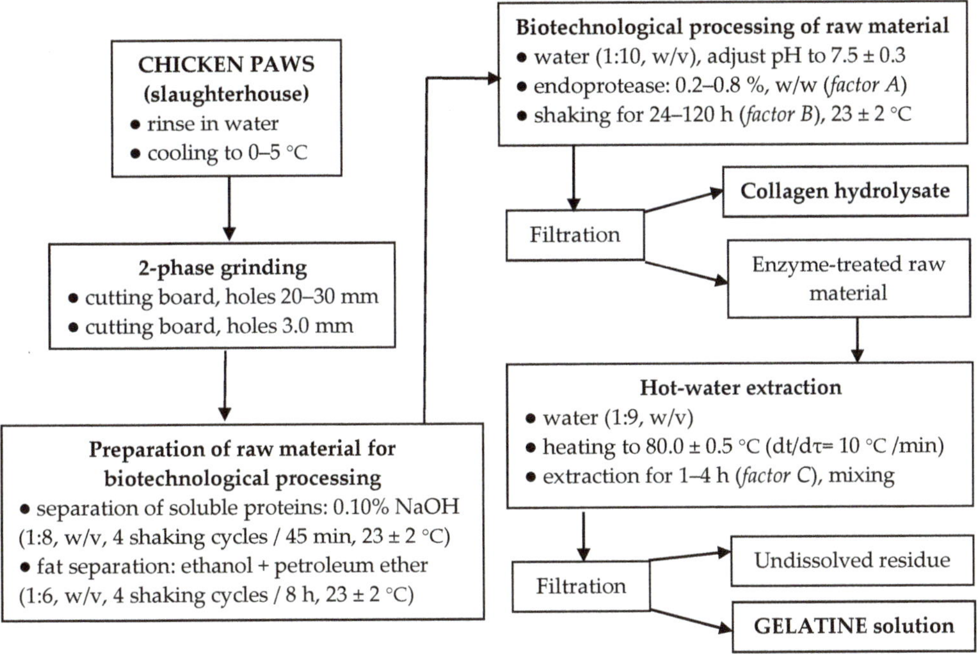

Scheme 1. A flow chart of preparation of gelatins from chicken feet.

2.1. Materials

Chicken feet were obtained in the Raciola poultry farm (Ltd., Uherský Brod, Czech Republic) and processed under strict hygiene conditions, see Section 2.3 and Section 2.4. Raw chicken feet composition: 35.0 ± 3.0% dry matter content; dry matter: 48.3 ± 0.4% protein content (82.8 ± 0.7% collagen content), 34.8 ± 0.8% fat content, 16.1 ± 0.2% ash content. The polarzyme 6.0 T-granulated endoprotease that hydrolyzes internal peptide bonds were manufactured by the fermentation of a microorganism that is not present in the final product (Novozymes, Copenhagen, Denmark), petrolether, ethanol, NaOH, 36% HCl. The procedure also used a SPAR Mixer SP-100AD-B industrial meat cutting machine with a four-arm knife (Panther T & H Industry, Taichung, Taiwan), Nabertherm muffle furnace (Lilienthal, Germany), WTW pH 526 pH meter (WTW, Oberbayern, Germany), Kavalier LT3 shaker (Sázava, Czech Republic), Kern 770 analytical and precision balances (Balingen, Germany), Kern 440-47 laboratory scales (Kern, Balingen, Germany), MEMMERT ULP 400 dryer (Schwabach, Germany), WTB Binder

E-28-TB1 dryer (Tuttlingen, Germany), IKA LABORTECHNIK RCT BASIC magnetic mixer with heating plate (Staufen, Germany), stainless steel sieve with 1.0-mm size, Stevens-LFRA Texture Analyser (Brookfield, UK).

2.2. Strategy

A two-stage biotechnological procedure was designed for the conversion of collagen to gelatines, which due to its variability of the process parameters would allow preparing the gelatines of the desired properties with the optimum utilization of the source material. In industrial practice, factor experiments are widely used to design experiments consisting of multiple technological variables, allowing maximum information to be obtained and optimal process conditions to be designed [23].

To study the influence of selected process factors on gelatine yield and the quality of prepared gelatines, 2^3 factor schemes with one central experiment and one repetition were used; then, in the optimization part of the process, 2^2 factor schemes with one central experiment and one repetition were used. The process factors studied were: factor A—enzyme addition: a minimum value of 0.2% (based on the protein dry matter, w/w), a mean value of 0.5% (w/w), a maximum value of 0.8% (w/w); factor B—enzyme treatment time: minimum 24 h, mean 72 h, maximum value 120 h; factor C—gelatine extraction time: minimum 1 h, mean 2.5 h, maximum value 4 h. The variables evaluated were gelatine yield (percentage conversion of the starting protein substrate to gelatine), the strength of the gelatine gels, the gelatine viscosity, and the ash content of the gelatines. On the basis of statistical evaluation of the influence of process parameters on the evaluated quantities, the process optimization was carried out. The process was based on monitoring the influence of the two most important process parameters on the evaluated variables and the design of optimal processing conditions for the conversion of collagen protein of chicken feet to high-quality gelatines.

2.3. Grinding and Preparation of Raw Material for Biotechnological Processing

When preparing the starting material for further biotechnological processing, it is necessary to set conditions to avoid the denaturation of the collagen protein, convert collagen to gelatine in biotechnological processing with optimum efficiency, and produce quality gelatines. Immediately after separating the feet, the raw material was rinsed in water and cooled to 0–5 °C to prevent negative microbial growth (the chilled raw material can be stored for a maximum of 36 h after slaughter).

Prior to two-phase grinding (on a meat cutter), the feedstock was slightly pre-frozen at −2 to −5 °C in the core of the feedstock. In the first milling process, kidney-shaped cutting plates with a hole size of 20.0–30.0 mm were used, the second was a hole size of 3.0 mm; during grinding, the temperature of the raw material increased to a maximum of 3 °C. If the ground and homogenized raw material is not processed within a maximum of 24 h, it must be wrapped in vacuum containers with a minimum wall thickness of 80 μm and deep-frozen at −36 ± 2 °C; the frozen raw material can be stored at −20 ± 2 °C for up to 24 months. Furthermore, the homogenized raw material needs to be purified from the accompanying non-collagenous components, pigments, and fat. The removal of albumins, globulins, and pigments was based on the earlier procedure [24] with slight modifications. The raw material was mixed with 0.10% NaOH solution in a ratio of 1:8 (w/v) and shaken for 45 min at room temperature; then, the solution was filtered through a 1.0-mm stainless steel sieve, and the raw material was washed with running water; the whole procedure was repeated four times. Then, the raw material was dried in an air circulating oven at 35.0 ± 0.5 °C (36 h).

The defatting of poultry tissues is described in the literature by several methods, such as the Soxhlet extraction process [21], using $NaHCO_3$ [24] or by mechanical separation after fat leaching in alkaline medium [25]. Defatting of the raw material was carried out according to the procedure previously described [26]. The dried raw material was mixed with a mixture of ethanol and petroleum ether (1:1, v/v) solvent in a ratio of 1:6 (w/v) and shaken for 32 h at room temperature; after 8 h, the solvent mixture was replaced with a new one (filtration was carried out through a stainless sieve with a size of 1.0 mm). The defatted raw material was spread on a metal sheet, and the solvent residues were evaporated in a fume hood at room temperature.

2.4. Biotechnological Processing of Raw Material and Gelatine Extraction

As there is a lack of biotechnological processing of the raw material prior to gelatine extraction in the available literature, such a procedure has been designed and tested in our workplace. We proceeded from the results of our previous research, which focused on the preparation of protein hydrolysates from chicken feet [27]. The defatted raw material was mixed with distilled water in a ratio of 1:10 (w/v); after 30 min, the pH of the mixture was adjusted to 7.5 ± 0.3 (optimal efficiency of the proteolytic enzyme used) by the addition of 10% HCl. After the addition of Polarzyme 6.0 T in an amount according to factor A (enzyme load based on the dry weight of the defatted raw material), the mixture was shaken at room temperature for a time according to factor B. After filtration (through a 1.0-mm stainless steel sieve), the protein hydrolysate solution was first brought to boil and was boiled for 10 min (inactivation of proteolytic enzyme). Then, it was dried (after pouring onto a stainless steel plate) in a thin film in an air-circulating oven at 45.0 ± 1.0 °C (48 h); afterwards, the dried film was scraped off and weighed. The enzyme treated raw material was thoroughly washed on the sieve under running water and then mixed in a beaker with distilled water in a ratio of 1:9 (w/v). The mixture was placed on a hot plate, stirred gently (magnetic stirrer), heated to 80.0 ± 0.5 °C ($dt/d\tau = 10$ °C/min), and after reaching this temperature, gelatine was extracted according to the factor C, time. After completion of the extraction, the gelatine solution was separated by filtration through a 1.0-mm stainless steel sieve equipped with three layers of polyamide fabric (300-µm pore size). The gelatine solution was brought to the boil and was boiled for 5 min; then, it was poured onto a thin-plate thin film and dried in an air-circulating oven at 45.0 ± 1.0 °C (48 h); afterwards, the dried film was scraped off and weighed. The undissolved residue of the raw material after gelatine extraction was dried in an air circulating oven at 103.0 ± 1.0 °C (for 16 h) and then weighed.

2.5. Analytical Methods

Dry matter, ash, fat, and protein were determined by conventional food methods [28–30]. The dry matter was determined by the indirect method of drying the sample for 18 h at 103.0 ± 2.0 °C; the ash was determined gravimetrically after burning and annealing the sample; fat was determined by Soxhlet extraction; nitrogen was determined by the Kjeldahl method, and the protein content was calculated from the determined nitrogen content by multiplying by a factor of 6.25. Collagen content was calculated from the hydroxyproline content (determined colorimetrically after sample hydrolysis in 6 mol·L^{-1} HCl) by multiplying by a factor of eight [31,32]. Gel strength, gelatine viscosity, and pH were determined according to the Official Procedure of the Gelatine Manufacturers Institute of America [33]. The pH of a 1.5% gelatine solution was determined by potentiometry at a temperature of 35 ± 0.5 °C using a pH meter. The gelatine gel strength was determined from a gel formed from a 6.67% solution prepared according to prescribed conditions by the measuring of force (weight) required to depress a prescribed area of the surface of the sample to a distance of 4 mm. The dynamic viscosity of a 6.67% gelatine solution was determined at 60 °C by measuring the flow time of 100 mL of the solution through a standard pipette; the viscosity was calculated from Equation (1). The yield of the hydrolysate was calculated from the weight of the hydrolysate prepared after the biotechnological treatment of the raw material, the yield of gelatine from the weight of the gelatine (both yields based on the weight of the defatted raw material); further, the total yield was calculated; see Equations (2, 3, 4). The mass balance error is expressed by the percentage difference of the dry matter mass balance between the input (defatted raw material) and the outputs (hydrolysate, gelatine, and undissolved residue); see Equation (5).

$$\eta = \left(A\tau - \frac{B}{\tau}\right) d \qquad (1)$$

$$HY = \frac{m_1}{m_0} \, 100 \tag{2}$$

$$GY = \frac{m_2}{m_0} \, 100 \, (3) \tag{3}$$

$$Y_\Sigma = HY + GY \tag{4}$$

$$MBE = \frac{\lfloor (m_1 + m_2 + m_3) - m_0 \rfloor}{m_0} 100 \tag{5}$$

where η is gelatine viscosity (mPa·s), A and B are pipette constants, τ is efflux time (s), d is solution density (for a 6.67% gelatine solution at 60 °C d = 1.003), HY is the hydrolysate yield (%), m_0 is the weight of the defatted raw material (g), m_1 is the weight of the hydrolysate, GY is the gelatine yield (%), m_2 is the weight of gelatine (g), m_3 is the weight of the undissolved residue (g), Y_Σ is the total yield (%), and MBE is a mass balance error (%).

2.6. Statistical Analysis

A two-level factorial design of experiments and evaluation of the results were carried out with Minitab® 17.2.1 software (Fujitsu Ltd., Tokyo, Japan). Statistical analyses (arithmetic means and standard deviations) were accompanied using Excel 2010 (Microsoft, Inc., Seattle, WA, USA) at the significance level of 5% ($P < 0.05$).

3. Results and Discussion

3.1. Study of the Influence of Process Factors on the Gelatine Yield and Quality of Prepared Products

A schedule of the experiments and summary results of the processing of chicken feet proteins into gelatine and hydrolysates by two-level factor schemes with three factors of concern are given in Table 1. The pH of the prepared gelatines ranged from 6.0 to 6.4, which corresponds to standards for food and pharmaceutical gelatines where pH 4.0–7.5 is prescribed. The pH of commercially produced porcine gelatines ranges from 5.5 to 6.5, with beef gelatines usually between 5.5–7.0, as well as fish gelatines. All the prepared gelatines were characterized by very low ash content (0.61–1.66%), and thus meet the stringent parameters for food and pharmaceutical gelatines (Food Chemical Codex 10; United States Pharmacopoeia 35 NF 30; European Pharmacopoeia; Japanese Pharmacopoeia 15).

Table 1. The experimental design and the results of processing of chicken feet into gelatines (study of the influence of process factors).

| | Factors under Study | | | Collagen Hydrolysate | | Gelatine | | | | | Summary of the Process | |
| | A | B | C | | | | | | | | | |
Exp. No.	Enzyme Addition (%)	Enzyme Treatment (h)	Extraction Time (h)	HY (%)	Ash$_H$[a] ± SD (%)	GY (%)	Ash$_G$[a] ± SD (%)	F ± SD (Bloom)	η ± SD (mPa·s)	pH ± SD	Y_Σ (%)	MBE (%)
1	0.2	24	1	9.8	20.6 ± 0.1	20.1	1.35 ± 0.02	295 ± 3	6.9 ± 0.1	6.2 ± 0.1	29.9	2.0
2	0.2	24	4	10.0	21.8 ± 0.2	27.4	0.61 ± 0.03	273 ± 3	6.5 ± 0.1	6.4 ± 0.1	37.4	4.9
3	0.2	120	1	10.7	23.9 ± 0.1	24.1	1.66 ± 0.01	266 ± 2	5.9 ± 0.2	6.4 ± 0.1	34.8	5.9
4	0.2	120	4	11.0	24.7 ± 0.3	33.5	0.88 ± 0.03	263 ± 2	5.2 ± 0.1	6.0 ± 0.2	44.5	3.8
5	0.8	24	1	11.2	27.2 ± 0.1	36.5	0.93 ± 0.02	241 ± 2	5.1 ± 0.2	6.4 ± 0.1	47.7	2.5
6	0.8	24	4	11.1	28.3 ± 0.2	37.9	0.77 ± 0.02	235 ± 3	4.7 ± 0.1	6.4 ± 0.1	49.0	2.6
7	0.8	120	1	11.4	28.6 ± 0.1	38.3	1.61 ± 0.01	228 ± 2	3.7 ± 0.1	6.3 ± 0.2	49.7	1.6
8	0.8	120	4	11.5	29.4 ± 0.1	39.1	1.32 ± 0.01	206 ± 2	3.1 ± 0.1	6.1 ± 0.1	50.6	0.7
9	0.5	72	2.5	10.9	25.7 ± 0.2	35.4	1.53 ± 0.02	249 ± 3	6.5 ± 0.3	6.1 ± 0.1	46.2	2.6

[a]—based on dry matter; HY—hydrolysate yield; GY—gelatine yield; Y_Σ—total yield; Ash$_H$—ash content in hydrolysate; Ash$_G$—ash content in gelatine; F—gelatine gel strength; η—gelatine viscosity; MBE—mass balance error

The statistical significance of the studied process factors in the observed limits was evaluated using the standard Fisher's significance test and the P-values for a 95% confidence level. For the factor schemes used by us, the critical value is $F = 10.13$, so the higher the F-value is above the critical value, the greater the influence of the process factor. Similarly, results for P-values were evaluated; factors with a value lower than $\alpha = 0.05$ have an effect on the evaluated variables with 95% probability, and the lower the P-value, the greater the influence of the process factor [34]. Table 2 shows the results of analysis of variance for gelatine yield, gelatine gel strength, and gelatine viscosity.

Table 2. Analysis of variance of the experimental design for gelatine yield, gelatine gel strength, and gelatine viscosity (study of the influence of process factors).

	Degree of Freedom	Sum of Squares	Mean Squares	F-Value	P-Value
Response: Gelatine yield, GY (%) = 16.35 + 19.46 A + 0.0341 B + 1.575 C; R^2 = 88.65					
Regression	3	338.71	112.905	13.01	0.008
Factor A (Enzyme addition)	1	272.61	272.611	31.42	0.002
Factor B (Enzyme treatment time)	1	21.45	21.451	2.47	0.177
Factor C (Gelatine extraction time)	1	44.65	44.651	5.15	0.073
Error	5	43.38	8.676		
Total	8	382.10			
Response: Gelatine gel strength, F (Bloom) = 315.85 − 77.92 A − 0.2109 B − 4.42 C; R^2 = 97.20					
Regression	3	5542.4	1847.46	57.87	0.000
Factor A (Enzyme addition)	1	4371.1	4371.13	136.92	0.000
Factor B (Enzyme treatment time)	1	820.1	820.12	25.69	0.004
Factor C (Gelatine extraction time)	1	351.1	351.13	11.00	0.021
Error	5	159.6	31.93		
Total	8	5702.0			
Response: Gelatine viscosity, η (mPa·s) = 8.37–3.29 A—0.0138 B—0.175 C; R^2 = 87.18					
Regression	3	11.87	3.955	11.33	0.011
Factor A (Enzyme addition)	1	7.80	7.801	22.35	0.005
Factor B (Enzyme treatment time)	1	3.51	3.511	10.16	0.025
Factor C (Gelatine extraction time)	1	0.55	0.551	1.58	0.264
Error	5	1.75	0.349		
Total	8	13.61			

3.1.1. Gelatine Yield

The results of the statistical evaluation showed that only factor A (enzyme addition) is statistically significant for the yield of gelatine (GY). The effect of the two most important process factors (enzyme addition and gelatine extraction time) on GY is represented by the contour graph in Figure 1. It is evident that with increasing enzyme addition and at the same time extending the extraction time, GY increases. This trend is particularly evident at lower enzyme additions (up to about 0.4%) and at shorter extraction times (up to about 2.5 h). There is no significant GY growth with any addition of enzymes above 0.5% or extraction time more than 2.5 h. The minimum GY ($\approx$ 21%) is reached under the lower limit of the factors of interest, i.e., 0.2% enzyme addition and 1.0-h extraction time. The maximum GY ($\approx$ 38%) then corresponds to approximately 0.7% enzyme addition and 2.5 h; further increasing the extraction time will no longer affect the increase of GY.

GY values are comparable or better compared to the available results of processing the proteinic poultry tissues into gelatines and hydrolysates. Using an acid extraction procedure and ultrasonic extraction, very low yields of gelatines (4% and 17%) from chicken feet are reported [35]. Almeida, Calarge, and Santana processed chicken feet by pressure extraction in water (120 °C for 20 min) and obtained a 36% yield of low-strength gelatine [36]. Du et al., by processing the chicken and turkey heads in an acidic environment and extracting in two stages (temperatures 50 and 60 °C), prepared high-quality gelatines with yields of 21% and 31% (chicken heads) and 25% and 38% (turkey heads) [24].

Sarbon, Badii, and Howell used the combined alkaline-enzyme processing of raw material to produce high-quality chicken skin gelatine, with a gelatine yield of 16% [21].

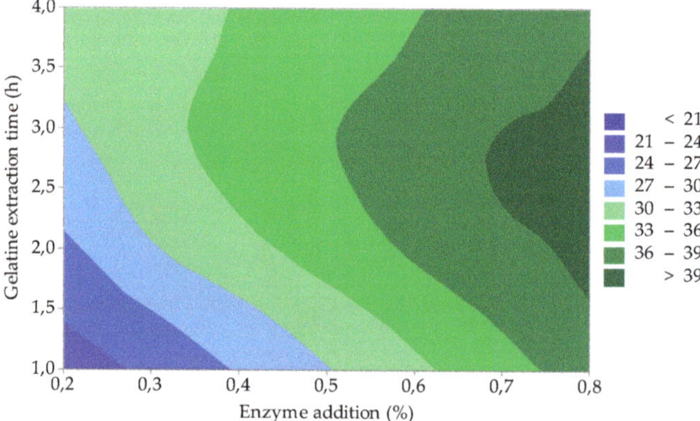

Figure 1. Effect of enzyme addition and gelatine extraction time on gelatine yield (study of the influence of process factors).

3.1.2. Gelatine Gel Strength

The results of the statistical evaluation showed that all three studied factors are statistically significant in the strength of gelatine gel (F); the influence of factor A (enzyme addition) and factor B (enzyme treatment time) on F is represented by the contour graph in Figure 2. It can be seen from the figure that high-quality gelatines (F = 220–280 Bloom) can be prepared within the limits of the process factors studied (0.2–0.8% enzyme addition and raw material enzyme processing for 30 to 120 h). Under the lower limit of the observed factors (0.2% enzyme addition and 30-h enzyme treatment of the raw material, gelatine with F > 280 Bloom can be prepared with a 21% yield of gelatine (see Figure 1). With an almost double yield of gelatine (GY = 38%, see Figure 1) under the upper limits of the observed factors (0.8% enzyme addition and 120-h enzyme treatment of the raw material), it is evident that there is no significant decrease in F: gelatine still has a high gel strength (F ≈ 220 Bloom). The standard specifications for food gelatines prescribe F = 150–280 Bloom, depending on the method of application. For the production of hard gelatine capsules (HGC), F = 200–280 Bloom is prescribed; gelatines with a gel strength of 130–200 Bloom are sufficient for soft gelatine capsules (SGC).

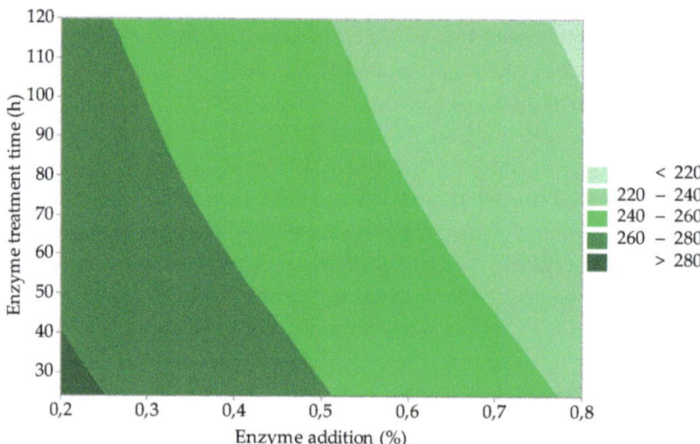

Figure 2. Effect of enzyme addition and enzyme treatment time on gelatine gel strength (study of the influence of process factors).

The gel strength of gelatines prepared under various process conditions is the same or higher than that of commercial high-strength gelatines (*F* > 200 Bloom) made from bovine and pork hides/skins and bones and fish. The gelatines prepared by us have a comparable or lower F compared to gelatines prepared from poultry by-products by other authors. Gelatine prepared from chicken feet by acid treatment of the starting material had a gel strength of 295 Bloom [20]. Du et al. reported *F* = 200–248 Bloom in gelatines prepared from chicken heads; however, higher F values (333–368 Bloom) were achieved in gelatines prepared from turkey heads [24]. The results are similar in high-quality gelatine (*F* = 355 Bloom) from chicken skins [21].

3.1.3. Gelatine Viscosity

The results of statistical evaluation showed that factor A (enzyme addition) and factor B (enzyme treatment time) are statistically significant for gelatine viscosity (η); the influence of these process factors on η is represented by the contour graph in Figure 3. It is evident from the figure that η is the highest (>6.5 mPa·s), and at the same time, its value is not influenced by the change of these technological factors at low enzyme addition (0.2–0.3%) and at short enzymatic treatment times (within 90 h). It is also evident from the graph that with increasing enzyme addition (>0.4%) and at short enzyme treatment times (up to ≈ 90 h), η gradually decreases to 5.0 mPa·s. Long enzyme treatment times (110–120 h) in combination with the highest enzyme additions (0.7–0.8%) mean an even more significant decrease of η (3.5–4.0 mPa·s). Gelatines with η 2.0–7.5 mPa·s cover a wide range of applications in the food industry. For the production of HGC, gelatines with a higher viscosity (4.5–6.5 mPa·s) are required. For SGC, it is 2.5–4.5 mPa·s, and for tablets, the viscosity of 1.7–3.5 mPa·s is sufficient. It can be seen from the viscosity measurement results that gelatines prepared under different process conditions meet the gelatine specifications for the production of HGC and SGC.

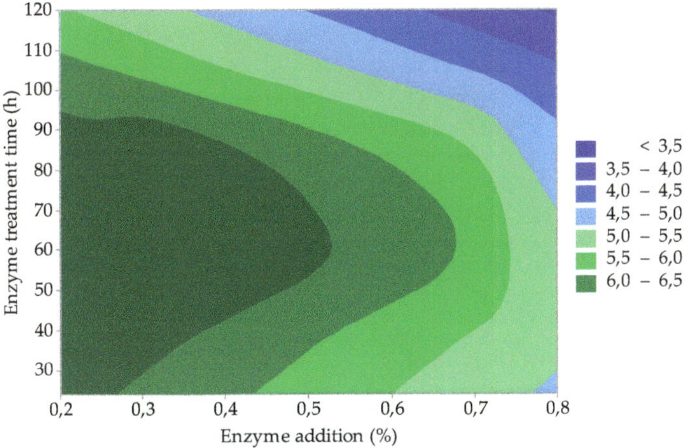

Figure 3. Effect of enzyme addition and enzyme treatment time on the gelatine viscosity (study of the influence of process factors).

3.2. *Process Optimization*

From the study of the influence of process parameters on the gelatine yields and the quality of the prepared gelatines, the following conclusions can be drawn: *a)* with increasing enzyme addition (factor A) and with increasing gelatine extraction time (factor C), the gelatine yield increases; *b)* with increasing enzyme addition (factor A) and with increasing enzyme treatment time (factor B), the strength of the gelatine gels decreases; *c)* the highest gelatine gel strength (≈ 295 Bloom) was recorded under the conditions of the lower process factors observed (0.2% enzyme addition, 24-h enzyme treatment time and 1-h gelatine extraction time); *d)* it can be assumed that the strength of the gelatine gels will grow at lower enzyme additions, shorter enzyme treatment times, and shorter extraction times.

When optimizing the process conditions for the preparation of gelatines from chicken feet, the goal was to prepare high-quality gelatines (with a gel strength of at least 300 Bloom) even at the cost of a lower gelatine yield. It was decided to enzymatically process the raw material for a constant time (20 h) and monitor the effect of enzyme addition (0.1–0.4%) and gelatine extraction time (15–45 min) on the quality of the prepared gelatines (gel strength, viscosity, ash content) and gelatine yield. A schedule of experiments (factor schemes 2^2) and summary results of the optimization part of processing chicken feet proteins to gelatines are given in Table 3.

Table 3. The experimental design and the results of processing of chicken feet into gelatines (process optimization).

Exp. No.	Factors under Study		Gelatine				
	A Enzyme Addition (%)	C Extraction Time (min)	GY (%)	Ash$_G$ [a] ± SD (%)	F ± SD (Bloom)	η ± SD (mPa·s)	pH ± SD
10	0.1	15	17.0	1.31	325	7.3	7.3
11	0.1	45	18.9	1.23	315	7.2	7.2
12	0.4	15	20.6	1.94	308	6.9	7.2
13	0.4	45	21.2	1.58	301	6.8	6.9
14	0.25	30	19.8	1.45	310	6.9	7.2

[a]—based on dry matter; GY—gelatine yield; Ash$_G$—ash content in gelatine; F—gelatine gel strength; η—gelatine viscosity

As shown in Figure 4, gelatines with a very high gel strength can be prepared by the appropriate choice of enzyme addition (factor A) and gelatine extraction time (factor C). In the case of low enzyme addition (up to 0.15%) and short extraction times (up to 20 min), the highest quality gelatine (F = 320–325 Bloom) is prepared; the gelatine yield is approximately 17.5–18.0% (see Figure 5). By increasing the enzyme addition to the upper test limit (0.4%) and extending the extraction time (up to 45 min), the gelatine yield increases to about 21.0% (Figure 5), but gelatine still has a very high gel strength (F ≈ 305 Bloom); see Figure 4.

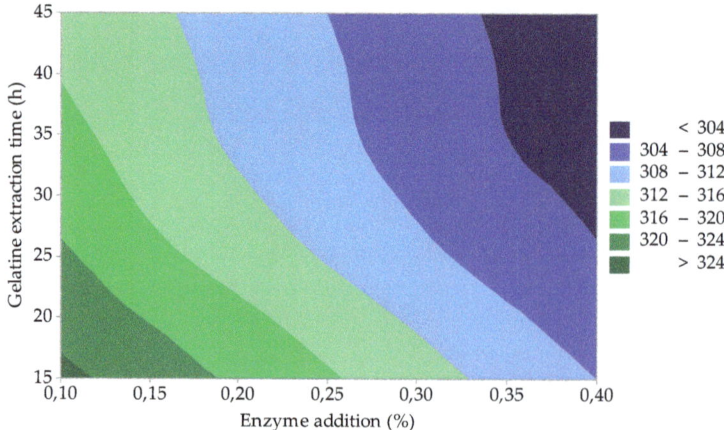

Figure 4. Effect of enzyme addition and gelatine extraction time on gelatine gel strength (process optimization).

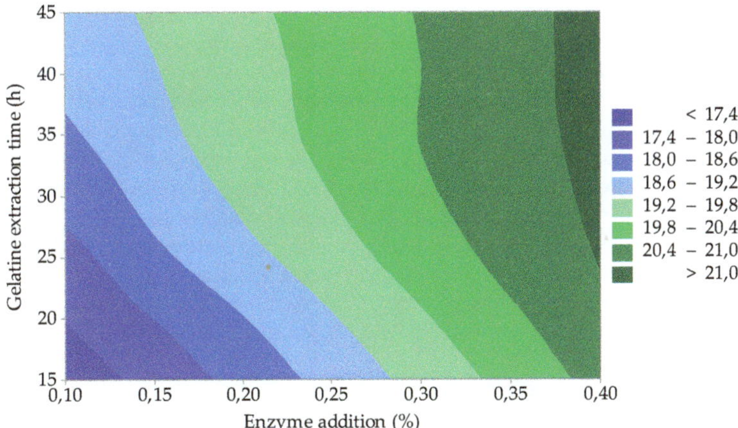

Figure 5. Effect of enzyme addition and gelatine extraction time on gelatine yield (process optimization).

The pH of the prepared gelatines ranged between 6.9–7.3 and the ash content was 1.23–1.94%, which meet both the food standards where the permitted ash content is up to 2.0% (Food Chemical Codex 10) and pharmaceutical standards where the permitted ash content is up to 3.0% (United States Pharmacopoeia 35 NF 30; European Pharmacopoeia; Japanese Pharmacopoeia 15). Gelatines have a higher viscosity (6.8–7.3 mPa·s) and are suitable, for example, in the food industry for the production of deposited marshmallows, or as a binder and gelling agent for meat industry products or for coatings.

The observed technological conditions in the preparation of gelatines have an analogous effect on the yield of gelatines and their quality, as is the case with the production of gelatines from traditional raw materials (beef and pork hides/skins and bones, and fish) by established technologies, i.e., by processing the raw material in an acidic or alkaline environment [1]. By properly selecting the amount and time of action of the enzyme in the processing of the raw material and the appropriate extraction time, gelatines with the desired gel strength and viscosity can be prepared. In industrial practice, it will be advantageous to utilize the multiple fractionation of the raw material at increasing extraction temperatures (typically 60–100 °C), which will result in the preparation of different quality gelatines with optimum utilization of the starting material (maximal gelatine yield). The undissolved residue after the gelatine extraction can be processed by hydrolysis with a suitable proteolytic enzyme into a collagen hydrolysate, which can be used in the food industry (e.g., as nutritional supplements, thickeners, etc.) or as an additive (humectant) in the cosmetic industry.

From the study of the influence of process factors and from the optimization part of the biotechnological preparation of gelatines from chicken feet, it can be stated that gelatines with a gel strength of 220–320 Bloom, viscosity of 3.5–7.3 mPa·s, and an ash content of up to 2.0% can be prepared by a suitable combination of process factors (amount of enzyme, time of raw material enzyme treatment, time and temperature of gelatine extraction). These are gelatines belonging to the category of high-quality food and pharmaceutical gelatines (gel strength of 220–300 Bloom, viscosity of 2.0–7.5 mPa·s) complying with the strict standards prescribed in the European Union, US or Japan (Directorate General for Health and Consumer Protection of the European Commission, USDA, FDA or MHLW).

Chicken gelatines with a high gel strength and a viscosity of about 4.0–5.5 mPa·s are suitable, for example, for the production of gelatine desserts, confectionery products (e.g., gummy bears, extruded marshmallows), and are also perfect for use in the meat industry (aspics, binder for meat emulsions, ham, jellies), in the manufacture of dairy products (low-fat butter spreads, panna cotta, aerated desserts, yogurt-based products), or on frozen semi-finished or finished products. Conversely, gelatines with a lower viscosity (<4.0 mPa·s) are suitable for making chewing gums or sugar sanded/oiled jelly items. For the preparation of hard gelatine capsules, all the prepared chicken gelatines with a viscosity of 4.5–6.5 mPa·s are suitable; for producing soft gelatine capsules, gelatines with a viscosity of 2.5–4.5 mPa·s are sufficient. Due to the high gel strength, chicken gelatines may find various applications in the biomedical field, similar to gelatines produced from pork and bovine feedstock. For example, they can be used as hydrogel carriers for delivering bioactive macromolecules, producing membranes, microspheres, and nanoparticles, encapsulating carriers for the controlled release of biologically active substances, or delivering cell transplants for tissue repairs [37–39]. A high Bloom value has a positive effect on the mechanical properties of gelatine films [40]. In addition, Lai suggested that the gelatine Bloom value has an influence on cellular responses to gelatine material as well [41]. The Bloom value of chicken gelatines is a critical parameter for preparing mixtures with other biopolymers, above all with polysaccharides, to adjust the gelling/thickening properties of such mixtures for food and pharmaceutical applications [42].

4. Conclusions

Most commercial gelatines are made from beef, pork, or fish tissues; industrial processing is based on the acidic or alkaline processing of the feedstock. The presented technology focuses on the preparation of gelatines from by-product collagen raw materials derived from the slaughter of chicken (chicken feet). The innovative technology element brings the biotechnological processing of (purified) feedstock by commercial food endoprotease, which, in contrast to acidic (type A gelatines) or alkaline (type B gelatines) processing, has a variety of economic, technological, and environmental advantages. The feedstock is treated with a small amount of enzyme (0.1–0.8%, based on the dry raw material weight) in neutral environment at room temperature; and the processing time, e.g., compared to alkaline beef processing, is significantly reduced (from weeks to months to tens of hours). After enzymatic treatment and washing of the feedstock in water, a standard hot-water extraction of gelatine follows. By selecting the technological conditions, gelatines with high gel strength (220–320 Bloom) with an ash content of less than 2.0% can be prepared to meet food and pharmaceutical standards. The gelatine yield is 18–38% for the one-step extraction process, which are very good values; a more optimal utilization of the feedstock can then be achieved in practice by a multi-step extraction process. The observed process parameters in the preparation of chicken gelatines also affects the gelatine viscosity (3.5–7.3 mPa·s). The strength of gelatine gels (a key parameter for the application of gelatines) is not significantly altered by the change in process parameters, as all of the gelatines prepared belong to the gelatine category with a high gel strength (>220 Bloom). Chicken gelatines can be an alternative to beef, pork, and fish gelatines, as they meet halal and kosher requirements. For their excellent properties, they are suitable in many food applications; for example, for confectioneries, sweet desserts, dairy products, or meat products. They can also be used in the pharmaceutical field for manufacturing hard gelatine capsules (HGC) and soft gelatine capsules (SGC), as well as various applications in the biomedical field (hydrogels, membranes, carriers, and films).

5. Patents

From the work reported in this manuscript, the following patent resulted: Patent CZ 307665 – biotechnology-based production of food gelatine from poultry by-products.

Author Contributions: Supervision, visualization, writing—original draft preparation, writing—review and editing: P.M. (Pavel Mokrejš); methodology, validation: P.M. (Pavel Mokrejš) and P.M. (Petr Mrázek;) formal analysis, resources: P.M. (Petr Mrázek;) funding acquisition, project administration: R.G.; software: J.P.; data curation, investigation: P.M. (Petr Mrázek) and R.G.

Funding: This research was funded by the Internal Grant Agency of the Faculty of Technology, Tomas Bata University in Zlín, ref. IGA/FT/2019/003.

Acknowledgments: The authors thank to David Dohnal (Olomouc, the Czech Republic) for professional editing of the manuscript.

Conflicts of Interest: The authors declare no conflict of interest. The funders had no role in the design of the study; in the collection, analyses, or interpretation of data; in the writing of the manuscript, or in the decision to publish the results.

References

1. Schrieber, R.; Gareis, H. *Gelatine Handbook—Theory and Industrial Practice*, 1st ed.; Wiley-VCH: Weinheim, Germany, 2007; pp. 45–117.
2. Gómez-Guillén, M.C.; Giménez, B.; López-Caballero, M.E.; Montero, M.P. Functional and bioactive properties of collagen and gelatin from alternative sources: A review. *Food Hydrocoll.* **2011**, *25*, 1813–1827. [CrossRef]
3. Sheela, A.K. Gelatin Market—Global Industry Analysis, Size, Share, Growth, Trends and Forecast, 2012–2018. Available online: http://www.transparencymarketresearch.com/gelatin.html/ (accessed on 22 March 2019).
4. Ferraro, V.; Anton, M.; Santé-Lhoutellier, V. The "sisters" α-helices of collagen, elastin and keratin recovered from animal by-products: Functionality, bioactivity and trends of application. *Trends Food Sci. Technol.* **2016**, *51*, 65–75. [CrossRef]
5. The National Agricultural Statistics Service (NASS), United States Department of Agriculture. Poultry Slaughter 2016 Summary. Available online: https://www.nass.usda.gov/Publications/Todays_Reports/reports/pslaan17.pdf/ (accessed on 22 March 2019).
6. Eurostat Statistics Explained, Meat Production Statistics for EU-28. Available online: https://ec.europa.eu/eurostat/statistics-explained/index.php/Meat_production_statistics#Poultry_meat/ (accessed on 22 March 2019).
7. Jayathilakan, K.; Sultana, K.; Radhakrishna, K.; Bawa, A.S. Utilization of by products and waste materials from meat, poultry and fish processing industries: A review. *J. Food Sci. Technol.* **2012**, *49*, 278–293. [CrossRef] [PubMed]
8. Ockerman, H.W.; Hansen, C.I. *Animal By-Product Processing and Utilization*, 1st ed.; CRC Press: London, UK, 2000; pp. 23–83.
9. Coutand, M.; Cyr, M.; Deydier, E.; Guilet, R.; Clastres, P. Characteristics of industrial and laboratory meat and bone meal ashes and their potential applications. *J. Hazard. Mater.* **2008**, *150*, 522–532. [CrossRef] [PubMed]
10. Pitk, P.; Kaparaju, P.; Vilu, R. Methane potential of sterilized solid slaughterhouse wastes. *Bioresour. Technol.* **2012**, *116*, 42–46. [CrossRef] [PubMed]
11. Gómez-Guilloen, M.C.; Montero, M.V.P. Extraction of gelatin from megrim (*Lepidorhombus boscii*) skins with several organic acids. *J. Food Sci.* **2001**, *66*, 213–216. [CrossRef]
12. Sinthusamran, S.; Benjakul, S.; Kishimura, H. Characteristics and gel properties of gelatin from skin of seabass (*Lates calcarifer*) as influenced by extraction conditions. *Food Chem.* **2014**, *152*, 276–284. [CrossRef]
13. Ahmad, M.; Benjakul, S. Characteristics of gelatin from the skin of unicorn leatherjacket (*Aluterus monoceros*) as influenced by acid pretreatment and extraction time. *Food Hydrocoll.* **2011**, *25*, 381–388. [CrossRef]
14. Norziah, M.H.; Al-Hassan, A.; Khairulnizam, A.B.; Mordi, M.N.; Norita, M. Characterization of fish gelatin from surimi processing wastes: Thermal analysis and effect of transglutaminase on gel properties. *Food Hydrocoll.* **2009**, *23*, 1610–1616. [CrossRef]
15. Nalinanon, S.; Benjakul, S.; Visessanguan, W.; Kishimura, H. Improvement of gelatin extraction from bigeye snapper skin using pepsin-aided process in combination with protease inhibitor. *Food Hydrocoll.* **2008**, *22*, 615–622. [CrossRef]
16. Kasankala, L.M.; Xue, Y.; Weilong, Y.; Hong, S.D.; He, Q. Optimization of gelatine extraction from grass carp (*Catenopharyngodon idella*) fish skin by response surface methodology. *Bioresour. Technol.* **2007**, *98*, 3338–3343. [CrossRef] [PubMed]
17. Cheng, F.Y.; Hsu, F.W.; Chang, H.S.; Lin, L.C.; Sakata, R. Effect of different acids on the extraction of pepsin-solubilised collagen containing melanin from silky fowl feet. *Food Chem.* **2009**, *113*, 563–567. [CrossRef]
18. Liu, D.C.; Lin, Y.K.; Chen, M.T. Optimum condition of extracting collagen from chicken feet and its characteristics. *Asian-Australas. J. Anim. Sci.* **2001**, *14*, 1638–1644. [CrossRef]
19. Chakka, A.K.; Muhammed, A.; Sakhare, P.Z.; Bhaskar, N. Poultry processing waste as an alternative source for mammalian gelatin: Extraction and characterization of gelatin from chicken feet using food grade acids. *Waste Biomass Valoriz.* **2017**, *8*, 2583–2593. [CrossRef]

20. Almeida, P.F.; da Silva Lannes, S.C. Extraction and physicochemical characterization of gelatin from chicken by-product. *J. Food Process Eng.* **2013**, *36*, 824–833. [CrossRef]

21. Sarbon, N.M.; Nazlin, F.B.; Howell, K. Preparation and characterisation of chicken skin gelatin as an alternative to mammalian gelatin. *Food Hydrocoll.* **2013**, *30*, 143–151. [CrossRef]

22. Xu, M.; Wei, L.; Xiao, Y.; Bi, H.; Yang, H.; Du, Y. Physicochemical and functional properties of gelatin extracted from Yak skin. *Int. J. Biol. Macromol.* **2017**, *95*, 1246–1253. [CrossRef]

23. Antony, J. *Design of Experiments for Engineers and Scientists*, 1st ed.; Butterworth-Heinemann: Oxford, UK, 2003; pp. 54–70.

24. Du, L.; Khiari, Z.; Pietrasik, Z.; Betti, M. Physicochemical and functional properties of gelatins extracted from turkey and chicken heads. *Poult. Sci.* **2013**, *92*, 2463–2474. [CrossRef]

25. Huda, N.; Seow, E.K.; Normawati, M.N.; Nik Aisyah, N.M. Preliminary study on physicochemical properties of duck feet collagen. *Int. J. Poult. Sci.* **2013**, *12*, 615–621. [CrossRef]

26. Mrázek, P.; Mokrejš, P.; Gál, R.; Krejčí, O. Preparation of collagen concentrate from chicken feet. *Waste Forum* **2018**, *4*, 444–451.

27. Mokrejš, P.; Gál, R.; Janáčová, D.; Plšková, M.; Brychtová, M. Chicken paws by-products as an alternative source of proteins. *Orient. J. Chem.* **2017**, *33*, 2209–2216. [CrossRef]

28. Nollet, L.M.L.; Toldrá, F. *Handbook of Food Analysis*, 3rd ed.; CRC Press: Boca Raton, FL, USA, 2015; pp. 357–754.

29. ISO 1443:1973. *Meat and Meat Products—Determination of Total Fat Content*; ISO: Geneva, Switzerland, 1973.

30. ISO 937:1978. *Meat and Meat Products—Determination of Nitrogen Content*; ISO: Geneva, Switzerland, 1978.

31. ISO 3496:1994. *Meat and Meat Products—Determination of Hydroxyproline Content*; ISO: Geneva, Switzerland, 1994.

32. Vázquez-Ortiz, F.A.; González-Méndez, N.F. Determination of collagen as a quality index in Bologna from Northwestern Mexico. *J. Food Compos. Anal.* **1996**, *9*, 269–276. [CrossRef]

33. Standard Testing Methods for Edible Gelatin. Official Procedure of the Gelatin Manufacturers Institute of America, Inc. Available online: http://www.gelatin-gmia.com/images/GMIA_Official_Methods_of_Gelatin_Revised_2013.pdf/ (accessed on 18 March 2019).

34. Stange, K. *Angewandte Statistik Teil 2—Mehrdimensionale Probleme*; Springer: Heidelberg, Germany, 1971; pp. 469–495.

35. Widyasari, R.; Rawdkuen, S. Extraction and characterization of gelatin from chicken feet by acid and ultrasound assisted extraction. *Food Appl. Biosci. J.* **2014**, *2*, 83–95.

36. Almeida, P.F.; Calarge, F.A.; Santana, J.C.C. Production of a product similar to gelatin from chicken feet collagen. *Eng. Agrícola* **2013**, *33*, 1289–1300. [CrossRef]

37. Lai, J.Y.; Lin, P.K.; Hsiue, G.H.; Cheng, H.Y.; Huang, S.J.; Li, Y.T. Low Bloom strength gelatin as a carrier for potential use in retinal sheet encapsulation and transplantation. *Biomacromolecules* **2009**, *10*, 310–319. [CrossRef]

38. Lai, J.Y.; Lu, P.L.; Chen, K.H.; Tabata, Y.; Hsiue, G.H. Effect of charge and molecular weight on the functionality of gelatin carriers for corneal endothelial cell therapy. *Biomacromolecules* **2006**, *7*, 1836–1844. [CrossRef]

39. Chou, S.F.; Luo, L.J.; Lai, J.Y.; Ma, D.H.K. On the importance of Bloom number of gelatin to the development of biodegradable in situ gelling copolymers for intracameral drug delivery. *Int. J. Pharm.* **2016**, *511*, 30–43. [CrossRef]

40. Bigi, A.; Panzavolta, S.; Rubini, K. Relationship between triple-helix content and mechanical properties of gelatin films. *Biomaterials* **2004**, *25*, 5675–5680. [CrossRef]

41. Lai, J.Y. The role of Bloom index of gelatin on the interaction with retinal pigment epithelial cells. *Int. J. Mol. Sci.* **2009**, *10*, 3442–3456. [CrossRef]

42. Wang, C.S.; Virgilio, N.; Wood-Adams, P.; Heuzey, M.C. A mechanism for the synergistic gelation properties of gelatin B and xanthan gum aqueous mixtures. *Carbohydr. Polym.* **2017**, *175*, 484–492. [CrossRef]

Article

Compositional and Temperature Effects on the Rheological Properties of Polyelectrolyte–Surfactant Hydrogels

Jiří Smilek *, Sabína Jarábková, Tomáš Velcer and Miloslav Pekař

Faculty of Chemistry, Brno University of Technology, Purkynova 464/118, 612 00 Brno, Czech Republic; xcjarabkova@fch.vut.cz (S.J.); xcvelcer@fch.vut.cz (T.V.); pekar@fch.vut.cz (M.P.)
* Correspondence: smilek@fch.vut.cz; Tel.: +420-541-149-483

Received: 25 April 2019; Accepted: 24 May 2019; Published: 27 May 2019

Abstract: The rheological properties of hydrogels prepared by physical interactions between oppositely charged polyelectrolyte and surfactant in micellar form were studied. Specifically, hyaluronan was employed as a negatively charged polyelectrolyte and Septonex (carbethopendecinium bromide) as a cationic surfactant. Amino-modified dextran was used as a positively charged polyelectrolyte interacting with sodium dodecylsulphate as an anionic surfactant. The effects of the preparation method, surfactant concentration, ionic strength (the concentration of NaCl background electrolyte), pH (buffers), multivalent cations, and elevated temperature on the properties were investigated. The formation of gels required an optimum ionic strength (set by the NaCl solution), ranging from 0.15–0.3 M regardless of the type of hydrogel system and surfactant concentration. The other compositional effects and the effect of temperature were dependent on the polyelectrolyte type or its molecular weight. General differences between the behaviour of hyaluronan-based and cationized dextran-based materials were attributed to differences in the chain conformations of the two biopolymers and in the accessibility of their charged groups.

Keywords: Hyaluronan; carbethopendecinium bromide; sodium dodecyl sulphate; diethylaminoethyl-dextran hydrochloride; rheology; hydrogels

---

1. Introduction

Interactions between polyelectrolytes and oppositely charged surfactants constitute an area of intensive research not only for theoretical reasons but also because of the practical applications of these systems, e.g., in cosmetic, pharmaceutical, and food products [1]. In this work the interactions of polyelectrolytes with surfactants in micellar form are of special interest. Kizilay et al. [1] overviewed the mechanistic and structural features of the complexation and coacervation of polyelectrolytes with oppositely charged colloids, including surfactant micelles. The first interaction step is the linking of the colloid and polyelectrolyte, which is described by two types of model—either the "condensation" of polyelectrolyte chains on the surface of colloidal particles or the binding of colloids as "ligands" to host polyelectrolytes. Generally, polyelectrolyte–colloid interactions depend mainly on the charge per polyelectrolyte repeat unit, the ionic strength, pH and the colloid charge density. The effect of the polyelectrolyte molecular weight is usually negligible [1]. Association typically continues in several subsequent steps, controlled also by the concentration of interacting species: non-interacting, individual polymers and colloids (e.g., due to the low concentration of interacting species or the subcritical colloid surface charge density); primary complexes of an intrapolymer type; soluble aggregates; and coacervates (the whole system is separated into two immiscible liquid phases, one of which—the coacervate—is relatively concentrated in macromolecules). Intrapolymer complexes are formed by individual polymer chains decorated with colloids and have radii similar to those of the maternal polyelectrolyte. Soluble aggregates are formed by the association of primary complexes. Coacervation arises from extended interactions among soluble aggregates (complexes) and is considered to be a true form of liquid–liquid phase separation. In many systems, maximum coacervation is found when the charges of the colloid and polyelectrolyte are neutralized [1]. However, particularly in polyelectrolyte–micelle systems, the coacervation region is broader than the electroneutrality point. Entropic effects, related to counter ion release or the formation of partially charged neutralized droplets also play a role in interactions between oppositely charged polyelectrolyte-colloids [1].

At surfactant concentrations well above the critical micelle concentration, interactions with oppositely charged polyelectrolyte can lead to the separation of a gel-like (viscoelastic) material. Kizilay et al. [2] attributed the viscoelastic properties of coacervate to a "disproportionation" process, during which poly-ions and their counter ions in systems where the poly-ion charge exceeds that of colloids are expelled to domains of 50–300 nm in size. Thus, proximal regions of more complete charge neutralization and higher density are created, which serve as crosslinks in the resulting viscoelastic material. The formation of these domains is driven by similar forces such as the formation of neutral aggregates in solution—the enthalpic contribution of interacting opposite charges—and the entropic contribution from both counter ion release and chain configuration change.

Polyelectrolyte–surfactant gels retain the amphiphilic character of the surfactant building blocks—inside the hydrophilic gel matrix they contain hydrophobic micelle-like domains capable of solubilizing non-polar species [2]. However, these materials are still underexplored and only limited information on their properties can be found in the literature. Perhaps the most information can be found on hyaluronan–cationic surfactant systems. Thalberg and Lindman [3] describe the separation of a few volume percent of a highly viscous gel-like phase after mixing hyaluronan with decyl-, dodecyl, tetradecyl- or cetyltrimethylammonium bromide. Just the gel phase formed by hyaluronan and tetradecyltrimethylammonium bromide was studied by Wong et al. using NMR techniques [4]. They observed that, in the gels, the surfactant molecules formed relatively small micellar aggregates which were bound to the polyelectrolyte chains. The surfactant aggregates seemed to be completely covered with polyelectrolyte chains at polyelectrolyte/surfactant ratios above 0.5 (by weight). Thalberg and Lindman [5] prepared the gels from hyaluronan of various molecular weights and decyl-, dodecyl, or tetradecyltrimethylammonium bromide. The formation of the gel phase was observed for the hyaluronan molecular weight below about 20,000 g·mol^{-1}. No crystalline phases were detected with small-angle X-ray scattering. The gels were able to solubilize a hydrophobic dye (in contrast to hyaluronan–water systems); thus they should contain hydrophobic domains. Proton NMR relaxation revealed the presence of micellar aggregates in the gels. Analogical self-diffusion studies showed that surfactants in micellar solutions had self-diffusion coefficients of the same order of magnitude as those measured in the gels. The surfactant self-diffusion was attributed both to the diffusion of its monomers within the micellar aggregates and to the monomer exchange between different aggregates. Water self-diffusion was reduced in the gels by obstruction and the presence of hydration water within the gel structure.

Buchold et al. [6] investigated hyaluronan–tetradecyltrimethyl ammonium bromide systems at high surfactant concentrations and in the presence of 160 mM NaBr. They also presented a phase diagram covering several decades of surfactant concentration in which a two-phase area can be observed. The authors reported that one of the phases was a turbid, dense gel which was not formed under salt-free conditions, a finding which is consistent with our previous study [7]. However, only single phase solutions with very high surfactant concentrations (245 mM and more) were further investigated in that study [8].

Recently, we reported on the rheological properties of hyaluronan–cetyltrimethylammonium bromide hydrogels and demonstrated how they can be controlled over a broad range of liquid-like to solid-like states by the polyelectrolyte molecular weight [7]. In this work, we focused in more detail on the effects of various processing parameters on the properties of similar gels prepared using a slightly different cationic surfactant (approved for use in pharmaceutical formulations) and also investigated a reversely charged system—a positively charged polyelectrolyte (cationized dextran) and an anionic surfactant.

2. Materials and Methods

The sodium form of hyaluronan (HYA) was purchased from Contipro (Dolní Dobrouč, Czech Republic) and used as anionic polyelectrolyte without further treatment. In this study, two types of hyaluronan were used, high and low molecular weight hyaluronan (HMW and LMW, respectively). Diethylaminoethyl-dextran hydrochloride (Sigma-Aldrich, Prague, Czech Republic, batch BCBQ8681 with 3.0% of nitrogen; DEAED) was used as received as the cationic polyelectrolyte. The exact molecular weights of all polysaccharides were checked by the SEC-MALLS technique and the results are shown in Table S1 (Supplementary Material).

Carbethopendecinium bromide (Septonex, Czech Pharmacopoeia quality), a pharmaceutical cationic surfactant, was purchased from GBNchem (Prague, Czech Republic). Sodium dodecyl sulphate (SDS, ≥99.0%), purchased from Sigma Aldrich (Czech Republic), and was used as received as the anionic surfactant.

The composition of the gel samples was selected on the basis of previous experience and preliminary experiments. Basically, when increasing the concentration of polyelectrolyte at a constant surfactant concentration, or vice versa, the gel started to form at a certain concentration, then its amount increased up to a maximum, after which it decreased upon a further increase in surfactant concentration. Therefore, we selected concentrations at which sufficient amounts of gel were formed and at which the viscosity of the polyelectrolyte solutions was not too high to complicate the mixing and preparation of the samples. The surfactant concentration should be well above its critical micelle concentration (cf. Table 1).

Table 1. Concentrations of initial solutions of hyaluronan (HYA), DEAED (Diethylaminoethyl-dextran hydrochloride), Septonex and SDS (Sodium dodecyl sulphate) used to prepare hydrogels by the solution method.

| Sample Name | Hyaluronan | | DEAED | Septonex | SDS | Charge Ratio |
	HMW (High Molecular Weight)	LMW (Low Molecular Weight)				
	% (*w/v*)			mM		
H1	2			200		4.0
H2	2			100		2.0
H3	2			50		1.0
H4		2		200		4.0
H5		2		100		2.0
H6		2		50		1.0
D1			4		400	4.7
D2			4		100	1.2

Hydrogels were prepared in two ways—the solution or powder method. The solution method was based on the simple mixing of the pre-prepared surfactant and polyelectrolyte solutions with the required concentrations. We found that gels could also be prepared by pouring the liquid solvent over the mixture of powdered polyelectrolyte and surfactant; this was called the powder method, which is patent pending [9].

The solution method was realized by mixing a stock solution of polysaccharide and a stock solution of surfactant in a volume ratio of 1:1 in a vial and leaving the vial on a shaker overnight to complete the gelation process and the separation of the gel phase. Stock solutions were prepared in 0.15 M NaCl using deionized water (Purelab Flex, ELGA system), because previous experiments showed that a non-zero ionic strength of the aqueous medium is important for obtaining gel-like materials [7]. The initial concentrations of stock solutions are listed in Table 1. The theoretical charge ratio was calculated supposing the existence of one charge on each surfactant molecule, one charge per hyaluronan basic dimeric unit (sodium form) with a molecular weight of 401.299 g·mol^{-1}, and one charge per each nitrogen atom in DEAED.

In the powder method, required amounts of biopolymer and surfactant in powder form were weighed directly into the vials. Then, an appropriate volume of 0.15 M NaCl aqueous solution was added. The final concentrations of the resulting mixtures were half of those listed in Table 1. The samples were vortexed briefly and left overnight to complete the gelation process and phase separation.

Alternatively, phase separation in both methods was accelerated by centrifugation. The polysaccharide and surfactant stock solutions were thoroughly mixed in a centrifuge tube and centrifuged at 4000 rpm for 10 min.

The effect of multivalent ions was investigated by preparing samples by the powder method and using corresponding salt solutions ($CaCl_2$, $MgCl_2$ and $FeCl_3$) at a concentration of 0.15 M (instead of NaCl). Similarly, the effect of pH was studied using appropriate buffer solutions of ionic strength 0.15 M. The composition of buffer solutions is summarized in Table S2. In addition, samples in the study of the effect of ionic strength and pH were prepared by the powder method using aqueous solutions of NaCl at different concentrations (0.05–1.0 M).

Rheological measurements were performed on an AR-G2 rheometer (TA Instruments, New Castle, DE, USA) using steel plate-plate geometry (a diameter of 25 or 8 mm, gap size of 100 μm). Experiments were carried out at least in duplicates, under a controlled temperature of 25 or 37 °C. Each sample was equilibrated for 3 min at the given temperature before measurement (conditioning). The maximum normal force used for compressing the sample did not exceed 5 N. During measurement, a solvent trap was used to prevent water evaporation and subsequent changes in the hydrogel structure. First, the linear viscoelastic region (LVR) was determined by strain sweep tests (deformation of 0.01–1000%, frequency of 1 Hz, 6 points per decade). Viscoelastic properties (elastic and viscous moduli, complex viscosity) as a function of oscillation frequency were then determined by frequency sweep tests in the range of 0.01–20 Hz, the deformation chosen within the LVR (the chosen amplitude of deformation was the same for all frequency sweeps). Steady state shear experiments were performed in the shear rate range of 0.02–200 s^{-1} (logarithmic sweep, 6 points per decade). Each experiment was conducted at least twice. The results are the average of all measurements. The standard deviation did not exceed 7% in either case—frequency or flow measurement.

The fundamental rheological parameters such as relaxation moduli (G) and relaxation time (λ) were calculated from oscillatory measurements with respect to the Maxwell model [10]. For each hydrogel, the mesh size was calculated from the frequency sweep measurements according to Equation (1) [10]

$$\xi = \sqrt[-3]{\frac{G_\infty}{k_B T}} \qquad (1)$$

where ξ represents the mesh size calculated in [m], G_∞ is the value of the storage modulus when the plateau has been reached (or, if the plateau is not reached, the value of the storage modulus at the maximum frequency of oscillation), k_B is the Boltzmann constant, and T represents the absolute temperature.

An alternative way of determining the mesh size was also tested. The mesh size was calculated according to Pescosolido et al. [11] On the basis of measurements and rheological investigation, it was concluded that the optimal number of Maxwell elements corresponds to four in order to fit the hydrogel viscoelastic curves. According to this finding, it is possible to calculate the hydrogel shear modulus as the sum of all relaxation moduli.

The mechanical properties of polyelectrolyte–surfactant hydrogels can be analysed using rubber elasticity theory [12]. The application of this theory on biopolymer hydrogels has been questioned. Nevertheless, recent studies [13] have shown that for hydrogels with elastic components, such as the polyelectrolyte–surfactant hydrogels described, this theory can be applied. One of the most important assumptions is that the mechanical properties of hydrogels are determined in the linear viscoelastic region (the amplitude of deformation in frequency sweep tests must always be chosen from the linear viscoelastic region). If this condition is met, a new parameter—crosslink density (ρ_x)—can be calculated according to Equation (2) [11].

$$\rho_x = \frac{G}{RT} \tag{2}$$

where R is the universal gas constant, T represents the absolute temperature, and G is the shear modulus. From the crosslink density, which can give us information about the density of the junction (unit: mol·m^{-3}), the mesh size can be calculated with respect to Equation (3) [11].

$$\xi = \sqrt[3]{\frac{6}{\pi \rho_x N_A}} \tag{3}$$

3. Results and Discussion

3.1. The Effect of the Preparation Method

Hydrogels of the same composition were prepared by different techniques, as described in Materials and methods. First, the solution and powder methods were compared. It was found that both methods gave materials with very similar rheological properties in both oscillatory and shear tests, which indicates that a final, equilibrium state was reached. An example can be found in Figure S1 (Supporting material), where the frequency sweep for H1 is shown. The independency of viscoelastic properties on the preparation method was observed for all tested hyaluronan and dextran hydrogels. This fact is important also from the point of view of potential applications—it would not be necessary to prepare a solution but just to mix powders and the liquid dispersion medium directly, which would be more cost-effective.

The separation of hydrogels by free standing, i.e., by gravitation, was rather time consuming (the phase preparation of hydrogels took at least 24 h), which is undesirable with respect to potential commercial production. Therefore, the preparation methods were modified by centrifuging the system after mixing the components. The accelerated separation of the gels did not affect their rheological properties (data not shown).

Thus, the various preparation methods did not affect the mechanical properties of the resulting hydrogels. For further investigations, gel samples were separated by the centrifugation method.

3.2. The Effect of Surfactant Concentration

3.2.1. Oscillatory Tests

Strain sweep measurements primarily give information on the extent of the linear viscoelasticity region, but can also be used to detect the effects of composition. Generally, the surfactant concentration demonstrated a very small effect on the shape of the strain sweep curves and corresponding moduli (an example is given in Figure S2 in Supplementary Material).

The storage modulus of HMW hyaluronan-based gels was higher than the loss modulus throughout the whole linear viscoelasticity region, stressing the gel-like character of all materials. The highest G'/G'' ratio was found for intermediate surfactant concentrations (samples H2, H5), which indicated the increased rigidity of densely crosslinked materials. In contrast, the loss modulus of LMW hyaluronan-based gels was higher than the storage modulus. The length of the linear viscoelasticity region of hyaluronan-based gels was dependent on the surfactant concentration—interestingly, the length increased with the concentration of surfactant for HMW hydrogels and decreased for LMW hydrogels (cf. Table S3 in Supplementary Material). Thus, the hyaluronan molecular weight influenced the structural destruction of the gels—longer chains are necessary to prepare materials with enhanced resistance to the mechanical destruction of their networks. The lowest value of the storage modulus in the linear viscoelasticity region was found for sample H3, where the charge ratio was 1:1. This means that micelles could be bound both in an electrostatic and steric manner in an excess of surfactants in the case of samples H1 and H2, which possessed the most rigid behaviour.

Hydrogels prepared from DEAED and a lower concentration of SDS (sample D2) exhibited higher absolute values of both moduli (elastic and viscous). The linear viscoelastic region was shorter for the D1 sample. Viscous moduli exceeded elastic moduli during the whole of each measurement and the difference between elastic and viscous moduli was the same for both concentrations of SDS (the example is given in Figure S3 in Supplementary Material).

The frequency sweep is the main method of oscillatory rheometry. The effect of surfactant concentration on frequency sweep results was dependent on the hyaluronan molecular weight. Gels prepared from HMW hyaluronan showed similar shapes of the measured curves (see the example on Figure S4 in Supplementary Material). Also, the modulus value at the crossover point was only weakly dependent on the surfactant concentration. However, the crossover frequency was shifted to higher values with increasing surfactant concentration, i.e., the corresponding relaxation time decreased in the same time (Table S4). The calculated mesh size slightly decreased when the surfactant concentration increased and was generally smaller for HMW hyaluronan gels. The smaller mesh size for HMW hyaluronan gels can be explained by the more dense coiled structure of HMW hyaluronan in comparison with LMW biopolymer, which is subsequently crosslinked by surfactant micelles [14] (see Table 2).

Table 2. Cross-over frequency and modulus, mesh size for all prepared hydrogels.

Sample Name	Cross-over Frequency (Hz)	Cross-over Modulus (Pa)	Mesh Size (nm)
H1	0.14	324.5	13.01
H2	0.10	287.8	14.56
H3	0.05	314.2	16.41
H4	0.01	65.6	20.4
H5	4.64	242.2	22.8
H6	2.15	199.1	21.6
D1	10.00	5765	8.94
D2	6.81	11430	7.12

The sample of HMW hyaluronan gels with the lowest concentration of surfactant (H3) had theoretically saturated all potential sites of electrostatic interactions (the charge ratio was 1). Of course, this supposes the total dissociation and accessibility of all charged groups, which, in real samples, will not likely be attained [6]. Nevertheless, at an increased surfactant concentration, at least some micelles are not bound electrostatically but physically entrapped in the gel network and function like a (nano-sized) filler. This could also explain why the viscoelastic properties change slightly with an increasing concentration of surfactant and why the mesh size is slightly decreased with an increasing concentration of surfactant.

In the case of LMW hyaluronan, almost the same general change in moduli with surfactant concentration as in HMW hyaluronan-based hydrogels was observed. However, due to the shorter chain length, gels with predominantly elastic moduli could be prepared only when using the highest Septonex concentration (200 mM), which means that excess surfactant (micelles), can participate in the network as an elastically active component. Lower concentrations of Septonex (samples H5, H6) caused the loss moduli of hydrogels to be higher over the whole range of frequencies. Materials formed from LMW hyaluronan generally behaved more like a viscoelastic liquid than a (soft) solid.

The values of the principal measured or calculated parameters are given in Table 2. Discrete relaxation spectra were obtained with five elements (see Table S4)—the relaxation spectra for HWM hyaluronan hydrogels practically overlapped and showed a plateau during the first three or four relaxation times followed by a noticeable decrease (Figure S5). In the case of LMW hyaluronan hydrogels, relaxation moduli decreased significantly with increasing relaxation time. This decrease was practically independent of the Septonex concentration.

In the case of DEAED-based hydrogels, increased surfactant concentration significantly changed their rheological behaviour. For samples with lower surfactant concentrations, both viscoelastic moduli were up to one order of magnitude higher than in the case of samples with higher surfactant concentrations; the crossover point was shifted to a somewhat lower frequency with a higher value of crossover modulus (Table 2). Higher moduli values for samples with a lower concentration of surfactant mean tougher hydrogels (Figure 1).

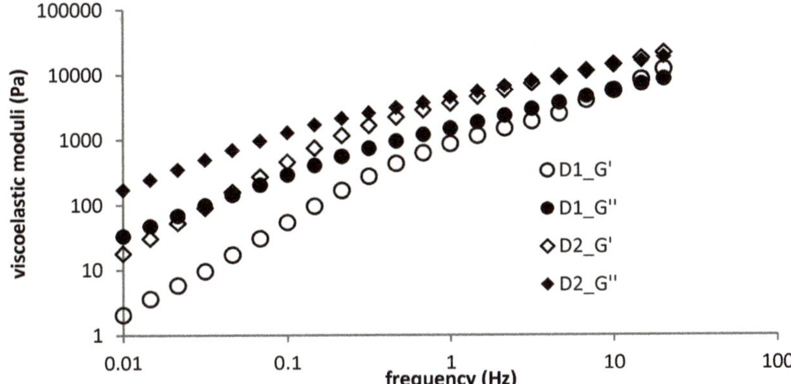

Figure 1. Frequency sweep for diethylaminoethyldextran hydrochoride (DEAE) hydrogels—concentration dependence.

The gels prepared with higher surfactant concentrations also demonstrated a much lower slope in their relaxation spectra (Figure S6 in Supplementary Material), i.e., slower relaxation. In this case, the excess of surfactant functioned much more like a "lubricant" than a reinforcing filler, causing a deterioration in mechanical properties.

3.2.2. Flow Curves

All hydrogels were sufficiently soft to be subject to flow curve measurements. In contrast to oscillatory tests, much bigger deformations are employed in these tests and the network structure is thus definitely disturbed. Generally, using an oscillatory technique, the structure of hydrogels should not be damaged within the linear viscoelasticity region.

All hydrogels demonstrated pseudoplastic behaviour with a Newtonian plateau at low shear rates. An increased surfactant concentration in hyaluronan-based gels resulted in increased zero-shear viscosity (see Table S5 in Supplementary Material); i.e., micelles worked as thickening fillers. LMW hyaluronan hydrogels had lower viscosities than their HMW counterparts, which is attributable to their shorter chain lengths [15,16]. Their HMW counterparts also possessed a much shorter Newtonian plateau and a steeper viscosity decrease after this plateau. DEAED-based hydrogels manifested increased zero-shear viscosity at the lower surfactant concentration, which is consistent with findings in oscillatory tests, whereas the width of the Newtonian region was weakly dependent on the surfactant concentration (Table S5 in Supplementary Material).

The observed pseudoplastic behaviour was a result of relatively weak bonding interactions, which could be relatively easy disrupted by low shear rates. These physical interactions were destroyed at high shear rates and the biopolymer chains in tattered hydrogels were oriented in the direction of flow, decreasing the apparent viscosity.

The differences between the rheological behaviour of hyaluronan- and DEAED-based gels may be attributed to the structural differences in their polysaccharide backbone [17]. Hyaluronan probably possesses more opportunities for hydrogen bonding and forms a single strain alpha helix [18], whereas dextran adopts a ribbon like conformation with two antiparallel chains [19–22]. Further, there should be differences in the accessibility of the charged groups amenable to electrostatic interactions. Whereas the positively charged groups protrude from the dextran chain, the negatively charged carboxyls on hyaluronan are positioned much closer to its backbone—see the structures in Figure 2. The impact of such accessibility on polyelectrolyte–surfactant interactions was reported by Buchold et al. [6].

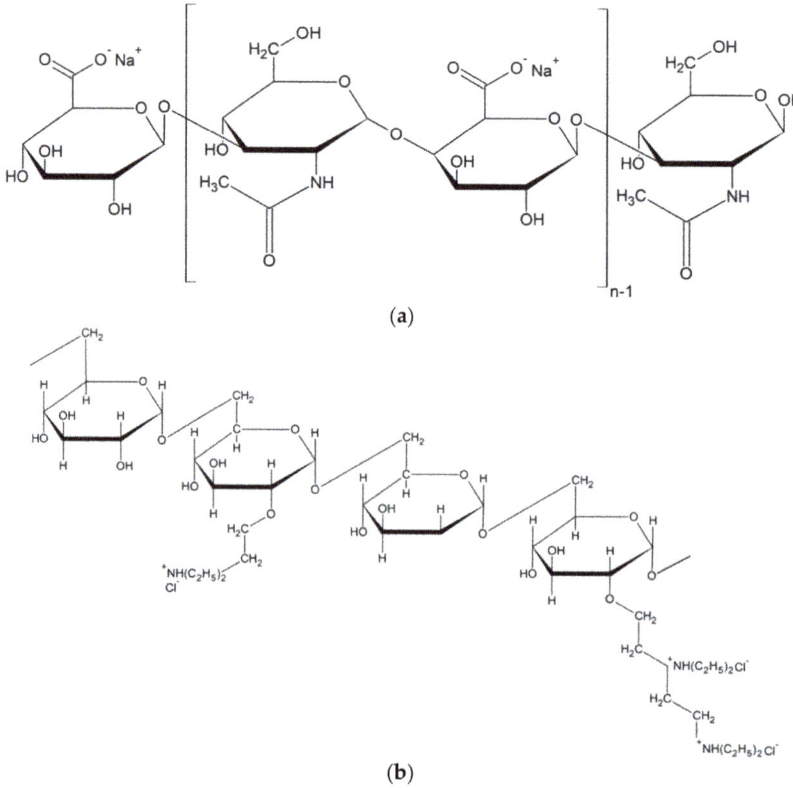

(a)

(b)

Figure 2. (a) Structure of sodium hyaluronate; (b) structure of modified dextran (DEAE—diethylaminoethyldextran hydrochoride).

3.3. Effect of Temperature

The viscoelastic properties of biopolymer–surfactant hydrogels were determined at 25 and 37 °C. These temperatures were chosen purposefully, because of the potential dermal applications of these hydrogels. Increased temperature had a very small effect on the strain sweep curves of both LMW and HMW hyaluronan-based hydrogels. The linear viscoelasticity region remained almost unchanged, and moduli very slightly increased with increased temperature in all hydrogels prepared from hyaluronan (for an example see Figure S7 in Supplementary Material).

In the case of DEAED-based hydrogels, the temperature effect on the moduli was rather stronger, while the linear viscoelasticity region also remained almost unchanged for both hydrogels with different concentrations of SDS (an example is shown in Figure S8 in Supplementary Material). Both viscoelastic moduli were slightly shifted to higher values at the elevated temperature.

The temperature effect on frequency sweep measurements was dependent on the hyaluronan molecular weight. In the case of HMW Ha, temperature increased the values of both moduli (more distinctly, the elastic modulus) and shifted the crossover point significantly to lower frequencies; also, the crossover modulus value was decreased (an example is shown in Figure 3a). The relaxation spectrum remained almost unchanged when the temperature increased (see Figure S9 in Supplementary

Material). Thus, to some extent, a higher temperature stiffened the structure of the hydrogel prepared from HMW hyaluronan.

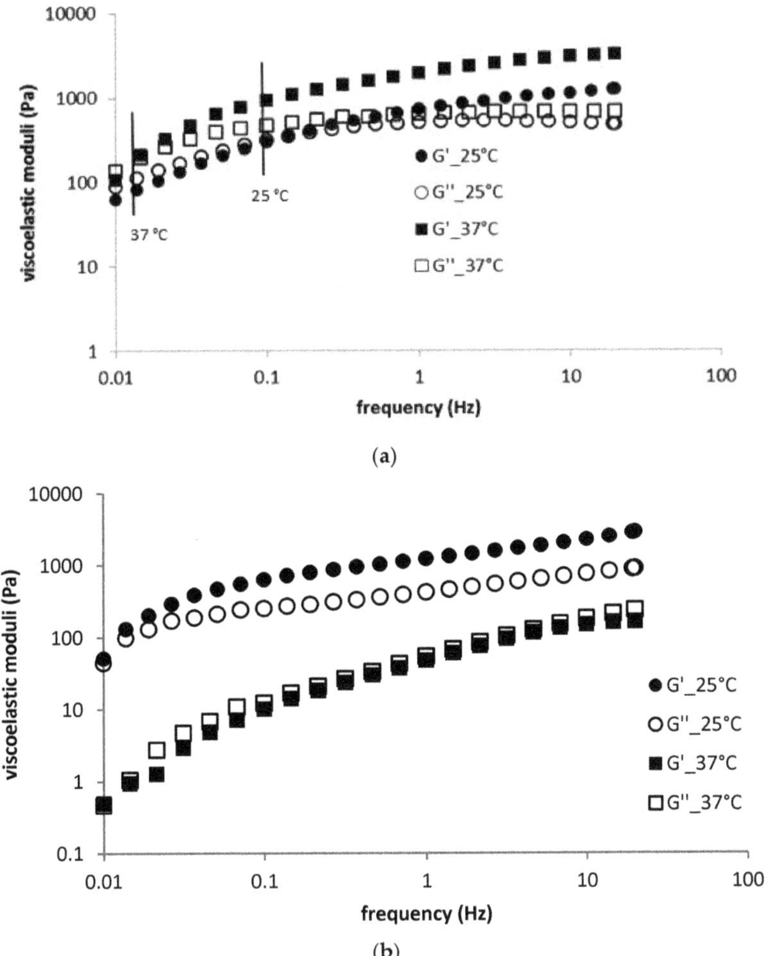

(a)

(b)

Figure 3. (a) Frequency sweep for H1 sample—temperature dependence; (b) frequency sweep for H4 sample—temperature dependence

In contrast, increased temperature significantly reduced the moduli of LMW hyaluronan-based hydrogels and both moduli exhibited very similar values at the higher temperature over the whole frequency range (an example is given in Figure 3b). The crossover point seemed to remain essentially unchanged from the point of view of frequency, whereas the crossover modulus value was probably significantly lower at the elevated temperature (determining the exact crossover location is problematic due to the almost completely overlapping corresponding curves). The relaxation spectrum was shifted downwards and to the left with temperature (cf. the example in Figure S9 in Supplementary Material).

The polymer molecular weight is thus crucial for the heat resistance of the structure and elasticity of hyaluronan polyelectrolyte–surfactant hydrogels. Short chains form a looser network structure which is released by a moderate increase in temperature. In contrast, the structure formed by long chains is stiffened, which suggests the accessibility of more crosslinking points at the elevated temperature, probably as a result of the higher molecular mobility and conformational rearrangements of biopolymer chains.

The temperature effect on the frequency sweep of DEAED–SDS hydrogels (at both concentrations of SDS) was manifested by decreased moduli values and increased crossover frequencies at the elevated temperature (example is shown in Figure 4). The crossover modulus value was not substantially changed by changes in temperature. The relaxation spectrum was shifted downwards with increased temperature for both systems (see the example in Figure S10 in Supplementary Material). The behaviour at the elevated temperature was thus similar to that of LMW hyaluronan-based gels.

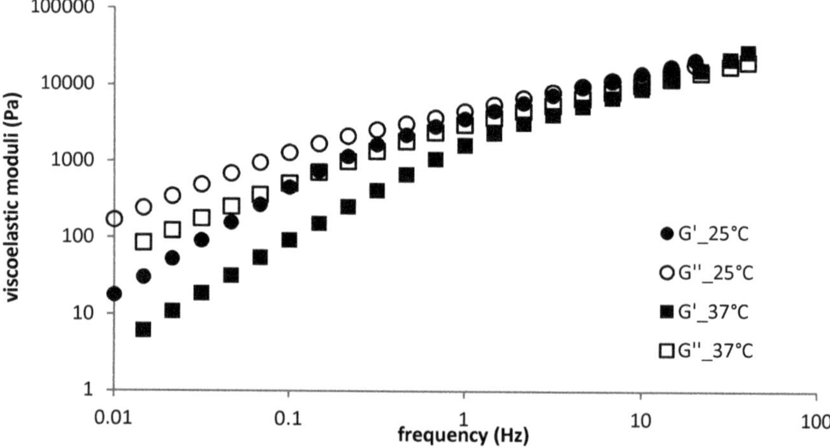

Figure 4. Frequency sweep for sample D2—temperature dependence.

Flow curves were not significantly affected by temperature in the case of HMW hyaluronan-based hydrogels and DEAED–SDS hydrogels. The disturbed network structure of the resulting fluid-like materials was thus not sensitive to the applied temperature increase. LMW hyaluronan-containing materials demonstrated increased viscosity and no Newtonian plateau at the higher temperature, as shown in Figure S11 in Supplementary Material.

3.4. Effect of Ionic Strength

The viscoelastic properties of HMW hyaluronan-based hydrogels were strongly dependent on ionic strength. Hydrogels with lower ionic strength (0.05 M NaCl) had a higher elastic modulus than viscous modulus at higher frequencies of oscillation. The crossover point was shifted to higher frequencies in comparison with a standard sample (0.15 M NaCl); moreover, both moduli exhibited lower absolute values. On the other hand, at a higher ionic strength (0.3 M NaCl), both moduli were distinctly shifted to lower absolute values in comparison with standard samples or hydrogels with a lower ionic strength. The viscous modulus exceeded the elastic modulus over the whole range of measurable frequencies, which means that materials with a higher ionic strength behave more liquid-like. Moreover, hydrogels at yet higher ionic strengths (>0.3 M NaCl) could not be prepared—there was no gelation process.

The same behaviour was also observed for LMW hyaluronan-based hydrogels. Furthermore, the critical ionic strength of the LMW hyaluronan-hydrogels gelation process was lower than for HMW hyaluronan-based hydrogels—when an ionic strength of 0.15 M NaCl was exceeded, the gelation process was not observed.

DEAED–SDS-based hydrogels with the highest ionic strength (0.5 M) could not be prepared due to surfactant solubility problems in this environment. Samples with lower (0.05 and 0.1 M) and higher (0.3 and 0.5 M) ionic strengths than the standard ionic strength (0.15 M) were prepared without problems. On the other hand, hydrogel based on DEAED–SDS in pure water without the adjustment of ionic strength could not be prepared. Non-zero ionic strength is required for the preparation of DEAED–SDS hydrogel (at least 0.05 M).

The linear viscoelastic region of DEAED–SDS hydrogels was independent of ionic strength and was comparable with that of the standard sample. The results from frequency sweep test measurements showed a very small effect of ionic strength on sample D2 and a toughening effect of increased ionic strength on sample D1.

Flow curves for DEAED–SDS-based hydrogels detected a very small effect of ionic strength in comparison with flow curves for standard samples. Flow curves at the changed ionic strength were close to the flow curve for the standard sample, especially in the Newtonian region. The biggest difference in flow curves was demonstrated by the hydrogel prepared with an ionic strength of 0.3 M, which had the highest viscosity values in the non-Newtonian region.

Gelation in both biopolymer systems was sensitive to ionic strength. Gels could not be prepared without the addition of low molecular weight electrolyte. Thalberg and Lindman [3] reported that a certain minimum concentration of surfactant is needed for the marked formation of complexes between hyaluronan and cationic surfactants. A too high surfactant concentration is known to suppress polyelectrolyte–surfactant interactions due to the screening of electrostatic forces. The same was observed in this work for DEAED systems. Ions of the added electrolyte affect biopolymer polyelectrolyte conformations and the aggregation of surfactants (by changing their critical micellar concentration and acting as additional counter ions to dissociated surfactant moieties). For the formation of hydrogel materials and achieving adequate rheological properties, there is an optimum ionic strength for both biopolymers, which lies around 0.15–0.3 M.

3.5. Effect of pH (Buffers)

Standard gel samples (i.e., samples without extraneous pH control) prepared from hyaluronan were of essentially neutral pH, whereas DEAED-based gels were slightly acidic (see Table 3).

Table 3. pH values of hydrogels and supernatants.

Sample Name	pH	
	Gel Phase	Supernatant
H1	6.88	7.09
H2	7.23	7.20
H3	7.15	7.00
H4	6.83	7.32
H5	7.05	7.25
H6	7.38	7.13
D1	4.58	6.74
D2	4.93	6.60

Generally, changes in pH decreased both moduli and suppressed the elastic behaviour of hyaluronan-based hydrogels during strain sweep in comparison to a standard sample. The effect of pH buffers has been studied only for phase-separated hydrogels based on HMW hyaluronan. In the case of LMW hyaluronan, the elastic modulus was lower than the viscous modulus and in some samples even unmeasurable. In the case of HMW hyaluronan, both moduli had comparable values, which decreased with increasing pH, except for the highest pH value where the moduli values were the highest.

No effect of pH was observed on the width of the viscoelastic region in DEAED hydrogels. D1 hydrogels could be prepared only in a pH range of 4.5–7, which is very close to the pH of standard sample; no gels were formed in more acidic or alkaline environments. In contrast, it was possible to prepare D2 hydrogels throughout the chosen pH range. More acidic buffer slightly toughened the D2 hydrogels.

Frequency sweep investigations resulted qualitatively in the same conclusions—LMW hyaluronan resulted in a liquid-like, viscous material with practically unmeasurable elastic moduli (in contrast to the corresponding standard samples). The curves measured for HMW hyaluronan gels were all located about two orders of magnitude below the curves for the standard sample. They were slightly shifted to lower moduli values with increasing pH, except for the highest pH curve, which gave the highest moduli (see Figure 5a,b). The crossover point frequency first increased with increasing pH, then dropped down for the highest pH. The effect of pH on the crossover modulus was small.

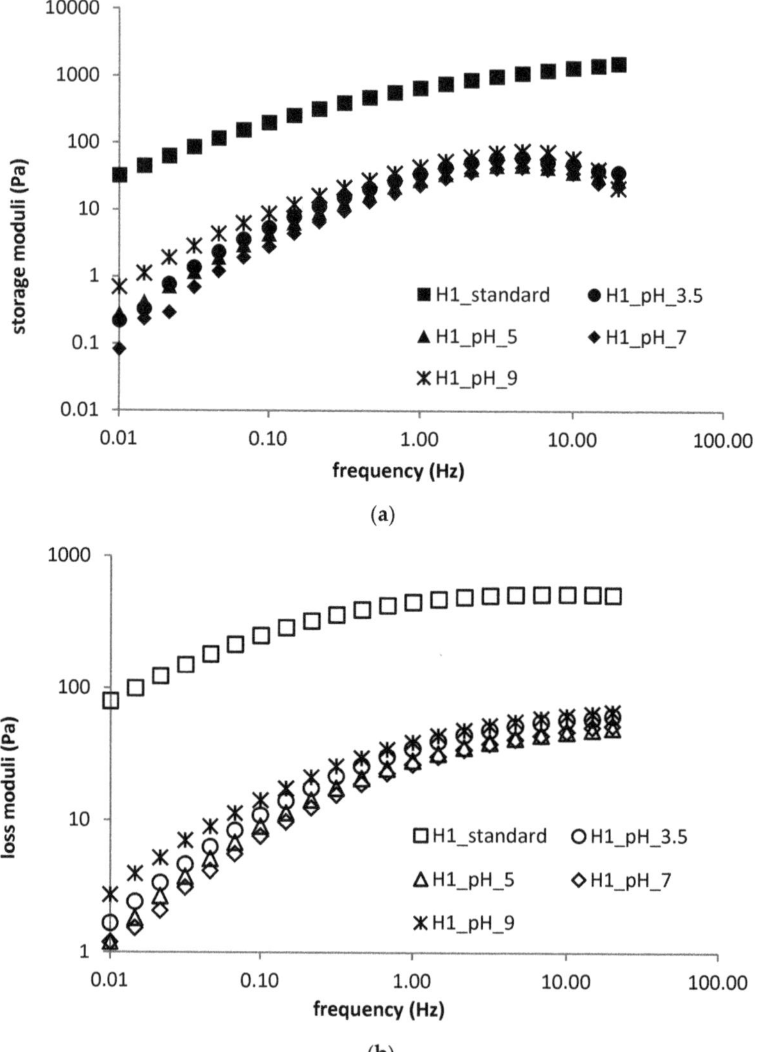

(a)

(b)

Figure 5. (a) Frequency sweep (storage moduli) for H1 sample—pH dependence; (b) frequency sweep (loss moduli) for H1 sample—pH dependence

The D1 standard sample achieved the lowest viscoelastic moduli in frequency sweep tests. Both moduli thus increased with increased pH and were almost identical for D1 gels prepared in buffers of pH 5 and 7. Crossover moduli and crossover frequencies were shifted to higher and lower values, respectively, in comparison to the standard sample. The effect of pH on D2-based gels was much smaller—decreasing pH somewhat toughened the gels, while the highest pH resulted in gels with the lowest moduli, which were very close to the moduli of the standard sample (see Figure S12). Whereas the crossover modulus was comparable for gels prepared at the three lower pH values, its value dropped two orders of magnitude for the gel prepared at the highest pH; similar behaviour was also observed for the crossover frequency (see Table 4). The effect of pH on the basic parameters of the frequency sweep tests is summarized in Table S7.

Table 4. Cross-over frequency and modulus, mesh size for all prepared hydrogels, pH influence.

pH	Sample Name	Cross-over Frequency (Hz)	Cross-over Modulus (Pa)	Mesh Size (nm)
3.5	H1	0.954	34.24	48.25
	H2	1.474	43.19	43.75
	H3	1.732	29.03	56.58
	H4	gel not observed		
	H5	gel not observed		
	H6	gel not observed		
	D1	gel not observed		
	D2	4.511	16320	3.93
5	H1	1.175	29.47	52.27
	H2	0.507	55.82	28.71
	H3	1.972	35.57	54.93
	H4	gel not observed		
	H5	gel not observed		
	H6	gel not observed		
	D1	11.660	16440	5.72
	D2	8.055	15220	5.46
7	H1	3.160	37.54	56.52
	H2	1.196	40.19	50.00
	H3	2.121	27.02	61.19
	H4	gel not observed		
	H5	gel not observed		
	H6	gel not observed		
	D1	1,468	4250	
	D2	7.853	13740	5.18
9	H1	0.650	34.14	57.67
	H2	18.070	78.23	38.34
	H3	1.674	19.79	68.41
	H4	gel not observed		
	H5	gel not observed		
	H6	gel not observed		
	D1	gel not observed		
	D2	0.248	871	6.82

The effect of pH on flow curves was also dependent on hyaluronan molecular weight. Flow curves measured for HMW hyaluronan-containing gels (see Figure 6) almost always showed lower apparent viscosities than the standard sample. Gels prepared at pH 9 possessed the highest viscosities whereas those prepared at pH 7 had the lowest. All flow curves started to overlap at the highest frequencies. In the case of LMW hyaluronan-containing materials, there was about a one order of magnitude decrease in apparent viscosities and a reduction in the Newtonian region compared with the standard sample. There was no significant effect of pH on the flow properties of hydrogels prepared from DEAED and SDS, which accords with the results of oscillatory tests. All hydrogels, independently of pH, displayed pseudoplastic behaviour, with mostly overlapping flow curves possessing a clear Newtonian plateau at low shear rates up to approximately $1.0\ \text{s}^{-1}$, with the exception of the standard sample, where the Newtonian plateau ended at approximately $0.2\ \text{s}^{-1}$.

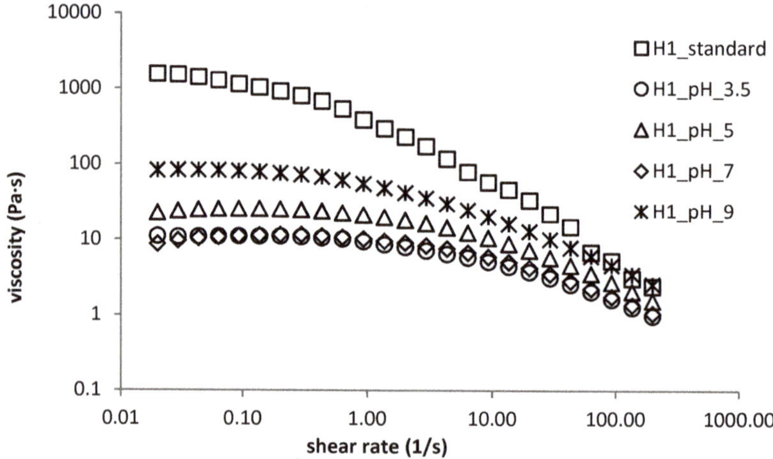

Figure 6. Flow properties of H1 sample—pH dependence.

The results obtained with hydrogels prepared in buffer solutions demonstrated that besides the effect of pH the effect of buffer ions should also be considered, despite the fact that the buffers were prepared at identical ionic strengths. Acidic pH should protonate hyaluronan and thus suppress its electrostatic interactions with surfactant micelles and, consequently, the formation of crosslinks. This was confirmed, particularly in the case of LMW hyaluronan-based gels. Samples prepared in phosphate buffer had approximately similar pH in comparison to the standard sample; however, different rheological properties and decreased acidic pH slightly toughened HMW hyaluronan-based hydrogels. Thus, differences in the size and charge of buffer ions seemed to affect the structures of the formed materials. In the case of DEAED-based gels, a much stronger influence of surfactant concentration on the effect of pH was observed. This could be a result of the better accessibility of DEAED charged groups together with the higher sensitivity of anionic surfactant to pH and buffer ions. Much higher sensitivity to the buffer environment was detected in the sample with higher surfactant concentration—both acidic and alkaline environments hindered the formation of gel material. Whereas alkaline conditions should suppress DEAED charging (the dissociation of amino groups), acidic environments should tend to neutralize the surfactant charge. At lower surfactant concentrations, weak gel was already formed in the absence of buffer, which was not very sensitive to the addition of buffer ions.

3.6. Effect of Multivalent Ions

The presence of multivalent ions affected the linear viscoelasticity region of hyaluronan-based hydrogels detected in strain sweep experiments. The effect of ions on moduli values was stronger in the case of LMW hyaluronan-based hydrogels. The highest moduli were obtained for hydrogels containing Fe^{3+} ions, the lowest for hydrogels with Ca^{2+} ions (HMW hyaluronan) or Mg^{2+} ions (LMW hyaluronan). All hydrogels containing multivalent ions had lower viscoelastic moduli in comparison with the standard sample (without the addition of multivalent ions), because all added cations had the same charge as the micelles and therefore could supress polymer–micelle interactions. On the other hand, Fe^{3+} ions are known to crosslink hyaluronan chains themselves, which could be the reason for the highest moduli observed among the tested ions [23]. The range of the linear viscoelastic region was independent of the presence of multivalent ions.

The linear viscoelastic region of DEAED gels was weakly dependent on the presence and type of multivalent ions (Ca^{2+} ions had the biggest effect) and was comparable with that of the standard sample without the addition of multivalent ions. The effect of multivalent ions on the viscoelastic properties of DEAED hydrogels was dependent on the surfactant concentration. In the case of D1 samples, the presence of multivalent ions caused an increase in both moduli. The highest increase was observed for Ca^{2+} ions. In contrast, sample D2 containing multivalent ions had lower viscoelastic moduli in comparison with the standard sample. Again, the greatest influence was observed for Ca^{2+} ions (more than one order of magnitude higher than standard sample moduli). In contrast to hyaluronan-based hydrogels, there was no significant influence of ferric ions on the values of viscoelastic moduli.

In DEAED-based samples the added multivalent cations had an opposite charge to the surfactant micelles and were thus supposed to affect the hydrogels' behaviour mainly through their interactions with micelles. This is in accord with the observed dependence of the effect of the cations on the surfactant concentration. In an excess of surfactant (a charge ratio greater than 1, D1 samples) the added cations were observed to have a stiffening effect, which could be attributed to the formation of new structures between the added cations and micelles which did not participate in the biopolymer–surfactant network crosslinks. In the case of D2 samples, a proportion of the micelles interacted with the added cations, which disturbed the biopolymer–surfactant network and lowered its rheological parameters. (The basic characteristics of the linear viscoelasticity region in the presence of multivalent ions are summarized in Tables S6 and S7 in Supplementary Material).

The frequency sweep of HMW hyaluronan-based gels showed a stiffening effect on the part of Fe^{3+} ions, whereas bivalent ions decreased the values of both moduli (for a representative example, see Figure 7a,b).

In particular, ferric ions increased the storage modulus at lower frequencies and shifted the cross-over point to very low frequencies (in fact, the storage modulus outweighed the loss modulus across the whole of the measured frequency range). The frequency sweep of LMW hyaluronan-based gels possessed the same behaviour as that of HMW hyaluronan-based gels in the environment of bivalent ions, where both moduli were slightly decreased. In the presence of ferric ions, no gelation process was observed. The effect of multivalent ions on the viscoelastic properties of both types of hydrogels is summarized in Table 5.

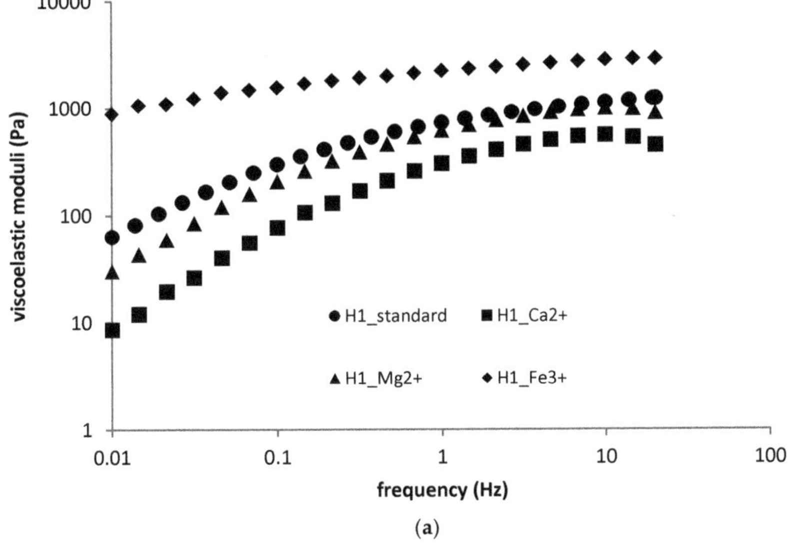

(a)

Figure 7. *Cont.*

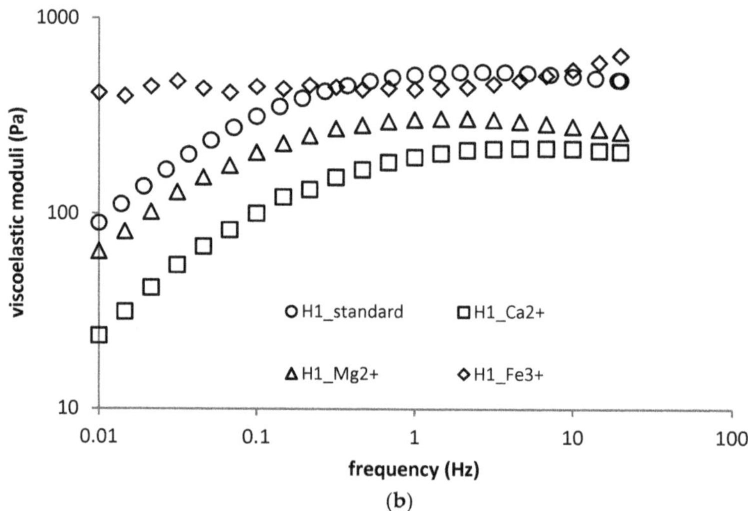

(b)

Figure 7. (a) Frequency sweep (storage moduli) for H1 sample—the effect of multivalent ions; (b) frequency sweep (loss moduli) for H1 sample—the effect of multivalent ions

Table 5. The effect of multivalent ions to cross-over point in samples, values of frequency and $G' = G''$.

Sample Name	Cross-over Point							
	Original Sample		$CaCl_2$		$MgCl_2 \cdot 6H_2O$		$FeCl_3$	
	Moduli (Pa)	Frequency (Hz)	Moduli (Pa)	Frequency (Hz)	Moduli (Pa)	Frequency (Hz)	Moduli (Pa)	Frequency (Hz)
H1	328	0.14	206	0.16	200	0.09	227	0.01
H2	278	0.10	207	0.07	246	0.14	-	<0.01
H3	338	0.06	255	0.05	154	0.05	-	<0.01
H4	69	0.01	477	19.85	90	0.11	75	0.12
H5	245	4.77	584	6.51	89	0.15	74	0.12
H6	192	2.04	522	2.55	73	0.11	76	0.05
D1	5765	10.00	9089	1.00	14670	10.00	17140	31.62
D2	11430	6.81	no cross-point		5142	3.16	1560	1.47

In the case of DEAED-based hydrogels, the results of frequency sweep investigations basically corresponded to those obtained from strain sweep measurements. Hydrogels with higher surfactant concentrations and containing multivalent ions had lower viscoelastic moduli, whereas those with lower surfactant concentrations had higher viscoelastic moduli in comparison with the standard sample. Hydrogels containing calcium ions had the biggest influence on viscoelastic moduli, which were more than one order of magnitude higher (D1) or lower (D2) than standard sample moduli, see Figure 8a,b.

Flow curves detected a very small effect of bivalent ions on hydrogels prepared from both HMW and LMW hyaluronan (Figures S13 and S14 in Supplementary Material); in the case of LMW biopolymer, the curves were shifted slightly upwards, i.e., the ions slightly increased viscosity. The effect of Fe^{3+} ions was stronger, particularly in the case of LMW hyaluronan, where viscosity increased by up to four orders of magnitude and the increase in viscosity was inversely proportional to the shear rate (Figure S14).

Multivalent ions somewhat increased the apparent viscosity within the Newtonian plateau of the standard sample D1, whereas a decrease was observed in the case of D2 samples (Figures S15 and S16). Also, with respect to D1 samples, the Newtonian plateau was disturbed by multivalent ions in the sense that the viscosity was not constant but slightly decreasing. In the presence of multivalent ions, the viscosity decrease in the non-Newtonian region was not so steep in D1 samples and almost unchanged in D2 samples, except in the Ca-containing sample. Calcium ions had generally the strongest effect on flow curves of DEAED-based hydrogels.

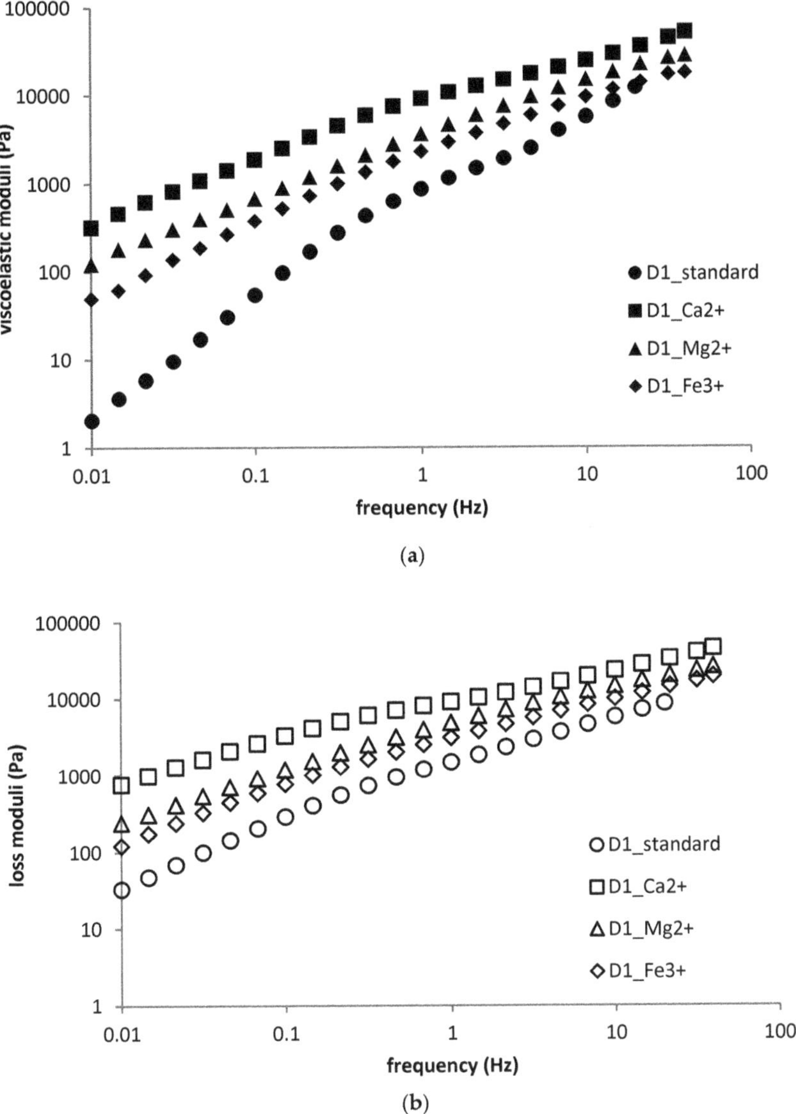

Figure 8. (a) Frequency sweep (storage moduli) for D1 sample—the effect of multivalent ions; (b) frequency sweep (loss moduli) for D1 sample—the effect of multivalent ions

4. Conclusions

Hydrogels were prepared from oppositely charged biopolymer polyelectrolyte and surfactant in micellar form at a surfactant:biopolymer charge ratio at least equal to one. Hyaluronan acted as a negatively charged biopolymer, whereas DEAED (amino-modified dextran) was used as a positively charged biopolymer. The former interacted with Septonex (carbethopendecinium bromide) whereas the latter interacted with sodium dodecylsulphate. The rheological properties of hyaluronan-based hydrogels were dependent mainly on the polymer molecular weight. Surfactant concentration (more precisely, the concentration of micelles and a surfactant:biopolymer charge ratio above 1) showed only a small effect. Surfactant concentration was found to have a much greater effect for DEAED-based hydrogels.

Gels were formed at an optimum ionic strength (set by NaCl solution), ranging from 0.15–0.3 M regardless of the type of hydrogel system or surfactant concentration. Outside the optimum range, the formation of either very weak gels (viscoelastic liquids) or viscous liquids was observed.

The effect of increased temperature (37 °C) was dependent on the hydrogel composition and biopolymer (hyaluronan) molecular weight. Both the stiffening and weakening of gels with increased temperature was observed.

Preparing gels in buffers of acidic, neutral, or alkaline pH generally deteriorated their mechanical properties in comparison to hydrogels prepared in NaCl solution with the optimum concentration (ionic strength). The best materials were thus prepared in model physiological solution (0.15 M NaCl) at natural pH. Besides controlling the pH, the buffers also affected the hydrogel properties due to the presence of their ions, which changed the environment of the ionically crosslinked hydrogels.

The addition of multivalent cations decreased the values of the rheological parameters of hyaluronan-based hydrogels probably due to their interference in the interactions between hyaluronan and micelles possessing the same charge as the added cations. In the case of DEAED-based materials, a stiffening effect on the part of the added cations was observed in the presence of excess surfactant charge as a result of their interactions with (excess) micelles which were not involved in the hydrogel network crosslinks.

General differences between the behaviour of hyaluronan and DEAED materials were attributed to the differences in the chain conformations of the two biopolymers and in the accessibility of their charged groups. To sum up our work, it was experimentally proved that the final flow and viscoelastic properties of phase-separated hydrogels based on surfactant:biopolymer can be certainly modulated by changing of chemical composition of the hydrogels as well as physico-chemical parameters (temperature, ionic strength, presence of multivalent ions). These findings are crucial for the potential applications of phase-separated hydrogels on a wider level.

Supplementary Materials: The supplementary materials are available online at http://www.mdpi.com/2073-4360/11/5/927/s1.

Author Contributions: The list of authors described in the introduction contributed to this study. Investigation and methodology were conducted by J.S., S.J. and T.V. with simultaneous supervision by M.P. The first and last of these authors also carried out the consequent formal analysis and original draft writing. Review writing & editing were made by M.P.

Funding: This work was supported by the Czech Science Foundation (project No. 16-12477S); the Materials Research Centre is supported by the Ministry of Education, Czech Republic (project No. LO1211).

Conflicts of Interest: The authors declare no conflict of interest.

References

1. Kizilay, E.; Kayitmazer, A.B.; Dubin, P.L. Complexation and coacervation of polyelectrolytes with oppositely charged colloids. *Adv. Colloid Interface Sci.* **2011**, *167*, 24–37. [CrossRef] [PubMed]

2. Kizilay, E.; Dinsmore, A.D.; Hoagland, D.A.; Sun, L.; Dubin, P.L. Evolution of hierarchical structures in polyelectrolyte-micelle coacervates. *Soft Matter* **2013**, *9*, 7320–7332. [CrossRef]

3. Thalberg, K.; Lindman, B. Interaction between hyaluronan and cationic surfactants. *J. Phys. Chem.* **1989**, *93*, 1478–1483. [CrossRef]

4. Wong, T.C.; Thalberg, K.; Lindman, B.; Gracz, H. Surfactant carbon-13 relaxation and differential line broadening in a system of a polyanion and a cationic surfactant. *J. Phys. Chem.* **1991**, *95*, 8850–8857. [CrossRef]

5. Thalberg, K.; Lindman, B. Gel formation in aqueous systems of a polyanion and an oppositely charged surfactant. *Langmuir* **1991**, *7*, 277–283. [CrossRef]

6. Buchold, P.; Schweins, R.; Di, Z.; Gradzielski, M. Structural behaviour of sodium hyaluronate in concentrated oppositely charged surfactant solutions. *Soft Matter.* **2017**, *13*, 2253–2263. [CrossRef] [PubMed]

7. Venerová, T.; Pekař, M. Rheological properties of gels formed by physical interactions between hyaluronan and cationic surfatants. *Carbohydr. Polym.* **2017**, *170*, 176–181.

8. Holmberg, K. *Surfactants and Polymers in Aqueous Solution*, 2nd ed.; John Wiley: Hoboken, NJ, USA, 2003.

9. Brno University of Technology. Process for preparing physically crosslinked hydrogel with at least one solubilized hydrophobic compound. Czech patent application No. 2014-946, 22 December 2014.

10. Gradzielski, M.; Hoffmann, I. Polyelectrolyte-surfactants complexes (PESCs) composed of oppositely charged components. *Curr. Opin. Colloid Interface Sci.* **2018**, *35*, 124–141. [CrossRef]

11. Pescosolido, L.; Feruglio, L.; Farra, R.; Fiorentino, S.; Colombo, I.; Coviello, T.; Matricardi, P.W.; Hennink, W.E.; Vermonden, T.; Grassi, M. Mesh size distribution determination of interpenetrating polymer network hydrogels. *Soft Matter.* **2012**, *8*, 7708–7715. [CrossRef]

12. Flory, P.J. *Principles of Polymer Chemistry*; Ithaca, NY, USA, 1953.

13. Gardel, M.L.; Shin, J.H.; MacKintosh, F.C.; Mahadevan, L.; Matsudaira, P.; Weitz, D.A. Elastic behaviour of cross-linked and bundled actin networks. *Science* **2004**, *304*, 1301–1305. [CrossRef]

14. Cowman, M.K.; Matsuoka, S. Experimental approaches to hyaluronan structure. *Carbohydr. Res.* **2005**, *340*, 791–809. [CrossRef]

15. Fouissac, E.; Milas, M.; Rinaudo, M. Shear-rate, concentration, molecular weight, and temperature viscosity dependences of hyaluronate, a wormlike polyelectrolyte. *Macromolecules* **1993**, *26*, 6945–6951. [CrossRef]

16. Krause, W.E.; Bellomo, E.G.; Colby, R.H. Rheology of sodium hyaluronate under physiological conditions. *Biomacromolecules* **2001**, *2*, 65–69. [CrossRef] [PubMed]

17. Dumitriu, S. *Polysaccharides: Structural Diversity and Functional Versatility*, 2nd ed.; Marcel Dekker: New York, NY, USA, 2005.

18. Li, S.; Kelly, S.J.; Lamani, E.; Ferraroni, M.; Jedrzejas, M.J. Structural basis of hyaluronan degradation by Streptococcus pneumoniae hyaluronate lyase. *EMBO J.* **2000**, *19*, 1228–1240. [CrossRef]

19. *Polysaccharides*; Dimitru, S. (Ed.) M.Dekker: New York, NY, USA, 1998.

20. Maréchal, Y. Observing the water molecule in macromoleculerared spectrometry: Structure of the hydrogen bond network and hydration mechanism. *J. Mol. Struct.* **2004**, *700*, 217–223.

21. Almond, A.; Sheehan, J.K. Predicting the molecular shape of polysaccharides from dynamic interactions with water. *Glycobiology* **2003**, *13*, 255–264. [CrossRef] [PubMed]

22. Almond, A. Towards understanding the interaction between oligosaccharides and water molecules. *Carbohydr. Res.* **2005**, *340*, 907–920. [CrossRef]

23. Garg, H.G.; Hales, C.A. *Chemistry and Biology of Hyaluronan*; Elsevier: Amsterdam, The Netherlands, 2004.

 *polymers*

Article

Anticoagulant Polyethylene Terephthalate Surface by Plasma-Mediated Fucoidan Immobilization

Kadir Ozaltin [1,*], Marian Lehocky [1], Petr Humpolicek [1], Jana Pelkova [2,3], Antonio Di Martino [1], Ilkay Karakurt [1] and Petr Saha [1]

[1]   Centre of Polymer Systems, Tomas Bata University in Zlín, Tr. Tomase Bati 5678, 76001 Zlín, Czech Republic; lehocky@post.cz (M.L.); humpolicek@utb.cz (P.H.); dimartino@utb.cz (A.D.M.); ykarakurt@utb.cz (I.K.); saha@utb.cz (P.S.)
[2]   Department of Hematology, Tomas Bata Regional Hospital, Havlickovo Nabrezi 2916, 76001 Zlin, Czech Republic; pelkova@utb.cz
[3]   Faculty of Humanities, Tomas Bata University in Zlín, Stefanikova 5670, 76001 Zlín, Czech Republic
*   Correspondence: ozaltin@utb.cz; Tel.: +4205-7603-1741

Received: 27 March 2019; Accepted: 25 April 2019; Published: 28 April 2019

Abstract: Biomaterial-based blood clot formation is one of the biggest drawbacks of blood-contacting devices. To avoid blood clot formation, their surface must be tailored to increase hemocompatibility. Most synthetic polymeric biomaterials are inert and lack bonding sites for chemical agents to bond or tailor to the surface. In this study, polyethylene terephthalate was subjected to direct current air plasma treatment to enhance its surface energy and to bring oxidative functional binding sites. Marine-sourced anticoagulant sulphated polysaccharide fucoidan from *Fucus vesiculosus* was then immobilized onto the treated polyethylene terephthalate (PET) surface at different pH values to optimize chemical bonding behavior and therefore anticoagulant performance. Surface properties of samples were monitored using the water contact angle; chemical analyses were performed by FTIR and X-ray photoelectron spectroscopy (XPS) and their anticoagulant activity was tested by means of prothrombin time, activated partial thromboplastin time and thrombin time. On each of the fucoidan-immobilized surfaces, anticoagulation activity was performed by extending the thrombin time threshold and their pH 5 counterpart performed the best result compared to others.

Keywords: polyethylene terephthalate; fucoidan; blood coagulation; anticoagulant; plasma treatment; surface coating

1. Introduction

Besides adequate mechanical performances of polymeric biomaterials, their surface properties (surface chemistry, hydrophilicity, energy and charge density) and their interactions with living tissue are equally essential to ensure biocompatibility. In the case of blood-contacting medical devices, such as vascular graft and stents, heart valves and intravascular catheters, regulating the blood–material interface is required to regulate hemocompatibility. The major drawback of such biomaterials (i.e., polyethylene terephthalate, polylactic acid, polycaprolactone and their nanofibers) is their insufficient hemocompatibility, which triggers thrombus formation on the material surface (namely, surface-induced thrombus formation) and may cause failure of the implanted material, vascular occlusion, heart attack and stroke [1–7].

The first response of the body to the implanted biomaterial is rapid protein adsorption within seconds, in accordance with the Vroman Effect, which leads to platelet adhesion and mediates thrombus formation [8–12]. This protein layer is recognizable by the integrin receptors of the cells; hence, protein adsorption is important for cellular interactions as well [13–17].

Platelet adhesion onto the protein layer is followed by platelet aggregation and activation of the intrinsic pathway of the blood coagulation cascade by blood protein factor XII (also called coagulation factor) to XIIa, fibrin network formation and finally complement system activation via erythrocytes and leukocytes interactions [18–21]. This activated blood coagulation cascade triggers thrombus formation on the biomaterial's surface (surface-mediated thrombosis) [22].

Hence, reducing the protein adhesion at the beginning onto the biomaterial's surface is the key point to prevent platelet adhesion, and therefore thrombus formation, for short- and long-term hemocompatibility of blood-contacting devices.

Modification of the biomaterial's surface can be considered for two approaches: first, tailoring the physical properties to change its surface topography, and therefore hydrophilicity and cellular interactions; second, chemical immobilization of a bioactive agent to obtain specific features according to the individual properties of the chemical agent. Immobilization of a bioactive agent onto a polymeric substrate is challenging due to the hydrophobic nature and lack of the bonding interfaces to create a covalent bond with the chemical agents for most of the polymeric materials. There are several methods to treat polymeric surfaces, such as wet chemistry, UV irradiation, corona discharge, flame treatment, ozone-induced treatment and plasma treatment [23–28]. Besides efficiency of the immobilization, desired bulk properties are equally important, especially for heat-sensitive and chemically unstable substrates. Therefore, plasma treatment is one of the best candidates for etching the surface to change its topography (therefore increasing its hydrophilicity and surface area) homogeneously and to introduce the beneficial functional hydroxyl, carboxyl and carbonyl groups on the surface for further chemical bonding with selected chemical agents. Due to the fact that plasma treatment is heat-free and its effect is limited to the top surface, the bulk properties remain unchanged.

Heparin is a well-known anticoagulant polysaccharide for avoiding blood thrombus, but its biggest drawbacks are hemorrhage and thrombocytopenia. Furthermore, it may cause animal-based infections since it is mostly obtained from animals [29–33].

Fucoidan (Figure 1) is a marine-sourced sulphated polysaccharide, extracted from brown algae, with a better anticoagulant effect than that of anticoagulant heparin due to its action mechanism on the coagulation cascade by interacting with the natural thrombin inhibitors of antithrombin (AT III), heparin cofactor II (HCII), activated factor II (thrombin) and activated factor X [30–38]. Fucoidan has numerous biological activities besides its anticoagulant activity, such as anticancer, antitumor, antivirus and antiinflammation activities [39–44]. Its effect is related to monosaccharide composition, sulfation degree and pattern and molecular weight to regulate binding properties [32–34,44,45]. Furthermore, fucoidan does not cause hemorrhage as heparin does, and it has no risk of inflammation as animal-sourced heparin does, due to its marine origin [30–32].

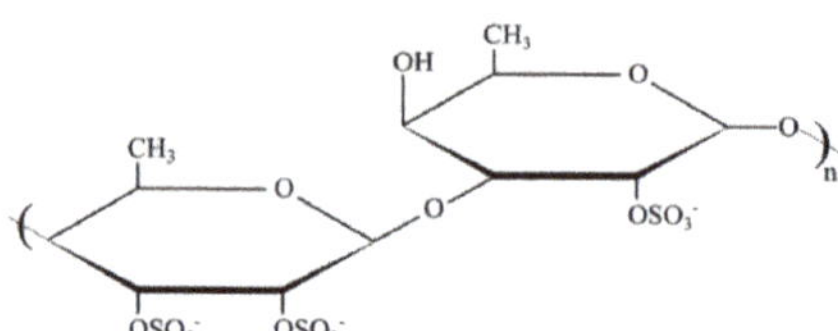

Figure 1. Chemical structure of fucoidan.

Vesel et. al. studied the fucoidan from *Fucus vesiculosus* immobilization onto a polyethylene terephthalate (PET) surface after radio frequency (RF) plasma treatment at different pH values and demonstrated the immobilization performance by X-ray photoelectron spectroscopy. They found that the highest immobilization level occurred at pH 5, but anticoagulation activity was not carried out [46].

In our previous study, fucoidan from *Fucus vesiculosus* was immobilized onto low density polyethylene at its natural pH value and it exhibited an anticoagulant activity slightly higher than the threshold [47]. In this study, polyethylene terephthalate (PET) was used as a substrate, due to its unique chemical and physical properties with highly crystalline structure, to create a functional anticoagulant surface using fucoidan from *Fucus vesiculosus* (FU). PET surfaces were activated by direct current (DC) plasma to form functional groups for FU immobilizations with pH levels of 3, 4, 5, 6 and 7 to compare its anticoagulation activity at various pH values, as it depicted in Figure 2. Surface characterizations were studied by wettability test, scanning electron microscopy, X-ray photoelectron spectroscopy and Fourier transform infrared spectroscopy. The anticoagulation activity of the samples was carried out for prothrombin time (PT), activated partial thromboplastin time (aPTT) and thrombin time (TT).

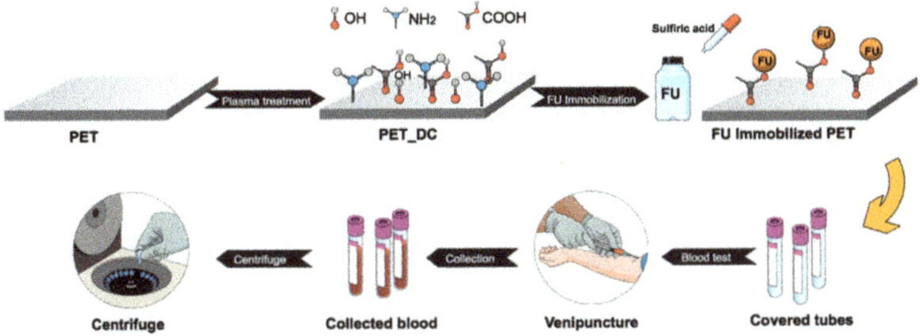

Figure 2. Experimental scheme.

2. Materials and Methods

2.1. Materials and Preparation of Anticoagulant Surfaces

In this work, polyethylene terephthalate (PET) sheets with $70 \times 30 \times 0.1$ mm dimensions were used as a polymeric substrate. All PET substrates were thoroughly cleaned with distilled water and dried at 30 °C for 24 h in an oven. PET substrates were exposed to direct current (DC) plasma at 50 W of reactor power and 40 kHz of frequency, generated by a PICO (Diener, Germany) plasma reactor under the chamber pressure of 50 Pa. As a discharge gas, air was used with 20 sccm flow rate, and both sides of PET sheets were treated for 60 s, hereafter referred to as PET_DC. Anticoagulant fucoidan from *Fucus vesiculosus* (FU) (Sigma Aldrich, St. Louis, MO, USA) solution was prepared as 1% (w/v) in distilled water and placed in the vials to change their pH. The total volume of the FU solutions in vials was 4 mL each. The pH values of the FU solutions of pH = 3, 4, 5, 6 and 7 were obtained by diluted H_2SO_4 addition. PET_DC samples were subsequently placed into each solution vial containing FU solution with pH = 3, 4, 5, 6 and 7 and placed in a rotational shaker for 24 hours at room temperature in order to immobilize anticoagulant fucoidan onto the functionalized surface of PET by plasma. Samples were labeled as FU3, FU4, FU5, FU6 and FU7, respectively. Finally, the samples were taken out of the solution vials, thoroughly washed with distilled water to eliminate non-immobilized fucoidan species and dried at room temperature for 24 h. Each sample was placed into a separated blood collection tube for further anticoagulation activity evaluation.

2.2. Surface Wettability Evaluation

The wettability behavior of the samples to determine its surface hydrophilicity was studied by the Sessile drop method via SEE System (Advex Instruments, Brno, Czech Republic) equipped with a CCD camera. Distilled water was used as the testing liquid at 22 °C and 60% relative humidity. Ten separate droplets with a volume of 5 μL were placed onto each sample surface for 30 s to obtain the average water contact angle value (Q_w).

2.3. Surface Morphology Investigation by SEM

Surface morphology of all samples was monitored by a NANOSEM 450 (FEI, Hillsboro, OR, USA) scanning electron microscope (SEM), equipped with a low vacuum detector and operated at 5 kV under 90 Pa pressure in a water vapor environment. Images were taken at the magnification of 10 k × with a spot size of 50 nm.

2.4. Attenuated Total Reflectance Fourier Transform Infrared Spectroscopy (ATR-FTIR) Analysis

For surface chemistry examination, in order to compare the changes in chemical compositions of the studied samples, a Nicolet iS5 (Thermo Scientific, Grand Island, NY, USA) single beam Fourier transform infrared spectroscopy (FTIR) equipped with iD5 attenuated total reflectance (ATR) was used. Collected spectra were recorded between 400 and 4000 cm^{-1} with a resolution of 2 cm^{-1} and 64 scans using a ZnSe crystal at an incident angle of 45°.

2.5. X-Ray Photoelectron Spectroscopy (XPS) Analysis

To reveal the changes in chemical composition on the surfaces of the samples, X-ray photoelectron spectroscopy (XPS) analysis was carried out on a bioemulsifier film deposited on a glass slide using ESCALAB 200A, VG Scientific (East Grinstead, UK), with PISCES software for data acquisition and analysis. An achromatic Al (K_a) X-ray source was operated at 15 kV (300 W), and the spectrometer was calibrated with reference to Ag $3d_{5/2}$ (368.27 eV) and operated in constant analyzer energy (CAE) mode with 20 eV pass energy. Data acquisition was performed with a pressure lower than 10^{-6} Pa. Gaussian–Lorentzian peak shape and Shirley type background subtraction were used for peak fitting of the spectral analysis.

2.6. Evaluation of Anticoagulation Activity

The blood was obtained by venous puncture from a healthy donor in accordance with the Helsinki Declaration, and placed into blood collection tubes (5 mL each) covered by prepared PET samples. The obtained human blood plasma was treated with 3.2% citric acid (109 mmol/L) and then centrifuged at room temperature, 3000 min^{-1}, for 15 min. Anticoagulant activity was determined by means of prothrombin time (PT), thrombin time (TT) and activated partial thromboplastin time (aPTT) using a SYSMEX CA 1500 (Siemens, Munich, Germany) instrument. Each of the samples was examined three times.

3. Results

3.1. Surface Wettability Behavior

Surface wettability results of studied samples, obtained by water contact angle, are given in Table 1. The water contact angle of 70.6° was monitored for the initial PET sample. This value of the water contact angle is referred to as a hydrophobic nature of the PET surface, which is not convenient for anticoagulant agent immobilizations. After DC plasma treatment (PET_DC), the water contact angle value sharply decreased to 22.09°, due to the introduced functional groups and the increased surface area by plasma, referring to the increased hydrophilicity and surface energy, which is convenient for further anticoagulant agent immobilizations. As can be seen in Table 1, FU-immobilized samples at various pH (samples of FU3 for pH 3; FU4 for pH 4; FU5 for pH 5; FU6 for pH 6 and FU7 for pH 7) showed almost the same water contact angle values within a range of 39.08° to 43.21°. Such an increase, compared to the plasma-treated sample, indicates the immobilization of FU onto PET_DC counterparts, as was demonstrated by SEM analysis.

Table 1. Water contact angle values of the samples.

Samples	Contact Angle (°)
PET	70.6 ± 0.60
PET_DC	22.09 ± 1.32
FU3	43.21 ± 2.67
FU4	39.08 ± 1.74
FU5	42.61 ± 2.92
FU6	39.74 ± 5.52
FU7	42.54 ± 7.45

3.2. Surface Morphology

The immobilization of fucoidan onto functionalized PET surface was investigated by scanning electron microscope, and the results are shown in Figure 3. As it is seen in Figure 3a, the reference PET surface is smooth without impurities; however, after DC plasma treatment it becomes relatively rough due to the increased surface area by plasma treatment (Figure 3b). The immobilization of fucoidan species on each sample was monitored, which revealed successful bonding to functional groups created by plasma. Nevertheless, the particle size and distribution of immobilized fucoidan species were different for each sample, due to the dissolution ability of fucoidan at different pH values. According to the SEM results, solubility and immobilization affinity of fucoidan seemed to be the best at pH 5 (Figure 3e) and pH 4 (Figure 3d). For both FU5 and FU4 samples, the particle size was smaller (this reduces the fraction of floating blood to decrease the chance of clustering blood particles such as fibrin and red blood cells, which trigger blood clot formation), and the distribution was more homogenous than that of the FU3 (Figure 3c), FU6 (Figure 3f) and FU7 (Figure 3g) counterparts. This was also confirmed by XPS results, showing that sulphur content was much higher for FU5 and FU4 compared to others. Furthermore, as expected, their anticoagulation activity was better, which is discussed later on.

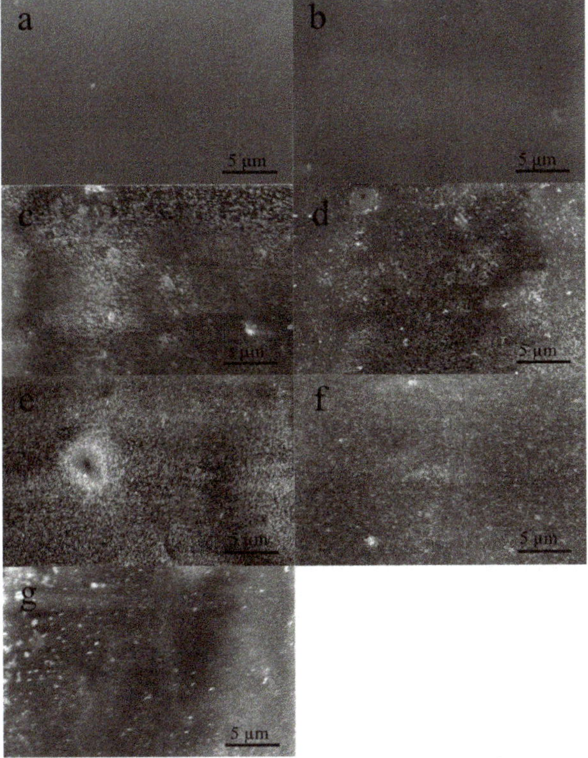

Figure 3. Surface morphology of fucoidan immobilized surfaces at various pH values: (**a**) PET; (**b**) PET_DC; (**c**) FU3; (**d**) FU4; (**e**) FU5; (**f**) FU6; (**g**) FU7.

3.3. Surface Chemical Anaylsis by ATR-FTIR

The chemical changes in the near-surface area of the samples were observed by attenuated total reflectance Fourier transform infrared spectra (ATR-FTIR) for the mid-IR range of 4000–400 cm^{-1}, and the collected data are shown in Figure 4. The major peaks of PET were observed at the wavenumbers of: 720 cm^{-1} related to CH$_2$ rocking vibrations; 795 cm^{-1} in the plane vibration of C–H; 850 cm^{-1} and 872 cm^{-1} corresponding to wagging vibration of O–CH$_2$ and out-of-plane C–H vibration, respectively; 975 cm^{-1} due to stretching of O–CH$_2$; 1015 cm^{-1} related to in-plane C–H vibration; 1095 cm^{-1} and 1243 cm^{-1} corresponding to asymmetric stretching vibration of C–O; 1340 cm^{-1} referring to CH$_2$ wagging; 1410 cm^{-1} related to C–H bending vibration; 1471cm^{-1} related to CH$_2$ bending; 1511 cm^{-1} referring to C–H bending vibration; 1718 cm^{-1} due to C=O stretching band [48–50]. The absorption peaks observed between 2850 and 3000 cm^{-1} were attributed to aliphatic C–H stretching vibrations [47,50].

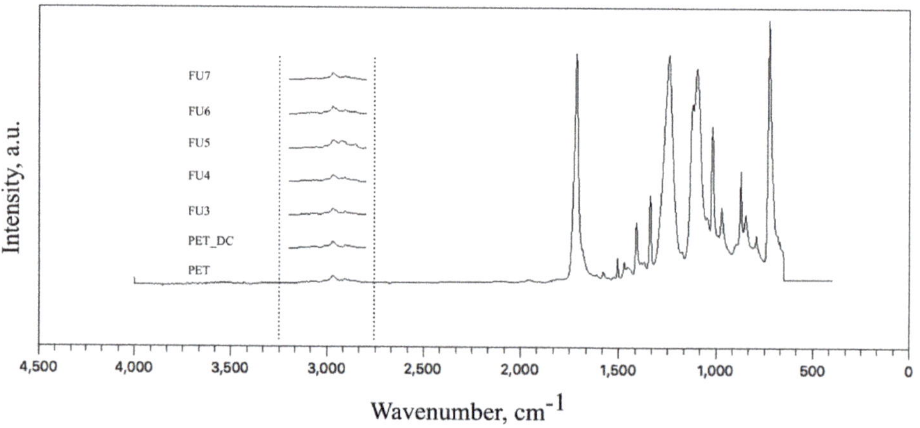

Figure 4. Attenuated total reflectance (ATR)-FTIR spectrum collected from the samples.

Due to the penetration depth of ATR crystals of ZnSe (0.6 μm), which was well above the thickness of the coated FU (assumed to be nanometer size), the collected data were mostly belongs to the PET itself. Nevertheless, between wavenumbers 3000 and 2850 cm^{-1}, especially for the FU5 samples, a slight intensity increase of peaks was observed, which refers to the immobilization of FU. The peak corresponds to the OH group vibration from the monosaccharide monomer. The peak at pH 5 showed the highest obtained intensity due to the highest amount of immobilized FU, as is discussed later on.

3.4. Surface Elemental Analysis by XPS

Surface chemical compositions of the samples were revealed by using X-ray photoelectron spectroscopy (XPS), and results are listed in Table 2. The XPS spectra of the reference PET sample consisted only of carbon and oxygen content with the levels of 69.7% and 30.3%, respectively. After DC plasma treatment, the oxygen level increased to 40.9% due to the oxidative group functionalities introduced by plasma, and the carbon level decreased proportionally. Likewise, nitrogen was detected after plasma treatment with a level of 0.7%, due to the amine groups introduced by the applied plasma. Oxygen and nitrogen levels decreased for all fucoidan-immobilized samples as an indicator of fucoidan bonding. Nevertheless, sulphur content was lowest for the FU3 sample and highest for the FU5 sample, which signifies fucoidan immobilization. The highest sulphur content was observed at pH 5. Therefore, maximum anticoagulation activity is expected for the FU5 counterpart, according to the XPS data.

Table 2. Collected elemental compositions in atomic percentages of the samples by X-ray photoelectron spectroscopy (XPS).

Samples	C1s	O1s	N1s	S2p	O1s/C1s	N1s/C1s	S2p/C1s
PET	69.7	30.3	0	0	0.435	0	0
PET_DC	58.4	40.9	0.7	0	0.700	0.012	0
FU3	63.4	35.7	0	0.2	0.563	0	0.003
FU4	63.7	34.5	0.2	1.6	0.542	0.003	0.025
FU5	63.6	34.4	0.2	1.8	0.541	0.003	0.028
FU6	63.7	35.7	0.1	0.5	0.560	0.002	0.008
FU7	63.8	35.6	0.1	0.5	0.558	0.002	0.008

3.5. Blood Coagulation Activity

The anticoagulation activity of the samples was examined to reveal prothrombin time (PT), activated partial thromboplastin time (aPTT) and thrombin time (TT). Therefore, three pathways of the blood coagulation cascade of the surface-mediated intrinsic pathway, the tissue-mediated extrinsic pathway and the common coagulation pathway were investigated in vitro and the results are given in Table 3.

Table 3. Anticoagulation activity results; PT: prothrombin time (s); aPTT: activated partial thromboplastin time (s); TT: thrombin time (s).

Samples	PT	aPTT	TT
PET	12.6	26.3	15.5
PET_DC	10.9	22	15.2
FU3	11.3	29.4	34.7
FU4	11	35.1	88.7
FU5	10.9	35.1	100+
FU6	11.1	30.8	58.6
FU7	11.2	31.5	60.7

PT measures the clot formation time in the extrinsic pathway, and the common coagulation pathway and PT range for the healthy donors is between 11 and 13.5 s [51]. According to the results of the PT assay, all samples are in the range with minor alteration (Figure 5a).

aPTT measures the coagulation activity of the intrinsic pathway and the common coagulation pathway, in contrast to PT. For the healthy donors, the time range of aPTT is between 25 and 32 s [52]. aPTT assay revealed that the PET surface is in the range by 26.3 s, but after DC plasma treatment it slightly decreased below the threshold by 22 s (Figure 5b). This is probably due to the increased surface roughness (which increases the hydrophilicity) that increases the fraction between the material surface and floating blood. The nitrogen content of PET_DC also needs to be taken into consideration. Fucoidan-immobilized samples FU3, FU6 and FU7 were within this range. By means of FU4 and FU5 samples, both were above the range by 35.1 s, which retarded the coagulation even further than that of upper anticoagulant limit of 32 s.

TT measures the thrombin formation time by means of the conversion rate of fibrinogen to fibrin in the common coagulation pathway. The threshold for the anticoagulation activity is 20 s, which means a TT over 20 s is considered to be anticoagulant (Figure 5c) [52–54].

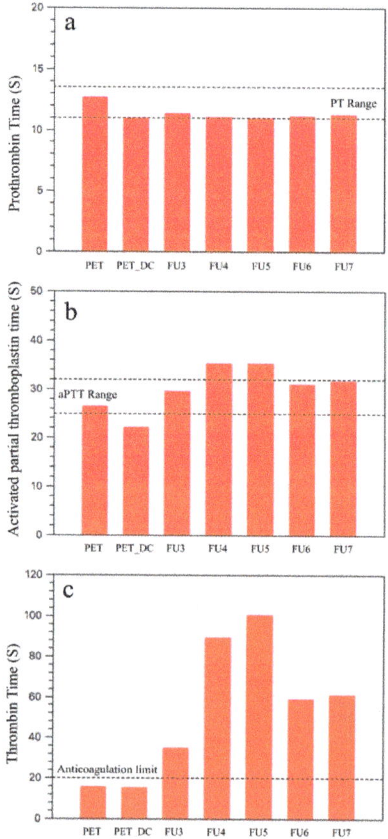

Figure 5. Anticoagulant activity results: (**a**) PT: prothrombin time (s); (**b**) aPTT: activated partial thromboplastin time (s); (**c**) TT: thrombin time (s).

The reference PET sample and its plasma-treated counterpart were found to have TT values of 15.5 and 15.2 s, implicating that neither the PET nor the PET_DC surfaces influenced the common coagulation pathway to retard blood thrombus formation. All fucoidan-immobilized samples had a TT value over 20 s, starting from 34.7 s of the FU3 sample. The TT value was increased by pH to 88.7 s and was >100 s for FU4 and FU5 samples, respectively. The further increase of pH to 6 and 7, for samples FU6 and FU7, meant the TT value was decreased to 58.6 and 60.7 s, respectively. Nevertheless, it was well above the threshold of 20 s, which implies that the FU species were successfully immobilized onto the surface at each pH value from 3 to 7 and that all of them exhibited anticoagulant activity. Hence, the pH effect on FU immobilization onto the plasma-treated PET surface and its anticoagulation performance can be seen from the TT results.

As it is also seen in Figure 3e, the FU species on the surface were denser and more homogenous, so that XPS collected the highest sulphur content of 1.8 for the FU5 sample, which increased the heparin cofactor II (HCII)-mediated antithrombotic activity [30,36]. The sulphur content of 1.6 was observed for the FU4 sample, which was the second highest among all the samples that refer to sufficient FU immobilization, and its anticoagulant activity showed the second highest performance. Even if all the FU-immobilized samples at different pH values can be considered as an anticoagulant, for the long-term usage of blood-contacting device, it could be considered that the FU5 sample is the most suitable since its TT value exceeds 100 s.

4. Conclusions

In order to achieve anticoagulant activity, a polyethylene terephthalate (PET) surface, for use in blood-contacting devices, was treated by DC air plasma and fucoidan from *Fucus vesiculosus* (FU) and was immobilized in different pH values with a range of 3–7. The FU immobilization onto the PET surface after plasma treatment was found to be best at pH 5, which it also correlated with FTIR, SEM and XPS results, and it exhibited the highest anticoagulant activity, more than 100 s, which indicates that it is a fully anticoagulant FU-immobilized sample suitable for blood-contacting PET devices. Nevertheless, FU immobilizations were successful for all samples and exhibited anticoagulant activity well above that of the TT threshold of 20 s.

Author Contributions: Conceptualization, K.O. and M.L.; Methodology, K.O.; Surface analysis and chemical analysis, K.O.; Blood coagulation test, J.P.; P.S. P.H, A.D.M. and I.K. contributed to the interpretation. Preparation of the manuscript, review and editing, K.O. and M.L. The manuscript was approved by all authors.

Funding: The authors would like to express their gratitude to the Czech Science Foundation, project number 17-10813S and the Ministry of Education, Youth and Sports of the Czech Republic, Program NPU I, LO1504. I.K. thanks the internal grant agency of Tomas Bata University in Zlin (IGA/CPS/2019/004).

Acknowledgments: The authors would like to express their gratitude to the Czech Science Foundation, project number 17-10813S.

Conflicts of Interest: The authors declare no conflict of interest.

References

1. Mozetic, M.; Ostrikov, K.; Ruzic, D.N.; Curreli, D.; Cvelbar, U.; Vesel, A.; Primc, G.; Leisch, M.; Jousten, K.; Malyshev, O.B.; et al. Recent advances in vacuum sciences and applications. *J. Phys. D Appl. Phys.* **2014**, *47*, 153001:1–153001:23. [CrossRef]

2. Mozetic, M.; Primc, G.; Vesel, A.; Zaplotnik, R.; Modic, M.; Junkar, I.; Recek, N.; Klanjsek-Gunde, M.; Guhy, L.; Sunkara, M.K.; et al. Application of extremely non-equilibrium plasmas in the processing of nano and biomedical materials. *Plasma Sources Sci. Technol.* **2015**, *24*, 015026:1–015026:12. [CrossRef]

3. Mozetic, M.; Vesel, A.; Primc, G.; Eisenmenger-Sittner, C.; Bauer, J.; Eder, A.; Schmid, G.H.S.; Ruzic, D.N.; Ahmed, Z.; Barker, D.; et al. Recent developments in surface science and engineering, thin films, nanoscience, biomaterials, plasma science, and vacuum technology. *Thin Solid Films* **2018**, *660*, 120–160. [CrossRef]

4. Hasebe, T.; Nagashima, S.; Kamijo, A.; Moon, M.W.; Kashiwagi, Y.; Hotta, A.; Lee, K.R.; Takahashi, K.; Yamagami, T.; Suzuki, T. Hydrophobicity and non-thrombogenicity of nanoscale dual rough surface coated with fluorine-incorporated diamond-like carbon films: Biomimetic surface for blood-contacting medical devices. *Diam. Relat. Mater.* **2015**, *38*, 14–18. [CrossRef]

5. Reviakine, I.; Jung, F.; Braune, S.; Brash, J.L.; Latour, R.; Gorbet, M.; Oeveren, W.V. Stirred, shaken, or stagnant: What goes on at the blood–biomaterial interface. *Blood Rev.* **2017**, *31*, 11–21. [CrossRef]

6. Alibeik, S.; Zhu, S.; Yau, J.W.; Weitz, J.I.; Brash, J.L. Surface modification with polyethylene glycol–corn trypsin inhibitor conjugate to inhibit the contact factor pathway on blood-contacting surfaces. *Acta Biomater.* **2011**, *7*, 4177–4186. [CrossRef] [PubMed]

7. Qi, P.; Maitz, M.F.; Huang, N. Surface modification of cardiovascular materials and implants. *Surf. Coat. Technol.* **2013**, *233*, 80–90. [CrossRef]

8. Zhang, Z.; Kuang, G.; Zong, S.; Liu, S.; Xiao, H.; Chen, X.; Zhou, D.; Huang, Y. Sandwich-Like Fibers/Sponge Composite Combining Chemotherapy and Hemostasis for Efficient Postoperative Prevention of Tumor Recurrence and Metastasis. *Adv. Mater.* **2018**, *30*, 1803217. [CrossRef] [PubMed]

9. Anderson, J.M.; Rodriguez, A.; Chang, D.T. Foreign body reaction to biomaterials. *Semin. Immunol.* **2008**, *20*, 86–100. [CrossRef]

10. Yamazoe, H.; Oyane, A.; Nashima, T.; Ito, A. Reduced platelet adhesion and blood coagulation on cross-linked albumin films. *Mater. Sci. Eng. C* **2010**, *30*, 812–816. [CrossRef]

11. Biran, R.; Pond, D. Heparin coatings for improving blood compatibility of medical devices. *Adv. Drug Deliv. Rev.* **2017**, *112*, 12–23. [CrossRef]

12. Tengvall, P. Protein Interactions with Biomaterials. In *Comprehensive Biomaterials*, 1st ed.; Hutmacher, D.W., Grainger, D.W., Ducheyne, P., Eds.; Elsevier Ltd.: Philadelphia, PA, USA, 2011; Volume 4, pp. 63–73.

13. Horbett, T.A. Adsorbed proteins on biomaterials. In *Biomaterials Science, an Introduction to Materials in Medicine*, 3rd ed.; Ratner, B.D., Hoffman, A.S., Schoen, F.J., Lemons, J.E., Eds.; Elsevier Inc.: Oxford, UK, 2013; pp. 394–408.

14. Xu, L.C.; Siedlecki, C.A. Heparin coatings for improving blood compatibility of medical devices. *Biomaterials* **2007**, *28*, 3273–3283. [CrossRef]

15. Rana, D.; Matsuura, T. Surface Modifications for Antifouling Membranes. *Chem. Rev.* **2010**, *110*, 2448–2471. [CrossRef]

16. Seyfert, U.T.; Biehl, V.; Schenk, J. In vitro hemocompatibility testing of biomaterials according to the ISO 10993-4. *Biomol. Eng.* **2002**, *19*, 91–96. [CrossRef]

17. Xu, L.C.; Bauer, J.W.; Siedlecki, C.A. Proteins, platelets, and blood coagulation at biomaterial interfaces. *Colloids Surf. B* **2014**, *124*, 49–68. [CrossRef]

18. Courtney, J.M.; Lamba, N.M.K.; Sundaram, S.; Forbes, C.D. Biomaterials for blood-contacting applications. *Biomaterials* **1994**, *15*, 737–744. [CrossRef]

19. Faxalv, L.; Ekblad, T.; Liedberg, B.; Lindahl, T.L. Blood compatibility of photografted hydrogel coatings. *Acta Biomater.* **2010**, *6*, 2599–2608. [CrossRef]

20. Cashman, J.D.; Kennah, E.; Shuto, A.; Winternitz, C.; Springate, C.M.K. Fucoidan film safely inhibits surgical adhesions in a rat model. *J. Surg. Res.* **2011**, *171*, 495–503. [CrossRef]

21. Ikada, Y. Surface modification of polymers for medical applications. *Biomaterials* **1994**, *15*, 725–736. [CrossRef]

22. Hsiao, C.R.; Lin, C.W.; Chou, C.M.; Chung, C.J.; He, J.L. Surface modification of blood-contacting biomaterials by plasma-polymerized superhydrophobic films using hexamethyldisiloxane and tetrafluoromethane as precursors. *Appl. Surf. Sci.* **2015**, *346*, 50–56. [CrossRef]

23. Lehocky, M.; Amaral, P.F.F.; Coelho, M.A.Z.; Stahel, P.; Barros-Timmons, A.M.; Coutinho, J.A.P. Attachment/detachment of *Saccharomyces cerevisiae* on plasma deposited organosilicon thin films. *Czechoslov. J. Phys.* **2006**, *56*, 1256–1262. [CrossRef]

24. Lehocky, M.; Lapcik, L.; Dlabaja, R.; Rachunek, L. Influence of artificially accelerated ageing on the adhesive joint of plasma treated polymer materials. *Czechoslov. J. Phys.* **2004**, *54*, C533–C538.

25. Lehocky, M.; Amaral, P.F.F.; Stahel, P.; Coelho, M.A.Z.; Barros-Timmons, A.M.; Coutinho, J.A.P. Preparation and characterization of organosilicon thin films for selective adhesion of *Yarrowia lipolytica* yeast cells. *J. Chem. Technol.* **2007**, *82*, 360–366.

26. Lehocky, M.; Stahel, P.; Koutny, M.; Cech, J.; Institoris, J.; Mracek, A. Adhesion of *Rhodococcus* sp. S3E2 and *Rhodococcus* sp. S3E3 to plasma prepared Teflon-like and organosilicon surfaces. *J. Mater. Process. Technol.* **2009**, *209*, 2871–2875. [CrossRef]

27. Goa, S.; Tang, G.; Hua, D.; Xiong, R.; Han, J.; Jiang, S.; Zhang, Q.; Huang, C. Stimuli-responsive bio-based polymeric systems and their applications. *J. Mater. Chem. B* **2019**, *7*, 709–729.

28. Ding, Q.; Xu, X.; Yue, Y.; Mei, C.; Huang, C.; Jiang, S.; Wu, X.; Han, J. Nanocellulose-Mediated Electroconductive Self-Healing Hydrogels with High Strength, Plasticity, Viscoelasticity, Stretchability, and Biocompatibility toward Multifunctional Applications. *ACS Appl. Mater. Interfaces* **2018**, *10*, 27987–28002. [CrossRef]

29. Chen, Z.; Wang, Z.; Fu, Q.; Ma, Z.; Fang, P.; He, C. Microstructure and surface state of plasma-treated high-density polyethylene elucidated by energy-tunable positron annihilation and water contact angle measurements. *JJAP Conf. Proc.* **2014**, *2*, 011202. [CrossRef]

30. Azevedo, T.C.G.; Bezerra, M.E.B.; Santos, M.D.G.D.L.; Souza, L.A.; Marques, C.T.; Benevides, N.M.B.; Leite, E.L. Heparinoids algal and their anticoagulant, hemorrhagic activities and platelet aggregation. *Biomed. Pharmacother.* **2009**, *63*, 477–483. [CrossRef]

31. Dore, C.M.P.G.; Alves, M.G.C.F.; Will, L.S.E.P.; Costa, T.G.; Sabry, D.A.; Rego, L.A.R.S.; Accardo, C.M.; Rocha, H.A.O.; Filgueira, L.G.A.; Leite, E.L. A sulfated polysaccharide, fucans, isolated from brown algae *Sargassum vulgare* with anticoagulant, antithrombotic, antioxidant and anti-inflammatory effects. *Carbohydr. Polym.* **2013**, *91*, 467–475. [CrossRef]

32. Hu, Y.; Li, S.; Li, J.; Ye, X.; Ding, T.; Liu, D.; Chen, J.; Ge, Z.; Chen, S. Identification of a highly sulfated fucoidan from sea cucumber *Pearsonothuria graeffei* with well-repeated tetrasaccharides units. *Carbohydr. Polym.* **2015**, *134*, 808–816. [CrossRef]

33. Mracek, A.; Varhanikova, J.; Lehocky, M.; Grundelova, L.; Pokopcova, A.; Velebny, V. The influence of hofmeister series ions on hyaluronan swelling and viscosity. *Molecules* **2008**, *13*, 1025–1034. [CrossRef]

34. Ustyuzhanina, N.E.; Bilan, M.I.; Gerbst, A.G.; Ushakova, N.A.; Tsvetkova, E.A.; Dmitrenok, A.S.; Usov, A.I.; Nifantiev, N.E. Anticoagulant and antithrombotic activities of modified xylofucan sulfate from the brown alga *Punctaria plantaginea*. *Carbohydr. Polym.* **2016**, *136*, 826–833. [CrossRef]

35. Zhu, Z.; Zhang, Q.; Chen, L.; Ren, S.; Xu, P.; Tang, Y. Higher specificity of the activity of low molecular weight fucoidan for thrombin-induced platelet aggregation. *Thromb. Res.* **2010**, *125*, 419–426. [CrossRef]

36. Fitton, J.H. Therapies from Fucoidan; Multifunctional marine polymers. *Mar. Drugs* **2011**, *9*, 1731–1760. [CrossRef]

37. Liewert, I.; Ehrig, K.; Alban, S. Effects of fucoidans and heparin on reactions of neutrophils induced by IL-8 and C5a. *Carbohydr. Polym.* **2017**, *165*, 462–469. [CrossRef]

38. Kim, J.M.; Bae, I.H.; Lima, K.S.; Park, J.K.; Park, D.S.; Lee, S.Y.; Jang, E.J.; Ji, M.S.; Sim, D.S.; Hong, Y.J.; et al. A method for coating fucoidan onto bare metal stent and in vivo evaluation. *Prog. Org. Coat.* **2015**, *78*, 348–356. [CrossRef]

39. Yang, Q.; Yang, R.; Li, M.; Liang, X.; Elmada, Z.C. Effects of dietary fucoidan on the blood constituents, anti-oxidation and innate immunity of juvenile yellow catfish (*Pelteobagrus fulvidraco*). *Fish Shelfish Immun.* **2014**, *41*, 264–270. [CrossRef]

40. Ale, M.T.; Maruyama, H.; Tamauchi, H.; Mikkelsen, J.D.; Meyer, A.S. Fucoidan from *Sargassum* sp. and *Fucus vesiculosus* reduces cell viability of lung carcinoma and melanoma cells in vitro and activates natural killer cells in mice in vivo. *Int. J. Biol. Macromol.* **2011**, *49*, 331–336. [CrossRef]

41. Pielesz, A.; Binias, W. Cellulose acetate membrane electrophoresis and FTIR spectroscopy as methods of identifying a fucoidan in *Fucus vesiculosus* Linnaeus. *Carbohydr. Res.* **2010**, *345*, 2676–2682. [CrossRef]

42. Zhao, X.; Dong, S.; Wang, J.; Li, F.; Chen, A.; Li, B. A comparative study of antithrombotic and antiplatelet activities of different fucoidans from *Laminaria japonica*. *Thromb. Res.* **2012**, *129*, 771–778. [CrossRef]

43. Tengdelius, M.; Lee, C.J.; Grenegard, M.; Griffith, M.; Pahlsson, P.; Konradsson, P. Synthesis and biological evaluation of fucoidan-mimetic glycopolymers through cyanoxyl-mediated free-radical polymerization. *Biomacromolecules* **2014**, *15*, 2359–2368. [CrossRef] [PubMed]

44. Rabanal, M.; Ponce, N.M.; Navarro, D.; Gomez, R.M.; Stortz, C. The system of fucoidans from the brown seaweed *Dictyota dichotoma*: Chemical analysis and antiviral activity. *Carbohydr. Polym.* **2014**, *101*, 804–811. [CrossRef]

45. Wu, L.; Sun, J.; Su, X.; Yu, Q.; Yu, Q.; Zhang, P. A review about the development of fucoidan in antitumor activity: Progress and challenges. *Carbohydr. Polym.* **2016**, *154*, 96–111. [CrossRef] [PubMed]

46. Vesel, A.; Mozetic, M.; Strnad, S. Improvement of adhesion of fucoidan on polyethylene terephthalate surface using gas plasma treatments. *Vacuum* **2011**, *85*, 1083–1086. [CrossRef]

47. Ozaltin, K.; Lehocky, M.; Humpolicek, P.; Pelkova, J.; Saha, P. A new route of fucoidan immobilization on low density polyethylene and its blood compatibility and anticoagulation activity. *Int. J. Mol. Sci.* **2016**, *17*, 908. [CrossRef] [PubMed]

48. Mracek, A.; Lehocky, M.; Smolka, P.; Grulich, O.; Velebny, V. The allyamine grafting on the plasma pre-treated polyester nonwowen fabric: Preperation, characterization and utilization. *Fiber Polym.* **2010**, *11*, 1106–1110. [CrossRef]

49. Almazan, M.C.A.; Paredes, J.I.; Mendoza, M.P.; Garcia, M.D.; Garzon, F.J.L.; Alonso, A.M.; Tascon, J.M.D. Surface characterisation of plasma-modified poly(ethylene terephthalate). *J. Colloid Interface Sci.* **2006**, *293*, 353–363. [CrossRef]

50. Andanson, J.M.; Kazarian, S.G. In situ ATR-FTIR spectroscopy of poly(ethylene terephthalate) subjected to high-temperature methanol. *Macromol. Symp.* **2008**, *265*, 195–204. [CrossRef]

51. Huang, L.Y.; Yang, M.C. Surface immobilization of chondroitin 6-sulfate/heparin multilayer on stainless steel for developing drug-eluting coronary stents. *Colloids Surf. B* **2008**, *61*, 43–52. [CrossRef]

52. Humpolicek, P.; Kucekova, Z.; Kasparkova, V.; Pelkova, J.; Modic, M.; Junkar, I.; Trchova, M.; Bober, P.; Stejskal, J.; Lehocky, M. Blood coagulation and platelet adhesion on polyaniline films. *Colloids Surf. B* **2015**, *133*, 278–285. [CrossRef]

53. Wijesinghe, W.A.; Jeon, Y.J. Biological activities and potential industrial applications of fucose rich sulfated polysaccharides and fucoidans isolated from brown seaweeds: A review. *Carbohydr. Polym.* **2012**, *88*, 13–20. [CrossRef]

54. Harris, L.F.; Lopez, V.C.; Killard, A.J. Coagulation monitoring devices: Past, present, and future at the point of care. *Trends Anal. Chem.* **2013**, *50*, 85–95. [CrossRef]

MDPI

St. Alban-Anlage 66

4052 Basel

Switzerland

Tel. +41 61 683 77 34

Fax +41 61 302 89 18

www.mdpi.com

Polymers Editorial Office

E-mail: polymers@mdpi.com

www.mdpi.com/journal/polymers